Sciences Basic to Psychiatry

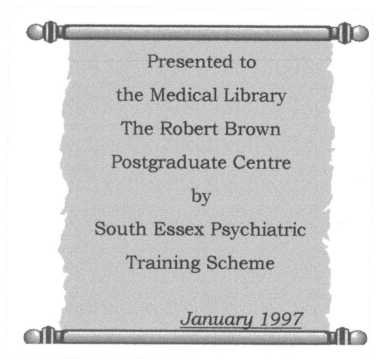

For Churchill Livingstone:
Publisher: Georgina Bentliff
Project Editor: Lucy Gardner
Editorial Co-ordination: Editorial Resources Unit
   Copy Editor: Paul Singleton
Production Controller: Nancy Henry
Design: Design Resources Unit
Sales Promotion Executive: Hilary Brown

Sciences Basic to Psychiatry

Basant K. Puri MA MB BChir MRCPsych
Senior Registrar in Psychiatry, Charing Cross Hospital,
London, UK

Peter J. Tyrer MD MRCP DPM FRCPsych
Professor of Community Psychiatry, St Mary's Medical School and St Charles' Hospital,
London, UK

CHURCHILL LIVINGSTONE
EDINBURGH LONDON MADRID MELBOURNE NEW YORK AND TOKYO 1992

CHURCHILL LIVINGSTONE
Medical Division of Longman Group UK Limited

Distributed in the United States of America by Churchill
Livingstone Inc., 650 Avenue of the Americas, New York,
N.Y. 10011, and by associated companies, branches and
representatives throughout the world.

First published 1992
  Reprinted 1995

ISBN 0-443-04478-3

British Library Cataloguing in Publication Data
A catalogue record for this book is available from the
British Library.

Library of Congress Cataloging in Publication Data
Puri, Basant K.
    Sciences basic to psychiatry/ Basant K. Puri, Peter J. Tyrer.
      p. cm.
    Includes bibliographical references and index.
    1. Neurosciences. 2. Psychiatry. I. Tyrer, Peter J. II. Title.
    [DNLM: 1. Neurosciences. 2. Psychiatry. WL 100 P985s]
RC341.P9 1992
616.8--dc20
DNLM/DLC
for Library of Congress   91–46616

The
publisher's
policy is to use
**paper manufactured
from sustainable forests**

Produced by Longman Singapore Publishers (Pte) Ltd.
Printed in Singapore

Preface

This book is a basic reference text for psychiatric trainees studying for Parts 1 and 2 of the MRCPsych examination. In short, it is a cookbook. Although comments about cookbooks in this context tend to be derogatory, the basic principles of a work of reference are no different from those of a good cookbook: to provide clear information in straightforward language, easy accessibility, accuracy and comprehensiveness. The pejorative use of cookbook comes about because its readers are implied to be dim-witted. The aspiring chef looks up the chosen recipe and follows the instructions carefully even though he may have no idea of the principles behind them. When unable to carry out the instructions exactly he is lost, because basically he is ignorant. This is a slur on cookery writing. Lists of recipes are no longer enough; the principles and techniques of food preparation are now given at least equal weight and options offered for the cook to impose his personal touch and talents on his culinary creations.

We have planned and written this book with the same objects in mind. The principles of the basic sciences related to psychiatry are described wherever possible, and we hope that this information makes their application more understandable to the reader.

This does not mean that factual details are skimped, but understanding their setting helps their retention in memory. Wherever possible we have used tables and diagrams to aid presentation of data and have made access easier by full indexing complete with cross-references. In keeping with the principles of a good cookbook we have confined ourselves to material that is relevant to the syllabus of the MRCPsych examination so that we have been able to present the information in one volume.

We should be grateful to receive any factual corrections to our psychiatric recipes, and feedback about areas of misunderstanding. These would have been greater but for the secretarial help of Ann Tyrer and the statistical assistance of Jonathan Tyrer. Although the final result is not a work of art we hope it will have the essence of a good cookbook and be of practical value both for revision and reference. An established textbook of surgery is now well-known to generations of medical students as 'pale, bulky and offensive'; we trust that if we introduce ours with a confident 'bon appetit' it will not have the same results.

London, September 1991

Basant Puri
Peter Tyrer

Dedicated to Professor James Gibbons, Chief Examiner, Royal College of Psychiatrists (1976-1981), who taught that simplicity in all things can even be true of psychiatry

Contents

1. Neuroanatomy

An understanding of the anatomy of the human nervous system, the most complex system of the body, is fundamental to the study of the sciences basic to psychiatry. In this chapter this subject is considered in the following order: the organization of the nervous system; the neurone; the neuroglia; the spinal cord; the developmental organization of the brain; the brain stem; the cerebellum; the diencephalon; the cerebral hemispheres; the organization of the cerebral cortex; the limbic system; the cranial nerves; the meninges; the cerebrospinal fluid; and the blood supply to the brain and spinal cord. The anatomy of the autonomic nervous system is considered in Chapter 2, and techniques of brain imaging are considered in Chapter 7.

ORGANIZATION OF THE NERVOUS SYSTEM

Central and peripheral nervous systems

The human nervous system can be divided into two basic parts: the **central nervous system**, which consists of the brain and spinal cord, and the **peripheral nervous system**, which consists of the cranial and spinal nerves and other neuronal processes and cell bodies lying outside the central nervous system. Although this conventional division is helpful for the purposes of description, it should be borne in mind that the central and peripheral nervous systems are interconnected. Indeed, parts of the same neurone may be found in both systems.

The central nervous system is well protected. Both the brain and the spinal cord are surrounded by three layers of connective tissue membrane, the **meninges**: the **dura mater** (outermost layer), the **arachnoid mater,** and the **pia mater** (innermost layer). The **subarachnoid space** contains cerebrospinal fluid, so that in effect the brain and spinal cord are suspended in fluid. Further protection is offered by the skull and vertebral column within which the brain and spinal cord are respectively housed.

In contrast, the peripheral nervous system is relatively unprotected and is more likely to suffer traumatic damage.

Autonomic nervous system

The nervous system can also be divided into two parts on a functional basis: the **somatic nervous system**, which is concerned primarily with the innervation of voluntary structures, and the **autonomic nervous system**, which is concerned primarily with the innervation of involuntary structures. The autonomic nervous system is subdivided into the **sympathetic** and **parasympathetic** parts, each of which has afferent and efferent fibres. Further details are given in the section on the autonomic nervous system in Chapter 2.

THE NEURONE

Figure 1.1 is a diagrammatic representation of a typical neurone or nerve cell. The cell body or soma of the neurone is known as the **perikaryon** or neurocyte. Projecting from the perikaryon are two types of neurite: the axon and the dendrite.

There is usually one **axon** per neurone, and it takes origin from the perikaryon or from one of the main dendrites. The axon may be ensheathed in a lamellated interrupted covering, the myelin sheath. These interruptions are the **nodes of Ranvier**, and the segments between the nodes are known as internodes. The proximal axon segment, the **axon hillock**, is unmyelinated and is, in many neurones, the site of initiation of the action potential. Myelination ceases at the distal end of the axon, which may be branched with terminal boutons. The axon conducts impulses away from the perikaryon with the presynaptic release of neurotransmitters taking place at the boutons. The Nissl substance, which is involved in protein synthesis, is absent in the axon.

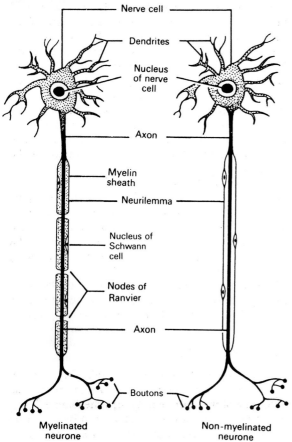

Myelinated   Non-myelinated
neurone   neurone

Fig. 1.1   The structure of a neurone. (Reproduced with permission from Wilson K J W 1990 Anatomy and physiology in health and illness 7th edn. Churchill Livingstone, Edinburgh.)

Dendrites differ from axons in that there is usually more than one per neurone, they conduct information to the perikaryon, and Nissl substance is present. Dendrites are usually branched and studded with dendritic spines, which are sites of synaptic contact.

Classification

Neurones can be classified on a morphological basis into three types: unipolar, bipolar and multipolar. In **unipolar** neurones the perikaryon only has one neurite which divides into two branches, as occurs for example in neurones of the posterior root ganglion. The perikaryon of a **bipolar** neurone has two neurites. Examples of bipolar neurones are found in the retina. The majority of neurones are **multipolar**, having one axon and more than one dendrite.

Neurones can also be classified on the basis of size into Golgi types I and II. **Golgi type I** neurones have a long axon. These axons form the long fibre tracts of the central nervous system, and the nerve fibres of peripheral nerves. Examples of Golgi type I neurones include the Purkinje cells of the cerebellar cortex and cerebral cortical pyramidal cells. **Golgi type II** neurones have a short axon which terminates close to the parent cell. They are much more numerous than Golgi type I neurones, being abundant, for example, in the cerebral and cerebellar cortex. Some central nervous system neurones do not appear to have an axon at all, and have been termed **amacrine**.

Structure

The neuronal **nucleus** is relatively large and usually possesses one prominent nucleolus. A Barr body occurs in the nuclei of females (see Ch. 6). The neuronal cytoplasm or neuroplasm is semifluid and viscous, and amongst its inclusions are the Nissl substance, Golgi apparatus, mitochondria, microfilaments, microtubules, lysosomes, centrioles, lipofuscin, melanin, glycogen and lipid.

The **Nissl substance** is composed of rough endoplasmic reticulum and synthesizes protein. This is transported to the synaptic terminals along the axon via axoplasmic flow. In a process known as chromatolysis there is a dispersion of the rough endoplasmic reticulum following neuronal damage or fatigue and the Nissl substance seems to disappear. During this process new protein biosynthesis takes place.

The **Golgi apparatus** is classically demonstrated under the light microscope following staining by osmium or silver impregnation methods. It has the appearance of a reticulum and is probably involved in synthetic activities.

Mitochondria are present throughout the neurone and are involved in energy production. Following section of a spinal nerve there is an increase in the mitochondria of the corresponding anterior horn cells. This reflects an increase in neuronal oxidative phosphorylation.

Microfilaments are arranged in bundles to form the neurofibrils demonstrated by light microscopy. Their function is uncertain.

Microtubules have been shown by electron microscopy to be present throughout the neurone. It has been suggested that they are involved in the transport of substances distally and in the process of axon growth.

Lysosomes are spherical membrane-surrounded bodies containing a number of powerful enzymes.

They may act as the neuronal digestive system or as internal scavengers. The inherited lack of a given lysosomal enzyme may lead to the accumulation of material and cause mental disorder. For example, in Niemann–Pick disease there is a deficiency of sphingomyelinase and the accumulation of sphingomyelin. Other so-called storage diseases include Gaucher's disease, Tay–Sach's disease and glycogen storage disease II. Further details are given in Chapter 6.

Lipofuscin is a lysosomal metabolic product which accumulates with age and with nervous system trauma or disease. It is thought to be harmless.

Melanin is a brown to black pigment found in the cytoplasm of neurones in, for example, the substantia nigra and the locus coeruleus.

NEUROGLIA

Also known as interstitial cells, the neuroglia make up most of the nervous tissue, outnumbering neurones five to ten times. There are four main types of neuroglial cells in the central nervous system: astrocytes, oligodendrocytes, microglia and ependyma. In the peripheral nervous system there are two types of neuroglia: Schwann cells and satellite cells.

Astrocytes or astroglia are multipolar cells which afford structural support to neurones. They are also the main type of cell in central nervous system neuroglial scar tissue. Some of their processes end as perivascular feet on capillaries. This gliovascular membrane helps to form the blood–brain barrier. It is thought that astrocytes may allow substances such as nutrients to pass between neurones and capillaries. Astrocytes are also involved in phagocytosis. There are two types: fibrous and protoplasmic astrocytes. Fibrous astrocytes have processes that are relatively long and thin. They are present mostly in the white matter of the central nervous system. By contrast, protoplasmic astrocytes have processes that are relatively short and thick and are found mostly in the grey matter.

Compared with astrocytes, **oligodendrocytes** or oligodendroglia have a smaller perikaryon, with fewer processes. They form the myelin sheaths of nerve fibres of the central nervous system. Like astrocytes, oligodendrocytes are also involved in phagocytosis.

The smallest neuroglial cells are the **microglia**. They are most abundant in the grey matter and when inactive they have many processes. Injury to the central nervous system activates the microglia which then migrate to the site of injury. Here they are phagocytic and remove the debris. Hence they are also known as scavenger cells.

The **ependyma** line the cavities of the central nervous system, that is, the ventricular system of the brain and the central canal of the spinal cord. Ependymal cells possess cilia whose beating contributes to the flow of the cerebrospinal fluid. Types of ependymal cells include ependymocytes, tanycytes and choroidal epithelial cells. Ependymocytes line the central canal of the spinal cord and the ventricles. Tanycytes line the floor of the third ventricle over the hypothalamic median eminence. Choroidal epithelial cells have tight junctions and cover the surfaces of the choroidal plexuses.

Schwann cells form the myelin sheaths of peripheral nerve axons. They also encircle some unmyelinated peripheral nerve axons. Schwann cells are also involved in the formation of the neurilemma. The latter is a membrane surrounding the myelin sheath that has a rôle in fibre regeneration following injury.

Satellite or capsular cells are flattened cells found in sensory and autonomic ganglia. Satellite cells have a supportive function for the neurones in those ganglia.

THE SPINAL CORD

The spinal cord is contained within the vertebral column and is continuous superiorly with the medulla oblongata of the brainstem through the foramen magnum of the skull. In the adult human it is on average approximately 45 cm long, terminating inferiorly at the level of the inferior border of the first lumbar vertebra. Prior to adolescence the spinal cord is relatively longer in comparison with the vertebral column, reaching the level of the third lumbar vertebra at birth, and being the length of the entire vertebral canal at three months after conception.

Thirty-one pairs of spinal nerves are attached to the spinal cord via the anterior or ventral motor roots and the posterior or dorsal sensory roots, as shown in Figures 1.2 and 1.3. The cell bodies of the sensory neurones are in the **posterior** (or **dorsal**) **root ganglia**. The cell bodies of the **lower motor neurones** which give rise to the anterior motor roots are in the **anterior grey columns** of the spinal cord itself. It can be seen from Figure 1.3 that the internal organization of the spinal cord demonstrates an inner H-shaped core of grey matter surrounded by white matter. The grey matter is divided into anterior, lateral and posterior columns or horns, and a central connecting grey commissure. The white matter is made up mainly of myelinated fibres, and is similarly divided into anterior, lateral and posterior columns.

The spinal cord is divided into five regions on the

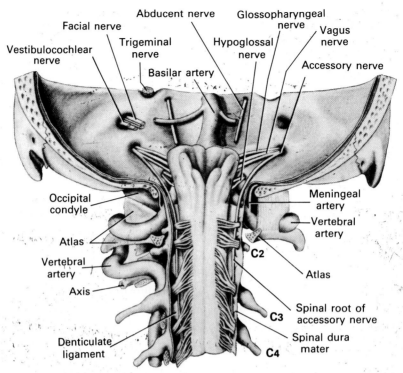

Fig. 1.2   The posterior cranial fossa and spinal canal opened from behind.
(Reproduced with permission from Last R J 1984 Last's anatomy 7th edn. Churchill
Livingstone, Edinburgh.)

basis of the site of exit from the vertebral column of
the 31 pairs of spinal nerves. Starting superiorly these
are: cervical (eight pairs), thoracic (twelve pairs),
lumbar (five pairs), sacral (five pairs), and coccygeal
(one pair).

The two most important functions of the spinal cord
are, firstly, the transmission of impulses to and from
the brain via the ascending and descending tracts, and
secondly, the mediation of spinal cord reflexes. The
anatomical basis of the first function is discussed
below. Details of the spinal cord reflexes are given in
Chapter 2.

Ascending white column tracts

Figure 1.4 shows the major ascending tracts or
fasciculi of the spinal cord. The ascending tracts
transmit sensory information from the spinal cord to
the brain and are usually named according to their
origin and termination.

It can be seen from Figure 1.4 that the anterior
white columns contain the **anterior spinothalamic
tracts**. These are concerned with the transmission of
light touch and pressure sense.

There are a number of important ascending lateral
white column tracts. The **anterior and posterior
spinocerebellar tracts** transmit proprioceptive,
pressure, and touch information. They enter the
cerebellum through the ipsilateral superior cerebellar
peduncles. The **lateral spinothalamic tracts** convey
pain and temperature sense. Before entering the
anterior and lateral spinothalamic tracts, the
appropriate sensory fibres, having entered the posterior
roots, ascend a short distance before crossing over to
ascend to the sensory cortex of the brain, via the
thalamus, in the contralateral tracts. The **spino-
olivary tracts** convey proprioceptive and cutaneous
information, whilst the **spinotectal tracts** are
concerned with spinovisual reflexes.

The ascending posterior white column tracts include
the **fasciculus cuneatus**, which conveys information
on discriminative touch and proprioception, and the
fasciculus gracilis, which conveys vibration sense.
These tracts pass, mostly uncrossed, to the cuneate
and gracile nuclei, respectively. These nuclei are in the
medulla oblongata and from them, following synapse,
fibres decussate and relay, via the thalamus, to the
sensory cortex. Fibres also pass from these nuclei

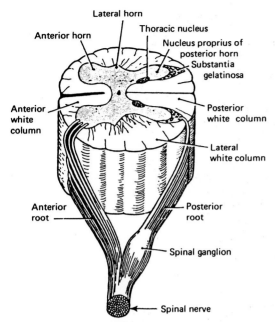

Fig. 1.3 A spinal segment from the thoracic region viewed from the left side. (Reproduced with permission from Emslie-Smith D, Paterson C R, Scratcherd T, Read N W 1988 Textbook of physiology 11th edn. Churchill Livingstone, Edinburgh.)

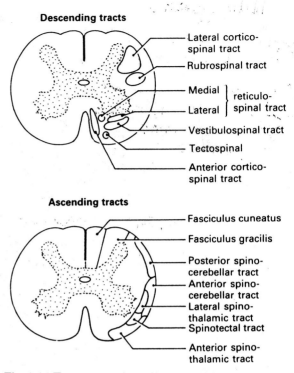

Fig. 1.4 Transverse sections of human spinal cord to show (top) the positions of the descending tracts and (bottom) those of the ascending tracts. Intersegmental tracts lie between the grey matter and the descending tracts. (Reproduced with permission from Emslie-Smith D, Paterson C R, Scratcherd T, Read N W 1988 Textbook of physiology 11th edn. Churchill Livingstone, Edinburgh.)

directly to the cerebellum, via the inferior cerebellar peduncles.

In summary, the major sensory pathways from general sensory endings to the thalamus and sensory cortex are as follows: light touch and pressure sensations are carried in the contralateral anterior spinothalamic tracts; pain and temperature sensations are carried in the contralateral lateral spinothalamic tracts; and discriminative touch, proprioception and vibration sensations are carried in the ipsilateral posterior white column.

Descending white column tracts

Figure 1.4 shows the major descending tracts of the spinal cord. They transmit motor information from the brain to the spinal cord.

The descending anterior white column tracts include the **anterior corticospinal tracts**, which are concerned with voluntary movement; the **reticulospinal fibres**, concerned with motor function; the **vestibulospinal tracts**, involved in the control of muscle tone; and the **tectospinal tracts**, which are involved in a head-turning reflex and the movement of the upper limbs in response to acoustic, cutaneous and visual stimuli.

The descending lateral white column tracts include the **lateral corticospinal tracts**, concerned with voluntary movement; the **rubrospinal and lateral reticulospinal tracts**, involved in muscular activity; **descending autonomic fibres** which are concerned with the control of visceral function; and the **olivospinal tracts**, which are probably concerned with muscular activity.

The corticospinal tracts provide the **upper motor neurones**. Their fibres originate from the pyramidal cells of the motor cortex and transmit voluntary motor information to the lower motor neurones that pass from the spinal cord to muscles, via the anterior spinal roots. The anterior corticospinal tracts, also known as the direct pyramidal or uncrossed motor tracts, are small tracts that are confined to the cervical and superior thoracic region of the spinal cord. The lateral corticospinal tracts are also known as the pyramidal or crossed motor tracts and, unlike the anterior corticospinal tracts, they decussate in the medulla oblongata and then pass inferiorly on the contralateral side of the spinal cord. At each spinal segment fibres leave the lateral corticospinal tracts to enter the anterior grey columns of the spinal cord and synapse with lower motor neurones. Thus the tracts diminish in size inferiorly.

To summarize, the corticospinal tracts are concerned with voluntary motor information, whilst the other descending tracts transmit involuntary motor information such as that concerned, for example, with muscle tone.

Segmental innervation of the skin

A **dermatome** is the area of skin supplied by the fibres of a posterior root. Various methods have been used to construct dermatome maps, a knowledge of which is useful in neuropsychiatric examination and diagnosis. For example, Foerster (1933) used surgical techniques, whilst Keegan (1943) studied areas of sensory impairment resulting from compression of individual posterior roots. Head (1920) constructed his map by studying cases of herpes zoster. In this condition, viral inflammation of cells in the posterior root ganglion leads to the eruption of vesicles in, and changes in the colour of, the corresponding dermatome.

The following is a useful guide: C5 to T1 — the upper limb; C7 — anterior and posterior aspects of the middle finger; T10 — the umbilical region; L1 — the inguinal region; L2 to L3 — anterior aspect of the upper leg; L4 to L5 — anterior aspect of the lower leg; S1 — posterior aspect of the lower leg and lateral aspect of the foot and sole. For more details the reader should refer to a dermatome map (for example, Williams et al 1989).

Sectioning of one posterior root will not cause clinically detectable sensory loss because there is considerable overlapping of adjacent dermatomes. There is greater overlap of fibres for pain, heat and cold than for fibres for light touch. Therefore, sectioning of spinal nerves will cause an area of tactile loss which is greater than the area of loss of sensations of pain, heat and cold.

Segmental innervation of muscles

Most skeletal muscles are innervated by more than one anterior motor root and hence by more than one spinal segment. Thus, pressure on one spinal segment, for example secondary to neoplasia, will lead to weakness in muscles receiving an innervation from that segment, but is unlikely to cause total paralysis of the muscles, unless adjacent segments are also affected.

In clinical practice it is possible to test routinely the

Table 1.1 The tendon reflexes

Reflex	Spinal segment
Biceps jerk	C5-6
Triceps jerk	C6-7
Supinator jerk	C5-6
Knee jerk	L2-4
Ankle jerk	S1-2

segmental innervation of certain muscles by eliciting the corresponding muscle tendon reflexes as shown in Table 1.1.

DEVELOPMENTAL ORGANIZATION OF THE BRAIN

At an early stage of fetal development, the midline neural tube differentiates into the following vesicles: the prosencephalon or forebrain, the mesencephalon or midbrain, and the rhombencephalon or hindbrain (see Fig. 1.5). The prosencephalon later differentiates into the telencephalon and the diencephalon. The rhombencephalon differentiates into the metencephalon and the myelencephalon.

The **telencephalon** gives rise to the cerebral hemispheres and contains the following structures: the

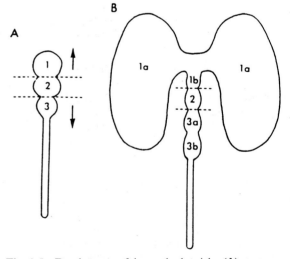

Fig. 1.5 Development of the cerebral vesicles (**A**) at an early, and (**B**) at a later stage. 1. Prosencephalon; 1a. telencephalon; 1b. diencephalon. 2. Mesencephalon. 3. Rhombencephalon; 3a. metencephalon; 3b. myelecephalon. (Reproduced with permission from Kendell R E, Zealley A K 1988 Companion to psychiatric studies 4th edn. Churchill Livingstone, Edinburgh.)

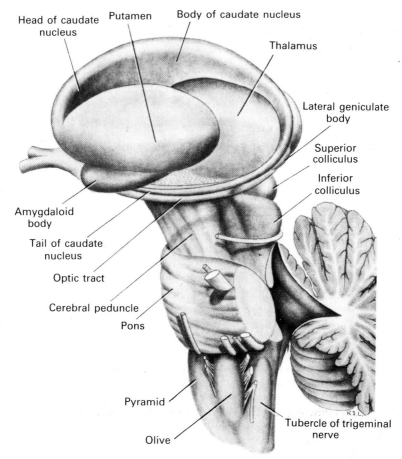

Head of caudate nucleus

Putamen

Body of caudate nucleus

Thalamus

Lateral geniculate body

Superior colliculus

Inferior colliculus

Amygdaloid body

Tail of caudate nucleus

Optic tract

Cerebral peduncle

Pons

Pyramid

Olive

Tubercle of trigeminal nerve

Fig. 1.6 The basal ganglia of the left hemisphere. The internal capsule has been removed. (Reproduced with permission from Last R J 1984 Last's anatomy 7th edn. Churchill Livingstone, Edinburgh.)

pallium or cerebral cortex, the rhinencephalon, the corpus striatum, and the medullary centre. The **rhinencephalon**, or 'nosebrain', consists of the olfactory mucosa, tracts and bulbs, and a strip of paleocortex from the temporal lobe uncus to the medial surface of the frontal lobe. The **corpus striatum** consists of the caudate and lentiform nuclei, the lentiform nucleus being divided into the putamen and globus pallidus (see Figs. 1.6 and 1.7). The corpus striatum is discussed in more detail later in this chapter, in the section on the basal ganglia. There are three types of nerve fibre in the medullary centre: fibres connecting cortical areas of the same cerebral hemisphere; fibres connecting cortical areas of both cerebral hemispheres by crossing the midline in the corpus callosum; and those passing in both directions between the cerebral cortex and subcortical centres.

The term **basal ganglia** is sometimes used as a synonym for the corpus striatum. According to Snell (1987) each of the basal ganglia of the telencephalon consists of the amygdaloid nucleus, the claustrum, and the corpus striatum (see Table 1.2). Another definition of the basal ganglia widely used in psychiatry and neurology includes the corpus striatum, claustrum, subthalamic nucleus and substantia nigra (Mettler 1968). Other definitions of the basal ganglia have also

Table 1.2 Components of the basal ganglia (after Snell 1987)

Corpus striatum
  caudate nucleus
  lentiform nucleus
    putamen
    globus pallidus

Amygdaloid nucleus

Claustrum

been proposed by some neuroanatomists, adding to the confusion on this topic.

The **diencephalon** consists of the following structures on each side: the thalamus, subthalamus, hypothalamus and epithalamus. The latter is made up of the habenular nucleus and the pineal gland.

The **mesencephalon** consists of the tectum, basis pedunculi and the tegmentum. The tectum, in turn, consists of the corpora quadrigemina, which are made up of the superior and inferior colliculi. Part of the substantia nigra, the pars compacta, is continuous with the globus pallidus. Indeed, as mentioned above, in some definitions the substantia nigra is included as being part of the basal ganglia. The tegmentum contains the red nuclei, fibre tracts and grey matter surrounding the cerebral aqueduct.

The **metencephalon** consists of the pons, the oral part of the medulla oblongata and the cerebellum.

The **myelencephalon** is the caudal part of the medulla oblongata.

BRAIN STEM

The brain stem consists of the medulla oblongata, pons and mesencephalon. The **medulla oblongata** is continuous inferiorly, through the foramen magnum, with the spinal cord, and is continuous superiorly with the pons. It is connected posteriorly, via the inferior cerebellar peduncles, with the cerebellum. The pons is continuous superiorly with the mesencephalon or midbrain. Posteriorly, the pons is connected with the cerebellum by way of the middle cerebellar peduncles. The mesencephalon lies in a gap in the tentorium cerebelli, and connects the pons and cerebellum to the diencephalon. Posteriorly, the mesencephalon is connected with the cerebellum via the superior cerebellar peduncles. The **pineal gland**, which may be seen in skull radiographs when calcified, lies between the superior colliculi of the mesencephalon.

In the midline of the brain stem is a loose network of neurones known as the **reticular activating system**, which is probably involved in sleep, consciousness, arousal, the activation and inhibition of movement, and motivation. It receives inputs from the spinal cord, cerebellum, basal ganglia, hypothalamus and cerebral cortex, and its axons project to the spinal cord, thalamus and hypothalamus.

Lying within the reticular system of the medulla oblongata are groups of neurones that make up the vital centres concerned with cardiovascular and respiratory functions. Compression, trauma or infection (for example, poliomyelitis) may particularly affect the latter, leading to respiratory failure. There is also a region sometimes referred to as the vomiting centre, which is involved in the control of gastrointestinal tract activity.

In the mesencephalon the **inferior colliculi** are involved in auditory reflexes, such as the reaction to a sudden loud noise. The **superior colliculi**, on the other hand, are concerned with visual reflexes, such as pupil constriction in response to bright light.

The brain stem also contains the **medial forebrain bundle** which consists of ascending noradrenergic, dopaminergic and serotinergic fibres terminating in higher brain areas. Further details of these and other biogenic amine tracts are given in Chapter 3.

CEREBELLUM

The cerebellum consists of two lateral **cerebellar hemispheres** and a median **vermis**. It occupies the greater part of the posterior cranial fossa and is separated from the cerebral hemispheres by the tentorium cerebelli, a fold of dura mater. As mentioned above, it is attached to the brainstem via the three pairs of cerebellar peduncles.

Its surface is divided into numerous narrow **folia**, so named because it bears some resemblance to leaves. Several deeper fissures divide the cerebellum into a number of lobules. In section it looks like a tree with many branches, the **arbor vitae**, on account of the fissuring. There is an outer **cortex** of grey matter where afferent fibres terminate. This covers an inner mass of white matter which constitutes the **medullary body**, in which lie several pairs of deep **nuclei** of grey matter. From medial to lateral in each cerebellar hemisphere, these nuclei are as follows: the fastigial, globose, emboliform and dentate.

The cerebellar cortex has a uniform structure in all parts of the organ and consists of three layers. The outer **molecular layer** has sparsely scattered neurones. Deep to this is a layer of one or two rows of large **Purkinje cells** associated with basket cells. The Purkinje dendrites ramify in the molecular layer with related stellate cells. The innermost **granular layer** is made up mostly of small closely packed granule cells and associated Golgi cells. Purkinje axons contribute to the white matter of the medullary body (see Fig. 2.7).

The cerebellum has three phylogenetically derived parts. The oldest is the **archicerebellum** or vestibulocerebellum, which receives inputs concerning the position and motion of the head from the inner ear. The **paleocerebellum** or spinocerebellum corresponds to the vermis and receives touch, pressure, proprioceptive and thermal inputs from ascending

Table 1.3 Connections of the cerebellum

Phylogenetic part	Afferents	Efferents
Archicerebellum	Vestibular neurones and nuclei	Vestibular nuclei
Paleocerebellum	Spinocerebellar tracts Cuneocerebellar tracts	Vestibular and reticular neurones
Neocerebellum	Pontine nuclei Pyramidal neurones	Thalamus

Table 1.4 Connections of the thalamus

Nucleus	Afferents	Efferents
Anterior	Mammillothalamic	Limbic association tract (cingulate) cortex
Dorsomedial	Prefrontal cortex Other thalamic nuclei Amygdala	Prefrontal cortex
Medial geniculate body	Auditory input via the lateral lemniscus	Primary auditory cortex
Lateral geniculate body	Visual input via the optic tract	Primary visual cortex
Ventral lateral	Superior cerebellar peduncle	Primary motor cortex
Ventral posterior	Medial lemniscus	Primary somaesthetic cortex

spinal cord pathways. The **neocerebellum** or ponto-cerebellum has its main input from nuclei in the pons, which in turn have received motor cortex information. Further details of these connections are given in Table 1.3, whilst details on the functions of the cerebellum are given in Chapter 2.

DIENCEPHALON

The diencephalon surrounds the third ventricle (see below), and as mentioned above, is made up of the thalamus, subthalamus, hypothalamus and epithalamus.

The **thalamus** is an ovoid shaped mass of grey matter that is reciprocally connected to the cerebral cortex and the limbic system. It lies above the hypothalamus and forms the lateral wall of the third ventricle. It consists largely of relay nuclei which project incoming sensory impulses widely via the thalamic radiations in the internal capsule. However, it should not be thought of as merely a relay station for sensory impulses as it is also involved in the processing and integration of such information. There are many nuclear divisions of the thalamus; a simplified summary of the main connections of the larger thalamic nuclei is shown in Table 1.4. The functions of the thalamus are considered further in Chapter 2.

As implied by its name, the **subthalamus** lies inferior to the thalamus. It contains a number of tracts and groups of neurones, for example, the cranial parts of the red nuclei and substantia nigra. The **epithalamus** is made up of the habenular nuclei, their connections, and the pineal gland. Further details of the pineal gland are given in the section on neuroendocrinology in Chapter 2.

The **hypothalamus** forms the floor of the third ventricle and is made up of a number of nuclei that are organized into three major structures: the supraoptic area superior to the optic chiasma, the tuber cinereum and infundibulum, and the mammillary area superior to the mammillary bodies. Another way of subdividing the hypothalamus is to consider the following three regions or zones on each side of the paramedian plane: the periventricular zone, the intermediate zone, and the lateral zone. The major outputs of the hypothalamus are to the pituitary gland, which lies inferior to the infundibulum and to the limbic system (see below), the autonomic nervous system, somatic motor system and mesencephalon. Together, the hypothalamus and pituitary gland have been termed the leader of the endocrine orchestra of the body, and they are conveniently described further in the section on neuroendocrinology in Chapter 2.

CEREBRAL HEMISPHERES

The two cerebral hemispheres lie on the anterior and middle cranial fossae and the tentorium cerebelli. They are connected to each other by the **corpus callosum**, being otherwise partly separated by the median **longitudinal fissure**. The cerebral hemispheres are made up of the cerebral cortex and its associated white matter, the basal ganglia, and the lateral ventricles. The lateral ventricles will be considered together with the rest of the ventricular system later in this Chapter.

Basal ganglia

The basal ganglia are also known as the basal or telencephalic nuclei, and historically this term referred loosely to large masses of grey matter found buried within the medullary body or white matter of each cerebral hemisphere. The components of the basal ganglia have been described above. For the purposes of this section, the definition given by Snell (1987) will

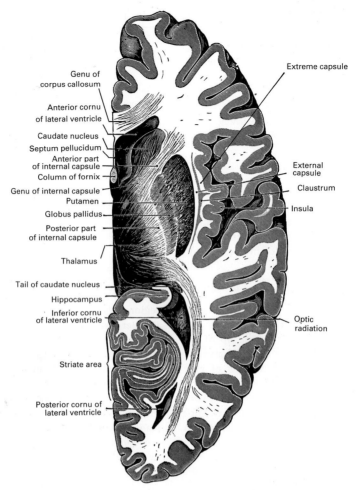

Fig. 1.7   Superior aspect of a horizontal section through the right cerebral hemisphere, showing the association of the corpus striatum with the internal capsule. (Reproduced with permission from 1989 Gray's anatomy 37th edn. Churchill Livingstone, Edinburgh.)

be followed. That is to say, the basal ganglia will be considered to consist of the **corpus striatum**, **amygdaloid nucleus** and **claustrum**. The reader is referred to Figures 1.6 and 1.7 and to Table 1.2.

Corpus striatum

The corpus striatum lies lateral to the thalamus and consists of the caudate and lentiform nuclei. The caudate nucleus is so named because of its resemblance to a C-shaped tail. The lentiform nucleus, which in turn consists of the putamen and globus pallidus (see Table 1.2), is named after its resemblance to a lens.

Laterally, the **caudate nucleus** is separated from

the putamen by the internal capsule. Medial to the caudate nucleus is the thalamus. As shown in Figure 1.6, the caudate nucleus can be divided, for descriptive reasons, into three regions: an expanded rounded head (rostrally), a narrower body and a tail. The head bulges medially into the anterior horn of the lateral ventricle, and inferiorly it is continuous with the putamen. Superior to the area of union of the head of the caudate with the putamen, strands of grey matter pass through the internal capsule. It is this that gives the region its striped appearance from which is derived the name corpus striatum, or striped body. The body of the caudate nucleus passes round the thalamus in the floor of the lateral ventricle. It is separated from the thalamus by a groove containing the stria

Table 1.5 Connections of the lentiform nucleus

	Afferents	Efferents
Putamen	Caudate nucleus Cerebral cortex	Globus pallidus
Globus pallidus	Caudate nucleus Putamen Substantia nigra	Hypothalamus Reticular formation Substantia nigra Subthalamus Thalamus ventroanterior nucleus ventrolateral nucleus

terminalis and the thalamostriate vein. The tail of the caudate nucleus terminates in the roof of the inferior horn of the lateral ventricle at the amygdaloid nucleus.

The **lentiform nucleus** is approximately lens-shaped with the more convex surface facing medially, adjacent to the genu of the internal capsule, and the less convex surface directed laterally. Related to the latter is the external capsule, which is a sheet of white matter separating the lentiform nucleus from another area of grey matter, the **claustrum**. The lentiform nucleus is subdivided into the darker lateral putamen, and the lighter more medial globus pallidus. The connections of the lentiform nucleus are summarized in Table 1.5.

Amygdaloid nucleus (or body)

The amygdaloid nucleus or body lies at the end of the tail of the caudate nucleus in the temporal lobe. It is part of the limbic system and will therefore be discussed in further detail in the section on this system later in this Chapter.

Claustrum

The claustrum is a thin sheet of grey matter with uncertain function; it lies lateral to the lentiform nucleus and is separated from it by the thin sheet of white matter known as the external capsule. The insula, which is a deeply situated area of cerebral cortex (see below), lies lateral to the claustrum.

Organization of the cerebral cortex

The cerebral cortex is the outermost layer of grey matter of the cerebral hemispheres. Its neuronal structure will be looked at first, followed by a consideration of the cortical layers and the cortical regions.

Cortical neurones

The main types of neurone found in the cerebral cortex are the relatively small stellate or granule cells, the cells of Martinotti, the horizontal cells of Cajal, and the larger fusiform cells and pyramidal cells. The horizontal cells of Cajal are found in the outermost cortical layers, and, as their name suggests, they are orientated 'horizontally' along the lie of the cortical layers. The stellate or granule cells have multiple branching dendrites and a short axon. The other three types of neurone have a more 'vertical' orientation with neurites that cut across the cortical layers. The pyramidal cells include the giant Betz cells, which are found in the motor precentral gyrus of the frontal lobe.

Cortical layers

Developmentally the pallium consists of the paleocortex, the archicortex, and the neocortex. Both the paleocortex or primary olfactory area and the archicortex or limbic formation have three cellular layers. The neocortex is phylogenetically the most recent cortex. It makes up about nine-tenths of the human pallium and has six identifiable cellular layers: (i) the molecular or plexiform layer, which is the most superficial layer; (ii) the external granular layer; (iii) the external pyramidal layer; (iv) the internal granular layer; (v) the ganglionic or internal pyramidal layer; (vi) the multiform or polymorphic cell layer.

Frontal cortex

The cortex of the frontal lobe constitutes the entire area anterior to the central sulcus and superior to the lateral sulcus on the lateral surface. Important cortical areas in this region include: the precentral area; the supplementary motor cortex; the frontal eye field or eye motor field; Broca's motor speech area; and the prefrontal cortex.

The **precentral area** comprises a posterior and an anterior region. The posterior region, or **primary motor area** contains the giant pyramidal cells of Betz and is concerned with voluntary movements. Experimental electrical stimulation of this area has been found to lead to the contraction of muscle groups causing either contralateral movements (for example in the limbs), or bilateral movements (for example in the upper face). It has been shown that movements are controlled by an area of cortex proportional to their complexity. Furthermore, there is a topographical representation of the body in the primary motor cortex, which has also been determined by electrical stimulation studies (see Penfield 1938). For example, electrical stimulation of the upper medial part of the left precentral area leads to right lower extremity movement.

The anterior region of the precentral area contains the **premotor** or **secondary motor area**. It lacks the giant pyramidal cells of Betz. Experimentally it has been found that, compared with the primary motor cortex, stronger electrical stimulation is required to give the same degree of movement.

The **supplementary motor cortex** lies on the medial surface of the cerebral hemisphere in the medial frontal gyrus. Again, compared with the primary motor cortex, stronger experimental electrical stimulation is required to give the same degree of movement.

Experimental electrical stimulation of the **frontal eye field** or **eye motor field** leads to conjugate eye movements.

Broca's motor speech area is situated in the dominant cerebral hemisphere. It is concerned with the motor aspect of speech.

The **prefrontal cortex** is a large area lying anterior to the precentral area. It is believed to be concerned with personality, depth of feeling, initiative and judgement.

Parietal cortex

On the lateral surface of the cerebral hemisphere the parietal cortex is bounded anteriorly by the central sulcus, and posteriorly by a line drawn from the preoccipital notch to the parieto-occipital fissure. Inferiorly it is bounded by both the lateral sulcus, and a line drawn from the midpoint between the preoccipital notch and parieto-occipital fissure to the lateral sulcus. On the medial surface, it is bounded by the calcarine sulcus, the corpus callosum, the frontal lobe, and the parieto-occipital fissure. Important cortical areas in this region include: the primary somaesthetic area; the secondary somaesthetic area; the somaesthetic association area; and the angular and supramarginal gyri.

The **primary somaesthetic area** is situated in the postcentral gyrus. The contralateral half of the body is represented in this area as an inverted sensory homunculus, corresponding very approximately to the pattern of cortical representation in the motor cortex. Whilst sensations from most of the body are represented in the contralateral primary somaesthetic area, some sensations from the oral area are represented in the ipsilateral primary somaesthetic area. Furthermore, sensations from the perineum, larynx and pharynx are represented bilaterally.

The **secondary somaesthetic area** lies in the superior lip of the posterior ramus of the lateral sulcus. It is believed to be associated with the spinothalamic tracts.

The **somaesthetic association area** lies in the superior parietal lobule. It is thought to be involved in the synthesis and elaboration of somatic sensory information and allows stereognosis.

The **angular and supramarginal gyri** lie posterior and inferior to the primary somaesthetic area. They are involved in the integration of somatic information with impulses of a visual and auditory nature.

Temporal cortex

On the lateral surface of the cerebral hemisphere the temporal cortex lies below the parietal cortex. Thus it is bounded above by both the posterior ramus of the lateral sulcus and a line drawn from the lateral sulcus to the middle of the imaginary line joining the preoccipital notch to the parieto-occipital fissure. The temporal cortex lies anterior to the occipital cortex and thus is bounded posteriorly by a line drawn from the preoccipital notch to the parieto-occipital fissure. The inferior surface is bounded posteriorly by a line drawn from the preoccipital notch to the anterior end of the calcarine sulcus. Important temporal cortical areas include: the primary auditory area; the secondary auditory area; the gustatory area; and the olfactory area.

The **primary auditory area** lies in the floor of the lateral sulcus, extending slightly on the lateral cerebral hemisphere. It receives afferent fibres carrying auditory impulses from the medial geniculate body; these form the auditory radiation of the internal capsule.

The **secondary auditory area** is situated adjacent and mainly posterior to the primary auditory area. Also known as the auditory association cortex, it is not a well defined area, and is probably involved in the interpretation of symbolic sound patterns.

The **gustatory or taste area** probably lies at the inferior tip of the postcentral gyrus of the parietal lobe. However, it is considered here as it is believed to be functionally related to the temporal lobe. Its functions include the reception and interpretation of taste impulses.

The **primary olfactory area** lies in the medial part of the temporal lobe and consists of the periamygdaloid and prepiriform areas. It receives olfactory impulses via the ipsilateral lateral olfactory stria and the contralateral medial olfactory stria. The **secondary olfactory area** consists of the entorhinal area of the parahippocampal gyrus (see the section on the limbic system, below). It receives impulses from the primary olfactory area. In addition to their reception, both areas are involved in the interpretation of olfactory impulses.

Occipital cortex

On the lateral surface, the occipital cortex is bounded anteriorly by a line drawn from the preoccipital notch to the parieto-occipital fissure. On the medial surface it is separated from the temporal lobe as described above. Important areas in this region include: the primary visual area; the parareceptive area; and the preoccipital area. The parareceptive area and the pre-occipital area are together known as the secondary visual area.

The **primary visual area** lies in the upper and lower lips of the calcarine sulcus on the medial surface of the occipital lobe, extending to adjacent areas on the lateral surface. Its afferent fibres are from the lateral geniculate body of the ipsilateral thalamus. It therefore receives visual information from the contra-lateral field of vision (see the section on the second cranial nerve later in this chapter), that is, from the ipsilateral temporal retina and the contralateral nasal retina. The superior wall of the calcarine sulcus receives visual information from the contralateral inferior field of vision, that is, from the superior retinal quadrants. The inferior wall of the calcarine sulcus receives visual information from the contralateral superior field of vision, that is, from the inferior retinal quadrants.

The **secondary visual area** surrounds the primary visual area and receives afferent fibres from the primary visual area, other cortical areas, and the thalamus. The **parareceptive area** immediately sur-rounds the primary visual area and is involved in the recognition and identification of visual responses. The **preoccipital area** immediately surrounds the para-receptive area and is involved in functions such as recall, orientation, and visual association. It is thought that the secondary visual area may contain the occipital eye field. Experimental electrical stimulation of this area leads to conjugate eye movements.

Insula

The insula, also known as the island of Reil, is an area of cerebral cortex which lies at the bottom of the lateral sulcus. It is normally hidden from view unless the lips of the lateral sulcus are separated. It may be bound to the corpus striatum during fetal develop-ment, and its functions and connections are not fully known.

Cerebral hemisphere fibres

There are three types of tract associated with the cerebral cortex: association fibres, commissural fibres and projection fibres. The association fibres are intrahemispheric fibres and the commissural fibres are interhemispheric.

Association fibres

Association fibres represent ipsilateral cortico-cortical axons. They allow a very large degree of intra-hemispheric interconnection between different parts of the cerebral cortex. They may be short or long, and diffuse or arranged in bundles.

The short association fibres connect vertical cortical neuronal columns lying adjacent to each other in either the same gyrus or in adjoining gyri. Types include the **intracortical association fibres** which remain cortical in their course and the **subcortical asso-ciation fibres** which enter the internal capsule (see below).

The long association fibres allow distant intra-hemispheric cortico-cortical connections. **Arcuate fibres** consist of long association fibres arranged in bundles in a curved shape running parallel to the cortical surface.

Commissural fibres

Commissural fibres represent axons connecting corti-cal areas with their contralateral homologous areas, and may enter the corpus callosum, the anterior commissure, or the commissure of the fornix. In addition there are further indirect connections between the cerebral hemispheres in the diencephalon and the brain stem.

The **corpus callosum** is the largest set of interhemispheric connecting fibres or commissures. It lies at the inferior end of the longitudinal fissure, superior to the diencephalon. It is divided, for des-criptive reasons, into the rostrum (the most rostral part), the genu (or knee), the body, and the splenium. It connects homologous areas of the neocortex.

The **anterior commissure** crosses the midline in the lamina terminalis. It connects homologous areas of the neocortex and of the paleocortex.

The **fornix** is made up of fibres representing the hippocampal efferent system. This system arches rostrally as the two posterior columns of the fornix, which lie inferior to the corpus callosum and superior to the thalamus. The **commissure of the fornix** is made up of transverse fibres that cross the midline from one posterior column to the other. It connects the archicortex of both cerebral hemispheres to each other.

Projection fibres

Projection fibres connect the cerebral cortex with subcortical nuclei. They ascend from or descend to the subcortical nuclei in major tracts, including the internal capsule and the fornix. The internal capsule serves the neocortex and is discussed in this section, while the fornix serves the archicortex and is considered in the section on the limbic system (see below).

The **internal capsule** is shown in horizontal section in Figure 1.7 from which it can be seen that it is V-shaped in this view and that it lies medial to the lentiform nucleus and lateral to the thalamus and caudate nucleus. The fibres of the internal capsule fan out superiorly as the **corona radiata**, which interdigitates with fibres of the corpus callosum. The internal capsule proper is divided into an anterior limb, a genu (or knee) that faces medially, and a posterior limb (see Fig. 1.7). The anterior limb lies between the caudate nucleus and the lentiform nucleus and contains the fibres listed in Table 1.6. The posterior limb lies between the thalamus and the lentiform nucleus and is divided into three portions according to the relationship of the fibres to the lentiform nucleus (Table 1.6). The lenticular portion passes superiorly to the lentiform nucleus; the retrolenticular portion passes caudal to the lentiform nucleus; and the sublenticular portion passes infe-

Table 1.6 Fibres of the internal capsule

Limb	Fibres
Anterior	Corticothalamic
	Frontopontine
	Thalamocortical
Posterior	
lenticular portion	Corticobulbar
	Corticospinal
	Thalamic radiation (somaesthetic)
retrolenticular portion	Posterior thalamic radiation
sublenticular portion	Auditory radiation
	Optic radiation (geniculocalcarine)
	Temporopontine

riorly to the lentiform nucleus, from which some of its fibres pass caudally.

LIMBIC SYSTEM

The structure and connections of the limbic system are considered in this section. Details of the functions of the limbic system, including an historical consideration of the Papez circuit, appear in Chapters 2 and 11.

The limbic system is not precisely defined anatomically, and has been implicated in functions such as emotion and memory. The term **limbic lobe** was used by Broca in 1878 to describe the arrangement of cortical structures around the diencephalon, forming a border on the medial side of each cerebral hemisphere between the neocortex and the rest of the brain. The name is derived from the Latin *limbus* meaning border. This definition includes the following cortical areas (the limbic cortex): the cingulate, parahippocampal, and subcallosal gyri. These structures surround the upper brain stem. Subcortical nuclear groups embraced in this system include: the amygdaloid body (nucleus) and the septal area.

Since 1878 the concept of the limbic system has been enlarged from that of the limbic lobe to include other structures that are related to the limbic cortex and subcortical nuclei given above. However, there is disagreement among neuroanatomists about precisely which additional structures should be included. In this book the definition used by Snell (1987) will mainly be followed. In this definition the following grey matter structures are the principal additional components of the limbic system: the hippocampal formation (made up of the hippocampus, dentate gyrus, and parahippocampal gyrus); the anterior nucleus of the thalamus; and the hypothalamus. The connecting pathways of the limbic system are shown in Figure 1.8. Note that in this book the septal area is also considered as part of the limbic system although it is not part of Snell's definition.

The thalamus has been discussed earlier in this chapter. Similarly the anatomy and connections of the hypothalamus have been considered above and its functions are taken up in detail in Chapter 2. The hippocampal formation, the amygdaloid body, and the septal area will be considered further in this section. The connecting pathways of the limbic system include the alveus, fimbria, fornix, mammillothalamic tract and stria terminalis. With the exception of the mammillothalamic tract, which as noted in the section on the thalamus allows connections between the mammillary body and anterior nucleus of the thalamus, these pathways are discussed, as appropriate, below.

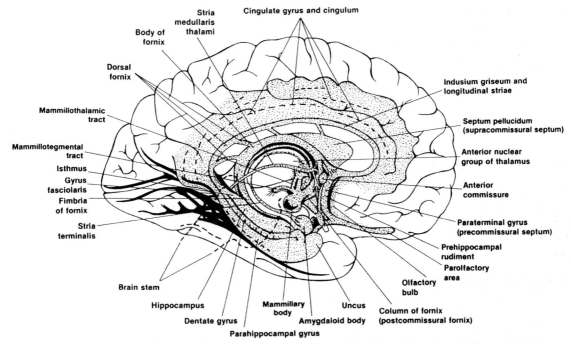

Fig. 1.8 Anatomy of the limbic system illustrated by the shaded areas of the figure. (Reproduced with permission from 1980 Gray's anatomy 35th edn. Churchill Livingstone, Edinburgh.)

Hippocampal formation

The hippocampal formation is made up of the hippocampus, dentate gyrus, and parahippocampal gyrus. Each of these structures will be considered in turn.

Hippocampus

The hippocampus is a cylindrical elevation of grey matter mainly lying in the floor of the inferior horn of the lateral ventricle. Anteriorly it is wide and forms the pes hippocampus. Posteriorly it narrows to end inferior to the splenium of the corpus callosum. The ependymal outer layer of the ventricular aspect covers a layer of white matter known as the **alveus**. Axons from the alveus of each hippocampus converge medially to form the **fimbria** and the crus of the **fornix**. Both crura converge inferior to the corpus callosum to form the body of the fornix, which in turn divides into the two columns of the fornix. The body of the fornix is connected anteriorly with the inferior surface of the corpus callosum via the midline triangular **septum pellucidum**. The latter forms a partition between the anterior horns of the lateral ventricles. The commissure of the fornix is formed by decussating fibres.

Histologically the hippocampus has a three-layered structure. The most superficial **molecular layer** contains nerve fibres and small neurones. The middle **pyramidal layer** is so named because it contains large pyramidal shaped neurones. The inner **polymorphic layer** has a structure similar to that of the polymorphic layer of the cerebral cortex described above.

The afferent and efferent connections of the hippocampus are shown in Tables 1.7 and 1.8 respectively. The **indusium griseum** mentioned in Table 1.7 refers to a layer of grey matter covering the superior surface of the corpus callosum. As mentioned previously fibres from each hippocampus pass to the

Table 1.7 Afferent connections of the hippocampus

Origin of fibres
Cingulate gyrus
Dentate gyrus
Hippocampus (opposite)
Indusium griseum
Parahippocampal gyrus
Secondary olfactory area (entorhinal area)
Septal area

Table 1.8   Efferent connections of the hippocampus

Destination of fibres in fornix
Anterior hypothalamus
Anterior nucleus of the thalamus
Habenular nucleus
Lateral preoptic area
Mammillary body (medial nucleus)
Septal nucleus
Tegmentum of the mesencephalon

other via the commissure of the fornix. The septal area connects with the hippocampi via the fornix.

Dentate gyrus

The dentate gyrus lies between the hippocampal fimbria and the parahippocampal gyrus. It is can be seen from Figure 1.8 that the dentate gyrus is continuous anteriorly with the hook-shaped **uncus**, and posteriorly with the indusium griseum.

Histologically the dentate gyrus, like the hippocampus, has a three-layered structure. The outer molecular layer and inner polymorphic layer are similar to those of the hippocampus. Instead of a pyramidal layer there is a middle **granular layer** of neurones which forms relays between parahippocampal structures and the hippocampus.

Parahippocampal gyrus

The parahippocampal gyrus adjoins the hippocampal fissure and is separated laterally from the remaining cerebral cortex by the longitudinal **collateral sulcus**. It is continuous anteriorly with the uncus, and along the medial border of the temporal lobe with the hippocampus. The parahippocampal gyrus includes the **secondary olfactory cortex** or **entorhinal area**. The part of the parahippocampal gyrus that adjoins the hippocampal fissure is the **subiculum**, and it is this structure that is considered by some neuroanatomists to be part of the hippocampal formation, rather than the whole parahippocampal gyrus.

The subiculum allows nerve fibres from the secondary olfactory cortex to pass to the dentate gyrus, from which they can pass to the hippocampus.

Amygdaloid body

The amygdaloid body is also known as the amygdala, amygdaloid nucleus and amygdaloid complex, and it consists of several nuclear groups. It is almond-shaped and situated at the tip of the tail of the caudate nucleus, with which it is continuous, and anterior and superior to the tip of the inferior horn of the lateral ventricle.

The amygdaloid body receives afferent connections from the lateral olfactory stria; the frontal and temporal association areas; the uncus; the septal area; the dopaminergic, noradrenergic, and serotonergic brain stem nuclei; and, via the anterior commissure, the opposite amygdaloid body. The most important efferent pathway is via the **stria terminalis**. This is an extensive set of nerve fibres that emerges from the posterior aspect of the amygdaloid body and passes to the hypothalamus and septal area. Efferents also pass to the lateral olfactory stria; the frontal and temporal association areas; the corpus striatum; and the thalamus.

Septal area

The septal area, also known as the septal nuclei, is situated inferior to the anterior portion of the corpus callosum, and like the amygdaloid body it comprises a subcortical group of nuclei.

It is connected with most of the other limbic cortical and subcortical structures including the amydaloid body; the hippocampus; the hypothalamus; the mammillary bodies; the medial forebrain bundle; the central grey matter of the mesencephalon; the thalamus; and the cerebral cortex.

CRANIAL NERVES

The 12 pairs of cranial nerves (see Table 1.9) all pass through foramina in the skull on leaving the brain. Their anatomy is considered in this section.

Olfactory nerve

The nerve fibres of the first cranial nerve represent the central processes of the olfactory receptor cells. They

Table 1.9   The cranial nerves

1. Olfactory nerve	7. Facial nerve
2. Optic nerve	8. Vestibulocochlear nerve
3. Oculomotor nerve	9. Glossopharyngeal nerve
4. Trochlear nerve	10. Vagus nerve
5. Trigeminal nerve	11. Accessory nerve
6. Abducent nerve	12. Hypoglossal nerve

pass from the olfactory mucosa, superiorly through the cribriform plate of the ethmoid bone, and synapse with the olfactory bulb mitral cells. From here the mitral cell axons pass in the olfactory tract, via the lateral olfactory striae, to the **primary olfactory cortex**. The latter is also known as the **periamygdaloid and prepiriform areas**.

Optic nerve

The optic nerve is formed by the axons of the retinal ganglion cells. In the optic chiasma, medial retinal fibres (carrying information from the temporal visual field) cross over to the contralateral optic tract. Lateral retinal fibres (carrying information from the nasal visual field) pass to the ipsilateral optic tract.

Most of the optic tract fibres synapse with neurones in the **lateral geniculate body** of the thalamus. A few optic tract fibres, concerned with pupillary and ocular reflexes, by-pass the lateral geniculate body and pass to the **pretectal nucleus** and the **superior colliculi**.

Axons leave the lateral geniculate body to form the **optic radiation** which passes, via the retrolenticular part of the **internal capsule** to the **visual cortex**.

The visual pathway is summarized in Figure 1.9.

Oculomotor nerve

The third cranial nerve has two motor nuclei: the **main oculomotor** or **somatic efferent nucleus** and the **accessory parasympathetic** or **Edinger–Westphal nucleus**. The main oculomotor nucleus lies in the grey

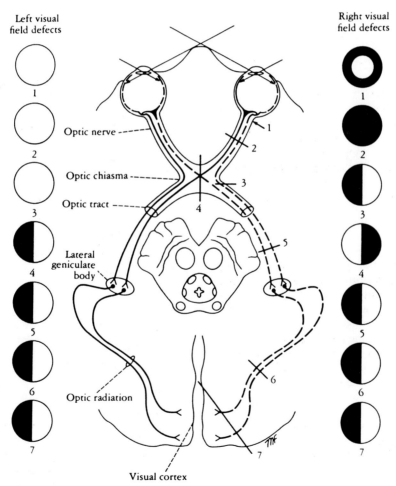

Fig. 1.9 The anatomical organization of the visual system and visual field defects that may result from lesions at different sites. (Reproduced with permission from Snell R S 1987 Clinical neuroanatomy for medical students. Little, Brown & Co, Boston.)

matter surrounding the cerebral aqueduct of the mesencephalon, at the level of the superior colliculi. It supplies all the **extrinsic ocular muscles** except the superior oblique (supplied by the trochlear nerve) and the lateral rectus (supplied by the abducent nerve). The accessory parasympathetic nucleus lies posterior to the main oculomotor nucleus and gives rise to **preganglionic parasympathetic** fibres to the **constrictor pupillae** and the **ciliary muscles**.

The oculomotor nerve divides into a superior and inferior branch prior to entering the superior orbital fissure. The superior branch supplies the levator palpebrae superioris and the superior rectus muscle. The inferior branch supplies the inferior oblique, the inferior rectus, and the medial rectus. The nerve to the inferior oblique carries the preganglionic parasympathetic fibres to the ciliary ganglion whence efferent fibres pass to the ciliary muscle and the constrictor pupillae of the iris, via about ten short ciliary nerves. Stimulation of this parasympathetic supply leads to pupil constriction and lens accommodation.

Trochlear nerve

The nucleus of the fourth cranial nerve lies in the grey matter surrounding the cerebral aqueduct of the mesencephalon at the level of the inferior colliculi and supplies one ocular muscle, the **superior oblique**.

Trigeminal nerve

Of the cranial nerves the fifth is the largest and possesses the following four nuclei: the **main sensory nucleus** lying in the posterior pons; the **spinal nucleus** continuous with the main sensory nucleus superiorly and passing through the medulla inferiorly to the level of C2; the **mesencephalic nucleus** lying in the grey matter surrounding the cerebral aqueduct and passing inferiorly into the pons; and the **motor nucleus** in the pons. In the remainder of this section the sensory and motor components of this nerve are considered in turn.

The trigeminal nerve is the main sensory nerve to most of the head and face and, as its name implies, there are three main divisions: the **ophthalmic nerve** (or division); the **maxillary nerve**; and the **mandibular nerve**. Their sensory fibres end in the inferior, middle and upper part of the spinal nucleus, respectively. The main branches and areas of innervation of these three divisions are shown in Table 1.10.

The motor component of the trigeminal nerve supplies the muscles of mastication, the anterior belly of the digastric, the mylohyoid, the tensor tympani and the tensor veli palatini.

Table 1.10 The sensory components of the trigeminal nerve

Division	Branch	Area of innervation
Ophthalmic	Frontal	Upper eyelid and scalp anterior to the lamboid suture (via the supraorbital and supratrochlear branches)
	Lacrimal	Lacrimal gland, lateral conjunctiva, and upper eyelid
	Nasociliary	Eyeball, medial lower eyelid, skin and mucosa of nose
Maxillary	Infraorbital	Skin of cheek
	Superior alveolar	Upper teeth
	Zygomatic	Skin of temple (via the zygomaticotemporal branch) Skin of cheek (via the zygomaticofacial branch)
	Branches from the sphenopalatine ganglion include:	
	greater and lesser palatine nerves nasal branches pharyngeal branch long and short sphenopalatine nerves	
Mandibular	Auriculotemporal	Skin of temple and auricle
	Buccal	Mucous membrane and skin of cheek
	Inferior alveolar	Lower teeth and skin of chin and lower lip
	Lingual	Anterior two-thirds of tongue and mucuous membrane of mouth

Both the ophthalmic and maxillary divisions are purely sensory, while the mandibular division is a mixed nerve, carrying all of the trigeminal nerve motor root.

Abducent nerve

The nucleus of the sixth cranial nerve lies in the upper pons inferior to the floor of the fourth ventricle and supplies one ocular muscle, the **lateral rectus**.

Facial nerve

The seventh cranial nerve has three nuclei: the **main motor nucleus** lying in the reticular formation in the lower pons; the **parasympathetic nuclei** lying posterolateral to the main motor nucleus; and the **sensory nucleus** which is the superior part of the tractus solitarius nucleus, lying near the main motor nucleus.

The main motor nucleus supplies the muscles of facial expression, the auricular muscles, the posterior belly of the digastric, the stapedius and the stylohyoid. Corticonuclear fibres from the contralateral cerebral hemisphere are received by the part of this nucleus that supplies lower face muscles, whilst corticonuclear fibres from both cerebral hemispheres are received by the part supplying muscles of the upper face.

The parasympathetic nuclei include the **lacrimal and superior salivary nuclei**. The lacrimal nucleus supplies the lacrimal gland. The superior salivary nucleus supplies the nasal and palatine glands, and the sublingual and submandibular salivary glands.

The sensory nucleus receives taste fibres, via the geniculate ganglion, from taste buds in the anterior two-thirds of the tongue, the floor of the mouth and the hard and soft palates.

The fibres of the motor root of the facial nerve pass from the motor nucleus posteriorly around the nucleus of the abducent nerve, forming the elevation of the floor of the fourth ventricle known as the **colliculus facialis**, and then pass anteriorly to emerge from the lateral aspect of the pons. The sensory root contains both the central processes of **geniculate ganglion** unipolar neurones and parasympathetic nuclei efferent preganglionic fibres.

The **chorda tympani** is a branch given off by the facial nerve before the latter passes through the stylomastoid foramen. It joins the lingual branch of the mandibular division of the trigeminal nerve, allowing its taste fibres to pass to the anterior two-thirds of the tongue.

Vestibulocochlear nerve

The eighth cranial nerve consists of two parts: the **cochlear nerve** which is concerned with hearing; and the **vestibular nerve** concerned with the maintenance of equilibrium.

The fibres of the cochlear nerve are the central processes of the spiral ganglion cells of the cochlea. They terminate in the anterior and posterior cochlear nuclei lying in the inferior cerebellar peduncle.

The fibres of the vestibular nerve are the central processes of neurones of the vestibular ganglion, which lies in the internal auditory meatus. They terminate in the lateral, medial, superior, and inferior vestibular nuclei, which lie in the floor of the fourth ventricle.

Figure 1.10 shows the central connections of the auditory pathway.

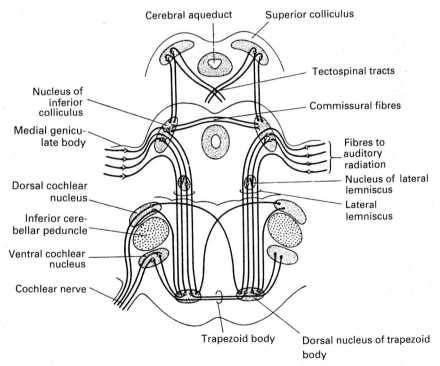

Fig. 1.10 A simplified diagram to show some of the central connections of the cochlear nerve and the auditory pathway through the brain stem. The fibres to the auditory radiation pass to the auditory cortex. Efferent fibres in the cochlear nerve and descending fibres in the auditory pathway have been omitted. (Reproduced with permission from 1989 Gray's anatomy 37th edn. Churchill Livingstone, Edinburgh.)

Glossopharyngeal nerve

The ninth cranial nerve has three nuclei: the **main motor nucleus** lying in the reticular formation in the medulla oblongata; the **parasympathetic or inferior salivary nucleus** lying inferior to the superior salivary nucleus and near the superior tip of the nucleus ambiguus; and the **sensory nucleus** which is part of the tractus solitarius nucleus.

The main motor nucleus receives corticonuclear fibres from both cerebral hemispheres and supplies the stylopharyngeus muscle.

The parasympathetic nucleus receives inputs from the hypothalamus, olfactory system, tractus solitarius nucleus and trigeminal sensory nucleus. Its pre-ganglionic fibres enter the tympanic branch of the glossopharyngeal nerve and reach the otic ganglion via the tympanic plexus and the lesser petrosal nerve. The postganglionic fibres supply the parotid gland via the auriculotemporal branch of the mandibular nerve.

The sensory nucleus receives taste information from the posterior one-third of the tongue.

Vagus nerve

The tenth cranial nerve has the most extensive distribution of the cranial nerves. It has three nuclei: the **main motor nucleus** lying in the reticular form-ation in the medulla oblongata; the **parasympathetic or dorsal nucleus** lying in the floor of the fourth ventricle; and the **sensory nucleus** which is the infe-rior part of the tractus solitarius nucleus.

The main motor nucleus is formed by the nucleus ambiguus. It receives corticonuclear fibres from both cerebral hemispheres. It supplies the intrinsic muscles of the larynx and the constrictor muscles of the pharynx.

The parasympathetic nucleus receives inputs from the hypothalamus, the glossopharyngeal nerve, the heart, the lower respiratory tract and the gastro-intestinal tract as far as the transverse colon. It supplies the involuntary muscle of the heart, the lower respira-tory tract and the gastrointestinal tract as far as the distal third of the transverse colon.

The sensory nucleus receives taste information from neurones of the inferior ganglion of the vagus nerve.

Accessory nerve

The eleventh cranial nerve consists of a small **cranial root** and a larger **spinal root**. The cranial root sup-plies, via the vagus nerve, muscles of the soft palate, larynx and pharynx. The spinal root is formed from fibres of neurones of the spinal nucleus, which lies in the anterior horn of cervical segments C1 to C5. It supplies the sternocleidomastoid and trapezius muscles.

Hypoglossal nerve

The nucleus of the twelfth cranial nerve lies in the floor of the fourth ventricle. It supplies all the extrinsic and intrinsic muscles of the tongue, with the exception of the palatoglossus muscle.

MENINGES

As mentioned above the brain and spinal cord are surrounded by the three meninges or membranes: the pia mater, the arachnoid mater and the dura mater.

Pia mater

The pia mater is the innermost of the meninges and closely ensheaths the brain and spinal cord. In the case of the brain it is carried into the sulci by cerebral arteries. A double fold of pia mater forms the **tela choroidea** of the root of the third and fourth ventricles. It fuses with the ventricular ependymal lining to form the **choroid plexuses** that are invaginated into the lateral, third and fourth ventricles (see the section on the ventricular system below).

In the spinal cord the pia mater is thickened on both sides between the nerve roots to form the **liga-mentum denticulatum**. This ligament is attached to the dura mater laterally. The ligamenta denticulata along both lateral aspects of the spinal cord help to keep the latter in place when the spine is moved. The pia mater is continued along spinal nerve roots. It continues inferiorly beyond the termination of the spinal cord as the **filum terminale** which is attached to the coccyx.

Arachnoid mater

The arachnoid mater is so named because of its web-like structure. It lies between the external dura mater and the internal pia mater. Between the arachnoid mater and dura mater lies a potential space, the **subdural space**, which normally contains a film of fluid.

The space between the arachnoid mater and pia mater is the **subarachnoid space** and it contains **cerebrospinal fluid**. The subarachnoid space is particularly large in some regions forming the **subarachnoid cisterns** which communicate freely with each other and the rest of the subarachnoid space.

The **cerebellomedullary space** or **cisterna magna** lies between the inferior surface of the cerebellum and the posterior aspect of the medulla oblongata. Cerebrospinal fluid passes directly into the cerebellomedullary space from the fourth ventricle via the **median aperture of Magendie**. The **pontine cistern** lies anterior to the pons and medulla oblongata and is traversed by the roots of the fifth to twelfth cranial nerves inclusive. Cerebrospinal fluid passes into this cistern from the fourth ventricle via the **lateral apertures of Luschka**. The **interpeduncular cistern** lies in the interpeduncular fossa between the two cerebral peduncles and is traversed by the third and fourth cranial nerves. In the spinal cord the subarachnoid space is enlarged inferior to the termination of the spinal cord. This allows cerebrospinal fluid to be obtained by lumbar puncture by a needle being passed, in the adult, between the third and fourth or fourth and fifth lumbar vertebrae.

Like the pia mater, the arachnoid mater is continued along the spinal cord nerve roots.

Dura mater

The dura mater is a firm fibrous membrane lining the inside of the skull. At the margins of the skull foramina and sutures it is continuous with the outer periosteum of the skull.

By convention the dura mater is divided into the **endosteal** and **meningeal layers** which are apparent as two layers only where they separate to form the **venous sinuses** and at the skull foramina (see Fig. 1.11). The inside of the vault of the skull contains a groove for the **superior sagittal sinus**. The **falx cerebri** is attached to the edges of this groove and consists of a vertical sickle-shaped fold of dura mater that lies in the longitudinal fissure between the cerebral hemispheres. The inferior concave free margin of the falx cerebri contains the **inferior sagittal sinus**. The latter passes posteriorly to the **straight sinus** which lies along the attachment of the falx cerebri to the tentorium cerebelli. The **tentorium cerebelli** is a crescent-shaped fold of dura mater that forms a partition between the posterior parts of the cerebral hemispheres and the cerebellum. The **transverse sinus** lies along the attachment of the tentorium cerebelli to the occipital bone.

CEREBROSPINAL FLUID

The cerebrospinal fluid is formed via an active process by the ependymal lining of the choroid plexuses of the

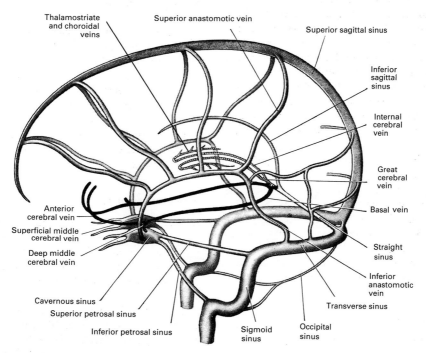

Fig. 1.11 Schema of the venous sinuses of the dura mater and their connections with the cerebral veins. (Reproduced with permission from 1989 Gray's anatomy 37th edn. Churchill Livingstone, Edinburgh.)

lateral, third and fourth ventricles. After being secreted into the ventricles it passes through the **foramina of Magendie and Luschka** (see above) of the fourth ventricle into the subarachnoid space. The cerebrospinal fluid flows from the region around the medulla oblongata superiorly around the cerebellum, pons, mesencephalon, and the cerebral surface. It also flows more slowly around the spinal cord.

It is re-absorbed, by a passive process, mainly by the **arachnoid villi** of the dural venous sinuses. The rest of the cerebrospinal fluid is absorbed by **spinal villi** and the **lymphatic system**.

The total volume of the cerebrospinal fluid in the ventricles and subarachnoid space is approximately 140 ml in the adult.

VENTRICULAR SYSTEM

Figure 1.12 is a diagrammatic representation of the ventricular system. It can be seen that each of the two lateral ventricles consists of an anterior horn, which is anterior to the interventricular foramen; a body; a posterior horn, which is in the occipital lobe; and an inferior horn, which reaches the temporal lobe.

The third ventricle lies between the thalami, as a slit-like cleft. Its floor is formed by the hypothalamus. The cerebrospinal fluid enters the third ventricle from the lateral ventricles via the **interventricular foramina (of Monro)**.

The fourth ventricle lies anterior to the cerebellum and posterior to the pons and the superior part of the medulla oblongata. It is rhomboid-shaped when viewed from above, and has a tent-shaped roof. The cerebrospinal fluid enters the fourth ventricle from the third ventricle via the **cerebral aqueduct (of Sylvius)** in the mesencephalon. As mentioned above,

the cerebrospinal fluid enters the subarachnoid space from the fourth ventricle via the **median aperture (foramen of Magendie)** and the **lateral apertures (foramina of Luschka)**.

Disorders involving the cerebrospinal fluid and the ventricular system are considered in Chapter 7.

BLOOD SUPPLY

The brain is supplied by two pairs of arteries: the **internal carotid arteries** and the **vertebral arteries**. They lie in the subarachnoid space and some of their branches anastomose to form the **circle or polygon of Willis**, also known as the **circulus arteriosus**. The spinal cord is supplied by the **posterior and anterior spinal arteries**, together with the **radicular arteries**, which are branches of the deep cervical, intercostal, and lumbar arteries.

The circle of Willis is shown in Figure 1.13. It surrounds the optic chiasma, and is formed by the following arteries: anterior communicating, anterior cerebral, middle cerebral, posterior communicating and posterior cerebral.

The anterior and middle cranial fossae are supplied by the **meningeal branches** of the internal carotid artery and the **middle meningeal branch** of the maxillary artery. The **ophthalmic branch** of the internal carotid artery supplies the contents of the orbit. Other branches of the internal carotid artery supply the pituitary gland and the trigeminal ganglion. The dura of the posterior cranial fossa is supplied by vertebral artery branches.

The **cortical branches** mainly supply the grey matter of the cerebral and cerebellar hemispheres. The **anterior cerebral artery** supplies the medial and superolateral aspects of the cerebral hemisphere, and

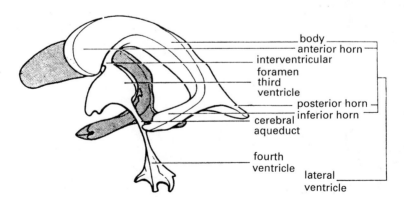

Fig. 1.12 The ventricular system. (Reproduced with permission from Aitken J T, Sholl D A, Webster K E, Young J Z 1971 A manual of human anatomy vol V: central nervous system 2nd edn. Churchill Livingstone, Edinburgh.)

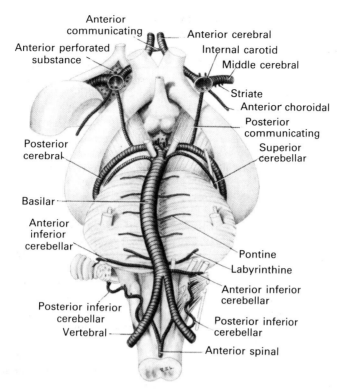

Fig. 1.13   The circle of Willis and the arteries of the brain stem. The arterial circle (horizontal) lies at right angles to the basilar artery (vertical). (Reproduced with permission from Last R J 1978 Last's anatomy 6th edn. Churchill Livingstone, Edinburgh.)

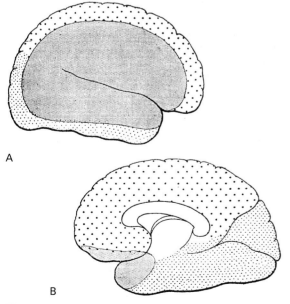

parts of the lentiform and caudate nuclei and the internal capsule. The **middle cerebral artery** supplies most of the lateral aspect of the cerebral cortex, the lentiform and caudate nuclei, and the internal capsule. The **posterior cerebral artery** supplies the inferolateral aspect of the temporal lobe, the lateral and medial aspects of the occipital lobe, parts of the thalamus and lentiform nucleus, the medial geniculate bodies, the midbrain, the pineal gland and the choroid plexuses of the lateral and third ventricles. The arterial supply of the cerebral cortex is shown in Figure 1.14. The **superior cerebellar**

	anterior cerebral artery
	middle cerebral artery
	posterior cerebral artery

Fig. 1.14   Diagram of (**A**) the lateral areas and (**B**) the medial areas, supplied by the cerebral arteries. (Reproduced with permission from Aitken J T, Sholl D A, Webster K E, Young J Z 1971 A manual of human anatomy vol V: central nervous system 2nd edn. Churchill Livingstone, Edinburgh.)

artery is a branch of the basilar artery and supplies the superior surface of the cerebellum, the pons, the pineal body and the superior medullary velum. The **anterior inferior cerebellar artery** is also a branch of the basilar artery and supplies the anterior and inferior parts of the cerebellum, the pons, and the medulla oblongata. The **posterior inferior cerebellar artery** is a branch of the vertebral artery and supplies the inferior surface of the cerebellum, the central nuclei of the cerebellum, the choroid plexus of the fourth ventricle and the medulla oblongata. Cerebral arterial syndromes are considered in Chapter 7.

The **central branches** are given off by the larger arteries and pierce the brain substance to supply the brain stem, the cerebellum, the pineal body, the medial geniculate body, the thalamus, the lentiform nucleus, and, via a branch of the ophthalmic artery, the retina.

The **anterior choroidal artery** is usually a branch of the internal carotid artery, and supplies the choroid plexus of the inferior horn of the lateral ventricle. The **posterior choroidal artery** is a branch of the posterior cerebral artery and supplies the choroid plexuses of the lateral and third ventricles.

The **anterior spinal artery** is formed by the fusion of a branch from each vertebral artery. It supplies the anterior two-thirds of the spinal cord. The **posterior spinal artery** arises from the posterior inferior cerebellar artery or the vertebral artery, and supplies the posterior one-third of the spinal cord.

The veins draining the brain open into the venous sinuses in the dura mater.

Further reading

Brodal A 1981 Neurological anatomy in relation to clinical medicine. 3rd edn. Oxford University Press, Oxford

De Groot J, Chusid J G 1991 Correlative neuroanatomy. 21st edn. Appleton & Lange, Los Altos

Snell R S 1987 Clinical neuroanatomy for medical students. 2nd edn. Little, Brown & Co, Boston

Williams P L, Warwick R, Dyson M, Bannister L H 1989 Gray's anatomy. 37th edn. Churchill Livingstone, Edinburgh

2. Neurophysiology

In this chapter the following areas in neurophysiology are considered: the physiology of neurones, synapses, and receptors; motor functions and reflexes of the spinal cord; motor functions of the brain stem; functions of the basal ganglia; functions of the cerebellum; functions of the thalamus; the cerebral cortex; the autonomic nervous system; neuroendocrinology; the limbic system; applied electrophysiology; and sleep and arousal. Neurotransmission is described in the next chapter.

NEURONES, SYNAPSES AND RECEPTORS

In this section basic concepts of the physiology of neurones, synapses and receptors are considered. Details of neurotransmission appear in Chapter 3.

Resting neurone

Neurones are excitable cells which can respond to a stimulus by undergoing a change in the resting electical potential that initiates a nerve impulse. The latter can then be conducted over relatively long distances. The **neuronal membrane** is a bimolecular lipid layer lined internally and externally with protein layers. The lipid is mainly phospholipid (see Chapter 3). The neuronal membrane has a high electrical resistance and electrical capacitance, and has different permeabilities for different ions. In the resting neurone the membrane is relatively permeable to potassium ions (K^+) and freely permeable to chloride ions (Cl^-). On the other hand, it is relatively impermeable to both sodium ions (Na^+) and organic anions (negatively charged ions such as protein ions).

Membrane potential

The resting membrane potential of a neurone is negative; for example that for a motoneurone is typically −70 mV. The extracellular concentration of sodium and chloride ions is greater than their intracellular concentrations. Similarly, the intracellular concentrations of potassium ions and organic cations is greater than their extracellular concentrations.

In considering the origin of the membrane potential, the intracellular organic anions are unimportant since the membrane is relatively impermeable to them. The gradient between the extracellular and intracellular sodium ion concentrations is maintained by a powerful **sodium pump** which, in an active process involving ATP (see Chapter 3), pumps sodium ions out of the neurone against their concentration gradient in exchange for potassium ions entering the neurone. It is relatively difficult for sodium ions to enter the intracellular fluid, however, because of the impermeability of the resting neuronal membrane to them. Although potassium ions enter the neurone as a result of the action of the sodium pump, a continuous build-up of the intracellular potassium ion concentration is avoided because the neuronal membrane is relatively permeable to the potassium ion. In spite of this permeability potassium ions are prevented from equilibrating with respect to concentration because of their positive charge; the active extrusion of sodium ions leaves a negative intraneuronal electrical charge which both attracts the intracellular positively charged potassium ions and repels the extracellular negatively charged chloride ions.

Active neurone

Action potential

When a neurone is stimulated, the membrane potential at the point of stimulation becomes less negative. This is known as **depolarization.** If the degree of depolarization is greater than a critical threshold level then a nerve impulse or **action potential** is generated, during

25

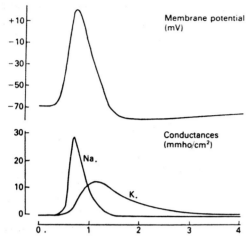

Fig. 2.1   Relation between the action potential and the changes in the conductances for Na$^+$ and K$^+$ in the squid axon. (Reproduced with permission from Emslie-Smith D, Paterson C R, Scratcherd T, Read N R 1988 Textbook of physiology 11th edn. Churchill Livingstone, Edinburgh; after Hodgkin A L, Huxley A F 1952 J Physiol 117: 500.)

which the membrane potential rapidly becomes positive, and then negative again, as shown in Figure 2.1.

The cause of the action potential is a change in the membrane permeability to sodium and potassium ions. It can be seen from Figure 2.1 that as the membrane potential increases (that is, becomes more positive) beyond the critical threshold level there is a rapid increase in the membrane permeability to sodium ions, which can be thought of as being caused by the opening up of membrane sodium ion channels. The rapid flow of sodium ions into the neurone continues until the membrane potential approaches the sodium ion equilibration potential whereupon the sodium ion channels can be thought of as closing. At the same time the membrane permeability to potassium ions increases above the resting level and there is increased movement of (positively charged) potassium ions out of the neurone, thereby restoring the membrane potential to a negative value. When this part of the membrane is again at rest the original ionic concentration gradients are restored by means of the energy-consuming sodium pump.

The process just described is repeated in adjacent parts of the neurone, allowing the conduction of the action potential along the length of an axon, for example. It is important to note that the number of ions actually involved in the production of an action potential is relatively small; most of the sodium and potassium ion gradients produced by the sodium pump remain intact.

All-or-none phenomenon

The passage of an action potential along a neuronal axon is an all-or-none phenomenon. The initiating stimulus to a neurone is either sufficient to cause a degree of depolarization that increases the membrane potential beyond the critical threshold level mentioned above, in which case an action potential results, or else this threshold is not reached and there is no action potential. It is not possible for there to be a fraction of an action potential. Since the action potential is regenerated at each stage in its conduction along a neurone, it does not undergo any diminution with conduction. Therefore this is an exceedingly effective mode of transmission of nerve impulses over long distances.

Refractory periods

When a part of the neuronal membrane is active it can be seen from Figure 2.2 that this results in a reversal of the polarity across it. During this period this part of the neurone is unable to conduct or initiate another nerve impulse. This is known as the **absolute refractory period.**

From Figure 2.1 it can also be seen that during the period of repolarization the membrane potential reaches a level more negative than the resting potential, owing to an increase in potassium ion conductance. This is known as **hyperpolarization** and during this time it is clearly more difficult for a stimulus to cause the membrane potential to reach the critical threshold level that it would need to in order to fire off another action potential. Hence the name **relative refractory period** for this time interval.

Because of these refractory periods it can be seen that a neuronal axon differs from an electrical wire that can carry continuous current. Rather, the 'current' can be thought of as being pulsed so that action potentials are carried singly or in bursts, but always as discrete units.

Conduction in unmyelinated fibres

In unmyelinated nerve fibres the velocity of passage of a wave of excitation is dependent on the **diameter** of the fibre. The greater the diameter, the greater the velocity. However, even with fibres of very large diameters the velocity is still much less than that in myelinated fibres of the same diameter.

Conduction in myelinated fibres

It will be recalled from Chapter 1 that myelinated nerve fibres are not covered with a homogeneous continuous myelin sheath. If this were the case the fatty myelin would prevent any conduction of action potentials. Instead, the myelin sheath is interrupted at the **nodes of Ranvier**. The result is that conduction appears to jump from one node of Ranvier to the next, in a process called **saltatory conduction** (from the Latin *saltare*, to leap). Therefore a myelinated axon has a much faster rate of conduction than an unmyelinated one of the same diameter.

A further advantage of myelination is that the entry of sodium into the neurone during conduction of a nerve impulse is confined to the nodes of Ranvier, rather than occurring along the whole length of the axon as is the case with unmyelinated fibres. Hence the relative amount of energy used by the sodium pump in restoring the intracellular ionic concentrations following the passage of an action potential is much lower for a myelinated fibre.

Synapses

Historically, it was observed that conduction usually occurred in one direction along a pathway of more than one neurone, in spite of the fact that, from the above discussion, one might expect a wave of excitation to be able to pass in either direction (Eccles 1964). The reason for this was found to be the fact that at the junction of two neurones, known as a synapse, there is no cytoplasmic continuity. Instead, there is a gap between the membrane of the presynaptic fibre and the membrane of the postsynaptic fibre. This gap is the **synaptic cleft**, and is often approximately 25 nm in width (one nanometre is 10^{-9} m).

Types

By far the commonest type of synapse in the human nervous system is the **chemical synapse**. In this type the transmission of a nerve impulse from one neurone to one or more others involves synaptic chemical neurotransmitters. Synapses can occur between most areas of the surface of neurones. They can also occur in a number of combinations, for example one neurone with another, one neurone with more than one other, or even one neurone with itself. Hence the various names for different types of synapses (presynaptic part first) such as: axo-dendritic; dendro-dendritic; axo-somatic; somato-somatic; and so on.

Compared with many lower vertebrates and the invertebrates, **electrical synapses** are relatively rare in the human nervous system. They are much faster than chemical synapses and, at least in invertebrates, sometimes allow bidirectional transmission. There is evidence that in humans they may occur in the cerebral cortex, the cerebellar cortex, and the superior olivary and vestibular nuclei.

Structure

As would be expected from the fact that conduction of nerve impulses occurs unidirectionally at chemical synapses, the structures of the presynaptic and postsynaptic membranes differ from each other. In the commonest type of synapse, the presynaptic axon divides into a large number of fine branches which terminate as expanded bulb-like regions known as **boutons** or **presynaptic knobs**. These boutons can also occur as rows along the surface of the distal part of the axon branches. The region of the bouton closest to the postsynaptic membrane contains rounded or flattened **vesicles** which contain neurotransmitter chemicals.

The postsynaptic area, particularly if it is part of a dendrite or the soma of the neurone, often has spine-like extensions known as **synaptic spines**. It is thought by some (e.g. Crick 1982) that these spines may be contractile. This would provide a way of altering the receptive area of the postsynaptic neurone, and may be important in the establishment of neural network memory (Koch & Poggio 1983).

Plasticity

In a process known as **long-term potentiation,** it has been found that periods of synaptic activity in regions such as the hippocampus, thought to be involved in memory, can strengthen the synaptic connection for relatively long periods. Moreover, cytoskeletal changes have been found in such synapses. This may mean that it is possible for **preferential conduction pathways** to be formed in some parts of the brain, particularly those involved in storing memory. Furthermore, experiments on newborn mammals in which one eye is covered for the first few weeks of postnatal life indicate that the normal increase in the number and size of synapses and of dendritic spines which occurs after birth is dependent on the amount of appropriate neuronal activity (Rothblat & Schwarz 1979).

Synaptic transmission

It has been mentioned that action potentials are all-or-none phenomena. Therefore, in order to transmit information about the intensity of the stimulus, the variables that can be brought into play are the frequency of firing of action potentials in the neurone, and the actual number of neurones in which action potentials are fired. Thus a stronger stimulus may initiate an increased frequency of firing in a larger number of neurones than a weaker stimulus might.

At **central excitatory synapses**, such as those that occur in the anterior or ventral horn of the spinal cord, stimulation of an afferent neurone leads to the release of an excitatory neurotransmitter, the generation of an **excitatory postsynaptic potential (EPSP)**, and a degree of depolarization in the efferent neurone. Further details of the process of neurotransmission are given in Chapter 3. If the EPSP is below the critical threshold for the initiation of an action potential then, although the latter is not generated, the efferent neurone is more sensitive to a subsequent EPSP for a short period of time. This phenomenon is known as **facilitation**. It has been found that this threshold for the membrane of the axon hillock (see Chapter 1) is lower than for the membrane of the dendrites or soma. Because the EPSP spreads through the soma of the neurone, if more than one afferent stimulus arrives at about the same time then the EPSPs can summate and may reach the threshold for an action potential. In this respect the EPSP is unlike an action potential in not being an all-or-none phenomenon. Figure 2.2 shows the effect on the membrane potential of the efferent neurone of an increase in the stimulus intensity, with the generation of an action potential.

At **central inhibitory synapses** stimulation of an afferent neurone leads to a hyperpolarization in the efferent neurone, that is, the membrane potential is made more negative and therefore changed in a direction away from the threshold. Hence this change in membrane potential is known as an **inhibitory postsynaptic potential (IPSP)** and this phenomenon is called **direct inhibition**. IPSPs can be caused by the release of an inhibitory neurotransmitter such as GABA (γ-aminobutyric acid), glycine and alanine (see Chapter 3). In a different type of inhibitory phenomenon known as **presynaptic inhibition** some small excitatory presynaptic fibres terminate on other presynaptic fibres. While the latter on their own generate an EPSP, when the former small fibres are stimulated they have the effect of diminishing the size of this EPSP.

A type of feedback inhibition known as **Renshaw** cell inhibition that is relevant to the functioning of the α-motoneurone is described later in this chapter.

Receptors

Sensory receptors

There are five major types of sensory receptor in the human body: mechanoreceptors, which detect mechanical deformation; thermoreceptors, which detect temperature change; light receptors in the retina; nociceptors, which detect tissue damage; and chemoreceptors. This section deals mainly with mechanoreceptors in order to illustrate the concepts discussed.

Many types of receptor, such as peripheral mechanoreceptors, are essentially **transducers** that convert detected stimulus changes into an electrical output. When the change detected is greater than a given threshold an action potential is generated which can be carried into the central nervous system from the peripheral nervous system. As mentioned above an increased stimulus is converted into an increased frequency of firing in the efferent neurone.

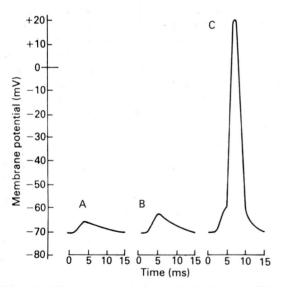

Fig. 2.2   Effect of an increase in the stimulus intensity of the membrane potential of the efferent neurone; in (**B**) the stimulus is greater than in (**A**) but still subthreshold; in (**C**) the stimulus has reached the threshold for the initiation of an action potential. (Reproduced with permission from Emslie-Smith D, Paterson C R, Scratcherd T, Read N W 1988 Textbook of physiology 11th edn. Churchill Livingstone, Edinburgh.)

Adaptation

Most sensory receptors exhibit the phenomenon of **adaptation** when they receive a continuous relatively prolonged appropriate stimulus, resulting in a progressively lowered firing frequency on the part of the receptor until for some types of receptor (**phasic receptors**) the firing stops altogether and for others (**tonic receptors**) it falls to a low maintained level. Examples of phasic mechanoreceptors include hair-follicle receptors and Pacinian corpuscles. Joint capsule receptors and muscle spindles (see below) are examples of tonic mechanoreceptors.

Chemoreceptors

Chemoreceptors contain specialized neurones which are able to respond to small changes in the extracellular concentration of chemicals. Types of chemoreceptor include taste bud receptors; olfactory receptors; various hypothalamic receptors which can detect blood osmolality, and blood concentrations of amino acids, fatty acids and glucose; and aortic and carotid body receptors able to detect changes in the blood oxygen and carbon dioxide concentrations.

Membrane receptors

From the description of the functioning of synapses it is clear that membrane receptors must exist on the postsynaptic membrane that are able to recognize and combine with neurotransmitters, with the subsequent electrical response outlined above. Membrane receptors also exist in other locations, such as the neuromuscular junction, where acetylcholine is an important neurotransmitter, and the neuroeffector junction of secretory glands.

MOTOR FUNCTIONS AND REFLEXES OF THE SPINAL CORD

Physiological aspects of the part played by the spinal cord in motor functions, including the integration of sensory information, are discussed in this section, together with a consideration of the stretch and tendon reflexes.

Spinal cord organization

Anterior motoneurones

Anterior motoneurones in the grey matter of the spinal cord give rise to the nerve fibres that exit the spinal cord via the anterior or ventral motor roots. There are two main types of anterior motoneurone: **alpha motoneurones** that give rise to large Aα **fibres** which innervate skeletal muscles; and **gamma motoneurones** that give rise to large Aγ **fibres** which innervate the intrafusal fibres of muscle spindles (see below). One Aα fibre and the muscle fibres supplied by it constitute a **motor unit.** The junction of a terminal branch of an Aα fibre and the voluntary muscle fibre it innervates is known as the **motor end-plate** or **neuromuscular junction** (see Fig. 2.3). Note that the axon loses its myelin sheath before its terminal branches each end on a muscle fibre.

Interneurones

Interneurones are present throughout the spinal cord grey matter. They interconnect with sensory fibres, with each other, and with anterior motoneurones. This allows for the integration of sensory and motor functions at the spinal cord level, including the functioning of spinal cord reflexes.

Renshaw cells are small inhibitory interneurones present in the spinal cord anterior horn. They produce an inhibitory negative feedback to the alpha motoneurones.

Sensory input to motoneurones and interneurones

Most sensory fibres to the spinal cord synapse first with interneurones. Hence spinal cord reflexes that involve such a pathway, from sensory fibre, through at least one interneurone, to anterior motoneurone, can be modified and can be complicated. An example of such a reflex is the flexor reflex discussed below. Because the pathway involves more than one synapse it is known as a **polysynaptic pathway.**

Some sensory fibres from muscle spindles synapse directly with anterior motoneurones without the intervention of interneurones. This forms the basis of the stretch reflex which, because it is a **monosynaptic pathway**, allows rapid feedback. A diagram of this monosynaptic pathway is shown in Figure 2.4.

Propriospinal fibres

Propriospinal fibres pass superiorly and inferiorly in the spinal cord and allow reflexes to occur that involve more than one spinal cord segment. An example of such multisegmental spinal cord reflexes is those involved in coordinating arm and leg movements while running.

Other inputs to motoneurones

In this section tracts that allow cortical and cerebellar control of motor functions are briefly discussed.

The **pyramidal** or **corticospinal tract** originates from the motor and somaesthetic areas of the cerebral cortex described in Chapter 1, and terminates mainly on interneurones in the dorsal horns of the spinal cord. A small proportion terminates directly on the anterior

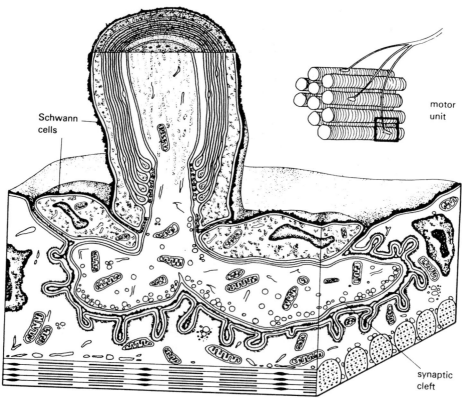

Schwann cells

motor unit

synaptic cleft

Fig. 2.3 The neuromuscular junction. (Reproduced with permission from Romero-Sierra C 1986 Neuroanatomy. A conceptual approach. Churchill Livingstone, Edinburgh.)

motoneurones. It will be recalled from the previous chapter that while most of the corticospinal tract fibres

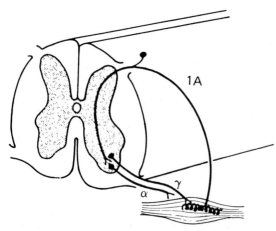

1A

γ

α

Fig. 2.4 The α and γ motor neurones and their relationships with the 1A afferent fibres in the monosynaptic stretch reflex. (Reproduced with permission from Chadwick N, Cartlidge N, Bates D 1989 Medical neurology. Churchill Livingstone, Edinburgh)

decussate to the opposite side of the body, some fibres provide an ipsilateral innervation. Stimulation leads to specific muscle contractions.

The **extrapyramidal tracts** are all the tracts that pass motor impulses from the cerebral cortex to the spinal cord, other than the pyramidal tract. They include the reticulospinal tract, rubrospinal tract, tectospinal tract, and vestibulospinal tract. A large proportion terminates directly on the anterior motoneurones while a small proportion terminates on neurones which inhibit the anterior motoneurones. Impulses transmitted by the extrapyramidal tracts cause less specific muscle contractions than is the case with the corticospinal tract, and lead to general facilitation and inhibition. These tracts also carry gross postural signals.

The muscle spindle and the stretch reflex

Physiological anatomy of the muscle spindle

Muscle spindles are specialized muscle receptors that are sensitive to changes in muscle fibre **length**. They also detect the **velocity of stretch**, that is, the rate of change of length. They occur abundantly in most

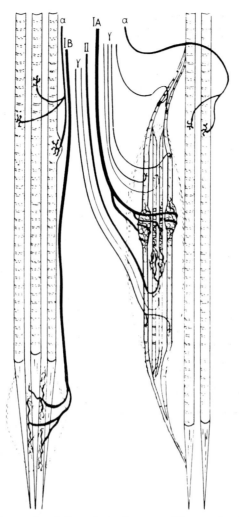

Fig. 2.5 Diagrammatic drawings of a muscle spindle and a tendon organ. The muscle spindle on the right is attached to extrafusal muscle fibres and tendon. It consists of small diameter intrafusal muscle fibres that are largely enclosed in a connective tissue capsule. Longitudinally the drawing is not to scale (the length of a spindle may be fifty times its width). Transversely in the drawing the width of the extrafusal muscle fibres represents a diameter of 40 μm; the intrafusal fibres are drawn to the same scale and represent diameters of about 20 μm for the two long fibres with nuclear bags at the equator of the spindle and about 10 μm for the two short fibres with nuclear chains at the equator. The group of nerve fibres shows the relative diameters of these fibres to each other. The largest nerve fibre, marked IA, supplies the main primary afferent ending lying over the nuclear bags and chains. Fibre II goes to a secondary afferent ending on the nuclear chain fibres adjacent to the primary ending. Six small γ fibres ('γ efferents', 'fusimotor fibres') of varying sizes supply motor endings on the intrafusal muscle fibres. The motor end-plates on the extrafusal muscle fibres are supplied by larger α nerve fibres. The remaining IB nerve fibre goes to the encapsulated tendon organ on the left; the branches of the afferent nerve ending lie between the tendons of a group of extrafusal muscle fibres. (By courtesy of Sybil Cooper.) (Reproduced with permission from Emslie-Smith D, Paterson C R, Scratcherd T, Read N W 1988 Textbook of physiology 11th edn. Churchill Livingstone, Edinburgh.)

voluntary muscles, a notable exception being the external muscles of the eye in which they occur in a much lower density.

Figure 2.5 is a representation of the structure of a muscle spindle with the motor and sensory innervation indication. Each muscle spindle contains an average of seven muscle fibres known as **intrafusal fibres** since they are encapsulated with fluid within a connective tissue sheath. The intrafusal fibres are attached at their ends to the surrounding extrafusal muscle fibres. When the ends of the intrafusal fibres contract, with the extrafusal fibres staying the same length, the central parts, because of their different structure, do not do so but instead are stretched by both ends. Similarly, when the whole muscle is stretched, this causes the attached ends of the intrafusal fibres to be placed under tension, with the result that the central parts of the intrafusal fibres are again stretched. These central parts comprise the sensory receptor part of the muscle spindle, while the ends of the intrafusal fibres receive motor efferent fibres.

It can be seen from Figure 2.5 that there are two types of sensory fibre from the central non-contractile sensory parts of the intrafusal fibres. The type or group Ia myelinated axons have a diameter of 12–20 µm and innervate the middle portion of the receptor. Their sensory endings spiral around this region of the intrafusal fibres, forming the **primary** or **annulospiral endings**. Because of their relatively large diameters the group Ia sensory fibres have a high speed of transmission of approximately $100 \, \text{ms}^{-1}$. The type or group II myelinated axons have a smaller diameter of 6–12 µm and innervate the intrafusal fibres adjacent to the primary ending, on both sides. Their sensory endings are known as **secondary** or **flower spray endings** because of their shape. Because these fibres have a smaller diameter they also have a lower speed of transmission of approximately $40 \, \text{ms}^{-1}$.

It can also be seen from Figure 2.5 that the contractile end portions of the intrafusal fibres receive a motor supply from small myelinated **gamma motoneurones**. In addition, each extrafusal muscle fibre has its own motor supply from an **alpha motoneurone**.

Comparator function

The two ways in which the central non-contractile part of the intrafusal fibres can be stretched, and therefore the muscle spindle stimulated, have been described above as being, firstly, contraction of the intrafusal fibres (following gamma efferent stimulation) with the extrafusal fibres remaining unchanged in length, and secondly, stretching of the whole muscle. It follows that the muscle spindle effectively functions as a **comparator** of the relative lengths of the extrafusal and intrafusal fibres. **Excitation** of the muscle spindle occurs when the extrafusal fibre length is greater than the intrafusal fibre length. Conversely, **inhibition** occurs when the intrafusal fibre length is greater than the extrafusal fibre length.

The stretch reflex

The neuronal pathway underlying the stretch reflex, also known as the muscle spindle reflex and the myo-

tatic reflex, is shown in Figure 2.4. There is a dynamic and a static component to the stretch reflex.

In the **dynamic stretch reflex**, when a voluntary muscle is suddenly and passively stretched, afferent impulses travel via type Ia fibres from muscle spindles to pass through the dorsal root of spinal cord. Since the fibres synapse directly with anterior alpha motoneurones in the spinal cord, the latter are stimulated and cause the same muscle to contract and regain its previous length before it was stretched. Because this is a monosynaptic pathway, and since type Ia fibres have a high rate of transmission, the dynamic stretch reflex occurs very rapidly indeed. It is the basis of **the tendon jerks** which are used clinically to test the functional integrity of the stretch reflexes. When the muscle tendon is struck with a tendon hammer the corresponding muscle is stretched. This initiates a dynamic stretch reflex. The segmental innervation of muscles has been described in Table 1.1.

In addition to the monosynaptic pathway just described, collaterals from the type Ia fibres together with type II afferents from the muscle spindles also transmit impulses to spinal cord interneurones and from them to propriospinal fibres. Thus, following the stretching of a muscle, sensory impulses from the muscle spindles also reach the brain stem, cerebellum and cerebral cortex. These in turn affect the degree of excitability of the efferent gamma motoneurones and so influence the overall muscle tone.

In the **static stretch reflex**, also called the **tonic stretch reflex**, steady stretch or vibration of a voluntary muscle leads to stimulation of all types of the muscle sensory receptor afferents. This results in more complicated stretch reflexes, including polysynaptic pathways, modulated by the descending pathways that control muscle tone and posture, so that the reflex muscle contraction continues for up to several hours while the muscle is being stretched or vibrated.

When a muscle is shortened suddenly, for example by the sudden loss of a load being carried, then both dynamic and static inhibition occur so that the muscle is not shortened further. This is known as the **negative stretch reflex** and is the opposite of the above positive stretch reflex.

The Golgi tendon organ and the tendon reflex

The Golgi tendon organ

Muscle tendons contain Golgi tendon organs, also known as neurotendinous endings, each connected in series with an average of 10 to 15 muscle fibres. The Golgi tendon organs detect muscle tension in the tendon caused by muscle contraction. They give rise to large myelinated type I axons that transmit information about changes in tension to the spinal cord and, via the spinocerebellar tracts, to the cerebellum.

The tendon reflex

The tendon reflex is an inhibitory reflex that occurs when a muscle is contracted, resulting in inhibition of the alpha motoneurones to the muscle. As well as being part of a feedback mechanism that controls muscle tension, this reflex acts as a protective mechanism for the muscle and its tendons allowing sudden relaxation of the muscle (the **lengthening reaction**) when the muscle is under very great tension, so preventing tendon avulsion or muscle tearing.

Spinal shock

Spinal shock refers to the phenomenon whereby following sudden transection of the spinal cord all the functions of the spinal cord, including the spinal cord reflexes, largely disappear. The spinal neurones gradually regain their normal functioning after a few days to a few months in many animals. However, this return may be delayed for much longer in humans, and sometimes it is not complete.

One of the effects in humans is abolition of the sacral reflexes involved in bladder and bowel evacuation. This may last several weeks but these reflexes usually return. Another effect is an immediate fall in the arterial blood pressure which may last a few days but which again usually returns to normal. Lost skeletal muscle reflexes may take several months to return to normal and some reflexes may eventually become hyperexcitable. The reflexes return in the following order: stretch reflex (returns first); flexor reflexes; postural reflexes; and parts of the stepping reflexes. The flexor reflex is a protective reflex that causes the withdrawal of a limb that has been subjected to a painful stimulus such as intense heat. The stepping reflex is a type of locomotive reflex of the spinal cord that causes rhythmic stepping movements in the limbs of, for example, spinal animals.

MOTOR FUNCTIONS OF THE BRAIN STEM

It will be recalled from Chapter 1 that the brain stem consists of the medulla oblongata, pons, and mesencephalon. So far as the anti-gravity support of the body is concerned, extensor muscle tone is maintained mainly by intrinsic excitation from the bulboreticular facilitatory area of the reticular formation and from the vestibular nuclei. These and other motor functions of the reticular formation and vestibular apparatus are considered in this section.

The reticular formation

The reticular formation is a loose network of neurones lying in the midline of the brain stem which is probably involved in sleep, consciousness, arousal, the activation and inhibition of movement, and motivation. It receives inputs from the spinal cord (including the spinoreticular tracts, spinothalamic tract collaterals and vestibular tracts), cerebellum, basal ganglia, hypothalamus, and cerebral cortex (particularly the motor cortex), and its axons project to the spinal cord, thalamus and hypothalamus.

The **bulboreticular facilitatory area**, making up most of the reticular formation, is excitatory. Stimulation of this region leads to an increase in muscle tone either generally or in local areas. Normally this excitatory activity is kept in balance because of inhibition by the basal ganglia, cerebellum and cerebral cortex. Experimentally, mammals that have these higher centres removed lose this inhibition and demonstrate rigidity of the anti-gravity extensor muscles.

The **bulboreticular inhibitory area**, consisting of a small part of the reticular formation in the inferior medulla, is inhibitory. Stimulation of this region leads to a decrease in muscle tone again either generally or in local areas. The inhibitory impulses from this region are not intrinsic. Rather, this region serves to channel the inhibitory signals mentioned above, from the basal ganglia, cerebellum, and cerebral cortex.

The vestibular apparatus

Structure

The vestibular apparatus consists of a **bony labyrinth** which contains the **membranous labyrinth**. The membranous labyrinth is shown in Figure 2.6 from which it can be seen to be made up of the cochlear duct, three semicircular canals, the utricle and the saccule. A sensory region known as the **macula** is located in the wall of the utricle and of the saccule, and is able to sense the orientation of the head relative to gravitational force and any forces of acceleration or deceleration to which the body is being subjected. The three semicircular canals, which contain the fluid **endolymph**, are arranged perpendicularly to each other, thus being able to detect changes in all three spatial dimensions. The **crista ampullaris** of each semicircular canal bears hairs that project into a jelly like mass known as the **cupula**. Movement of the

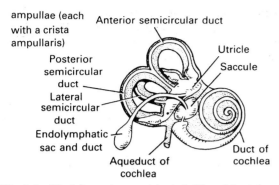

ampullae (each with a crista ampullaris)
Anterior semicircular duct
Utricle
Saccule
Posterior semicircular duct
Lateral semicircular duct
Endolymphatic sac and duct
Aqueduct of cochlea
Duct of cochlea

Fig. 2.6 The left membranous labyrinth. Viewed from the medial side. (Reproduced with permission from Last R J 1984 Last's anatomy 7th edn. Churchill Livingstone, Edinburgh.)

endolymph in a semicircular canal in one direction causes these hairs to bend and impulses to be transmitted from sensory afferents into the vestibular nerve. When these hairs are bent in the opposite direction by the endolymph their sensory afferents are inhibited.

The anatomy of the **vestibular nerve** has been described in Chapter 1. Its fibres are the central processes of neurones of the vestibular ganglion, which lies in the internal auditory meatus. They terminate in the lateral, medial, superior and inferior vestibular nuclei which lie in the floor of the fourth ventricle.

Functions

The cochlear duct is involved with hearing, the neuroanatomical pathways for which have been described in Chapter 1. The utricle and saccule are important in the maintenance of static equilibrium. In particular, the utricle is involved in the detection of linear acceleration.

The semicircular canals are also important in the maintenance of equilibrium. Their particular functions include the detection of angular acceleration; the detection of angular velocity; and the prediction that mal-equilibrium is going to occur before it happens. In addition, they play an important part in causing the stabilization of eye gaze. When the head is moved, the semicircular canals send signals that allow the angle of gaze to change by an amount equal and opposite to the change in angle of the head. Reflexes allowing this to occur are transmitted from the semicircular canals to the ocular nuclei via the vestibular nuclei, the cerebellum and the medial longitudinal fasciculus.

Nystagmus, a disturbance of eye posture in which there is almost rhythmical ocular oscillation, can occur when the semicircular canals are stimulated. Stimulation of the horizontal semicircular canals can lead to horizontal nystagmus, while stimulation of the vertical semicircular canals can cause rotary nystagmus. There are many other causes of nystagmus besides labyrinthine lesions — for example, retinal causes, cerebellar lesions, alcohol intoxication and intoxication with certain drugs such as some anticonvulsants. In addition, nystagmus may be congenital and indeed there are cases of familial nystagmus. Hysterical ocular movements resembling nystagmus have been described. Finally, it should be remembered that nystagmus may be a normal phenomenon occurring while trying to follow moving objects; this is known as optokinetic nystagmus.

Among the methods used clinically for testing the integrity of vestibular apparatus functioning are a test for positional nystagmus, the balancing test, the Barany chair test and the ice water test.

FUNCTIONS OF THE BASAL GANGLIA

It will be recalled from Chapter 1 that historically the basal ganglia have been described loosely and refer to large masses of grey matter found buried within the medullary body or white matter of each cerebral hemisphere. The components of the basal ganglia have been described in that chapter and are shown in Figure 1.6. The principal connections of the basal ganglia are also described in Chapter 1.

In essence, impulses converge into the basal ganglia from the cerebral cortex, thalamus, subthalamus and parts of the brain stem. After being 'processed' via the complex interconnections of the basal ganglia, impulses are then transmitted back to the same major regions of the central nervous system. In this way the basal ganglia are able to form an important subcortical link between, on the one hand, the motor cortex, and, on the other, the remainder of the cerebral cortex.

While it is clear that the basal ganglia play an important part in the functioning of the motor system of the central nervous system, it has so far proved difficult to allocate specific functions to each part of this structure. It is known that the corpus striatum is important in the control of posture. It is possible that the globus pallidus may be involved in the initiation of movement; bilateral ablation of this structure in the monkey causes poverty of movement including a diminished exhibition of manipulative movements. Since disease of the basal ganglia can lead to disturbances of muscle tone, including the onset of rigidity, and the development of tremor and athetoid, ballistic or choreiform movements, it is probable that the basal ganglia are involved in the balance between alpha and gamma motoneurone activity.

Clinically, neurosurgical operations directed at destroying the globus pallidus or the connected ventro-lateral nucleus of the thalamus have sometimes been successful in reducing contralateral rigidity or tremor. Neuropharmacological research has shown that dopamine, a neurotransmitter in the corpus striatum, substantia nigra, and the nigrostriatal pathway, is reduced in level in some cases of parkinsonism. Indeed their antidopaminergic effect in this region is probably the cause of phenothiazine-induced parkinsonism. Conversely, the therapeutic use of the dopamine precursor L-dopa in cases of parkinsonism can lead to increased levels of dopamine in this region, and can in turn cause athetoid movements as a side-effect.

The corpus striatum is also known to be rich in cholinergic neurones. Increased cholinergic activity can worsen parkinsonism, while anticholinergic drugs (actually antimuscarinic in their action — see Ch. 3) are used in clinical psychiatric practice for their anti-parkinsonian effect in patients treated with phenothiazines. Thus at the neurotransmitter level it may be that some of the motor disorders caused by abnormalities of the basal ganglia may be related to a disturbance of the balance between dopamine and acetylcholine, and between dopaminergic and cholinergic receptors and neurones.

FUNCTIONS OF THE CEREBELLUM

The structure and connections of the cerebellum have been considered in Chapter 1 (see also Fig. 2.7). The cerebellar cortex receives two types of afferent fibre: **climbing fibres**, which are the terminal fibres of the olivocerebellar tracts, and **mossy fibres**, which are the

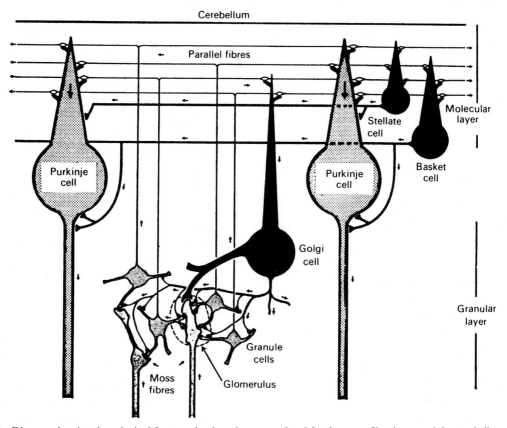

Fig. 2.7 Diagram showing the principal features that have been postulated for the moss-fibre input and the cerebellar glomerulus. The Golgi, stellate and basket cells, shown in black, are all inhibitory in action. The broken line represents the glial lamella that ensheathes a glomerulus. The diagram is drawn as for a section along the folium and the main distribution of the basket and stellate cells would be perpendicular to the plane of the diagram, but they are also distributed as shown to a band of several Purkinje cells along the folium. The arrows indicate the direction of impulse propagation. (Reproduced with permission from Eccles J C 1969 The inhibitory pathways of the central nervous system. Liverpool University Press, Liverpool.)

terminal fibres of all the other cerebellar afferent tracts. There is one type of efferent fibre from the cerebellar cortex, the Purkinje cell axons which release GABA (see Ch. 3) as an inhibitory neurotransmitter. The basket cells, Golgi cells and stellate cells (see Ch. 1) have an inhibitory action. The final output from the deep nuclei of the cerebellum is a function of the excitatory signals transmitted to them from the afferent fibres and the inhibitory signals transmitted to them from the Purkinje cells.

The main functions of the cerebellum are involved with the motor system, and these and other functions are now considered.

Motor functions

The cerebellum receives information about balance from the vestibular nerve. Information concerning involuntary movements is received mainly from the inferior olive. Visual information comes from the tectocerebellar tract. Information regarding voluntary movements is obtained from the cerebral cortex, muscle spindles and tendon organs. Regions of the motor system to which efferent signals from the cerebellum are sent include the motor cortex, the basal ganglia, the red nucleus, the reticular formation and the vestibular nuclei.

The cerebellum may act as a comparator, co-ordinating precise movements by comparing the somatosensory information, such as proprioceptive information from muscles, with the motor cortex and brain stem output, allowing the coordination of precise movements.

Motor abnormalities caused by cerebellar dysfunction include ataxia, disequilibrium (including asthenia), disturbance of reflexes, dysarthria and other speech defects caused by muscular incoordination, dysdiadochokinesis, dysmetria, hypotonia, intention tremor, nystagmus, past pointing, postural changes and alteration of gait. It has often been found to be difficult to map given motor abnormalities with distinct cerebellar regions. For example, midline lesions affecting the archicerebellum cause truncal ataxia (dysequilibrium without vertigo), in which patients stagger and sway and have a tendency to fall backwards, sometimes referred to as the flocculonodular syndrome. Such lesions may also cause positional nystagmus. However, truncal ataxia may also be caused by lesions affecting the paleocerebellum.

Other functions

Other functions that have been proposed for the cerebellum include the modulation of sensory transmission, autonomic homeostasis, and a role in motor memory and the acquisition of conditioned responses. It has also been suggested that since there is a regularity in the structure of the cerebellum with, for example, regularly spaced Purkinje cells that can be excited in sequence at relatively regular intervals, the cerebellum may act in some way as a timer or biological clock (Braitenberg 1977).

Cerebellar tumours may present as psychiatric disorders. There are also reports of atrophy of the vermis being found in CT brain scans of patients suffering from bipolar mood (affective) disorder and schizophrenia.

FUNCTIONS OF THE THALAMUS

The anatomy and connections of the thalamus have been presented in the previous chapter. The principal role of the thalamus is as the main sensory relay station of the nervous system. Neuroelectrical recordings have demonstrated that, like the somatotopic representation in the primary somaesthetic area described in Chapter 1, a similar somatotopic representation exists in the thalamus. Although somatosensory perceptions are mainly the result of sensory impulses reaching the cortical somaesthetic areas via the thalamus, it has been found that cortical lesions and even subcortical lesions that interrupt all thalamocortical fibres do not result in a total loss of the ability to perceive contralateral sensory information. Instead, the ability to perceive contralateral sensations of pain, temperature and crude touch remain relatively intact. On the other hand the perception of more delicate sensations such as discriminative touch is often very poor. This phenomenon has been interpreted as being indicative of the ability of the brain to allow cruder sensations to register in consciousness at the subcortical thalamic level.

Thalamic lesions can cause severe pain as part of the thalamic syndrome, which also leads to contralateral hemianaesthesia and hemianalgesia. The pain is very distressing and most commonly affects the face, arm and foot. Regions of the brain that can inhibit the transmission of pain signals to the thalamus include the nucleus raphe magnus of the medulla oblongata and the periaqueductal grey matter of the mesencephalon. As both these regions are rich in endogenous opiate receptors it has been postulated that endogenous opioid peptides acting in these regions are responsible at the neurotransmitter level for the inhibition of pain, including the inhibition caused by analgesics, placebos, and possibly even acupuncture (Martin et al 1976, Clement-Jones et al 1980).

THE CEREBRAL CORTEX

In addition to the functions of the cerebral cortex detailed in Chapter 1, such as its motor and sensory (olfaction, vision, hearing) roles, this is probably the part of the central nervous system most, although not necessarily solely, concerned with higher cognitive functions such as abstract thought and mental self-awareness.

The localization of function in the cerebral cortex has been discussed in Chapter 1. The role of the cerebral cortex in memory is looked at in Chapter 11. In this section the higher functions of the frontal lobe and cortical functional laterality are considered further.

Higher functions of the frontal lobes

In 1848 Phineas Gage, an American railway worker, suffered a major bifrontal lesion as the result of a crow-bar entering his skull. Following this injury he suffered a severe change in personality, becoming disinhibited and demonstrating diminished social control. Further-more, animal experiments showed that damage to the frontal lobes resulted in a reduction in problem-solving ability. Other indications of the higher functions of the frontal lobes have come from the results of psycho-surgical operations, particularly prefrontal leucotomy and lobotomy. Freeman and Watts, who in 1936 began using these psychosurgical techniques in the United States of America, concluded from the results that the human prefrontal regions were concerned with the adjustment of the personality as a whole to future contingencies, including the functions of foresight, imagination, and the apperception of the self (Brain & Strauss 1945).

Damage to the frontal lobes has not been found to be necessarily accompanied by loss of intellectual functioning. Moreover, when a deterioration in performance in intelligence tests has been found, it has not always proved possible to determine the extent to which this has been clouded by the greater distractibility that is known to occur in the frontal lobe syndrome. Further aspects of the frontal lobe syndrome are presented in Chapter 7.

Functional laterality

In 1861 the French anthropologist Broca found left hemispheric cortical damage in the area now known as Broca's motor speech area was associated with expressive aphasia. Since that time it has been found that damage in the equivalent region of the right cerebral hemisphere is not usually associated with such a speech defect. A similar situation is true for Wernicke's area, in the temporal lobe, which is also usually unilaterally represented in the left hemisphere, and damage to which is associated with receptive aphasia. The cerebral hemisphere associated with the expression of language is known as the **dominant hemisphere**. In almost all right-handed people the left hemisphere is dominant. In left-handed people the left hemisphere is dominant in approximately 60%, the rest having either a right dominant hemisphere or a bilateral representation of language functions. From studies on **split brain** patients, in whom the corpus callosum has been surgically severed so preventing the normal communication between both cerebral hemispheres, as well as from studies of patients with unilateral cerebral lesions and epilepsy, it has been concluded that the dominant hemisphere is associated with the more analytical, mathematical and 'logical' functions, in addition to its language functions.

Although the hemisphere that is not involved in the expression of language, usually the right side, is called the **non-dominant hemisphere**, this is really a misnomer. It is true that the non-dominant hemisphere has been found in split brain experiments to respond to only simple language and to be capable of only very simple arithmetic (Nebes & Sperry 1971). However, there is good evidence that the non-dominant hemisphere is in its own turn 'dominant' for certain other functions including visuospatial perception, and artistic and musical abilities. It also appears to be more involved in imagery, dreaming, emotion, impulsiveness and the perception of social cues.

Overall there is evidence that the more analytical dominant (usually left) hemisphere focuses on details while the non-dominant cerebral hemisphere has a holistic perceptual and problem-solving approach.

AUTONOMIC NERVOUS SYSTEM

It will be recalled from the previous chapter that the nervous system can also be divided into two parts on a functional basis: the somatic nervous system, which is concerned primarily with the innervation of voluntary structures, and the autonomic nervous system, which is concerned primarily with the innervation of involuntary structures and which is subdivided into the **sympathetic** and **parasympathetic** parts, each of which has afferent and efferent fibres.

Physiological anatomy of the sympathetic system

The preganglionic cell body lies in the intermediolateral horn of segments T1 to L2 of the spinal cord. It sends a fibre to the spinal nerve via the anterior root of the spinal cord. From here it passes, via the white ramus, to the sympathetic chain ganglion. The post-ganglionic nerve usually passes to the organ being innervated. Note that in the case of the suprarenal or adrenal medullae there is direct innervation by sympathetic preganglionic fibres.

Physiological anatomy of the parasympathetic system

Approximately three-quarters of all the parasympathetic fibres are in the **vagus nerve**. The parasympathetic or dorsal nucleus of the vagus, which is in the floor of the fourth ventricle, receives inputs from the hypothalamus, the glossopharyngeal nerve, the heart, the lower respiratory tract, and the gastrointestinal tract as far as the transverse colon. The vagus supplies the involuntary muscle of the heart, the lower respiratory tract, and the gastrointestinal tract as far as the distal one-third of the transverse colon.

Preganglionic parasympathetic fibres of the accessory parasympathetic or Edinger–Westphal nucleus of the **oculomotor nerve** pass to the ciliary ganglion. Ciliary ganglion efferent fibres in turn pass to the ciliary muscle and the constrictor pupillae of the iris, by way of about ten short ciliary nerves. Stimulation leads to pupil constriction and lens accommodation.

The lacrimal nucleus of the **facial nerve** supplies the lacrimal gland. The superior salivary nucleus of the facial nerve supplies the nasal and palatine glands, and the sublingual and submandibular salivary glands.

The parasympathetic or inferior salivary nucleus of the **glossopharyngeal nerve** receives inputs from the hypothalamus, olfactory system, tractus solitarius nucleus and trigeminal sensory nucleus. Preganglionic fibres from this nucleus enter the tympanic branch of the **glossopharyngeal** nerve and reach the otic ganglion via the tympanic plexus and the lesser petrosal nerve. Postganglionic fibres supply the parotid gland via the auriculotemporal branch of the mandibular nerve.

The **sacral** parasympathetic fibres, from S2-3, and sometimes also S1 and S4, form the **nervi erigentes** which supply the external genitalia, lower ureters, bladder, rectum and descending colon.

With the exception of some cranial nerves, the pre-ganglionic parasympathetic fibres pass to the wall of the organ being innervated, where they synapse with postganglionic neurones.

Neurotransmitters

Two neurotransmitters are involved in the autonomic nervous system: **acetylcholine** (released by **cholin-**

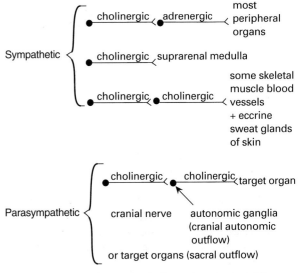

Fig. 2.8   The cholinergic and adrenergic neurones of the autonomic nervous system.

ergic fibres) and **noradrenaline** or norepinephrine (released by **noradrenergic fibres**). This is summarized in Figure 2.8.

All preganglionic neurones, both sympathetic and parasympathetic, are cholinergic. Of the sympathetic postganglionic neurones, only a minority supplying blood vessels in skeletal muscle involved in exercise and fainting, and eccrine sweat glands of the skin, are cholinergic. All parasympathetic postganglionic neurones, including the outflow carried by the cranial nerves, are cholinergic.

It can be seen from Figure 2.8 that the majority of sympathetic postganglionic neurones are noradrenergic.

Autonomic effects

The sympathetic nervous system is that part of the autonomic nervous system that can be thought of as being involved in emergency functions and the response to stress — the 'fight–flight' response. The parasympathetic nervous system, on the other hand, is involved in more 'vegetative' functions, such as the conservation of body resources and the preservation of normal resting functions. When these general concepts are borne in mind, the effects of sympathetic and parasympathetic stimulation are relatively easy to work out from first principles.

So far as the **eye** is concerned, parasympathetic stimulation leads to contraction of the ciliary muscle and the circular muscle of the iris. Ciliary muscle contraction results in a relaxation of the tension on the lens, allowing the latter to become more convex and thereby enabling near objects to be more easily focussed on the retina. Contraction of the circular muscle of the iris has the effect of constricting the pupil. On the other hand, sympathetic stimulation causes contraction of the radial muscle of iris, thereby dilating the pupil. Sympathetic stimulation also leads to contraction of the smooth muscle of the eyelids and the nictitating membrane.

Parasympathetic stimulation causes secretion from, and sympathetic stimulation vasoconstriction in, the following **glands**: the gastric, lacrimal, nasal, pancreatic, parotid and submaxillary.

In the **heart**, sympathetic stimulation leads to an increase of activity overall, which is consistent with the 'fight–flight' response. There is stimulation of the SA node, leading to an increase in the heart rate. Excitation of the AV node and conducting tissue of the heart takes place. The force of contraction of the cardiac muscle is increased, and the blood supply to the heart itself is increased through vasodilatation of the coronary arteries. Parasympathetic stimulation causes the opposite effects to occur; that is, inhibition of the SA node, depression of the AV node and conducting tissue, decreased force of atrial muscle contraction, and constriction of the coronaries.

Since the **lung airways** are poorly innervated by the autonomic nervous system, the effects of the latter are relatively mild. Nevertheless, again consistent with the 'fight–flight' response, sympathetic stimulation leads to an opening up of more airways by causing relaxation of the smooth muscle in the lungs. Parasympathetic stimulation, on the other hand, leads to the contraction of this smooth muscle, as well as causing secretion from glands in the lungs.

The effect of sympathetic stimulation on the **gut** is to decrease peristalsis and increase the tone of the lumen and sphincters. Parasympathetic stimulation leads to the opposite effects.

Important sympathetic effects on **hepatic function** include the promotion of glycogenolysis, which can be thought of as a way of helping to provide energy in a crisis situation, and the inhibition of bile secretion. The parasympathetic effects, on the other hand, are to stimulate both glycogenesis (thus helping to conserve body resources) and the secretion of bile.

The contraction of the detrusor muscle of the **urinary bladder** is a parasympathetic effect. The parasympathetic nervous system also sends inhibitory fibres to the internal sphincter. It would clearly be inappropriate for a strong desire to micturate to be stimulated during the 'fight–flight' response. In fact, the sympathetic nervous system sends inhibitory fibres to the detrusor muscle, and motor fibres to the internal

sphincter, the muscle of the upper urethra and the trigone. Similarly, while parasympathetic stimulation causes excitation in the **ureter**, sympathetic stimulation leads to inhibition in the ureter and a decreased **renal output** of urine.

With respect to **systemic blood vessels** sympathetic stimulation constricts abdominal vessels. This is useful in the 'fight–flight' response, allowing the blood to be diverted to where it may be needed more: the skeletal muscles, heart and brain, for example. Stimulation of the *adrenergic* postganglionic sympathetic neurones to skin and muscle blood vessels results in constriction. Conversely, stimulation of the *cholinergic* sympathetic fibres to skin and muscle blood vessels results in dilatation. The parasympathetic nervous system also sends inhibitory fibres to blood vessels in the skin.

Other effects of sympathetic stimulation on the **skin** include contraction of the pilomotor muscles, making the hairs on the surface of the body appear to stand on end, and secretion from the eccrine sweat glands, leading to sweating and the loss of excess body heat.

Erection of the **penis** results from parasympathetic stimulation, while sympathetic stimulation causes ejaculation.

Sympathetic stimulation leads to an increase in the **coagulability of the blood**, which can be thought of as a preparation for potential injury during the 'fight–flight' response.

Other effects of sympathetic stimulation include an increase in **basal metabolism**, **suprarenal (adrenal) cortical secretion**, and **mental activity**.

Suprarenal medulla

Stimulation of the medulla of the suprarenal or adrenal gland causes the release of relatively large quantities of the hormones adrenaline and noradrenaline into the bloodstream. This results in similar effects on the body as those of direct sympathetic stimulation, with the exception that these effects last much longer in the case of the bloodborne hormones.

NEURO-ENDOCRINOLOGY

The last section ended with a consideration of the effects on the body of hormones released by the suprarenal medulla. In this section the psychoneurological effects of these and other hormones are examined.

The pituitary gland both controls the release of hormones from many other endocrine organs, and is also an endocrine organ releasing hormones in its own right. The pituitary is under the influence of the hypothalamus, and so this section begins with a consideration of the physiology of the hypothalamus and pituitary gland. Indeed, as mentioned in Chapter 1, together, the hypothalamus and pituitary gland have been termed the master endocrine gland of the body. Historically these two structures together with the suprarenal cortex, the hypothalamic-pituitary-suprarenal cortex axis, have played a central part in neuroendocrine research for many years, beginning with studies into the endocrinological aspects of stress. This section ends by considering hormonal compounds of importance in neuro-endocrinology.

The influences of the limbic system on the hypothalamus are described in the next section of this chapter.

Hypothalamus

Hypothalamic nuclei

There are a number of methods of subdividing the hypothalamus. One such method, mentioned in Chapter 1, consists of organizing these nuclei into three major structures: the supraoptic area superior to the optic chiasma, the tuber cinereum and infundibulum and the mammillary area superior to the mammillary bodies. Another way of subdividing these is to consider the following three regions or zones on each side of the paramedian plane.

The **periventricular zone** is a subependymal medial region which contains the following nuclei: part of the preoptic nucleus, the suprachiasmatic nucleus, the paraventricular nucleus, the infundibular nucleus and the posterior nucleus. Of these the large **paraventricular nucleus** is the most important from a neuro-endocrinological viewpoint, and is discussed further below, as is the **suprachiasmatic nucleus**, which probably acts as an endogenous neural pacemaker or biological clock.

The **intermediate zone** also lies in the medial region and is alternatively referred to as the **medial zone**. It contains: another part of the preoptic nucleus, the anterior nucleus, the dorsomedial nucleus, the ventromedial nucleus and the premammillary nuclei.

The **lateral zone** lies lateral to the above two zones and contains: another part of the preoptic nucleus, the supraoptic nucleus, the lateral nucleus, the tubero-mammillary nucleus and the lateral tuberal nuclei. Like the paraventricular nucleus, the **supraoptic nucleus** is involved in the neurosecretion of hormones and is discussed below. It should be noted that because of their involvement in neurosecretion both the supraoptic and paraventricular nuclei receive a good blood supply. The suprachiasmatic nucleus

receives retinal photosensory information from the **retinolateral hypothalamic tract** via the **lateral nucleus** (see below).

Connections

The hypothalamus receives widespread afferent connections from the cerebral cortex, limbic system, diencephalon, mesencephalon, retina, olfactory pathways and ascending spinal cord pathways. The efferent pathways from the limbic system include the stria terminalis, fornix, medial forebrain bundle and ventral amygdalofugal tract.

The hypothalamus has efferent connections with the autonomic nervous system, somatic motor system, limbic system and mesencephalon. The hypothalamus also has neurovascular connections with the pituitary gland. The neuronal part consists of axons of hypothalamic neurones that end in the median eminence, infundibular stem and **posterior lobe** (pars posterior) of the pituitary, together known as the **neurohypophysis**. The vascular part consists of the long and short **hypophyseal portal veins** between sinusoids in the infundibulum and median eminence and capillary plexuses in the **adenohypophysis** (anterior pituitary). This **hypothalamopituitary portal supply** allows hormone releasing factors and hormone release-inhibiting factors to reach the adenohypophysis from the hypothalamus and median eminence.

Functions

The **endocrine control** of pituitary gland hormone release takes place by means of very small amounts of hypothalamic releasing factors and release-inhibiting factors being secreted into the hypothalamopituitary portal system. They thereby reach the adenohypophysis where they have a potent effect on the rate of synthesis and release of hormones. The latter may affect target cells directly as in the case of prolactin. Alternatively, some of the pituitary hormones act as tropic (trophic) hormones that act via a second endocrine organ. For example, thyrotropin (thyrotrophin) or thyroid stimulating hormone (TSH) acts on the thyroid gland, corticotropin or adrenocorticotropic hormone (ACTH) acts on the suprarenal cortex, and luteinizing hormone (LH) and follicle stimulating hormone (FSH) act on the gonads. Table 2.1 shows the anterior pituitary hormones and the corresponding hypothalamic releasing factors and release-inhibiting factors. Note that these factors are also commonly called hormones once their structure has been determined. For example, thyrotropin releasing factor is

Table 2.1 Anterior pituitary hormones and their corresponding hypothalamic releasing factors and release-inhibiting factors

Anterior pituitary hormones	Hypothalamic releasing factors and release-inhibiting factors
Corticotropin (adrenocorticotrophic hormone, ACTH)	Corticotropin releasing factor (CRF)
Follicle stimulating hormone (FSH)	Gonadotropin releasing factor (GnRF)
Luteinizing hormone (LH)	Gonadotropin releasing factor (GnRF)
Melanocyte stimulating hormone (MSH)	MSH release inhibitory factor (MIF)
Prolactin	Prolactin releasing factor (PRF)
	Prolactin release inhibitory factor (PIF) (dopamine)
Somatotropin (growth hormone, GH)	Growth hormone releasing factor (GRF; somatocrinin)
	Growth hormone release inhibitory factor (somatostatin)
Thyrotropin (thyroid stimulating hormone, TSH)	Thyrotropin releasing factor (TRF)

also known as thyrotropin releasing hormone. Note also that prolactin release inhibitory factor is now known to be dopamine.

The hypothalamus is responsible for the **neurosecretion** of the two posterior pituitary hormones **oxytocin** and **arginine vasopressin** (AVP) (also known as **argipressin** or antidiuretic hormone (ADH)). These two hormones are synthesized mainly in two nuclei of the hypothalamus, the paraventricular and the supraoptic. While both nuclei are involved in the synthesis of both hormones, it has been found that the **paraventricular nucleus** is predominantly responsible for the production of oxytocin, and the **supraoptic nucleus** is predominantly responsible for the production of arginine vasopressin. After combining with binding proteins known as **neurophysins**, these hormones travel down hypothalamopituitary (hypothalamohypophysial) tract axons to storage cells of the neurohypophysis. The rate of their release from these storage cells into the blood stream is influenced by a number of factors. In the case of oxytocin, release is increased by stimulation of the genitalia and nipples. The release of arginine vasopressin is increased by increased osmolarity and reduced volume of the blood. The former is

detected by **osmoreceptors** in the hypothalamus itself, and the latter is detected by volume receptors in walls of the large veins and the atria of the heart. Arginine vasopressin acts to increase renal reabsorption of water, making the urine more concentrated. It is considered further later in this chapter.

In addition to the above hormones, the paraventricular neurones are also rich in other peptides including angiotensin II, cholecystokinin (CCK), corticotropin releasing factor (CRF), dopamine, leu- and metenkephalin, neurotensin, peptide histidine isoleucine, somatostatin and substance P. Similarly, the supraoptic nucleus has also been found to be rich in other peptides. A third hypothalamic nucleus that is also particularly rich in peptides is the arcuate nucleus. It is clear that the workings and interactions of the hypothalamus are complex, and at the time of writing the neurophysiological implications of the presence of so many different peptides in the hypothalamus is still far from clear.

The hypothalamus has been shown, in stimulation and ablation studies, to influence the **autonomic nervous system**. Generally speaking, the anterior hypothalamic zones appear to influence mainly the parasympathetic side of the autonomic nervous system, and the posterior zones the sympathetic side. However, this distinction is not absolute and it is probably incorrect to talk of a parasympathetic centre and a sympathetic centre as has been the case in the past.

Since it controls the release of pituitary gonadotropins and prolactin, and is responsible for the neurosecretion of oxytocin, it is clear that the hypothalamus has a role in **sexual behaviour**. Mammalian experiments, including neuro-electrical stimulation, imply that some regions of the hypothalamus, particularly the periventricular zone, are concerned with reproductive endocrine reflexes, female receptivity, and copulatory movements in both males and females. Moreover, receptors for circulating sex steroids are found in the paraventricular and supraoptic nuclei. It has already been mentioned that the release of oxytocin, produced mainly by the paraventricular nucleus, is increased by nipple and genital stimulation. Such stimulation occurs during suckling and sexual intercourse and allows oxytocin to exert its effects of causing mammary muscular contraction and uterine muscular contraction during these events. The latter effect of this hormone also plays a role in parturition.

The medial and lateral hypothalamus have opposite effects so far as food intake is concerned. There appears to be a medial **satiety centre**; experimental stimulation of the medial zone in mammals leads to a reduction in eating, while ablation of this region results in hyperphagia. Conversely, the lateral zone appears to contain a **hunger centre**; stimulation of this region causes an increase in eating, while hypophagia or aphagia results from ablation. The mechanism by which the hunger or feeding centre operates may be related to the monitoring and regulation of metabolic functions which take place in this region; for example, it contains neurones that are sensitive to the circulating blood glucose concentration. It is probable that the hunger centre uses both the neuroendocrine and autonomic connections of the hypothalamus in carrying out its functions. Thus lateral stimulation in an experimental mammal in which hyperphagia is prevented still leads to increased deposition of adipose tissue in the body, presumably as a result of endocrine activity.

In addition to the hypothalamic osmoreceptors mentioned above, the lateral zone appears to contain a **thirst centre**. Stimulation of this region leads to increased drinking, even to the point of hyperhydration.

The hypothalamus has been shown to play a key role in **thermoregulation**. As in the case of satiety and hunger, it has been proposed that there are two hypothalamic centres with opposing actions on body temperature. The centre that acts against a rise in body temperature is thought to be in the anterior hypothalamus while its opposing 'antitemperature drop' centre is thought to be located more posteriorly. They act via the autonomic and endocrine systems to control the rate of loss of heat from the body by, for example, acting on the level of peripheral vasoconstriction or vasodilatation and stimulating either shivering or sweating. Receptors that send signals to these proposed hypothalamic thermoregulatory centres include peripheral thermodetectors, hypothalamic thermodetectors that detect the temperature of the blood reaching the brain, and hypothalamic neurones that respond to the presence of viruses and certain substances including drugs in the blood.

It was mentioned earlier in this chapter that there is evidence that the cerebellum may act in some way as a timer or biological clock. The evidence that the **suprachiasmatic nucleus** of the hypothalamus acts as, or is intimately involved with, a **biological clock**, is more compelling. For example, ablation experiments in mammals, including primates, have demonstrated that bilateral destruction of the suprachiasmatic nucleus leads to the disruption of many biological **circadian rhythms**, that is, regular endogenous bodily rhythms with a period of between 20 and 28 hours. The circadian rhythms interfered with include those concerned with drinking and eating, the sleep–wake cycle (which

is considered further in the section on sleep and arousal later in this chapter), and the heart rate. Neuroendocrine circadian rhythms that are disrupted include those for suprarenal cortical (adrenocortical) secretion, glucocorticoid secretion, somatotropin (growth hormone) secretion, and pineal N-acetyltransferase activity. It should be noted that so far lesions in no other part of the brain have been found to have a disruptive effect on such a wide variety of circadian rhythms. Since there tends to be a modifying environmental input to the period of many circadian rhythms, related in particular to the varying daily day–night cycle, it could be expected on theoretical grounds that there might be a neuro-anatomical pathway to the suprachiasmatic nucleus that provides sensory information about the amount of light in the environment. Such a pathway has indeed been identified. The **retinohypothalamic tract** is a monosynaptic pathway from the retina to the ipsilateral suprachiasmatic nucleus. Furthermore, the suprachiasmatic nucleus also receives retinal photosensory information from the **retinolateral hypothalamic tract** via the lateral nucleus of the hypothalamus. Photostimulation has been found experimentally to increase suprachiasmatic nucleus neuronal activity which in turn is probably responsible for reducing the secretion of N-acetyltransferase from the pineal gland, possibly by inhibiting sympathetic input to the pineal (see Nishino et al 1976). Further evidence of the role of the suprachiasmatic nucleus in initiating or regulating circadian rhythms comes from microelectrode studies on single neurones in this nucleus, which show spontaneous rhythmic activity, and from studies with radiocarbon-labelled deoxyglucose which demonstrate a rhythm to the metabolic activity of the nucleus. It has been proposed that the suprachiasmatic nucleus may act as a **neural pacemaker** that passes impulses to other structures, perhaps including other hypothalamic regions, which can act as secondary oscillators (Moore 1982).

The hypothalamus plays an important part in the elaboration of **emotions** such as aggression, fear, pleasure and reward. In this the hypothalamus acts closely with other parts of the nervous system including the prefrontal cortex, the autonomic nervous system, the reticular formation, the nucleus accumbens, the thalamus, and the limbic system (see Trimble 1981). The integrated actions of these structures is also involved in the control of behaviour and memory (see Ch. 11). So far as the hypothalamus itself is concerned various 'centres' have been proposed on the basis of focal ablation and stimulation experimental results. Thus it has been proposed that there is a centre concerned with **sham rage** in the posterior hypothalamus, stimulation of which leads to the elaboration of features compatible with the emotion of rage in experimental animals; for example, pupil dilatation, piloerection, arching of the back, growling and hissing, and tail lashing. Such rage was termed sham because it was first observed in decorticate dogs (Goltz 1892) and decorticate cats (Cannon & Britton 1925) which were thought at the time to be incapable of feeling true emotions without intact cortical functioning. A **pleasure centre** or **positive reward centre** has been proposed to exist in the medial forebrain bundle of the hypothalamus. Stimulation of this region has been found to elicit features compatible with pleasurable feelings in experimental animals. Moreover, when the experimental set-up is such that the animal is able to self-stimulate this region, for example by pressing a lever, then it has been found that such animals will continue to self-stimulate to the exclusion of other activities such as eating and drinking. Indeed they are often observed to stop self-stimulation only when they can no longer continue owing to physical exhaustion. For such self-stimulation to continue in this way it has been found that the dopaminergic pathways from the pars compacta of the substantia nigra need to be intact (Routtenberg & Santos-Anderson 1977). Other regions of the central nervous system which similarly appear to be pleasure centres include the orbitofrontal and parahippocampal or entorhinal neocortical areas, and the catecholaminergic complex of the brain stem (see Ch. 3). Stimulation of these pleasure centres in humans has been found to result in emotions of pleasure and well-being, often accompanied by erotic thoughts and feelings. Conversely, **negative reward centres** have also been proposed, stimulation of which leads to effects opposite to those noted for the positive reward centres.

The role of the mammillary area of the caudal hypothalamus and the mammillary bodies that lie inferior to this area in **memory** is still controversial. In Chapter 7 it is mentioned that there is often damage to this region in Korsakoff's psychosis, a condition in which there is impaired recent memory.

From the above discussion it can be seen that the hypothalamus has a wide variety of roles and that there are many influences on its functioning from other structures and systems of the nervous system such as the autonomic nervous system and the limbic system. The integrated actions of these various parts of the nervous system play a role in regard to learning, memory, basic drives, motivation and emotion, and the endocrine response to stress. Some aspects of these functions are considered further in this chapter and in the penultimate chapter on Psychology.

Pituitary gland

The pituitary gland or **hypophysis cerebri** lies inferior to the hypothalamus with which it is continuous by way of the hollow process of the tuber cinereum known as the infundibulum. It lies in a fossa of the sphenoid bone and is covered by the diaphragma sellae. The latter has a central aperture to allow the infundibulum to pass through it.

Components

There are many ways of naming the subdivisions of the pituitary gland. From an embryological (and functional) viewpoint the hypophysis cerebri can first be divided into two parts: the **adenohypophysis** or anterior pituitary, which is an ectodermal structure that develops from Rathke's pouch; and the **neurohypophysis** or posterior pituitary, which develops from the diencephalon as a downgrowth taking hypothalamic neuronal axons with it including those from the paraventricular and supraoptic nuclei (see above). The infundibulum is divided between the adenohypophysis and neurohypophysis; the parts of the hypophysis cerebri included in both divisions are shown in Table 2.2.

Adenohypophysis

Hormones biosynthesized and released by adenohypophysis are shown in Table 2.1 which also gives the corresponding hypothalamic releasing factors and release-inhibiting factors. Most of the hormones can be seen to be tropic (trophic), stimulating the target organs to secrete an amount of hormone whose mass is greater than that of the stimulating tropic hormone by a factor of at least one thousand. Likewise, the hypothalamic releasing factors stimulate the adenohypophysis to secrete an amount of hormone whose mass is greater than that of the releasing factor by a factor of at least one thousand.

Various staining techniques have been used to iden-

Table 2.2 Components of the pituitary gland

Adenohypophysis
    pars tuberalis
    pars intermedia
    pars anterior (or pars distalis or pars glandularis)

Neurohypophysis
    median eminence
    infundibular stem
    pars posterior (or pars nervosa)

Table 2.3 Classification of chromophilic endocrine cells of the pars anterior

	Endocrine cell	Hormone
Acidophils (α cells)	Mammotrope	Prolactin
	Somatotrope	Somatotropin (growth hormone)
Basophils (β cells)	Corticotrope	Corticotropin (adrenocorticotropic hormone)
	Gonadotrope (δ basophil)	Follicle stimulating hormone
		Luteinizing hormone
	Thyrotrope (β basophil)	Thyrotropin (thyroid stimulating hormone)

tify the different types of endocrine cells in the adeno-hypophysis. From early studies looking at whether cells were found to stain strongly or poorly with the dyes used, adenohypophysis cells have been classified as being **chromophilic** and **chromophobic** respectively. The chromophilic cells are further classified as being **acidophils** or α **cells** if they stain strongly with acidic dyes such as Orange G, and **basophils** or β **cells** if they stain strongly with basic dyes such as aldehyde fuchsin.

The types of chromophilic endocrine cells found in the pars anterior of the adenohypophysis are shown in Table 2.3 which also gives the corresponding hormones secreted by them. The pars anterior also has granules containing endorphin.

Secretory cells in the pars intermedia contain corticotropin, α-endorphin, β-endorphin, and α-MSH (melanocyte stimulating hormone).

Neurohypophysis

It has already been mentioned above that, following synthesis in the hypothalamus, the hormones oxytocin and arginine vasopressin combine with neurophysins and travel down hypothalamopituitary (hypothalamo-hypophysial) tract axons to storage cells of the neurohypophysis from which they are released. These axons also contain cholecystokinin (CCK) and met-enkephalin (M-ENK).

Hormones

The hormones considered in this section can be classified on the basis of their structure into **neuropeptides** such as corticotropin, **phenol derivatives**

such as thyroxine, and **steroids** such as the sex steroids. In addition to acting as blood borne hormones, neuropeptides can also act as modulators of neurotransmitters, and indeed as neurotransmitters themselves. The classical neurotransmitters, such as dopamine and noradrenaline, are considered in Chapter 3. In this section those substances of importance in psychoneuro-endocrinology are discussed.

Arginine vasopressin (argipressin)

The hypothalamic synthesis of arginine vasopressin (AVP) (also known as **argipressin** or antidiuretic hormone (ADH)) has been discussed above. Stimuli that lead to the release of arginine vasopressin include those related to the **volume of blood** in the body, such as reduced blood volume and increased plasma osmolarity. **Emotional stress** and **pain** can also act as stimuli for arginine vasopressin release, as well as causing the release of corticotropin (ACTH). The latter may possibly occur partly through the release of arginine vasopressin or oxytocin.

The well-established renal actions of arginine vasopressin result in **water retention** and an increase in the concentration of the urine. Hence the therapeutic use of intranasal administration of the analogue lysine vasopressin in the treatment of diabetes insipidus. There is also evidence that arginine vasopressin is involved in facilitating both the incorporation into and the accessing of **memory**, while conversely oxytocin appears to have an attenuating effect on memory processes (de Wied 1984). As a result, clinical trials have taken place in which analogues of arginine vasopressin have been administered to subjects with memory difficulties. Trials involving patients suffering from dementia and depression have yet to yield conclusive results.

The usual ratio of cerebrospinal fluid to plasma concentration of arginine vasopressin has been found to be inverted in patients with severe **anorexia nervosa** (Gold et al 1983). This ratio has been found to return towards the normal range with weight gain in such patients.

Chronic ethanol ingestion in humans has been found to increase the plasma concentration of arginine vasopressin (Beard & Sargent 1979). Animal studies have also shown that arginine vasopressin plays a role in the development of tolerance following chronic ethanol ingestion.

In control subjects provocative endocrine testing by administering arginine vasopressin, in the **vasopressin stimulation test**, leads to an increase in the circulating plasma concentrations of corticotropin (ACTH) and cortisol. In patients suffering from **depressive dis-**

order the corticotropin response has been found to be consistently suppressed while the cortisol response has tended to be normal (Kathol et al 1989). Similar results demonstrating disturbances in the hypothalamic-pituitary-suprarenal cortex axis (see below) have been found in depressed patients subjected to provocative endocrine testing with insulin hypoglycaemia and corticotropin releasing factor. These results are consistent with the hypothesis that hypothalamic or supra-hypothalamic overactivity occurs in depression. This is considered further later in this chapter.

Bombesin

Bombesin is a neuropeptide that is found in relatively high concentrations in parts of the central nervous system, particularly the hypothalamus, the gastrointestinal tract, and the lungs. At the time of writing its central actions appear to include the modulation of food intake and thermoregulation.

The cerebrospinal fluid levels of bombesin have been reported to be reduced in patients with schizophrenia, although the significance of this finding is not clear (Gerner 1984).

Cholecystokinin

The peptide cholecystokinin (CCK) was first located as a gastrointestinal and pancreatic hormone. It derives its name from the fact that the entry of food into the duodenum acts a stimulus for its release into the bloodstream, as a consequence of which gallbladder contraction occurs.

It is now known that cholecystokinin is located in high concentrations in the mammalian central nervous system, particularly in the cerebral cortex, hypothalamus, and parts of the limbic system (Baile et al 1986). Cholecystokinin has been found in mammalian studies to coexist with classical neurotransmitters. For example, cholecystokinin coexists with dopamine in the mesencephalon (Hökfelt et al 1985), with GABA in the cerebral cortex (Somogyi et al 1984), and with 5-hydroxytryptamine in the medulla oblongata (Mantyh & Hunt 1984). It appears that one of the most important cholecystokinin-like peptide fragments in the brain is CCK-8, the terminal octapeptide fragment of cholecystokinin.

Cholecystokinin is released following food intake, as mentioned above, and then has a **satiety action** on the brain which is probably indirect since cholecystokinin does not cross the blood-brain barrier. This action was discovered by Gibbs et al (1973a and b) who found that peripheral administration of cholecysto-

kinin inhibits food intake in rats. There is evidence that this action may take place via the vagus nerve since ablation, particularly of its gastric and abdominal branches, abolishes it. It has been suggested that once satiety signals reach the brain, cholecystokinin is again involved in the central satiety action, perhaps in relation to signals reaching the paraventricular nucleus of the hypothalamus. Microinjection of cholecystokinin into this nucleus has been found to have a satiety action (Faris 1985). Furthermore, it has been shown that injection of the CCK-8 antagonist proglumide into the third ventricle antagonizes the satiety effect of exogenous CCK-8 (Inui et al 1988). In spite of these results, clinical trials with intravenous cholecystokinin have not yet proved effective in reducing the binge eating found in bulimia nervosa.

Because cholecystokinin coexists with dopamine in two of the three major dopamine systems (see Ch. 3), namely the mesocortical and mesolimbic systems (but not the nigrostriatal system), it has been suggested that this may have important implications with respect to those psychiatric conditions in which these dopamine systems are involved. Indeed early studies in the 1980s provided evidence that it might be possible to use CCK-8 as an **antipsychotic drug** if an effective delivery system could be devised (see, for example, Moroji et al 1982). However, more recent controlled clinical trials have yielded essentially negative results (for example: Vanderhaeghen & Crawley 1985, Tamminga et al 1986). The potential use of cholecystokinin in schizophrenia has been reviewed by Montgomery and Green (1988).

Corticotropin, corticotropin releasing factor, cortisol

It has long been known that the hypothalamic-pituitary-suprarenal cortex axis is involved in the **endocrine response to stress**. In response to a stressor the secretion from the hypothalamus of corticotropin releasing factor (CRF) into the hypothalamopituitary portal system is stimulated. The corticotropin releasing factor in turn regulates the release of pro-opiomelanocortin (POMC) derived peptides, including corticotropin, from the adenohypophysis. POMC is considered further below in the section on the endogenous opioids. Corticotropin in turn stimulates the release of cortisol from the suprarenal cortex. The stressful stimuli can be both physical, such as infection and pain, and psychological. An example of the latter is the greatly elevated plasma cortisol levels sometimes found in candidates taking examinations.

In addition to the hypothalamus, high central concentrations of corticotropin have been found in many other parts of the brain such as the mesencephalic part of the limbic system. The corticotropin molecule itself is a 39 amino acid peptide which can be partitioned into four main segments, each of which has its own functions (Gispen 1982).

The administration of both corticotropin related peptides and vasopressin related peptides to mammals has been found to affect both forms of **associative learning** (see Ch. 11), namely classical conditioning and operant conditioning (de Wied & Jolles 1982). For example, the acquisition of conditioned responses has been shown to be enhanced following administration of these peptides. Conversely, Applezweig and Baudry (1955) had previously demonstrated that hypophysectomy leads to a deterioration in both the acquisition and retention of avoidance learning. These findings may be related to the observation that in humans the recall of events that have taken place at a time when an endocrine response to stress may have occurred would appear to be better than that of events occurring just a few hours anterior or posterior to the index events. Indeed, studies in humans have demonstrated the ability of corticotropin related peptides to enhance attention (see, for example, Sandman et al 1975). Pharmacological studies on the effects of these peptides on cognitive functioning in the elderly, and in demented and mentally retarded patients, have not so far yielded consistent positive results. Since cognitive dysfunction occurs in patients with alcohol dependence, the cerebrospinal fluid levels of corticotropin and corticotropin releasing factor have been studied in a group of alcoholics by DeJong et al (1990); no significant differences were found between this group and a group of normal controls.

Experiments in rats with corticotropin and the corticotropin fragment $ACTH_{1-24}$ (consisting of the first 24 amino acids of corticotropin) have shown that these substances can have potent effects on **sexual behaviour**. In male rats tested alone copulatory movements, penile erections and ejaculations may be stimulated, with alterations also occurring in male rats allowed to copulate with females (Serra & Gessa 1984). In female rats lordosis and increased progesterone levels have been found to be stimulated (Adler 1981).

There are a number of psychopathological effects of **Cushing's syndrome**, in which there is an excess of circulating glucocorticoids, including Cushing's disease, in which the raised corticotropin level is secondary to a pituitary adenoma (Jeffcoate et al 1979). Over half of such patients have been found to have a mood disturbance, with psychotic phenomena reported in up to one fifth (Jefferson & Marshall 1981). As is discussed below, disturbances in the hypothalamic-pituitary-

suprarenal axis have in turn been found in patients suffering from depressive disorders from causes not directly related to Cushing's syndrome. The psychopathological effects of **Addison's disease**, in which there is a decrease in the suprarenal steroids, show some similarity to those of hypopituitarism and include apathy and tiredness, reduced drive and initiative, decreased concentration, social withdrawal, sleep impairment, and depressed mood.

Yehuda et al (1990) have found low urinary cortisol excretion in patients suffering from the **post-traumatic stress disorder** when compared with control subjects. This finding is compatible with a model in which there is a physiological adaptation of the hypothalamic-pituitary-suprarenal axis to chronic stress.

Following **electroconvulsive therapy** there is a rise in the levels of corticotropin and cortisol without tolerance occurring (Johnstone et al 1980, Aperia et al 1984).

A reduction in the concentration of corticotropin releasing factor with a reciprocal increase in the density of corticotropin releasing factor receptors in the cerebral cortex has been found in **Alzheimer's disease** (de Souza et al 1986).

A large number of studies have looked at disturbances of the hypothalamic-pituitary-suprarenal cortex axis in **depression**. In normal human subjects the secretion of cortisol is episodic and follows a **circadian rhythm**. The peak cortisol secretion occurs in the morning usually just before awakening. In the hours between noon and four in the morning the secretion usually remains low, being lowest usually just after the onset of sleep (Suda et al 1979). In many depressed patients increased secretion has been reported of the following hormonal substances: corticotropin and β-endorphin (both POMC derivatives), and cortisol. Sachar et al (1973) reported that in melancholia (endogenous depression) there was a disruption in the normal circadian rhythm of cortisol secretion, with the morning peak being increased and lasting longer (see also Sachar et al 1980). A phase shift in the circadian rhythm with the morning peak occurring earlier has also been reported in some studies. That increased cortisol levels occur in elderly patients with depression has been confirmed in the study of Leake et al (1990).

Studies of patients with **mania** have yielded conflicting results, variously indicating that the plasma cortisol concentration is decreased (Gibbons & McHugh 1962), increased (Stokes et al 1984), or in the normal range (Sachar et al 1972). Manic patients have also been found to be less likely to show phase shifts in the cortisol circadian rhythm than are depressed patients.

In the consideration of the vasopressin stimulation

test (see above) it is mentioned that the results of its use in depressed patients lends support to the hypothesis that hypothalamic or supra-hypothalamic over-activity occurs in depression. There are a number of other provocative endocrine tests allowing stimulation at relatively localized points in the hypothalamic-pituitary-suprarenal cortex axis that also give results in depressed patients that are in harmony with this hypothesis. In the **CRF stimulation test** the administration of corticotropin releasing factor to normal human subjects leads, as would be expected, to the release of corticotropin. In depression it has been found that there is a consistent reduction of corticotropin response (see, for example, Gold et al 1984, Kathol et al 1989). **Insulin induced hypoglycaemia** and **bacterial pyrogens** have been used to stimulate pituitary release of corticotropin. A reduction in the corticotropin response in depressed patients has also been confirmed with the insulin hypoglycaemia test (Kathol et al 1989). The bacterial pyrogen test tends to be little used in human studies, compared with corticotropin releasing factor and insulin hypoglycaemia, owing to its lower precision of action and the greater discomfort it can cause to subjects.

The finding that depressed patients tend to have an elevated cortisol concentration is exploited in the **dexamethasone suppression test** (DST) in which the plasma cortisol levels are measured at one or more times following the administration of the long-acting potent synthetic steroid dexamethasone the previous evening or night. In normal control subjects the dexamethasone leads to a reduction in the level of cortisol over the next 24 hours through negative feedback, possibly at the level of corticotropin in the hypothalamic-pituitary-suprarenal cortex axis. In patients with melancholia, however, Carroll et al (1981) claimed that non-suppression of cortisol occurred in over 60% of cases, so that the DST could in effect be used as a specific laboratory test for the diagnosis of melancholia, with non-suppression being considered a positive result (see also Carroll et al 1968). The DST was also found to have a high specificity of up to 95% (Carroll 1982). Since the initial studies, the DST has been extensively evaluated and has not been found to live up to the earlier optimistic expectations. A relatively high level of positive results (cortisol non-suppression) has been found in psychiatric patients with diagnoses other than melancholia, such as mania, schizophrenia, types of depression other than melancholia, dementia, obsessive– compulsive disorder (see Berger et al 1984), and normal body mass bulimia nervosa (Hudson et al 1983, Mitchell & Bantle 1983). Claims that the DST might prove to be a predictor of

suicidal behaviour in depression have not found support in recent studies (Ayuso-Gutierres et al 1987, Schmidtke et al 1989). It has been found that the DST can be affected by factors such as age, changes in body mass, drugs, electroconvulsive therapy and endocrine disorders. The DST has not been found to be a useful predictor of treatment response in depression (Gitlin & Gerner 1986, Arana & Baldessarini 1987).

Endogenous opioids

The finding that opiate receptors with marked stereo-specificity occurred in brain tissue gave support, in the early 1970s, to the notion that there exist endogenous opioids. The first demonstration that this was indeed the case was by Hughes et al (1975) who identified two related pentapeptides from porcine brain tissue extracts which have potent opiate antagonist activity consistent with their having a strong affinity for opiate receptors. The structures of these pentapeptides, known as **enkephalins**, are shown in Figure 2.9. The term **endorphin** refers to all endogenous peptides with morphine-like action, and includes the enkephalins. It has been found that the five amino acid sequence of met-enkephalin shown in Figure 2.9 also occurs within the amino acid sequence of some of the other larger endogenous opioids such as the pituitary peptide β-lipotropin (β-lipotrophin) and its derivative β-endorphin. This is related to the fact that met-enkephalin is itself a derivative of β-endorphin.

All the endogenous opioids so far isolated from the central nervous system have been found to be derived from three peptide precursor molecules: **POMC** (pro-opiomelanocortin), from which is derived corticotropin (see above) and β-lipotropin; **proenkephalin**, which gives rise to met-enkephalin and leu-enkephalin; and **prodynorphin.** The peptide derivatives of β-lipotropin are shown in Figure 2.10 which is in the form of a family tree showing common amino acid sequences of peptide derivatives of POMC. The three peptides dynorphin A, dynorphin B, and α-neoendorphin, are derived from prodynorphin and each includes within its amino acid sequence the pentapeptide sequence of leu-enkephalin shown in Figure 2.9.

As might be expected from the fact that POMC, which is synthesized in the arcuate nucleus of the hypothalamus and the tractus solitarius, is a pituitary pep-

(A)   tyrosine–glycine–glycine–phenylalanine–methionine

(B)   tyrosine–glycine–glycine–phenylalanine–leucine

Fig. 2.9 Amino acid sequences of (**A**) met-enkephalin, and (**B**) leu-enkephalin.

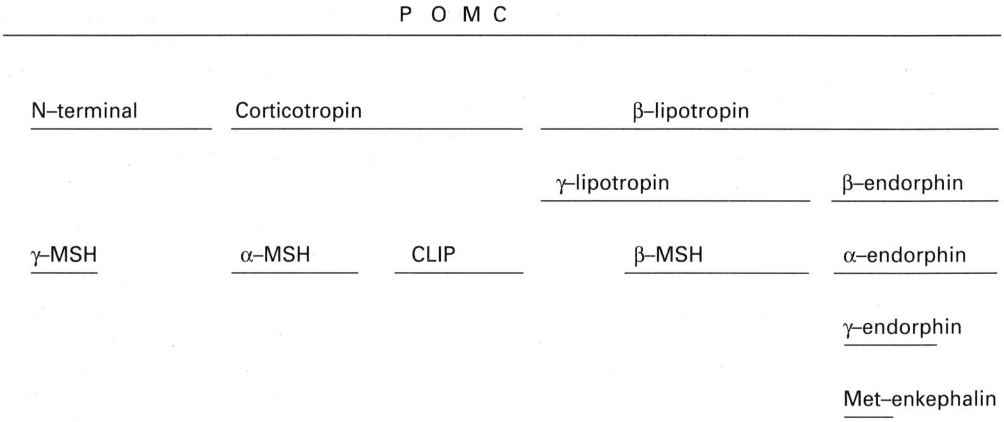

P O M C

N–terminal	Corticotropin		β–lipotropin	
			γ–lipotropin	β–endorphin
γ–MSH	α–MSH	CLIP	β–MSH	α–endorphin
				γ–endorphin
				Met–enkephalin

(CLIP = corticotropin–like immunoreactive peptide)

Fig. 2.10   Derivatives of POMC.

tide, endogenous opioids derived from POMC, and in particular β-endorphin which has relatively potent opiate activity and appears to be the stablest molecule in body fluids, are relatively highly concentrated in the pituitary and the arcuate nucleus. The latter gives rise to fibres that project to the locus coeruleus, the nucleus accumbens, and the mesencephalic periaqueductal grey matter. The derivatives of proenkephalin and prodynorphin are more widely distributed. The enkephalins are found in relatively high concentrations in the pituitary, the substantia nigra, parts of the limbic system, the globus pallidus, the cerebral cortex, and parts of the spinal cord such as the substantia gelatinosa. The ratio of met-enkephalin to leu-enkephalin varies in different locations, and both are also found outside the central nervous system, for example in the gastrointestinal tract and suprarenal medulla. Relatively high concentrations of pro-dynorphin derived peptides are found in the hypothalamus and pituitary, basal ganglia, and parts of the spinal cord. As in the case of cholecystokinin (see above), the phenomenon of coexistence occurs of endogenous opioids with classical neurotransmitters. For example, enkephalin has been found in noradrenergic neurones.

Three separate endogenous opioid **receptors** have been identified in the central nervous system: μ (mu) receptors for POMC derivatives including β-endorphin; δ (delta) receptors mainly for proenkephalin derivatives; and κ (kappa) receptors mainly for prodynorphin derivatives. There is evidence that β-endorphin can bind to all three receptor types and not just the μ receptors. Similarly, at least one type of dynorphin

binds to μ as well as to κ. There is also evidence from experimental binding data that the μ receptors consist of more than one type of binding site. These have been named μ_1 receptors, to which both morphine and the enkephalins bind, and μ_2 receptors which are morphine selective. If this subclassification is correct, then it will be necessary to determine the identity of the endogenous ligand(s) to which the μ_2 receptors are selective. In the classification of Martin et al (1976) it was proposed that there existed another receptor type known as σ (sigma), but at the time of writing there is little further evidence to support this. Non-μ β-endorphin selective receptors have been distinguished in the ductus deferens of the rat; these are known as ε (epsilon) receptors.

It is believed that agonists acting at μ receptors cause **analgesia**. There is also some evidence that the μ receptors may mediate **placebo analgesia** (see, for example, Grevert et al 1983). The effects mediated by the δ and κ receptors are probably more complicated. It has, for example, been suggested that δ receptors may be involved with limbic functions and that κ receptors may be associated with both analgesia and sedation (Chang et al 1980, Snyder 1986).

In addition to mediating drug induced analgesia and placebo analgesia, it is possible that endogenous opioids are responsible for causing the pain-free episodes sometimes seen in sudden injury or associated with acute shock. For example, there are numerous accounts of soldiers who feel no pain after sustaining injury during active duty. The administration of corticotropin releasing factor leads to the release not only of corticotropin, but also β-lipotropin, another POMC derivative, and therefore in turn a number of β-lipotropin derived endogenous opioids such as β-endorphin. Therefore it has been postulated that there is an increase in the endogenous opioids following stress. This may account for the phenomenon of analgesia associated with injuries occurring at times of great stress. It may also account for the feeling of exhilaration and well-being that can follow physical exercise. It has also been suggested that endogenous opioid action underlies **acupuncture**-induced analgesia, since the latter may be reversed by the opioid receptor antagonist naloxone (Pomeranz & Chiu 1976). Moreover, increased cerebrospinal fluid levels of endogenous opioids have been reported following electro-acupuncture (Sjolund et al 1977).

Regular administration of opioid receptor agonists can lead to tolerance and **dependence**. The cessation of the regular drug administration leads to a withdrawal syndrome. Whereas this may involve a craving for the drug in the case of, for example, the μ agonist morphine, only the physical symptoms and signs occur in the case of the withdrawal of κ agonists such as nalorphine in humans, with essentially no real craving. Some parts of the withdrawal syndrome seem to be associated with noradrenergic activity, particularly in the locus coeruleus; it has been found that such activity can be suppressed with clonidine. Vining et al (1988) have reported on the clinical utility of using clonidine in this way while withdrawing patients addicted to exogenous opiates by using naloxone. It has also been proposed that dependence on alcohol may be associated with changes in endogenous opioid activity (Jeffcoate 1981).

There is evidence that endogenous opioids have **neuroimmunological actions**. For example, dynorphin and enkephalins have been shown to alter natural killer cell activity in vitro (Oleson & Johnson 1988).

The results of experiments on rats published in 1976 were extrapolated to suggest that either opiate antagonists or, in contrast, opiate agonists may have a therapeutic use in **schizophrenia**: Bloom et al (1976) suggested that β-endorphin caused a catatonic mental illness in rats that was reversible with naloxone, while in the very next paper in the same journal Jacquet and Marks (1976) suggested that β-endorphin acts as an endogenous neuroleptic or antipsychotogen in the rat. The validity of these extrapolations to a human psychiatric disorder have since been questioned, and trials on (human) schizophrenics using either opiate antagonists or agonists have essentially been unsuccessful overall. For example, Pickar et al (1982) were unable to demonstrate consistent therapeutic benefit from nal-

oxone in schizophrenic patients. One year earlier Pickar et al (1981) had found similar results using β-endorphin in schizophrenic patients. A double-blind controlled study by Ruther et al (1980) using a synthetic analogue of met-enkephalin known as FK 33-824 which had previously been held, on the basis of less rigorously designed studies, to have beneficial effects in schizophrenia, also failed to demonstrate any consistent therapeutic benefit. Similarly, clinical trials with the γ-endorphin derived opioid peptide (des-tyr)-γ-endorphin have also proved disappointing. For example, Manchanda and Hirsch (1981) found that most of their group of schizophrenic patients failed to show any improvement with (des-tyr)-γ-endorphin.

Consistent changes in the levels of various endogenous opioids in **mood disorders** have not been demonstrated (see, for example, Nemeroff 1990). The double-blind study, mentioned above, by Pickar et al (1981) also included a group of depressed patients to whom β-endorphin was administered, while the above-mentioned double-blind study by Pickar et al (1982) likewise included a group of manic patients to whom naloxone was administered; no consistent therapeutic benefit was shown in either group of patients.

Gastrointestinal hormones

Ever since the discovery of the gastrointestinal hormones secretin and gastrin in the first decade of this century over 30 hormonal substances have been discovered in the gastrointestinal tract and pancreas. Many of these, including secretin and gastrin, have been shown to be present in the central nervous system since their initial discovery. The gastrointestinal hormone, cholecystokinin, has already been considered earlier in this section on Neuroendocrinology. Similarly, other gastrointestinal hormones that are individually discussed in this section include neurotensin, somatostatin, and substance P.

For the gastrointestinal hormones that are peptides it is possible, on the basis of similarities in their amino acid sequences, to group some of them into three families. One family consists of peptide hormones that all share the same terminal five amino acids; it includes gastrin, cholecystokinin, and caerulein.

Pancreatic polypeptide (PP), peptide YY (the YY refers to the fact that the amino acid represented by the single letter Y, namely tyrosine, is found at the amino and carboxy ends of the sequence) and neuropeptide Y (NYP) can also be grouped together as a family in which there are similarities in the amino acid sequences of at least ten regions of the peptides. Neuropeptide Y is believed, on the basis of mammalian experiments involving central administration, to stimulate appetite, influence circadian rhythms, and inhibit copulation. Neuropeptide Y has also been shown to stimulate the pituitary release of follicle stimulating hormone, luteinizing hormone, and somatotropin.

A third family includes gastric inhibitory peptide, glucagon, growth hormone releasing factor (GRF), peptide HI (the amino acid H or histidine is found at the amino end, and I or isoleucine at the carboxy end), secretin, and vasoactive intestinal polypeptide (VIP). Peptide HI potentiates the action of corticotropin releasing factor, stimulates the release of prolactin, and inhibits the action of vasoactive intestinal polypeptide.

Vasoactive intestinal polypeptide (VIP) has a number of functions in the gastrointestinal tract, including vasodilation, inhibiting gastric secretion, and stimulating bile flow, pancreatic secretion and lipolysis. It also appears to act on lymphocytes, which possess VIP receptors. In the central nervous system VIP is found in relatively high concentrations in the cerebral cortex, amygdala, hippocampus and hypothalamus. It stimulates the release of a number of hormones centrally, including luteinizing hormone, prolactin, somatostatin and somatotropin. Its other functions in the central nervous system and any role it may play in the pathophysiology of psychiatric disorders are as yet relatively unknown.

For many of the gastrointestinal hormones that are also found in the central nervous system, for example gastric inhibitory peptide and pancreatic polypeptide, again relatively little is known at the time of writing of their neuroendocrinological functions and they are not considered further in this section.

Gonadotropins, gonadotropin releasing factor and sex steroid hormones

The gonadotropins considered in this section are the two glycoprotein hormones follicle stimulating hormone (FSH) and luteinizing hormone (LH). The glycoprotein hormones also include thyrotropin (thyroid stimulating hormone, TSH), which like follicle stimulating hormone and luteinizing hormone is a pituitary hormone, and human chorionic gonadotropin which is derived mainly from the placenta although it is additionally found in much smaller concentrations in other organs including the pituitary. Gonadotropin releasing factor has been mentioned earlier in this chapter and is secreted in a pulsatile manner, possibly with the involvement of the classical inhibitory neurotransmitter γ-aminobutyric acid (GABA). This in turn is reflected in a normally pulsatile pattern of release of follicle stimulating hormone and luteinizing hormone from the adenohypophysis

which is itself superimposed on circadian rhythms and on variations dependent on the stage of sleep. The main targets of the adenohypophyseal gonadotropins are the gonads which in turn are stimulated to produce sex steroid hormones. The latter include oestrogen, progesterone, and testosterone, and a number of their metabolites. The sex steroid hormones in turn can modulate the pulsatile gonadotropin releasing factor secretion. The precise mechanism of this neuromodulation is still unknown; it may be direct, or it may occur indirectly via an intermediate neuromodulator such as, perhaps, β-endorphin.

The most important functions of these hormonal substances in the adult human are concerned with **reproduction**, including the onset of puberty, the menstrual cycle, pregnancy and childbirth, and the menopause. Gonadotropins are also synthesized by the fetal pituitary and are associated with sexual differentiation in utero.

Given their functions, the sex steroid hormones are likely to be implicated as playing a role in the **premenstrual syndrome** or premenstrual tension. To date it has not been possible to formulate a rigorous pathophysiological model entirely consistent with experimental findings. Indeed considerable methodological difficulties have been encountered in the study of the aetiology of this condition. It has been suggested that the premenstrual syndrome is associated with a reduced progesterone to oestrogen ratio in the luteal phase of the menstrual cycle. However, a similar change normally occurs in the follicular phase, during which the clinical features of the premenstrual syndrome are very unlikely to present. Other substances that have been suggested as having a role in the aetiology of this condition include the endogenous opioids, mineralocorticoids, prolactin, and the prostaglandins.

The results of mammalian experiments with gonadotropin releasing factor suggests it has a role in **sexual behaviour**. Following administration to rats, there is a decrease in the mounting and ejaculation latencies in the male, while increased lordosis occurs in females.

That the levels of these hormonal substances may be altered in **anorexia nervosa** is suggested by the finding that amenorrhoea is a feature of the condition that can occur independently of loss of body mass while the restoration of the lost body mass is not necessarily accompanied by a restoration of normal menstrual function (Fairburn & Hope 1988). It has indeed been shown that the plasma concentrations of gonadotropins and sex steroid hormones are reduced in anorexia nervosa (Mitchell 1986). Other endocrine abnormalities seen in this condition include reduced plasma triiodothyronine and increased plasma cortisol and somatropin levels.

The cerebrospinal fluid concentration of gonadotropin releasing factor has been found to be reduced in **Alzheimer's disease** (Rogers et al 1986) and there are uncontrolled studies that suggest that the administration of oestrogen may lead to some improvement in cognitive functioning in affected women.

There are reports of gonadotropin secretion abnormalities in **schizophrenia**. In chronic schizophrenia low plasma concentrations of follicle stimulating hormone and luteinizing hormone have been reported with loss of the normal pattern of secretion (Ferrier et al 1982).

Melatonin, melanocyte stimulating hormone, MSH release inhibitory factor

Modulation of the rhythmic secretion of the indoleamine melatonin, or *N*-acetyl-5-methoxytryptamine, by the **pineal gland** is believed to be the means by which changes in the **photoperiod** lead to changes in circadian and seasonal physiological rhythms. Apart from light there appear to be no other stimuli that alter the secretion of this hormone (Lewy 1984). In addition to the pineal gland, melatonin has been found to be located in many other parts of the central nervous system and also in the gastrointestinal tract. The biosynthesis of melatonin from its precursor serotonin, or 5-hydroxytryptamine, occurs via *N*-acetylation followed by *O*-methylation. The step involving the enzyme serotonin *N*-acetyltransferase is probably rate limiting and is stimulated at night in a process involving the neurotransmitter noradrenaline. As mentioned earlier in this chapter the hypothalamic **suprachiasmatic nucleus** acts as a biological clock, and there is evidence that this nucleus acts as the endogenous pacemaker for the nocturnal biosynthesis. Under appropriate conditions it is possible, therefore, to use the plasma concentration of melatonin as a biological marker of the activity of this pacemaker. The normal nocturnal production of melatonin can, however, be altered by light (Arendt 1988), although this fact could be utilized by using melatonin as a psychiatric probe in the study of the light sensitivity of patients with bipolar mood disorders (Arendt 1989).

As mentioned earlier in this chapter (alpha-)melanocyte stimulating hormone ((α-)MSH), like corticotropin, with which it shares a sequence of 13 amino acids, is an adenohypophyseal peptide hormone derived from POMC. In addition to controlling the release of melatonin, melanocyte stimulating hormone acts via the neurotransmitter dopamine to inhibit the release of luteinizing hormone and prolactin.

In light of the seasonal distribution of **depressive illness** and hypotheses suggesting an abnormality in the hypothalamus or in a central noradrenergic system, it has been suggested that altered melatonin secretion may occur in this condition. Indeed such an alteration, with a reduction of melatonin secretion, has been reported in a number of studies (see, for example, Wetterberg et al 1979). Further support for an association between low melatonin levels and depression is found in the report by Stanley and Brown (1988) of low melatonin concentrations in the pineal glands of suicide victims. However, most of these studies have been poorly controlled with respect to factors such as psychotropic medication which are known to influence melatonin secretion. In a well-controlled comparison of depressed patients versus normal controls Thompson et al (1988) failed to demonstrate a reduction in nocturnal melatonin secretion or a change in the timing of its secretion in depression. Although phototherapy in **seasonal affective disorder** (SAD) has proved beneficial (Lewy et al 1982, Rosenthal et al 1984) the implication that this occurs by means of an effect on melatonin secretion does not appear to be correct (Wehr et al 1986, Checkley et al 1989).

Melatonin administration has been found to alleviate some of the effects of the **jet lag** resulting from flying east. In particular, it can lead to a shift in the endogenous circadian rhythm of secretion and improve the mood of the subject (Arendt et al 1987). Arendt (1989) has suggested that melatonin may also have therapeutic potential in the treatment of insomnia.

Studies with contradictory results have found the nocturnal concentration of melatonin to be either raised (Brown et al 1981) or lowered (Birau et al 1984) in patients with **anorexia nervosa**. The significance of these results is not yet clear.

As mentioned earlier in this section, under 'corticotropin corticotropin releasing factor, cortisol' (where further details are given), the administration of both corticotropin related peptides, such as melanocyte stimulating hormone, and vasopressin related peptides to mammals has been found to affect both forms of **associative learning**.

Like corticotropin and corticotropin fragments such as $ACTH_{1-24}$, melanocyte stimulating hormone has been found to affect **sexual behaviour** in rats. In male rats tested alone copulatory movements, penile erections and ejaculations may be stimulated, with alterations also occurring in male rats allowed to copulate with females (Serra & Gessa 1984). In female rats lordosis has been found to be stimulated (Adler 1981).

Ehrensing and Kastin (1974, 1978) have reported that melanocyte stimulating hormone release inhibitory factor (MIF) demonstrated antidepressant effects in their double-blind studies. Since these results were published further studies have provided contrasting results, either against (see Levy et al 1982) or in favour (see van der Welde 1983) of MIF having antidepressant effects.

Neurotensin

Neurotensin is a gastrointestinal peptide hormone consisting of 13 amino acids. It has a number of inhibitory actions in the gastrointestinal tract, for example, on the motility of the tract and the secretion of gastric acid and pepsin, and its release is stimulated by the presence of fatty acids and mixed meals in the tract. Neurotensin is also found throughout the brain, being particularly concentrated in the hypothalamus, limbic system, substantia nigra and the brain stem. In addition to coexisting with dopamine in many parts of the brain, neurotensin also appears to have a functional relationship with dopaminergic pathways (see Seutin et al 1989) and is able to form a molecular complex with dopamine (Adachi et al 1990). Centrally neurotensin inhibits neuronal activity in the locus coeruleus, stimulates the release of somatostatin while inhibiting the release of somatotropin, and has been found to cause hypothermia.

Because neurotensin can affect the release of dopamine in the brain it has been suggested that the concentration of this peptide may be raised in **schizophrenia**. However, although there is evidence that the concentration of neurotensin may be raised in the caudate nucleus and nucleus accumbens by neuroleptic drugs (see, for example, Kilts et al 1988), there is no consistent evidence for a raised concentration otherwise being found in the brains of schizophrenic patients. For example, in their comparison of neuropeptide concentrations in limbic areas of the brain in schizophrenia with those of control brains, Ferrier et al (1984a) found no change. Furthermore, the influence of neurotensin on dopaminergic activity can vary, being stimulating in the dopamine perikarya and inhibitory in the nucleus accumbens. That neurotensin can antagonize both dopaminergic activity and the behavioural changes resulting from increased dopaminergic activity in the nucleus accumbens has also led to the contrasting suggestion that neurotensin may be able to act as an endogenous neuroleptic.

The concentration of neurotensin has generally been found to be normal in **Alzheimer's disease**, although Perry and Perry (1982) did report a low level in this condition. Its concentration generally appears to be raised in **Down's syndrome** and **Huntington's chorea** (see Whalley 1987). Neurotensin has also been found to increase **thirst** and to be a potent **analgesic**.

Oxytocin

The release of the neurohypophyseal neuropeptide oxytocin is discussed earlier in this chapter, where it is noted that this release is increased by stimulation of the genitalia and nipples. Oxytocin has also been found to be present in the pineal gland but the functional significance of this is not yet clear. Structurally the amino acid sequence of oxytocin differs from that of arginine vasopressin in respect to just one amino acid out of the nine that each neuropeptide consists of. Oxytocin is known to cause **uterine contractions** during childbirth, and contraction of mammary myoepithelial cells during suckling, leading to **milk ejection**. It is likely that oxytocin has a role in **sexual behaviour**. Mammalian experiments indicate that the act of copulation leads to the release of this hormone and subsequent uterine contractions, which in turn may facilitate the movement of spermatozoa, while intracerebroventricular infusion has been shown to increase lordosis.

Following **electroconvulsive therapy** the release of oxytocin-related neurophysin, a peptide with which the neurohypophyseal neuropeptides are both synthesized and released, is immediately stimulated (Whalley et al 1982). Scott et al (1986) found a positive correlation between the level of oxytocin-related neurophysin following electroconvulsive therapy and the degree of recovery from depression as measured on the Hamilton Rating Scale for Depression and the Montgomery and Asberg Depression Rating Scale. Other forms of **stress** can also stimulate oxytocin release.

Some of the effects of arginine vasopressin on **learning and memory** appear to be antagonized by oxytocin (see above). This antagonism is of interest in view of the structural similarity of the two neuropeptides. Another example of antagonism occurs with respect to the effects in neurophysiological experiments of the two neuropeptides on hypothalamic neuronal electrical activity.

Parathormone

Parathormone, or parathyroid hormone, is a peptide hormone consisting of 84 amino acids. It is released by the parathyroid glands and has the effect of increasing the plasma level of **calcium** and reducing the plasma level of **phosphate**. This takes place by means of its actions on gastrointestinal calcium absorption, renal calcium excretion, osteolysis and osteoclasts.

Changes in the plasma level of calcium can be associated with neurological and psychiatric manifestations. **Hypocalcaemia**, for example as a result of hypoparathyroidism, can lead to tetany, agitation, disorientation and poor memory. Mood disorders, hallucinations, and delusions are also recognized effects which can usually be treated successfully by increasing the plasma level of calcium.

Hypercalcaemia may result from a number of causes in addition to hyperparathyroidism. One important cause in psychiatric clinical practice is the administration of **lithium** which can lead to an increase in the level of parathormone. Manifestations include sedation, poor concentration, and apathy. Depressed mood, suspiciousness, and psychotic symptomatology may occur, and, again, usually improve following appropriate treatment that results in a calcium level within the normal range.

Prolactin

The adenohypophyseal hormone prolactin promotes mammary tissue development and lactation. Prolactin has also been found to increase the testosterone production stimulated by the action of luteinizing hormone on testicular interstitial Leydig cells. As indicated in Table 2.1, the hypothalamic control of the release of prolactin from the pituitary takes place via both prolactin releasing factor and prolactin release inhibitory factor (PIF). The latter is believed to be **dopamine**. The release of prolactin is facilitated by 5-hydroxytrytamine (serotonin), which also facilitates the release of corticotropin and somatotropin.

Hyperprolactinaemia is associated with gynaecomastia and galactorrhoea. In males a reduction in the sperm count, erectile dysfunction, failure of ejaculation, and reduced libido may also occur, while females may suffer menstrual disturbances. Hyperprolactinaemia may result from pituitary tumours. A much more common cause of hyperprolactinaemia in clinical psychiatric practice is as a side-effect of **neuroleptic medication**, since the latter has dopamine antagonistic effects which in turn diminish the tonic inhibition of adenohypophyseal prolactin release that normally occurs via dopamine or prolactin release inhibitory factor. Clinically this side-effect can be made use of by checking the prolactin level to check compliance with neuroleptic medication. Furthermore, from their study of plasma prolactin levels during treatment with chlorpromazine, Kolakowska et al (1979) found that the development of extrapyramidal side-effects from the neuroleptic was associated with higher prolactin levels. On the other hand, there are, to date, no

consistent results to indicate that prolactin levels might be used in schizophrenia as a reliable marker of treatment response.

In **depression** the prolactin level tends to be normal. Blunting of the prolactin responses to methadone and apomorphine have been found in depression and schizomood (schizoaffective) disorders (Judd et al 1982, Meltzer et al 1984).

Following **electroconvulsive therapy** there is a rise in the plasma concentration of prolactin. This may occur as a result of increased 5-hydroxytrytamine and β-endorphin release, both of which are known to facilitate the release of prolactin, and the release of both of which has been found to be increased in the rat brain following electroconvulsive therapy (Lebrecht & Nowak 1980, Dias et al 1981). Abrams and Schwartz (1985) reported an inverse relationship between the prolactin level following electroconvulsive therapy and clinical improvement. However, this finding has not been replicated in the studies by Whalley et al (1982) and Aperia et al (1984). Another argument against the prolactin level being a reliable marker of likely improvement following electroconvulsive therapy is the fact that in the well-constructed Northwick Park trial (Johnstone et al 1980) it was found that simulated electroconvulsive therapy also resulted in an increase in the prolactin level.

In addition to seizures induced by electroconvulsive therapy, those associated with **epilepsy** are also usually followed by raised plasma concentrations of prolactin. As this is not the case with hysterical seizures, the measurement of the plasma prolactin level 15 or 20 minutes after a seizure can be of help clinically in helping to differentiate epilepsy from hysteria (Trimble 1978).

Experimentally induced chronic hyperprolactinaemia in rodents tends to be associated with a reduction in **sexual behaviour** in both males and females. The effect on human male sexual behaviour has been noted above. Schwartz et al (1982) reported raised prolactin levels in approximately 6% of men attending a clinic for sexual disorders. Of the various sexual disorders it has been suggested that hyperprolactinaemia may be particularly associated with **paedophilia** (Gaffney & Berlin 1984, Harrison et al 1989). Hyperprolatinaemia is a recognized feature of Klinefelter's syndrome (in which males have more than one X chromosome per nucleus, for example, XXY) (Schiavi et al 1978); the suggestion, however, that there may be an association between this syndrome and paedophilia is the subject of some debate (see, for example, Lancet 1988).

As mentioned earlier, prolactin is one of the substances that has been suggested as possibly having a role in the aetiology of the **premenstrual syndrome**;

this would be consistent with the direct actions of prolactin on the mammary glands, mentioned above. In support of this possible role of prolactin is the fact that some studies have demonstrated increased prolactin concentrations in patients suffering from the premenstrual syndrome, particularly during the luteal phase (see, for example, Halbreich et al 1976, Vekemans et al 1977). Indeed, there is some evidence that the dopamine agonist bromocriptine can be of help in this condition (Carroll & Steiner 1978). However, these findings have not been consistently replicated, with many studies failing to demonstrate prolactin level elevation in premenstrual syndrome patients (for example, Harrisson & Letchworth 1976, O'Brien & Symonds 1982).

Somatotropin and somatostatin

The adenohypophyseal hormone somatotropin, also known as somatotropic hormone (SH) and growth hormone (GH), is a protein molecule consisting of 191 amino acids that affects all the tissues of the body that have the potential of growth, generally stimulating mitosis and growth of cells. These actions are not always direct. For example, somatotropin stimulates the growth of bone and cartilage indirectly via the hepatic somatomedins. Its metabolic effects include the promotion of increased cellular protein biosynthesis throughout the body, the inhibition of glucose uptake and utilization by cells, and the stimulation of glycogen storage and fatty acid mobilization and utilization. Somatotropin is secreted in response to stress related stimuli, hypoglycaemia, and reduced blood fatty acid concentrations. In normal individuals, therefore, the plasma level of somatotropin may vary widely. At night, the peak plasma somatotropin level is usually achieved at the end of about two hours of sleep (Takahashi et al 1968).

The release of somatotropin from the adenohypophysis is controlled by the hypothalamic factors listed in Table 2.1. Somatotropin release is stimulated by somatocrinin or growth hormone releasing factor (GHRF), and it is inhibited by somatostatin or growth hormone release inhibiting factor. Somatocrinin is not discussed further in this chapter. There are a number of forms of the neuropeptide somatostatin which are derived from a larger precursor peptide; the following have been identified: the 116-amino acid molecule preprosomatostatin, the 92-amino acid prosomatostatin, the 28-amino acid somatostatin-28, somatostatin-14, and somatostatin-12. Of these the 14-amino acid molecule somatostatin-14 appears to be the most important functionally. Somatostatin is widely distributed in the central nervous system, with the highest concentrations being in the hypothalamus. In addition to inhibiting the release of somatotropin, somatostatin also inhibits the release of cholecystokinin (CCK), glucagon, prolactin, thyrotropin (thyroid stimulating hormone, TSH) and vasoactive intestinal peptide (VIP). Hormones that are known to stimulate the hypothalamic release of somatostatin include glucagon, neurotensin, secretin, somatomedin A, somatotropin, substance P and triiodothyronine (T_3), while vasoactive intestinal peptide has been found to have the opposite action.

A number of **behavioural effects** occur following intracerebroventricular injection of somatostatin. Sleep disturbances occur which include reduced slow wave, rapid eye movement (REM) and total sleep (see below). There is a dose-related biphasic effect on locomotor activity, with low doses causing stimulation and higher doses leading to inhibition including the development of incoordination and rigidity. Food intake is also disturbed, being reported as either increasing or being associated with a dose related biphasic action. There is also evidence that analgesia and limbic system stimulation may occur.

Somatotropin is secreted in response to stress related stimuli as mentioned above. However, studies of basal levels of somatotropin in **anxiety disorders** have generally failed to show any consistent increase when compared with normal control subjects (Uhde et al 1986), although this may partly be the result of the wide normal variation of the basal level noted above. In endocrine provocative tests on patients with panic disorder (with or without agoraphobia), both the last study and that of Charney and Heninger (1986) reported a blunted somatotropin response to **clonidine stimulation** when compared with normal controls, clonidine being a relatively selective central α_2-adrenoceptor agonist that normally causes the release of somatotropin.

The study by Uhde et al (1986), just mentioned, also reported no consistent difference between the basal levels of somatotropin in patients with **depression** and normal controls. On the other hand, this study did report a blunting of the somatotropin response to **clonidine stimulation** in the depressed group, a finding that has been reported by other groups (see, for example, Checkley et al 1981a). A blunted somatotropin response has also been found to stimulation with the indirectly acting alpha receptor agonist desipramine (Checkley et al 1981b).

The cerebrospinal fluid levels of somatostatin have also been studied in patients with mood disorders. Beginning with Gerner and Yamada (1982), a number of groups have reported that depressed patients have a lower CSF concentration compared with normal controls and, in some studies, manic patients (see Rubinow 1986, Bissette et al 1986, Davis et al 1988).

The last study, by Davis et al (1988) also confirmed a finding reported by almost all other groups to the effect that there also tends to be a reduction in the CSF somatostatin concentration in **Alzheimer's disease**, a finding that has also been confirmed in post mortem studies of somatostatin concentrations and receptor densities. Studies of sleeping plasma somatotropin levels in Alzheimer's disease have given inconsistent results. It has been suggested that a reduced somatostatin level is associated with **cognitive dysfunction**, an hypothesis for which there is some experimental evidence (see, for example, Cook et al 1989). This would be consistent not only with the finding in Alzheimer's disease, but also the low CSF concentration of somatostatin found in depression. Other neurological disorders associated with cognitive dysfunction in which CSF and/or post mortem somatostatin concentrations have been found to be reduced include multiple sclerosis (Sørensen et al 1980), Parkinson's disease (Dupont et al 1982), and multi-infarct dementia (Beal et al 1986); similar studies in Huntington's chorea have yielded inconsistent results. Although alcoholism is associated with cognitive dysfunction, CSF somatostatin levels in a group of alcoholics have been reported by Roy et al (1990) as not being significantly different from those in a group of normal controls.

According to the dopamine hypothesis of **schizophrenia**, this condition is associated with dopaminergic overactivity in parts of the central nervous system (this is discussed more fully in the next chapter). Since dopamine and dopamine agonists stimulate somatotropin secretion, a number of studies have been carried out to examine whether there is any abnormality of this process in schizophrenia, the dopamine agonist often used to stimulate somatotropin secretion in these studies being apomorphine. In general, the somatotropin response to apomorphine has been found to be greater in acute schizophrenia than in control subjects and patients with chronic schizophrenia (see Pandey et al 1977, Ferrier et al 1984b, Meltzer et al 1984). In the study by Ferrier et al (1984b) patients with chronic schizophrenia were found to show a blunted somatotropin response. This was also reported eight years earlier by Ettigi et al (1976), although some groups have found no difference between the responses of chronic schizophrenia and normal controls. Acutely exacerbating chronic schizophrenic patients have been reported as showing somatotropin responses that are both increased and blunted (Rotrosen et al 1976).

Moreover, Whalley et al (1984) found that patients with Schneiderian first rank symptoms had a greater somatotropin response to apomorphine than psychotic patients without these symptoms. These results are compatible with a model in which dopamine receptors have increased sensitivity in acutely ill schizophrenics with positive symptoms, and decreased sensitivity in chronic schizophrenics. It should be noted, however, that not all investigators have reported results consistent with those mentioned, and that even with the above reports it may be necessary to take account of other factors, such as neuroleptic pharmacotherapy.

Substance P

Substance P is a neuropeptide consisting of 11 amino acids that is found in relatively high concentrations in a number of regions of the brain including the basal ganglia and hypothalamus, and outside the brain in the substantia gelatinosa of the spinal cord and in posterior or dorsal root ganglia. In the substantia gelatinosa substance P is believed to act as a neurotransmitter involved in the perception of **pain**, and it may have a similar function in other regions, such as the trigeminal nuclei, that are also associated with the perception of pain (Iversen & Iversen 1981). Both hyperalgesia and analgesia have been reported as following the administration of substance P.

Of special interest in view of the postulated role of the nigrostriatal dopaminergic pathway in the dopamine theory of **schizophrenia** is the existence of a large substance P striatonigral pathway. Another piece of evidence linking substance P with schizophrenia is the finding in post mortem brains that the concentration of substance P is increased in the caudate nucleus (Kleinman et al 1983) and hippocampus (Ferrier et al 1984a) in schizophrenic patients compared with controls. Moreover, substance P is known to act as an excitatory neurotransmitter for certain dopaminergic neurones. For example, experimentally it has been found that application of substance P to nigrostriatal dopaminergic neurones results in activation, again supporting the view that substance P might be involved in schizophrenia (Takeuchi et al 1988). Kaiya et al (1981) found raised plasma levels of substance P in patients with schizophrenia, although only in association with neuroleptic pharmacotherapy. It has been suggested that if it were indeed the case that substance P is involved in the pathophysiology of schizophrenia then its function with respect to pain modulation may be the means by which the perception of pain is sometimes found to be altered in schizophrenic patients.

Another result of the interaction of substance P with

dopamine has been found in the ventral tegmental area. Experimental administration of the neuropeptide into this region is associated with stimulation of prefrontal cortical dopamine metabolism and with locomotor hyperactivity (Kalivas & Miller 1984).

Thyrotropin releasing factor, thyrotropin, tri-iodothyronine, and thyroxine

Thyrotropin releasing factor (TRF), also known as thyrotropin releasing hormone (TRH), was the first hypothalamic releasing factor to be isolated and have its chemical structure worked out, independently, by both Guillemin (1978) and Schally (1978) who found it to be a tripeptide amide. The action of thyrotropin releasing factor in causing the release of thyrotropin from the adenohypophysis has been considered with the other hypothalamic releasing and release inhibiting factors earlier in this chapter. Although there is no known selective hypothalamic release inhibiting factor for thyrotropin, there is evidence that somatostatin may play this role in addition to inhibiting the release of somatotropin. It has been found that cold temperatures act as potent stimuli for the release of thyrotropin releasing factor and therefore also of thyrotropin.

An important effect of thyrotropin releasing factor administration in normal human subjects is **mood elevation**. Wilson et al (1973) reported beneficial effects of relaxation and a sense of well-being in normal subjects following administration of thyrotropin releasing factor, a finding that has been confirmed in other studies (see, for example, Betts et al 1976). Although Prange and his colleagues (Prange & Wilson 1972, Prange et al 1972) reported partial temporary improvement in mood in a series of female depressed euthyroid patients following intravenous administration of thyrotropin releasing factor, in general later studies have not consistently pointed to this releasing factor as being one which should be used therapeutically in depression. Prange et al (1979) also reported positive results from the administration of thyrotropin releasing factor to patients with schizophrenia. The results of clinical trials with thyrotropin releasing factor in schizophrenia have, however, been disappointing on the whole, with reports that symptoms in paranoid schizophrenia in particular may be worsened (see the review by Loosen & Prange 1984).

An important use of thyrotropin releasing factor in psychiatry is in the form on a provocative **TRF stimulation test** in which the thyrotropin response to intravenous administration of thyrotropin releasing factor is measured. The thyrotropin releasing factor is usually administered in the morning and the thyrotropin concentrations are measured in venous blood samples usually taken every half hour for two or three hours. Abnormalities in the results of this test are mentioned, as appropriate, below. Blunting of the thyrotropin response to TRF stimulation is estimated to occur in approximately 4% of normal subjects (Loosen & Prange 1980).

Thyrotropin, or thyroid stimulating hormone (TSH), is an adenohypophyseal glycoprotein that is released into the systemic circulation and thereby reaches the thyroid gland where it stimulates glandular cells. One of the important results of this is the stimulation of thyroid cellular proteolysis of prestored follicular thyroglobulin, which in turn leads to the release of the thyroid hormones L-tetraiodothyronine, or thyroxine, and L-tri-iodothyronine. Thyrotropin also stimulates the synthesis of further amounts of thyroid hormones by the gland. There is evidence that thyrotropin administration can be of benefit in depression (Prange et al 1984) but this may be only an indirect effect acting through the thyroid hormones.

Although more thyroxine (T_4) than tri-iodothyronine (T_3) is secreted by the thyroid gland, the tri-iodothyronine is the more potent of the two hormones, and following secretion there is considerable peripheral tissue enzymatic conversion of thyroxine into tri-iodothyronine. If the amount of iodine in the body is insufficient, then it is conserved by the thyroid gland secreting more tri-iodothyronine at the expense of the less potent hormone. Thyroxine can also be converted peripherally into a molecule similar to tri-iodothyronine known as reverse tri-iodothyronine (rT_3) which is believed to lack metabolic potency. The thyroid hormones thyroxine and tri-iodothyronine exert feedback control on the hypothalamic secretion of thyrotropin releasing factor and on the pituitary secretion of thyrotropin. The major actions of the thyroid hormones thyroxine and tri-iodothyronine are to stimulate the metabolic rate and to stimulate growth in children. The thyroid hormones cause increases in protein biosynthesis, mitochondrial activity, cellular enzymatic activity, and carbohydrate and fat metabolism. They cause widespread effects throughout the body which are made more evident by the somatic effects of hyperthyroidism and hypothyroidism.

Hyperthyroidism or thyrotoxicosis is often associated with psychological and psychiatric features. These include anxiety, overactivity, fatigue, emotional lability (including depressive swings) and irritability. Delirium is also recognized, and, more rarely, major psychotic disorders. The thyrotropin response to TRF stimulation is usually blunted in early hyperthyroidism. In clinical practice it is important to note that although pyrexia, autonomic dysfunction and tremors may be a presentation of hyperthyroidism, they may also occur because of neuroleptic pharmacotherapy as part of the neuroleptic malignant syndrome. In **hypothyroidism** or myxoedema there may be apathy, impaired cognitive functioning, depression, and fits. Delusions and auditory hallucinations may be part of a paranoid psychosis — the so-called myxoedematous madness (Asher 1949). In early primary hypothyroidism the thyrotropin response to TRF stimulation is usually greater than that seen in normal subjects. In psychiatric clinical practice it is important to bear in mind that hypothyroidism is a side effect of pharmacotherapy with lithium (see Chapter 5). In clinical psychiatric practice thyroid hormone preparations have been used, with benefit, in the treatment of **rapid cycling bipolar mood disorder** and **obesity**.

Since hyperthyroidism is associated with anxiety, studies have been carried out to discover whether elevated levels of thyroid hormones are found in anxiety disorders. In their study of patients suffering from **panic disorder**, Matuzas et al (1987) did indeed find such an abnormality in approximately 7% of the patients. On the other hand, Fishman et al (1985) and Roy-Byrne et al (1986) reported normal levels of thyroid hormones and reduced thyrotropin levels in similar patients. In the last study a blunted thyrotropin response to TRF stimulation was also found. In their comparison of patients suffering from social phobia with normal controls, Tancer et al (1990) found essentially no difference in the levels of thyroid hormones, thyrotropin, or antithyroid antibodies, or in the thyrotropin response to TRF stimulation, suggesting that thyroid abnormalities are not requisite neuroendocrine correlates of this condition.

In their study of female patients with **depression** Prange et al (1972) found no difference in the basal levels of thyrotropin when compared with normal controls, as would be expected since the exclusion criteria for the study included any historical, physical, or biochemical evidence of thyroid disorder. However, the thyrotropin response to TRF stimulation was blunted in the depressed patients. This finding has been replicated in many other studies which on average indicate that at least a quarter of depressed patients show blunting (see the review by Loosen & Prange 1982). It is of interest that blunting is also found in panic disorder (see above) as this may be a function of some underlying common pathophysiological condition in both depression and panic disorder. TRF stimulation studies in depression have also indicated that approximately 15% of patients show a raised thyrotropin response (see, for example, Gold &

Pearsall 1983). Many of these patients have been found to have antimicrosomal thyroid antibodies and antithyroglobulin antibodies, indicating that depression can be associated with symptomless autoimmune thyroiditis (see also Haggerty et al 1987). Changes in thyrotropin release and cerebrospinal fluid thyrotropin releasing factor levels have also been found to be associated with depression. Weeke and Weeke (1978) found that the normal circadian rhythm of thyrotropin release is altered in depressed patients, with a reduction of the normal raised nocturnal levels. This finding has since been confirmed by other groups. Using purer thyrotropin releasing factor antiserum Banki et al (1988) have confirmed the earlier reports of raised CSF concentrations of TRF in drug-free depressed patients compared with normal controls, and have also found normal levels in the two drug-free manic patients they studied.

Abnormalities of the hypothalamic-adenohypophyseal-thyroid axis have also been found in studies of patients with eating disorders. In addition to **anorexia nervosa** being associated with a reduced plasma concentration of tri-iodothyronine (Mitchell 1986), an abnormality that also occurs in emaciated states secondary to other causes, a blunted thyrotropin response to TRF stimulation has been found in a majority of studies (see, for example, Gerner & Gwirtsman 1981, Gold et al 1983). The fact that blunting also occurs in depression may be evidence of an underlying biological link between anorexia nervosa and depressive disorders. On the other hand, it may simply be related to the fact that lowered body mass is common to both conditions and has itself been found to be associated with blunting of the thyrotropin response to TRF stimulation (Fichter et al 1986). Although thyrotropin releasing factor does not usually stimulate the release of somatotropin in normal subjects, it has been found to have this effect in depression and anorexia nervosa (Kamijo et al 1983), making this another provocative endocrine test abnormality that is common to both psychiatric disorders. The thyrotropin response to TRF stimulation has been found to vary widely between different studies of normal body mass patients with **bulimia nervosa**. Some investigators have found rates of blunting as high as 80% (Gwirtsman et al 1983), while others have found no blunting (for example, Kaplan et al 1989).

In a study of post mortem brains of patients with **schizophrenia** compared with brains of normal controls, Nemeroff et al (1983) found decreased concentrations of thyrotropin releasing factor in the frontal cortex, and in Brodmann's areas 12 and 32, the former area also showing a decrease in the concentration of somatostatin. The significance of this finding is not yet clear, and, in general, significant abnormalities in the hypothalamic-adenohypophyseal-thyroid axis have not been found in schizophrenia. Results from several groups suggest that thyrotropin releasing factor is more likely to stimulate somatotropin release in adolescent schizophrenics, particularly if there is a positive family history, than in adult schizophrenics, suggesting that any hypothamic–pituitary abnormality in schizophrenia may be related to age (compare, for example, Gil-Ad et al 1984 and Weizman et al 1982 with Prange et al 1979).

In **Alzheimer's disease** in general no change has been found in cortical thyrotropin releasing factor levels, although Oram et al (1981) have found a reduction in the CSF level of this hormone. Blunting of the thyrotropin response to TRF stimulation has been reported in **alcoholism** (Loosen & Prange 1979) and in patients with **borderline personality** (Garbutt et al 1983).

LIMBIC SYSTEM

The anatomy of the limbic system has been presented in Chapter 1. In this section its functions, particularly those of importance from a psychiatric viewpoint, are considered.

In 1937 James Papez introduced the concept of the **Papez circuit** which he termed the limbic system and which consisted of the hippocampus, hypothalamus, anterior thalamus and cingulate gyrus. He believed that this constituted a reverberating circuit that formed the neuronal mechanism of emotion. In the years since 1937 the concept of the limbic system and the understanding of its functions have advanced considerably. It is now believed to play a major role not only in the elaboration of emotion, but also in perception, memory and certain types of behaviour. In the words of the late K. E. Livingston, the limbic system provides 'bridges that link brain, mind and behaviour in a functional continuum' (Livingston 1978).

Triune brain

Of fundamental importance in understanding the functions of the limbic system is the concept that MacLean has put forward of the **triune brain** in which the primate brain can be thought of as being a synthesis of three brains in one (MacLean 1967, 1970). The phylogenetically oldest part is the **reptilian brain** of reptiles and birds. This is surrounded by the younger **paleomammalian brain**, which is the limbic system. This first appears in early mammals; areas with a primitive

resemblance to the limbic system that exist in some reptilian and avian brains are only rudimentary. Covering this is the phylogenetically youngest part, the **neomammalian brain**, found only in late mammals. These three parts differ in their neurochemical structures and functions. Functions of the limbic system can be surmised from comparative ethological studies. Thus cerebral functions that occur in early mammals with a limbic system (paleomammalian brain) but not in reptiles, that have just a reptilian brain and lack a limbic system, may be attributable to the limbic system. An example of this approach is given in the subsection below on behaviour.

Emotion, perception and memory

The limbic system is believed to be involved in the elaboration of emotions, perception and memory. The hippocampal region is thought to be perhaps particularly involved with short-term memory and motivation. Evidence for the role of the limbic system in these functions is given in this section, while further aspects of these functions appear in Chapter 11.

Historically, one way of understanding the role of the limbic system has been the study of the effects of ictal discharge involving the temporal lobe in humans. Indeed, Hughlings Jackson used epilepsy as a display of brain function (MacLean 1986). A study of cerebral blood flow and metabolism in temporal lobe epilepsy using positron emission tomography and radiolabelled oxygen demonstrated that there is a large region of influence in temporal lobe epilepsy with spread to the limbic system (Gallhofer et al 1985). Indeed, some researchers now refer to **temporolimbic epilepsy** rather than temporal lobe epilepsy. The phenomena experienced include emotional, perceptual and mnemonic features and may appear subjectively to have the quality of occurring as real events although they are rarely mistaken as such. They can also be elicited in conscious humans through **electrical stimulation** of the temporal lobe.

Emotion

Of the emotional features that occur as a result of electrical stimulation of or epileptic discharge from the temporal lobe the commonest is a feeling of **fear**. Others include anger, disgust, guilt, strangeness, unhappiness, a feeling of unreality and a wish to be alone (Maclean 1962). Positive emotions, such as feeling elated, have also been reported. **Erotic ictal manifestations** have been found to predominate in women; they occur only very rarely in males. The erotic feelings in women have the same subjective quality as those elicited by sexual intercourse or masturbation, and indeed may result in orgasm (Rémillard et al 1983).

As has been mentioned earlier in this chapter, there exists a **lateralization of function** in the brain, with the dominant (usually left) cerebral hemisphere being responsible for different functions to those of the non-dominant (usually right) cerebral hemisphere. Since there are hemispheric differences so far as emotional functions are concerned, it has been suggested that lateral specialization in corticolimbic connections related to emotional function also exists. It has been proposed that the part of the limbic system in the dominant cerebral hemisphere together with its cortical connections are responsible for mediating intellectual reflection and the affective investment of words and verbal concepts (Bear & Fedio 1977). For the contralateral hemisphere the suggested functions include affective and autonomic responses within each sensory modality, the decoding and transmission of affective signals, and emotional drive-related surveillance associated with the recognition of threats and the pursuit of goals (Bear 1986).

Perception

The perceptual phenomena that occur in temporal lobe seizures, and upon electrical stimulation of the temporal lobe, include both sensory distortions and sensory deceptions. The latter include hallucinations (false sensory perceptions in the absence of real external stimuli) such as **complex auditory hallucinations**, for example hearing music or speech, and **complex visual hallucinations**, for example seeing a face or scene. Hallucinations involving taste and smell, gustatory and olfactory hallucinations respectively, may also occur, although less frequently. Somatosensory hallucinations have not been reported so far.

Memory

The mnemonic phenomena that occur in seizures and upon stimulation of the temporal lobe include **memory flashbacks** which may involve complex auditory and visual hallucinations and emotional feelings that are appropriate to the memories being recalled. These phenomena were first elicited by electrical stimulation of the temporal lobe by Penfield who concluded that there exists in the human brain a 'permanent record of the stream of consciousness' (Penfield 1951, 1955). This phenomenon can be illustrated by an example observed on electrical stimulation of the right amygdala of a 40-year-old woman who suffered from spon-

taneous seizures: 'Upon stimulation... the patient looked perplexed...she experienced a pleasant feeling in her vulva and on the inner surface of her thighs, as if she were having sexual intercourse. She did not see her partner, but knew it was 'X', the boyfriend with whom she had had her first sexual intercourse at age 16, and subsequently many times more. She affirmed that the experience was an evocation of an old memory, and although she subsequently had had intercourse with other men, she was positive that the feeling evoked by amygdaloid stimulation was the feeling of having sexual relations with her old boyfriend 'X'. ... She volunteered that she had the same sexual experience at the onset of her spontaneous seizures.' (Gloor 1986).

Another mnemonic phenomenon that may occur is **déjà vu**, a type of paramnesia in which occurs the illusion that what is taking place currently is identical, or almost identical, with some previous experience.

In the **Wernicke–Korsakoff syndrome**, described further in Chapter 7, patients suffering from the more severe Korsakoff's psychosis typically have impairment of recent recall. This poverty of short-term memory is associated with pathology of the hippocampal region of the limbic system. For example, Korsakoff's psychosis can be caused by bilateral hippocampal damage following neurosurgery.

With respect to **lateralization of function** the part of the limbic system in the dominant cerebral hemisphere together with its cortical connections are thought to be possibly responsible for mediating learning based on sequentially processed stimuli such as is involved in operations involving language and arithmetic (Bear & Fedio 1977). For the contralateral hemisphere the suggested functions include mediating learning based on topographical associations of stimuli, incidental learning and memory for complex stimuli (Bear 1986).

Behaviour

In this section aspects of the role of the limbic system in aggressive behaviour, maternal nursing behaviour, audiovocal communication and play are considered. The role of the limbic system in the emotional aspects of sexual behaviour have been mentioned above. That the limbic system has a role in the behavioural aspects of this phenomenon is indicated by the fact that hyposexuality is a common feature of temporolimbic epilepsy.

Aggressive behaviour

It might be expected on theoretical grounds that disturbances of functions such as behaviour with which the limbic system is involved may be caused by 'limbic pathoneurophysiology' (Koella 1984). The evidence that that is so in the aetiology of some cases of violent behaviour comes from a number of sources.

From a neuro-anatomical viewpoint it has been shown that the limbic system is closely interconnected with parts of the **basal ganglia**. This would indicate that limbic pathoneurophysiology may lead to both motor disorders and disorders of other limbic system functions in association with each other. It has, for example, been postulated that in severe depressive disorders limbic pathoneurophysiology may be associated with both the depressed mood, a disorder of emotion, and psychomotor retardation, a motor disorder.

Studies into aggression have shown that there is a significant, although weak, association with **temporolimbic epilepsy**. A number of possibilities, such as impaired learning and sociocultural deprivation, have been put forward, mainly in the 1960s and 1970s, to account for this association in such a way that there is no direct causal link between temporal lobe epilepsy and aggressive behaviour. In the 1980s, however, following the demonstration of definite cases of epileptic aggressivity by means of depth electrode recordings (St Hilaire et al 1980), a number of possible direct mechanisms of aggressivity in temporolimbic epilepsy were formulated. These include temporary disconnection of medial limbic structures from cortical control and interference with higher cerebral functions (Ferguson et al 1986). Furthermore, intensive monitoring of patients with temporal lobe epilepsy who have a history of episodic aggressive behaviour has demonstrated that such behaviour is usually an interictal phenomenon, ictal aggressive behaviour being rare (Ramani & Gumnit 1981). This may be associated with the fact that in temporolimbic epilepsy both epileptic foci and distant brain regions, including contralateral regions, have an interictal hypometabolism and hypoperfusion (Valmier et al 1987).

Placidity is a prominent feature of the **Klüver–Bucy syndrome**, which is caused by bilateral excision of the temporal lobes. Other features include hyperorality, hypersexuality, hypermetamorphosis and hyperphagia. It is believed that it is the bilateral destruction of the amygdaloid bodies of the limbic system, situated in the dorsomedial temporal lobes, which is of aetiological significance in giving rise to this syndrome, thus adding further weight to the notion that the limbic system is involved in the elaboration of aspects of aggressive behaviour. Clinically, aspects of the Klüver–Bucy syndrome are seen in Alzheimer's disease and Pick's disease (see Ch. 7).

Maternal nursing behaviour, audiovocal communication and play

Maternal nursing behaviour, audiovocal communication between mother and offspring, and playful behaviour in the offspring are important components of the mammalian family group; they are not evident in extant reptiles. That the limbic system may be involved in the elaboration of this mammalian behavioural triad was suggested by comparative neuro-anatomical findings. Thus Clark and Meyer (1950) demonstrated that the reptilian brain lacks the cingulate gyrus of the mammalian brain and concluded that this part of the mammalian brain, together with its subcortical connections, was related to types of cerebral activity found exclusively in mammals. Subsequent ablation experiments have shown that ablation of the cingulate cortex abolishes this mammalian behavioural triad (see for example Murphy et al 1981).

Dreaming

It has been mentioned above that one possible function of the part of the limbic system in the non-dominant cerebral hemisphere and its cortical connections may be emotional surveillance. Experimental evidence would tend to locate the relevant emotional surveillance circuits to the dorsal 'non-dominant' (usually right) limbic system of the parietofrontal system. Bear (1986) has suggested that in dreaming there occurs an activation or priming of these dorsal surveillance circuits. If this hypothesis were correct then prevention of this activation or priming would be expected to lead to a decrease in attention and emotional control. This is indeed the result of dream (REM sleep) deprivation (see the section on sleep, below). Also consistent with this hypothesis is the finding that damage to the non-dominant parietal lobe, but not to the temporal lobe, is associated with the cessation of dreaming (Humphrey & Zangwill 1951).

APPLIED ELECTROPHYSIOLOGY

In this section the following types of electrical data, based on applied electrophysiological techniques, are discussed: the electromyogram (EMG), the electrooculogram (EOG), the electroencephalogram (EEG), evoked potentials (EPs) and brain electrical activity mapping (BEAM).

Electromyography

Electromyography is essentially a technique used to investigate the functioning of the motor unit. The physiology of the latter is discussed earlier in this chapter. Different types of EMG can be recorded by altering the number and type of electrodes used and by using electronic filters. The EMG provides useful information in the clinical management of conditions such as amyotrophic lateral sclerosis (motor neurone disease), primary myopathies, and disorders of neuro-muscular transmission such as myasthenia gravis.

Sleep studies

From a psychiatric viewpoint electromyography is of use in recording the degree of muscle tone when studying human sleep. A common arrangement is the submental application of two electrodes so that the potential difference between them can be recorded. This is known as **bipolar recording** and gives a measure of the muscle tone in the submental facial muscles during the different stages of sleep (see below).

In **polysomnography**, a complex type of recording procedure used in the diagnosis of sleep disorders, **anterior tibialis electromyography** is commonly used. As the name implies two electrodes are applied to the anterior tibialis muscle of the leg. This allows the detection of periodic leg movements, such as ankle dorsiflexion, which occur in nocturnal myoclonus. Increased anterior tibialis EMG activity may also be observed prior to sleep in the restless leg syndrome. Polysomnography is discussed further in the section on sleep and arousal, below.

Electroconvulsive therapy

It has been suggested that, given the widespread use of muscle relaxants in electroconvulsive therapy, EMG recordings from the masseter muscles provide a more sensitive means than simple observation for gaining information about induced seizures (Sørensen et al 1981). A more common and easier method to use consists of applying an inflated sphygmomanometer cuff to a limb before the intravenous introduction of the muscle relaxant. The muscle relaxant is thereby prevented from reaching the corresponding limb extremity (for example the forearm) which therefore is subject to unmodified convulsions during an induced seizure.

Neonates

Submental electromyography is used as part of poly-graphic studies in neonates who may be at risk of mental retardation following perinatal problems. Other recordings made in such studies include electro-encephalography, electrocardiography, and waking and sleeping electro-oculography.

Depression

Facial electromyographic studies in depressed patients have revealed differences in the data obtained compared with normal controls (Greden et al 1986).

Electro-oculography

In electro-oculography electrodes are applied to the skin close to the lateral corner of each eye, thereby providing two eye channels. The differences between the signals obtained and a reference lead are amplified and recorded, and vary with ocular movements. Because of the mutually contralateral positioning of the index electrodes, conjugate eye movements result in recordings from the two eye channels that are reflections of each other.

Sleep studies

Like the EMG, the EOG is used in sleep studies including polysomnography. Electro-oculography is particularly useful in studying rapid eye movement (REM) sleep and REM latency since it reveals the number of eye movements per unit time. These are discussed later in this chapter in the section on sleep and arousal.

Schizophrenia

Electro-oculography, and in particular **visual tracking**, has revealed ocular movement abnormalities in schizophrenic patients and their relatives (Holzman et al 1977). Two types of abnormality have been found to be significantly commoner in schizophrenia: an increased frequency of **blinking** and abnormal saccadic eye movements. The administration of neuroleptic medication has been found to lead to a reduction in the rate of blinking. Therefore it is possible that the rate of blinking may be associated with central dop-aminergic pathways. **Saccades** are rapid ocular movements that allow the line of sight to be re-aligned during the visual tracking of moving objects. Between approximately 50% and 80% of the samples of schizo-phrenic patients tested have been found to have abnormal saccades when attempting smooth-pursuit visual tracking of moving objects. The corresponding figure for normal controls is about 8%. However, for first-degree relatives of schizophrenic patients the corresponding figure is significantly higher at approximately

40%. Holzman et al (1988) have found that by treating schizophrenic symptomatology and abnormal saccades as two independent manifestations of a latent trait then data relating to abnormal saccades in the offspring of mono- and dizygotic twins discordant for schizophrenia can be accounted for by a genetic model that proposes a single dominant high-penetrance gene for the latent trait. Genetic aspects of schizophrenia are discussed further in Chapter 6.

Increased saccadic abnormalities have been found in those schizophrenic patients with **tardive dyskinesia** when compared with schizophrenic patients without tardive dyskinesia, the latter in turn having increased abnormalities than controls (Thaker et al 1989). It has been proposed that this result is caused by dysfunction of subcortical GABA circuits involved with the frontal eye fields. The result is consistent with a model of schizophrenia put forward by Early et al (1989) that is discussed further in Chapter 3.

Other conditions in which abnormal saccades have been found include mood disorders and drug-induced states. An association has also been found between visual tracking dysfunction and thought disorder in psychosis (Solomon et al 1987).

Neonates

As mentioned above, waking and sleeping electro-oculography are used as part of polygraphic studies in neonates who may be at risk of mental retardation following perinatal problems.

Electroencephalography

The electroencephalogram (EEG) is a recording of the electrical potential activity of the brain. The basic resting EEG alpha rhythm (see below) was first described by the Austrian psychiatrist Hans Berger in 1929 and electroencephalography was further developed by Adrian and Yamagiwa (1935). It was initially hoped that specific functional psychiatric disorders would be found to be associated with pathognomonic EEGs. Although this has not proved to be the case the EEG has continued to be used in clinical psychiatry as a safe technique for investigating brain function.

Conventional recording techniques

In the conventional EEG recording electrodes are placed on the scalp, making this a safe non-invasive investigation procedure. This method can be used in conjunction with the activation procedures considered later in this section. The EEG can also be recorded more directly from either the surface of the brain (electrocorticography) or from electrodes placed inside the brain (depth electroencephalography).

In the conventional EEG recording procedure the scalp electrodes are usually positioned according to the **International 10–20 System**. This entails the placement of the electrodes at positions defined by measurements from the following scalp landmarks: the nasion, the inion, and the right and left auricular depressions. These measurements are proportional to the dimensions of the skull. For example, for the midline (z) electrodes, the occipital electrode (O_z) is placed on the midline joining the nasion and inion such that its distance from the inion is 10% of the distance from the inion to the nasion. Similarly the corresponding proportions for the parietal midline electrode (P_z) and the central midline or vertex electrode (C_z) are 30% and 50% respectively. Therefore the International 10–20 System allows the standardization of electrode placement over a range of skull sizes including those of children. The proportions between electrodes are 10% or 20%, hence the name of this system.

The output from the electrodes, after passing through electronic filters and amplifiers, can be viewed on a visual display unit or on a paper recording. It can also be stored electronically. In the case of **ambulatory electroencephalography** the output is stored on cassette tapes in small portable cassette recorders. By using a number of electronic filters and amplifiers it is possible to obtain simultaneous multi-channel recordings, each channel having two electrode inputs. The recording for each channel represents the electrical potential difference between its two inputs. The recording can be either **monopolar**, in which case one input to each channel is from a scalp electrode and the other from a relatively neutral distant reference electrode such as the combined output of electrodes attached to the earlobes, or it can be **bipolar**, in which case both inputs to each channel are from scalp electrodes. The way in which the electrodes are connected for recording purposes is usually in one of a number of preset ways which can be switched between electronically.

Specialized recording techniques

Nasopharyngeal leads are inserted through the nasal orifices to allow the electrodes to lie in the superior part of the nasopharynx. Nasopharyngeal electrodes are used to obtain recordings of electrical activity from the inferior and medial temporal lobe.

Sphenoidal electrodes are inserted between the mandibular coronoid notch and the zygoma and like nasopharyngeal electrodes can also be used to obtain recordings from the inferior temporal lobe. Compared with nasopharyngeal electrodes, sphenoidal electrodes require more skill to be positioned, but obtain recordings that are probably less subject to unwanted artefactual electrical potentials.

In **electrocorticography** the electrodes are placed directly on the surface of the brain. Because the electrical resistance of the scalp is bypassed, the recordings of electrocorticography are much larger than those of electroencephalography.

In **depth electroencephalography** the electrodes are placed inside the brain itself. Again, the recordings are much larger than those from conventional recording techniques. Electrocorticography and depth electroencephalography may be used, for example, in mammalian experiments and in human neurosurgery.

Normal EEG rhythms

Normal EEG rhythms are believed to arise from activity in the superficial layers of the cerebral cortex that is synchronized by the thalamus (see Fig. 2.11). By convention, these normal EEG rhythms are classified according to their frequencies into the four classes shown in Table 2.4.

Alpha activity represents the resting normal EEG rhythm of an awake adult human with closed eyes. It was recorded from the occipital region by Berger who noted that it disappeared when the eyes were opened. Hence the previous name of Berger rhythm. Although strongest over the occipital region, alpha activity can also sometimes be recorded from other regions of the brain of a subject resting peacefully.

When the awake resting adult human subject opens his or her eyes or is otherwise stimulated out of a state of quiet cerebration, the alpha activity is largely replaced by **beta** activity which has a higher frequency of greater than or equal to 13 Hz. This replacement of alpha activity is known as **alpha blocking** and also as the alerting or arousal response. Compared with alpha

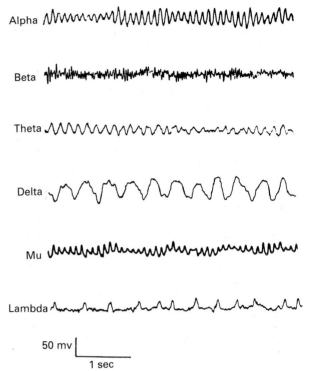

Fig. 2.11 Single channel tracings of various forms of EEG activity from normal adult individuals. Alpha, characteristic of the awake, eyes closed state is seen over the parieto-occipital region usually at 20–50 µV. It ranges in frequency between 8–13 cycles per second and is 'blocked' by eye opening. Beta activity, 13 cycles per second and over, is present in small amounts in the fronto-central regions. It is markedly increased by drug administration such as benzodiazepines or barbiturates. Theta, 7.5–4 cycles per second activity occurs in drowsiness, in adults, youth and abnormally in a variety of abnormal conditions. Delta activity, below 4 cycles per second occurs in deep sleep in normals, also in coma and many disorders. Mu rhythm is seen over the rolandic regions and is 'blocked' by movement of the contralateral limb. Lambda waves occur over the occipital regions in subjects with their eyes open and scanning patterned visual fields. (Reproduced with permission from Swash M, Kennard C 1985 Scientific basis of clinical neurology. Churchill Livingstone, Edinburgh.)

activity, beta activity consists of waves of smaller amplitude and greater irregularity. Different cortical sites simultaneously give rise to beta waves that are out of phase and differ in amplitude and frequency with each other. For this reason the EEG is said to exhibit **desynchronization**.

Low-frequency, high-amplitude **theta** activity and **delta** activity do not normally occur in waking adult EEGs but are normal features of sleep (see below).

Lambda activity occurs only over the occipital region in subjects with opened eyes. It is related to ocular movements occurring during visual attention.

Table 2.4 Classification of EEG rhythms according to frequency

Type of rhythm	Frequency (f) of rhythm[1]
Delta (δ)	$f < 4$ Hz
Theta (θ)	4 Hz $\leq f < 8$ Hz
Alpha (α)	8 Hz $\leq f < 13$ Hz
Beta (β)	$f \geq 13$ Hz

[1]Hertz = Hz = cycles per second.

Mu activity occurs over the motor cortex and is related to motor activity, being abolished by movement of the contralateral limb.

Normal EEG rhythms in humans vary with **age**. EEG recordings from newborn full-term babies are characterized by the relative absence of electrical activity. In infants, the presence of desynchronized low frequency activity is common. As a child ages, waking delta and theta activity are gradually reduced and the adult alpha-dominant EEG is usually present by the end of the second decade. In old age, normal changes include a decrease in the amplitude and average frequency of alpha activity, diffuse slowing, and the presence of brief runs of frontotemporal, mainly left-sided, low-frequency activity.

The average frequency of normal EEG rhythms also varies with the level of **alertness**. High-frequency beta activity occurs during a high state of alertness, for example, when engaged in intense studying. As the subject becomes more relaxed the average frequency tends to decrease, so that when the adult subject is awake but relaxing with closed eyes, alpha activity is present. During drowsiness and sleep low-frequency theta activity appears. With increasing depth of sleep lower frequency delta activity becomes present. Delta activity is also a characteristic feature of loss of consciousness from other causes such as general anaesthesia.

Spikes and sharp waves

Spikes are transient high peaks lasting less than 80 milliseconds that can clearly be differentiated from the general amplitude of the background EEG recording. **Sharp waves** are conspicuous sharply defined wave formations that rise rapidly but fall more slowly and last more than 80 milliseconds. Although spikes and sharp waves usually represent EEG abnormalities, occurring for example in epilepsy, this is not always the case. Thus the occipital lambda activity referred to above consists of sharp waves. Similarly, another type of sharp wave occurring over the vertex during drowsiness, sometimes known as V-waves, is also a normal phenomenon.

Activating procedures

EEG activating procedures are used to expose EEG abnormalities that are either not manifest in the normal EEG recording or are present but not clearly defined.

Hyperventilation for two or three minutes may induce high amplitude delta activity in epileptics. It is particularly effective in children, in whom it may reveal the EEG pattern seen in petit mal (see below). The mechanism of activation is believed to be cerebral hypoxia leading to cortical hyperexcitability.

Photic stimulation with repetitive light flashes of variable frequency, produced for example by means of a stroboscope, can lead to **photic driving** in which the EEG rhythm, particularly alpha activity over the occipital region, synchronizes with the stimulus frequency. High-amplitude abnormal EEG seizure patterns can spread to other parts of the cerebral cortex and lead to a photoconvulsion. The latter may also be seen clinically in some epileptics when sitting close to a flickering television screen.

Drugs are also used in activating procedures. A drug that lowers the seizure threshold, for example chlorpromazine, can be administered in order to allow seizure patterns to emerge. The induction of sleep, for example with pentobarbitone, can also be used to allow seizure patterns to emerge as the subject passes through the different stages of sleep. Intravenous thiopentone may be administered to induce high-frequency beta activity.

An alternative non-pharmacological method of sleep induction consists of **sleep deprivation**. The sleep-deprived subject is likely to fall asleep during the EEG recording and, as in the case of drug-induced sleep, while passing through the stages of sleep seizure patterns may emerge.

In cases of seizures occurring in response to **known or suspected stimuli** such as reading, these stimuli can be used for activation.

Clinical uses

As mentioned above, specific functional psychiatric disorders have not been found to be associated with pathognomonic EEGs, thereby reducing the diagnostic power of electroencephalography initially hoped for (see Kiloh et al 1981). Another difficulty with the diagnostic use of the EEG is the occurrence of false-positive and false-negative results. Therefore, a normal EEG cannot be used on its own to exclude a diagnosis, including epilepsy, without further investigation. On the other hand, an abnormal EEG may sometimes be recorded from a normal subject.

A further limitation to the clinical use of electroencephalography is the fact that the EEG can be influenced by physiological changes. The effects of age and the level of alertness (or drowsiness) have been mentioned above. In addition to the changes associated with the stages of sleep, it has been found that the average frequency of alpha activity has its own circadian rhythm and also varies in women according to the time in the menstrual cycle. Hypothermia leads to a reduction in the amplitude of the EEG, while hyperthermia again causes changes in alpha activity. The EEG is also altered by hyperventilation, other causes of changes in the acid-base balance, and hypoglycaemia. The consumption of alcohol can also alter the EEG — sometimes, for example, causing changes in the EEG of normal subjects that are usually associated with epilepsy.

Limitations of electroencephalography that are particularly important to note in clinical psychiatric practice include the occurrence of variable EEG abnormalities in a proportion of patients with functional psychiatric disorders that are not believed to be organic in aetiology (Fenton 1974). This situation is analogous to the occurrence of false positives in up to 15% of normal subjects, except that the proportion of false positives in psychiatric patients has been reported as being higher (Lishman 1987). Other limitations are the changes in the EEG caused by the use of psychotropic medication and electroconvulsive therapy, and following psychosurgery. These changes are discussed further below.

On the positive side, the EEG can be helpful in the localization of focal lesions. It has been mentioned above that, in normal adults, low-frequency theta and delta activity normally occurs only during drowsiness and sleep. When such activity is present in the waking adult it may be caused by cerebral changes in tissues adjacent to a cerebral lesion such as a tumour, abscess, infarction or haemorrhage. In the past this made the EEG a valuable tool for localization. However, the source of low-frequency activity may not always be in close proximity to the actual site of the lesion. Therefore it is much more useful in such cases to utilize more modern and accurate methods of brain imaging, such as computerized tomography.

Absence seizures (petit mal) are usually associated with the relatively specific EEG feature of generalized compound wave-and-spike discharges occurring at a frequency of 3 Hz. Creutzfeldt–Jakob disease is also associated with a relatively specific EEG feature: repeated generalized irregular spike and slow wave complexes may be recorded. Subdural haematomas may cause ipsilateral reduction of alpha activity and ipsilateral low-frequency activity. Herpes simplex encephalitis and subacute sclerosing panencephalitis are also associated with relatively specific EEG recordings.

Details concerning the EEG features seen in psychiatric disorders are not dealt with in this book. The interested reader is referred to the review by Fenton (1989).

Effects of psychiatric treatment

There have been a number of studies of the effects of psychotropic medication on the EEG. In general it has been found that **anxiolytics** including both barbiturates and benzodiazepines cause increased beta activity and sometimes reduced alpha activity (Itil 1974). **Antidepressants** lead to an increase in delta activity (Herrmann 1982). **Antipsychotic drugs** generally lead to a decrease in beta activity and an increase in low-frequency delta and/or theta activity (Itil 1977). The EEG effects of **lithium** at therapeutic levels are small and likely to be missed on visual analysis of routine recordings (Small et al 1972, Reilly et al 1973).

When receiving **electroconvulsive therapy** (ECT) the EEG recording is that seen during tonic–clonic seizures (grand mal), possibly as a result of either subcortical synchronization or cortical spread (Berrios & Katona 1983). Between treatments during a course of ECT, the most commonly reported change in the EEG is the build-up of diffuse irregular low-frequency activity which is usually intermittent and which increases with each treatment and then decreases with time following the completion of the course (Fink 1979). This change is usually approximately symmetrical but has occasionally been reported as showing a left-sided predominance even following bilateral ECT (see, for example, Stromgren & Juul-Jensen 1975). Furthermore, this EEG slowing has been found to be enhanced by barbiturates, relaxation and hyperventilation (Roth 1951). The EEG slowing usually disappears by one month after the end of a course of ECT. It is only infrequently seen after three months have elapsed (Weiner 1980).

Bifrontal delta activity has been reported following the **psychosurgical operations** of both prefrontal leucotomy in the past and stereotactic subcaudate tractotomy performed more recently. Indeed in the latter group it has been found that the presence of this EEG change is a good prognostic factor (Evans et al 1982).

Evoked potentials

The repeated presention of a stimulus in a given sensory modality can give rise to electrical responses, known as evoked potentials, in the cerebral cortex, the brainstem and neuronal tracts. The sensory modalities used include visual, auditory, somatosensory and olfactory. Evoked potentials can be recorded by means of EEG electrodes and electrodes positioned close to the spinal cord tracts.

Analysis

Evoked potentials (EPs) probably mainly represent slow synaptic activity. As with the EEG, the electrodes record the average totalled electrical potential changes caused by large numbers of neurones. When using the EEG to record EPs it is usually the case that the small EPs superimposed on the EEG tracings cannot be discerned. This problem has been resolved by use of the technique of **signal averaging**. The responses to repeated stimuli are added so that the combined response has a much larger amplitude than an individual response, while the EEG background tracings that are not related to the stimuli are averaged out.

Typically an EP consists of a number of positive and negative changes of electrical potential occurring in succession with respect to time and known individually as **components**. A diagrammatic example of a brainstem auditory EP is shown in Figure 2.12 in which successive positive components are labelled with increasing Roman numerals, beginning with I. The **latency** of a given component is the period of time elapsing from the time of onset of the stimulus to the time that the maximum deflection in the component occurs. The time interval between the maximum deflections of two components of an EP is known as the interpeak latency. For a given EP the different components may arise as a result of more than one site of origin and the interpeak latencies may represent the time taken for neuronal conduction to occur between different sites.

Types

The types of light used in **visual evoked potentials** (VEPs) include unpatterned and patterned stimuli. The use of chessboard type patterned stimuli has proved useful clinically in the investigation of cases of possible multiple sclerosis. In definite cases of this condition it is found that there is likely to be an increased latency for the positive component that normally has a latency of approximately 100 ms and which is therefore known as P100 (P for positive).

Auditory evoked potentials (AEPs) can be used clinically in audiometric studies and in investigating the possible presence of lesions such as acoustic neuromas. Thus AEPs can be used to determine the level of hearing ability in subjects such as neonates, infants, and some psychiatric patients who are unable to give an appropriate response to more conventional testing.

Somatosensory evoked potentials (SEPs) can be recorded after the stimulation of peripheral sensory or mixed neurones or receptors. They are used in the assessment of neurological trauma, for example to the spinal cord.

Evoked potential changes in psychiatric disorders

EPs may vary with age, gender, the level of alertness and the use of psychotropic drugs. Thus these factors, in addition to the occurrence of artifacts, have to be taken into account in the analysis of EP changes occurring in psychiatric disorders.

EP studies of **schizophrenia** outnumber those of any other psychiatric diagnosis. They have demonstrated relatively consistent changes in the SEP amplitude, AEP amplitude and latency, temporal and spatial wave-shape variability, amplitude recovery, latency recovery, and P300 in schizophrenic patients compared with normal controls and depressed non-psychotic patients (Shagass 1983). VEP studies have tended to give variable results.

Similarly, relatively consistent changes have been found in the SEP, AEP, VEP, and temporal wave-shape variability in **depressive disorders** (Shagass 1983). Differences have been found between psychotic and non-psychotic depressed patients (von Knorring et al

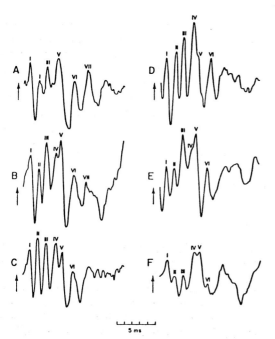

Fig. 2.12 Brain-stem auditory evoked potentials: waveform variations within the normal range. The responses were recorded from a single ear of six subjects (**A**–**F**). Positivity at the vertex produces a deflection upwards. (Reproduced with permission from Swash M, Kennard C 1985 Scientific basis of clinical neurology. Churchill Livingstone, Edinburgh.)

1974), and between those with and without a genetic loading (Satterfield 1972).

Studies on **mania** have proved more difficult owing to the greater occurrence of artifacts from less compliant subjects. Nevertheless, Shagass et al (1978) have demonstrated changes in the SEP, AEP and VEP in mania. These results cannot be taken as being as consistent as those in the previous two groups owing to the smaller numbers studied.

Levy et al (1971) demonstrated changes in the SEP in patients with **dementia** compared with patients of a similar age who were depressed. Changes in the VEP of demented patients compared with control subjects of similar age have been found by Straumanis et al (1965). Similarly, changes have been found in the AEP.

Attempts to find relatively consistent changes in non-psychotic functional psychiatric disorders such as neuroses and personality disorders have mostly been unsuccessful. However differences have been found between non-psychotic and psychotic patients (Shagass et al 1978). In particular, in the latter group the amplitude of components of an EP occurring with a latency of greater than 80 ms tends to be smaller than in non-psychotic patients.

Since EP studies allow the analysis of cerebral processing to take place, they can be used in the study of rapid normal information processing and possible related cognitive disorders including learning disabilities.

Brain electrical activity mapping

Brain electrical activity mapping (BEAM) was developed by Duffy et al (1979) as a method of allowing electrical data from the brain to be displayed in a computerized topographic form on a visual display unit. The electrical data input can be in the form of the EEG or EP. The display is a dynamic one, so that real-time changes can be noted following the presentation of various stimuli, for example.

EEG data

In the case of EEG data, the two-dimensional display represents different EEG frequency ranges (delta, theta, alpha and beta activity) by different colours or shades of grey. It is possible to represent given three-dimensional regions of the scalp by equivalent two-dimensional areas on the brain electrical activity map.

EP data

When the input in BEAM is from EP data, the different regions on the map represent different ranges of EP electrical potential differences. This can be superimposed on the use of two different colour ranges that represent positive and negative values. By electronically averaging the EP data over given time intervals or epochs, the brain electrical activity map for successive epochs can be displayed on the visual display unit.

Use in psychiatry

The use of BEAM in psychiatry is in a relatively early stage. It has been applied to patients with dyslexia, who have been found to show consistently higher mean alpha activity per region than occurs in normal subjects (Duffy et al 1980a, b), and it has also been used in studies of schizophrenia (Morihisa & McAnulty 1985).

As a functional brain imaging technique BEAM has a number of potential advantages over regional cerebral blood flow and positron emission tomography, including its lower cost, and the fact that BEAM does not require the use of radioactive markers. Moreover, with EP data, BEAM allows events occurring over much shorter time intervals to be mapped (see Morihisa 1987).

SLEEP AND AROUSAL

Sleep has been defined by Anch et al (1988) as being 'a recurring state of existence characterized by 1. reductions in awareness of and interaction with the environment, 2. lowered motility and muscular activity, and 3. partial or complete abeyance of voluntary behavior and consciousness.' Although normally linked to the 24-hour cycle, the circadian sleep–wakefulness rhythm follows a rather longer cycle of approximately 25 hours when human subjects are freed from the link with the external day–night cycle, for example by living experimentally underground without access to the time (Aschoff 1969).

Stages of sleep

Sleep has been found through applied electrophysiological methods such as the EEG, EOG and EMG to consist of two types: **rapid eye movement** (REM) sleep during which the eyes undergo rapid movements, and **non-REM sleep** during which the EEG shows low-frequency activity. The staging of non-REM sleep based on EEG recordings was first published in 1937 (Davis et al 1937, Harvey et al 1937), with the discovery of REM (rapid eye movement) sleep being made over a decade later by Aserinsky and Kleitman (1953).

Figure 2.13 is a diagrammatic representation of the stages of sleep based on the EEG with typical EEG

Awake   low voltage – random, fast

50 μV

1 sec

Drowsy – 8 to 12 cps – alpha waves

Stage 1 – 3 to 7 cps – theta waves

Theta Waves

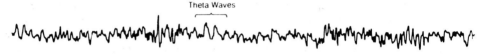

Stage 2 – 12 to 14 cps – sleep spindles and K complexes

Sleep Spindle

K Complex —

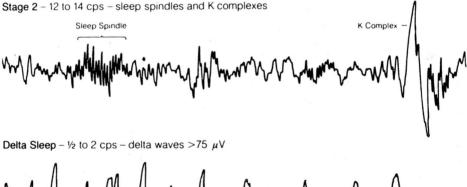

Delta Sleep – ½ to 2 cps – delta waves >75 μV

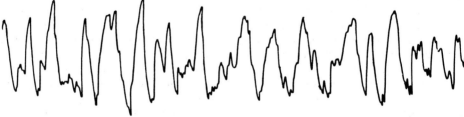

REM Sleep – low voltage – random, fast with sawtooth waves

Sawtooth Waves   Sawtooth Waves

Fig. 2.13   Electroencephalograms of the different states and stages of human sleep. (Reproduced with permission from Hauri P 1977 The sleep disorders. Upjohn, Kalamazoo, M I.)

tracings shown. During quiet wakefulness with shut eyes, the EEG shows alpha activity; this can be called

stage 0. In stage 1 the subject is falling asleep and the EEG has a low amplitude. There is decreased alpha activity and the appearance of low-voltage theta activity. With further deepening of sleep stage 2 occurs, typically during light sleep. There is a further reduction in the frequency to 2–7 Hz. Stage 2 low-frequency activity is characteristically broken by occasional **sleep spindles** and **K complexes**. Sleep spindles are transient runs of waves with a frequency of 12 to 14 Hz lasting at least 0.5 s. K complexes each last approximately 0.5 s and have a marked positive deflection immediately preceded by a marked negative one on the EEG. Stages 3 and 4 correspond to deep sleep and their EEG patterns show increasing levels of delta activity (frequency less than 4 Hz); the proportion of delta activity in stage 3 is 20% to 50%, and for stage 4 it is greater than 50%. Stages 3 and 4 are together known as **slow-wave sleep**.

The basic EEG pattern of REM sleep is similar to that of stage 1 sleep, with superimposed **sawtooth waves**, shaped as their name implies. REM sleep is also known as dreaming sleep. This is because sleep studies have consistently demonstrated that when a subject is woken while known from polysomnographic recordings to be in REM sleep, the subject is more likely to report that he or she is in the middle of a dream, than is the case when the subject is woken from non-REM sleep. However, dreaming is by no means confined to REM sleep. Therefore it is misleading to use the synonym dreaming sleep for REM sleep. What can be said with confidence is that a subject is more likely to remember that they were dreaming if woken during REM sleep. Non-REM sleep dreams are more likely to be simpler than REM dreams. For example, non-REM dreams may involve little or no movement in the dream scenes (Anch et al 1988).

Other names used for REM sleep include desynchronized sleep and paradoxical sleep. Non-REM sleep is also known as synchronized sleep. Some authors use the term slow-wave sleep to refer to non-REM sleep in general and not just to stages 3 and 4.

With increasing age the total length of time spent sleeping in each 24-hour cycle diminishes from an average of approximately 16 hours per day in the newborn full-term baby, through eight hours per day during adolescence, to less than six hours per day in old age. Within the sleeping time the proportion of REM sleep to non-REM sleep also diminishes, from approximately 50% in the newborn full-term baby to less than 20% by middle age.

During normal adult sleep the length of time spent in REM sleep in each sleep cycle tends to increase with successive sleep cycles (see Kales & Kales 1974). In total, REM sleep normally takes up about one-fifth to a quarter of the total time spent asleep. The average sleep cycle lasts approximately 90 minutes, with the first cycle tending to be relatively short and the second cycle usually being the longest.

Polysomnography

Polysomnography is a technique involving a battery of simultaneous tests that allows physiological variables to be measured during sleep. The tests carried out can include electroencephalography, electro-oculography, electromyography, electrocardiography and the monitoring of respiration, movement, body temperature, galvanic skin conductance and penile tumescence. Other variables such as the blood oxygen level, gastric juice pH and vaginal blood flow can also be measured, as appropriate.

Polysomnography is used in the scientific study of sleep. For example, the physiological correlates of REM sleep and non-REM sleep mentioned below have been found or confirmed using polysomnographic studies.

Polysomnography is also useful in the clinical investigation of sleep disorders and other medical conditions such as certain respiratory and cardiovascular disorders, impotence, gastrointestinal disorders such as sleep related oesophageal refux, nocturnal enuresis and epilepsy. Sleep changes in psychiatric disorders, for example mood disorders, can also be studied in more detail with this technique.

Physiological correlates

In addition to the difference in the recall of dreaming noted above, REM sleep and non-REM sleep have been found to differ in a number of other ways physiologically. Thus, so far as autonomic functions are concerned, REM sleep is dominated by sympathetic activity while non-REM sleep is dominated by parasympathetic activity (Snyder et al 1964).

As mentioned above, REM sleep is characterized by the occurrence of transient runs of conjugate ocular movements. Other features of REM sleep not already mentioned include maximal loss of muscle tone, increased heart rate and systolic blood pressure, increased respiratory rate, increased cerebral blood flow, occasional myoclonic jerks, and either penile erection or increased vaginal blood flow. REM sleep is also associated with increased protein synthesis in the rat brain (Levental et al 1975) while experimental inhibition of protein synthesis leads to an inhibition of REM sleep (Drucker-Colin et al 1982). It has been suggested that this may be associated with other evidence that links REM sleep with learning and memory (Webb 1981, Mirmiran et al 1983). That REM sleep is probably associated with important functions is indicated by the occurrence of REM rebound following REM deprivation.

During non-REM sleep many of the changes that occur are the opposite of those during REM sleep. These include decreased heart rate and systolic blood pressure, decreased respiratory rate, decreased cerebral blood flow and upward ocular deviation with few or no movements. The penis is not normally erect during non-REM sleep. Abolition of tendon reflexes also occurs.

In addition to the neuroendocrine circadian rhythms mentioned earlier in this chapter, changes in some hormone levels have been found that are closely associated with the stage of sleep. Nocturnal peaks of somatotropin occur during slow-wave sleep (stages 3 and 4 of sleep) (Takahashi et al 1968, Quabbe 1977). Prolactin levels increase during the first sleep cycle and peak during the last two hours of normal sleep (Sassin et al 1972). No consistent association has been found between prolactin secretion and the stage of sleep (van Cauter et al 1982). Testosterone levels have been found to increase soon after the onset of sleep and generally to continue increasing during most of the sleep time (Evans et al 1971). A trough in the level of melatonin occurs during REM sleep (Birkeland 1982). Similarly, renin levels also decrease during REM sleep (Mullen et al 1980). Parathormone levels have also been found to be associated with the stage of sleep (Kripke et al 1978).

Theories of the causes of the sleeping–waking cycle

There are two main theories that relate to the causes of the sleeping–waking cycle. These have variously been named the monoaminergic, biochemical, two-stage, or Jouvet's model, and the cellular, or Hobson's model.

The **monoaminergic model** was first put forward by Jouvet (1965; 1967; 1972; 1977) and essentially attributes non-REM sleep to serotonergic neuronal activity and REM sleep to noradrenergic neuronal activity. The serotonergic neurones postulated as being involved are in the **raphe complex**. The evidence for this includes the finding that insomnia results from the inhibition of 5-hydroxytryptamine (serotonin) and also from the destruction of serotonergic neurones of the raphe complex. The 5-hydroxytryptamine precursor L-tryptophan would be expected, on the basis of this model, to have a hypnotic effect. This is indeed the case, and for many years it was used clinically as an hypnotic until its withdrawal (in 1990 in Britain) owing to a strong association of its use with the potentially fatal eosinophilia–myalgia syndrome (Drugs

and Therapeutics Bulletin 1990). The noradrenergic neurones involved are believed to be in the **locus coeruleus**, destruction of which leads to REM sleep suppression without non-REM sleep being suppressed. Jacobs and Jones (1978) have modified Jouvet's model, proposing that the above central serotonergic and noradrenergic neurones are not involved in mediating the onset of non-REM sleep and REM sleep, respectively, but rather that they are involved with the modulation of those respective parts of the sleep cycle.

Hobson (1974) and McCarley and Hobson (1975) put forward an alternative **cellular model** which involves three groups of central neurones: the gigantocellular tegmental field of the pons (the nucleus reticularis pontis caudalis), dorsal raphe nuclei, and the locus coeruleus; and three corresponding neurotransmitters: acetylcholine, 5-hydroxytryptamine and noradrenaline, respectively. According to this model, the gigantocellular tegmental field or 'on cells', which are inhibited by the dorsal raphe nuclei and the locus coeruleus (the 'off cells'), are responsible for causing the onset of REM sleep, during which a gradual increase in the activity of the off cells leads to an inhibition of the on cells and the restoration of non-REM sleep or wakefulness. Evidence in support of this model includes the finding that the injection of very small quantities of cholinergic agents into the gigantocellular tegmental field leads rapidly to the onset of REM sleep (see, for example, Silberman et al 1980). On the other hand, ablation experiments have failed to lend support to this model. For example, Sastre et al (1979) have found that, following chemical destruction of the gigantocellular tegmental field of the feline pons, REM sleep persists, whereas Hobson's model would predict that the opposite should occur. Moreover, in light of the evidence that dopamine, in addition to noradrenaline, is associated with wakefulness (Jouvet 1972), this model requires further modification.

Arousal thresholds

There is a large body of evidence from mammalian experiments that suggests that noradrenergic central neurones, and in particular those from the locus coeruleus, are associated with arousal (Brodal 1981). Their activity has been found to be highest, usually around two spikes per second, during active wakefulness and lowest, at almost zero spikes per second, during REM sleep, which is consistent with the models of both Jouvet and Hobson considered above. Arousal stimuli have been found to be associated with very high levels of activity, of as much as 20 spikes per second (see, for example, Aston-Jones & Bloom 1981, and Rasmussen et al 1986), as well as increased sympathetic activity. On the other hand, during waking activities associated with low arousal, such as repetitive behaviour and defaecation, there is a low level of locus coeruleus noradrenergic neuronal activity (Rasmussen et al 1986). The noradrenergic pathways are described in the next chapter and psychological aspects of arousal are discussed in Chapter 11. In the rest of this section arousal thresholds are considered.

It has proved difficult to obtain consistent data on arousal thresholds from human sleep experiments. One of the problems is that the intensity of arousal stimuli may vary widely. For example, whereas a very loud noise may usually be needed to waken a mother, she may also wake after hearing just a very quiet sound from her baby. It is clearly important to take into account the meaning of a stimulus to the subject, as well as controlling for more objective factors such as the stimulus intensity and the stage of sleep at the time of stimulus presentation. Another problem arises when attempting to define when arousal has actually taken place in a previously sleeping subject. It has been found that the time interval between the onset of alpha activity as monitored with the EEG and awakening is variable (Anch et al 1982). One way around this is to define arousal in terms of a change in the EEG without awakening necessarily occurring; this is known as an EEG arousal (Anch et al 1988).

FURTHER READING

Carpenter R H S 1989 Neurophysiology. 2nd edn. Edward Arnold, London

De Groot J, Chusid J G 1991 Correlative neuroanatomy. 21st edn. Appleton & Lange, Los Altos

Doane B K, Livingston K E (eds) 1986 The limbic system: functional organization and clinical disorders. Raven Press, New York

Emslie-Smith D, Paterson C R, Scratcherd T, Read N W (eds) 1988 Textbook of physiology. 11th edn. Churchill Livingstone, Edinburgh

Guyton A C 1991 Textbook of medical physiology. 8th edn. Saunders, Philadelphia

Kandel E R, Schwartz J H, Jessell T M 1991 Principles of neural science. 3rd edn. Elsevier, New York

3. Neurochemistry

This chapter begins by looking at proteins, including a consideration of enzymes and amino acid metabolism. Lipids, including cerebral lipids and neuronal membranes, are described next, followed by a consideration of energy metabolism. Nucleic acids are conveniently described briefly in Chapter 6 (Genetics). The chapter ends with a section on neurotransmission which includes a description of neurotransmitters not already considered in this book.

PROTEINS

Structure

All proteins consist of polymers of **amino acid** building blocks. The **common** amino acids are those for which DNA nucleotide triplet sequences, known as **codons**, exist (see below). A total of 20 common amino acids are known, and are shown in Figure 3.1, which also includes the usual three-letter abbreviations used for each of them. It can be seen that, in general, amino acids consist of a carbon atom, known as the **alpha** carbon atom, to which is attached both an **amino** group (NH_3^+ at pH 7) and a **carboxylic acid** group (COO^- at pH 7). Hence the name amino acid.

Amino acids are polymerized into proteins by means of a dehydration reaction that causes the formation of **peptide bonds**. This is shown for two given amino acids, with side chains R_1 and R_2 in Figure 3.2. The sequence of amino acids of a protein is known as the **primary structure** of the protein, and determines the way in which the polypeptide chain it comprises folds; in other words, the one dimensional amino acid sequence information determines the three dimensional protein structure, and, therefore, its function. The forces that cause the folding of a protein and maintain its structure include a number of non-covalent forces such as hydrophobic bonds, hydrogen bonds, electrostatic bonds and van der Waals–London forces.

Proteins with low relative molecular masses made up of less than 50 or so amino acids are also termed **peptides**. For example, it will be recalled from the last chapter that thyrotropin releasing factor (TRF) consists of just three amino acids; it can be therefore be termed a neuropeptide.

Functions

Approximately two-fifths of the dry mass of the human brain consists of protein. Proteins in the central nervous system, like proteins elsewhere in the body, have a variety of structural and dynamic functions. The **structural** functions give support and form to tissues. **Dynamic** functions include enzymatic action (see below) and transport of substances such as hormones and drugs (discussed in the next chapter). In the preceding chapter it has been seen how many neuro-hormones are themselves proteins. Proteins also play an important role in neuro-immunology, with immuno-globulin (antibody) molecules each consisting of four polypeptide chains. Complex molecules consisting of both protein and lipid, called lipoproteins, and consisting of both protein and carbohydrate, called glycoproteins, also occur. An example of the latter is the hormone thyrotropin (thyroid stimulating hormone, TSH). Other types of complexes also exist in the central nervous system. For example, proteins associated with copper, such as cerebrocuprein, have been found in the brain.

There are a number of cerebral peptides and proteins which occur only or mainly in the central nervous system. Omitting peptide hormones, which have been discussed in the section on neuro-endocrinology in the previous chapter, some of these peptides and proteins are now briefly described.

Of all the organs of the body the brain is the richest in levels of the amino acid derivative **N-acetylaspartic acid**. Its biosynthesis from aspartic acid, via a reaction catalyzed by aspartate N-acetyl transferase, probably

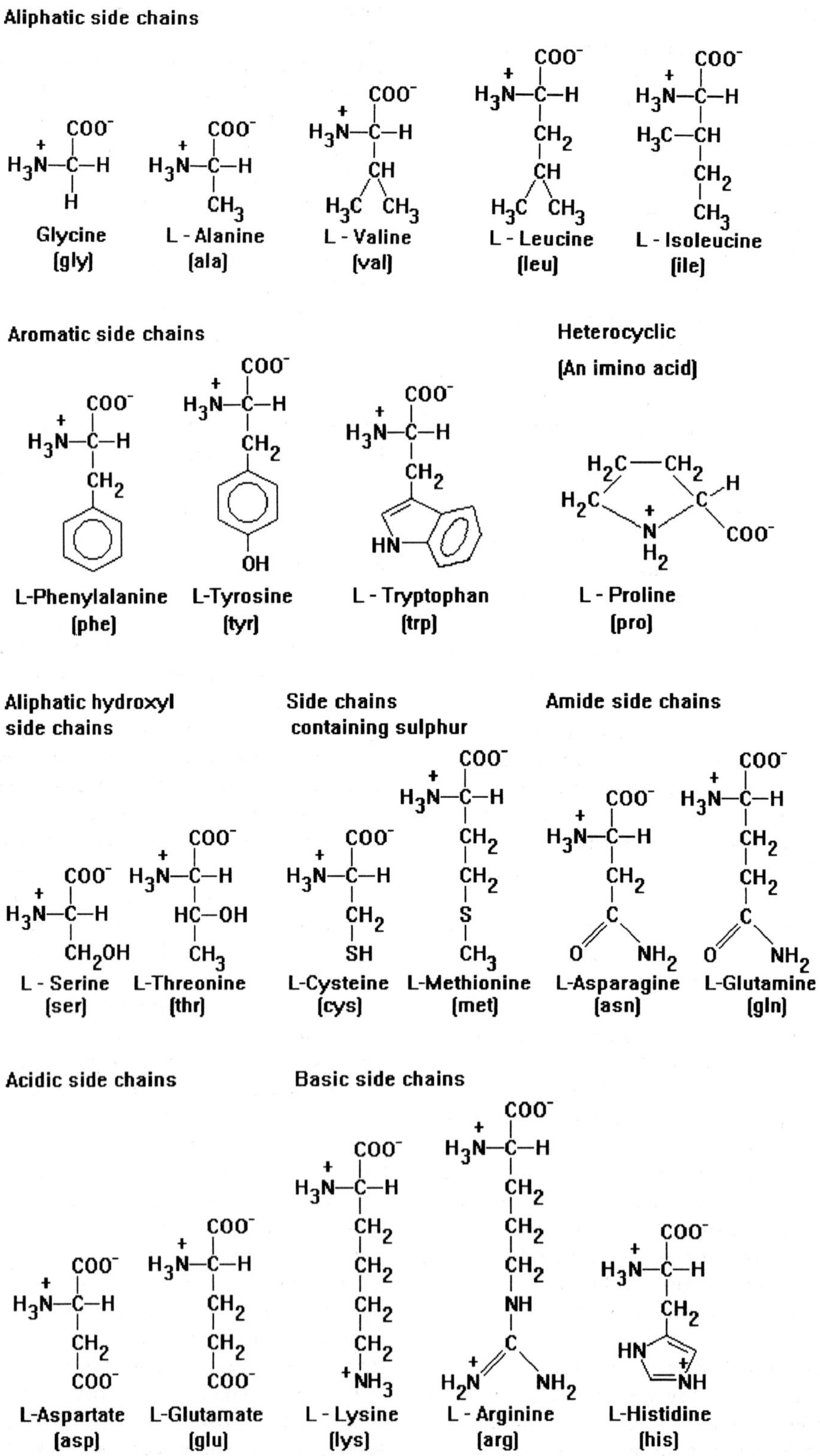

Aliphatic side chains

Glycine (gly)

L - Alanine (ala)

L - Valine (val)

L - Leucine (leu)

L - Isoleucine (ile)

Aromatic side chains

Heterocyclic

(An imino acid)

L-Phenylalanine (phe)

L-Tyrosine (tyr)

L - Tryptophan (trp)

L - Proline (pro)

Aliphatic hydroxyl side chains

Side chains containing sulphur

Amide side chains

L - Serine (ser)

L-Threonine (thr)

L-Cysteine (cys)

L-Methionine (met)

L-Asparagine (asn)

L-Glutamine (gln)

Acidic side chains

Basic side chains

L-Aspartate (asp)

L-Glutamate (glu)

L - Lysine (lys)

L - Arginine (arg)

L-Histidine (his)

Fig. 3.1 The structures of the 20 common amino acids.

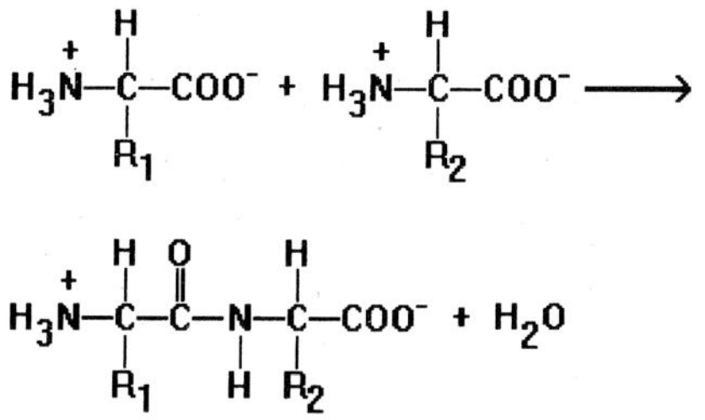

$$H_3\overset{+}{N}-\overset{\overset{\displaystyle H}{|}}{\underset{\underset{\displaystyle R_1}{|}}{C}}-COO^- + H_3\overset{+}{N}-\overset{\overset{\displaystyle H}{|}}{\underset{\underset{\displaystyle R_2}{|}}{C}}-COO^- \longrightarrow$$

$$H_3\overset{+}{N}-\overset{\overset{\displaystyle H}{|}}{\underset{\underset{\displaystyle R_1}{|}}{C}}-\overset{\overset{\displaystyle O}{\|}}{C}-\overset{\displaystyle H}{\underset{\displaystyle H}{N}}-\overset{\overset{\displaystyle H}{|}}{\underset{\underset{\displaystyle R_2}{|}}{C}}-COO^- + H_2O$$

Fig. 3.2 Polymerization of two amino acids.

does not usually take place in non-neural tissues, so that it may be that this substance is transferred to other parts of the body following its biosynthesis in the central nervous system. A number of related peptides is also found in the central nervous system. Their functions have yet to be determined.

The peptide **carnosine** or β-alanyl-L-histidine is present in the central nervous system, and is particularly abundant in olfactory neurones, so that it is possible it has a role in the olfactory system, possibly as a neurotransmitter. The related peptides **homo-carnosine** or γ-aminobutyryl-L-histidine and **homo-anserine** or γ-aminobutyryl-L-methylhistidine are also cerebral peptides whose detailed functions have yet to be elucidated.

Although it also occurs in the cells of other tissues, the tripeptide **glutathione** can be considered to be a cerebral peptide since the brain is particularly rich in it. Its biosynthesis from free amino acids occurs as part of the **γ-glutamyl cycle**, described below in the section on amino acid metabolism. The small number of patients who have been found to be deficient in the enzyme glutathione synthetase have been found to variously develop haemolytic anaemia, neurological abnormalities, psychiatric disorder, and mental retardation.

Encephalitogenic or myelin **basic proteins** are found exclusively in **myelin**. They have a relative molecular mass of approximately 15 000 to 20 000 and, as their name implies, lead to encephalomyelitis when administered to mammals.

Although first discovered in the brain, where they occur in a relatively high concentration, the **calmodulins** have been found to exist throughout the body. They bind calcium ions and are associated with cyclic nucleotide metabolism and ATPase activation. Calmodulin binding proteins have also been discovered.

Glial fibrillary acidic protein (GFAP) is an astrocytic protein with a relative molecular mass of approximately 54 000. Central nervous system damage leads to astrocytic hypertrophy and hyperplasia, in a process known as astrocytosis or gliosis. This has been found to be accompanied by increased production of glial fibrillary acidic protein. Indeed, glial fibrillary acidic protein was first discovered in the plaques of multiple (disseminated) sclerosis. Thus it is possible to use glial fibrillary acidic protein as a cytochemical marker for such damage, in addition to being used as a marker for astrocytes in undamaged nervous tissue. For example, antibodies to glial fibrillary acidic protein can be used as markers for tumours such as astrocytomas. The surface membrane protein **antigen I** is also believed to be glial.

Neurostenin is similar to the muscle protein actinomyosin and may be involved with calcium ion mediated neurotransmission.

S 100 has a relative molecular mass of approximately 21 000 and derives its name from the fact that it was found to be a soluble acidic protein. S 100 is found particularly in glia, in which it is believed to be a cytoplasmic protein, and in postsynaptic neuronal membranes, where it may be involved in calcium ion gating functions. However, S 100 is also found in non-nervous tissue, for example, in malignant melanomas. It has been used as a marker for tissue of neural crest origin, but has not proved as useful as was first expected as a neuro-epithelial tumour marker because of its occurrence in tumours of non-neuroepithelial origin such as melanomas. S 100 can be divided into two types electrophoretically, one of which, S 100b, may have a role in RNA synthesis. Another soluble acidic brain-specific protein is **14-3-2**, thought to be the same as **antigen α**, which has a relative molecular mass of 40 000 to 50 000. It is located in neuronal cytoplasm and is believed to be involved in the function of D-2-phosphoglycerate hydrolase, or enolase, an enzyme that catalyses the dehydration of 2-phosphoglycerate to phosphoenolpyruvate. The latter reaction occurs in glycolysis, which is considered further below.

As indicated by its name, the brain-specific protein **synapsin** is found particularly in association with synaptic vesicles. It has a relative molecular mass of approximately 45 000 and is believed to be associated with the functioning of protein kinases. Related proteins include the cerebellar Purkinje cell **P 400**, so-named because of its location and relative molecular mass of approximately 400 000, and the synaptic membrane proteins D_1, D_2 and D_3.

Tubulin is a soluble glycoprotein which forms microtubules. It is therefore relatively common in the central nervous system, comprising up to one tenth of the soluble proteins of mammalian brains. It has a relative molecular mass of approximately 110 000, and binds with colchine and guanosine triphosphate (GTP). Other brain-specific glycoproteins include the glial protein **glycoprotein** α_2 and the neuronal protein

glycoprotein 350 which has a relative molecular mass of approximately 12 000.

Amino acid transport

The concentration of cerebral amino acids is related to the transport of amino acids into and out of the central nervous system, which is discussed in this section. It is also related to their metabolism, described later in this chapter.

In general the uptake of amino acids into the brain is relatively rapid and involves active transport systems. Hence there can be transport of amino acids into the brain against electrochemical gradients. Similarly, following experimental injection of relatively large amounts of amino acids into the mammalian brain, efflux can occur against electrochemical gradients. The active transport systems are relatively specific, as is indicated by competition for entry into the brain between certain amino acids (for example, between aspartic acid and glutamic acid) and also between amino acids and other substances such as amino acid analogues. There appear to be separate types of transport systems for acidic, basic, and neutral amino acids. Because they are generally energy dependent, the transport systems are sensitive to changes in brain energy metabolism. They sometimes also require the presence of cations, with sodium ions in particular having been shown to be necessary for the transport of large amino acid molecules by the 'L system', an amino acid carrier mechanism that preferentially transports histidine, methionine, and large amino acids including those with aromatic side chains and branched side chains.

Enzymes

Enzymes are proteins which catalyze chemical reactions. They are therefore involved in the metabolic reactions discussed later in this chapter. The substance upon which an enzyme acts is known as the enzyme **substrate**. The latter binds to the enzyme substrate binding site, the affinity of which gives the enzyme its specificity, and the reaction is then catalyzed by the groups within the **active site** of the enzyme. A third site that occurs in some enzymes is the **allosteric site**, binding of small non-substrate molecules to which can alter the substrate binding site affinity or the active site functioning.

Cofactors and **coenzymes** are molecules that need to be present to allow certain enzymes to be functionally active. For example, thiamine or vitamin B_1 is converted into thiamine pyrophosphate which acts as a cofactor in two of the reactions of the tricarboxylic acid cycle or Krebs cycle (see below).

Nomenclature

Enzymes can be categorized, according to the International Union of Biochemistry, into six classes. These classes are then subdivided twice. This classification provides the first three digits of a numbering system of the form a.b.c.d, with the last digit being a number allocated to each specific enzyme. Enzymes can also be named in a form that includes the substrate name and the type of chemical reaction catalyzed. The later is suffixed with 'ase'. Many non-systematic names also exist for enzymes, some of which are recognized officially and used widely.

Classes

Class 1 of the International Union of Biochemistry system consists of the **oxidoreductases**. Enzymes of this class catalyze oxidation and reduction of substrates. Types of oxidoreductases include dehydrogenases, hydroxylases, oxidases, oxygenases and peroxidases. The action of the enzyme catalase is also included in this class.

Class 2 enzymes, the **transferases**, are associated with the transfer of functional chemical groups from donor molecules to acceptor molecules. Aminotransferases, also known as transaminases, catalyze the transfer of an amino group from one donor amino acid to an acceptor keto acid, in the process changing the amino acid into another keto acid, and the acceptor keto acid into another amino acid. Other examples include acyltransferase, glucosyltransferase, methyltransferase, phosphoryltransferase, kinases and phosphomutases.

Hydrolases, class 3, are a form of transferase in which the acceptor molecule is water. This class includes the amidases, deaminases, esterases, glycosidases, peptidases, phosphatases, phospholipases, ribonucleases and thiolases.

Class 4 comprises enzymes that catalyze the addition or removal of carbon dioxide, water or ammonia elements. They are known collectively as **lyases**, and include the aldolases, decarboxylases, dehydratases, hydratases, lyases and synthases. Isomerization reactions are catalyzed by the class 5 **isomerases**, which include the epimerases and racemases, as well as some mutases. Class 6 comprises the **ligases** which are involved in synthetic reactions requiring energy derived from adenosine 5^1-triphosphate (ATP). It includes the carboxylases and synthetases.

Amino acid metabolism

It has been seen that the structure of an amino acid is such that attached to the α-carbon is an amino group (NH_2 or NH_3^+), known as the α-amino group, a carboxyl group (COOH or COO^-), a hydrogen atom, and a side-chain (usually referred to by the letter R). Amino acids surplus to body requirements cannot be stored as such, and are degraded in such a way that the α-amino group is removed to form the ammonium ion (NH_4^+), most of this degradation being hepatic. Although some of the ammonium ions formed in this way are used in the biosynthesis of new amino acids, most enter the urea cycle, shown in a simplified form in Figure 3.3, to form urea. The remaining carbon skeletons of the degraded amino acids are transformed into acetyl-CoA or one of a number of carbohydrate intermediates, and can thence be formed into fatty acids, ketone bodies, glucose or carbon dioxide and water with the release of energy.

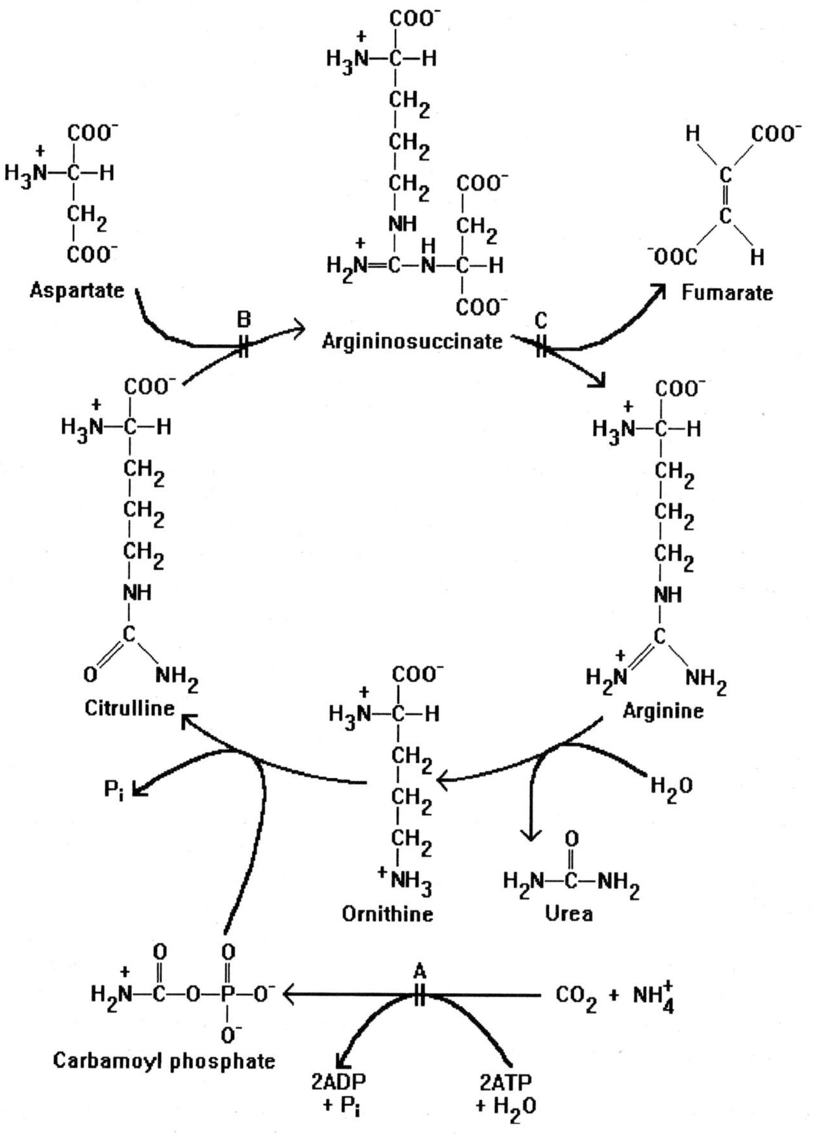

Fig. 3.3 The urea cycle. Enzyme deficiences at the sites labelled **A** to **C** are associated with the following inborn errors of metabolism: (**A**) hyperammonaemia; (**B**) citrullinaemia; (**C**) arginosuccinic aciduria.

Because the urea cycle is the only known urea bio-synthetic pathway in mammals, an inherited incomplete deficiency of one or more of the enzymes of the urea cycle leads to an **inborn error of metabolism** and can be associated with mental retardation. The commonest such disease involves a deficiency of ornithine transcarbamoylase, the enzyme that catalyzes the transfer of the carbamoyl group from carbamoyl phosphate to ornithine to form citrulline, and affects males in greater numbers than females owing to the location of the corresponding gene on the X chromosome. A complete deficiency of such an enzyme is fatal. In Figure 3.3 the sites of enzyme deficiencies associated with the inborn errors of urea metabolism known as **hyperammonaemia, citrullinaemia** or citrullinuria, and **arginosuccinic aciduria** are shown. The urea cycle disorders are mostly inherited as autosomal recessive conditions.

The normal metabolic fates of the carbon skeletons of the three branched-chain amino acids **leucine, isoleucine,** and **valine** are similar to each other. Following initial transamination to the α-keto acid, **oxidative decarboxylation** results in the formation of isovaleryl-CoA from leucine (which in turn is converted into acetyl-CoA and acetoacetate), acetyl-CoA and propionyl-CoA from isoleucine, and methylmalonyl-CoA from valine. In **maple syrup urine disease**, so-named because of the characteristic odour of the urine of patients with this disorder of amino acid metabolism, there is a block of the oxidative decarboxylation reactions by branched-chain keto acid decarboxylase for these three amino acids. Elevated plasma and urinary levels of the three amino acids and their corresponding α-keto acids results. This disease is inherited as an autosomal recessive trait and results in early death unless a diet low in the three amino acids is instituted. This condition is associated with epilepsy and mental retardation.

The hydroxyamino acids **serine** and **glycine** are interconvertible. The metabolism of glycine occurs by two oxidative steps to form oxalate. In **non-ketotic hyperglycinaemia** there is a deficiency of glycine decarboxylase with resulting elevated plasma levels of glycine. The condition is often fatal early in life, and is associated with mental retardation.

Homocysteine is a precursor of the sulphur-containing amino acid cysteine. It also takes part as a precursor of the sulphur-containing amino acid methionine in the activated methyl cycle, shown in Figure 3.4. In the biosynthesis of cysteine, homocysteine combines with serine in the reactions shown in Figure 3.5. In the autosomal recessive condition **homocystinuria** there is a deficiency of the enzyme **cystathionine synthetase** so that the first reaction shown in Figure 3.5 is affected, leading to an accumulation of homocysteine and the interconvertible compound homocystine, the structure of which is shown in Figure 3.6. Owing to the activated methyl cycle shown in Figure 3.4 homocystinuria also results in raised plasma methionine levels. This inborn error of metabolism is associated with mental retardation, epilepsy, skeletal abnormalities, thin fair hair, and cardiovascular and ocular disorders. A deficiency of the enzyme that catalyzes the second reaction in the biosynthesis of cysteine, shown in Figure 3.5, namely **cystathionase**, results in the inborn error of urea metabolism known as **cystathioninuria**. Apart from the accumulation of cystathionine and its urinary excretion, cystathioninuria is associated with few clinical features.

The amino acids **phenylalanine** and **tyrosine** both have side-chains that contain an aromatic ring. The degradation of phenylalanine commences via its

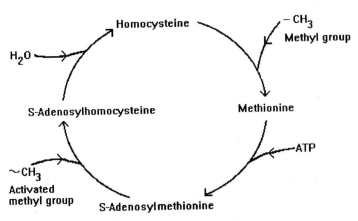

Fig. 3.4   The activated methyl cycle.

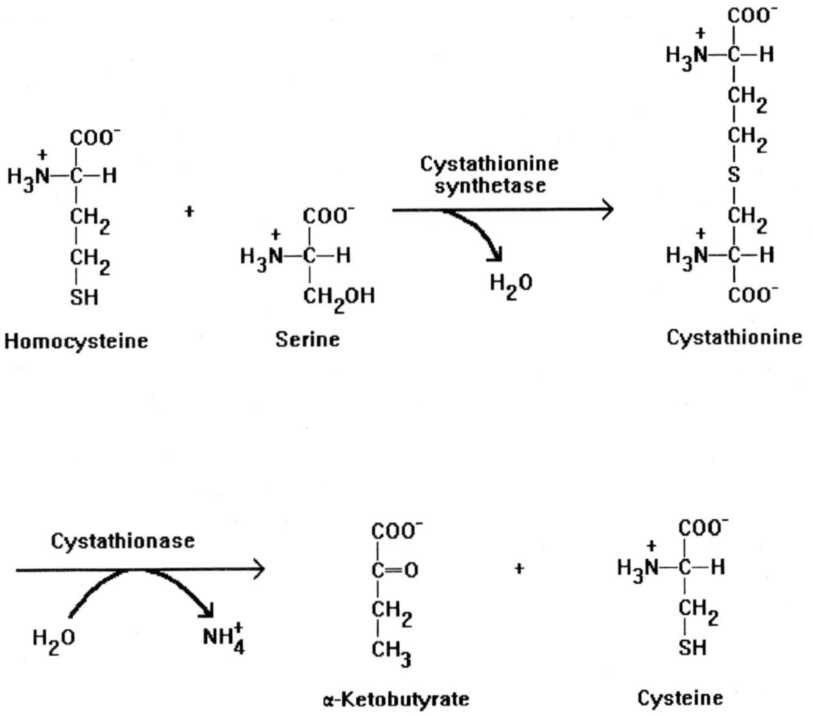

Fig. 3.5 Biosynthesis of cysteine from homocysteine and serine.

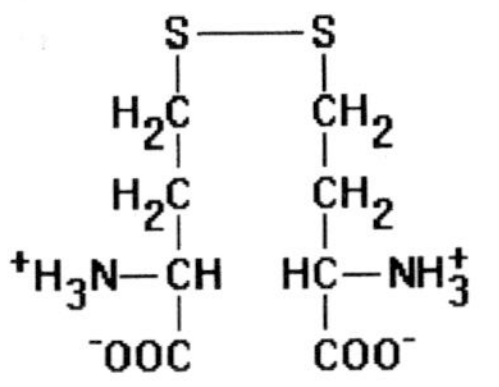

Fig. 3.6 Homocystine.

hydroxylation to tyrosine by means of a reaction catalyzed by **phenylalanine hydroxylase**, also known as **L-aromatic amino acid hydroxylase**, and involving the oxidation of the coenzyme **tetrahydrobiopterin** to dihydrobiopterin (see Fig. 3.7). In addition to being the first reaction in the degradation of phenylalanine, this reaction clearly also serves a biosynthetic role in the biosynthesis of tyrosine. The degradation pathway for both phenylalanine and tyrosine then proceeds via a number of steps, including reactions catalyzed by oxidases (oxygenases) that use oxygen to break the aromatic ring, to result in the formation of fumarate and acetoacetate. In untreated **phenylketonuria**, the commonest inborn error of amino acid metabolism, there is an accumulation of phenylalanine because of a block in the reaction shown in Figure 3.7, leading to mental retardation and epilepsy. Other features of this

autosomal recessive condition include decreased height and head size, kyphosis, and increased upper incisor spacing. Because the biosynthesis of the pigment melanin occurs via a pathway commencing with the hydroxylation of tyrosine, there is decreased melanin formation; patients have decreased pigmentation of the hair, skin (with café-au-lait patches), and iris (blue eyes are therefore common). Phenylketonuria usually results from a deficiency of phenylalanine hydroxylase, but can also be caused by a deficiency of tetrahydrobiopterin, which is also required for the conversion of phenylalanine into tyrosine. Patients with this inborn error of phenylalanine metabolism are treated with a phenylalanine restricted diet. This must be commenced soon after birth if severe brain damage is to be avoided. Hence it is important to identify babies with this condition soon after birth; this is carried out by screening with the Guthrie test in which the bacterial detection of plasma phenylalanine is used. Following maturation, the central nervous system does not appear to be as sensitive to the presence of elevated levels of phenylalanine. Although heterozygotes appear normal, they can be tested for the presence of the autosomal recessive gene by means of a phenylalanine loading test, the result showing that they have a reduced ability to carry out the reaction of Figure 3.7. The inborn error of metabolism known as **type II**

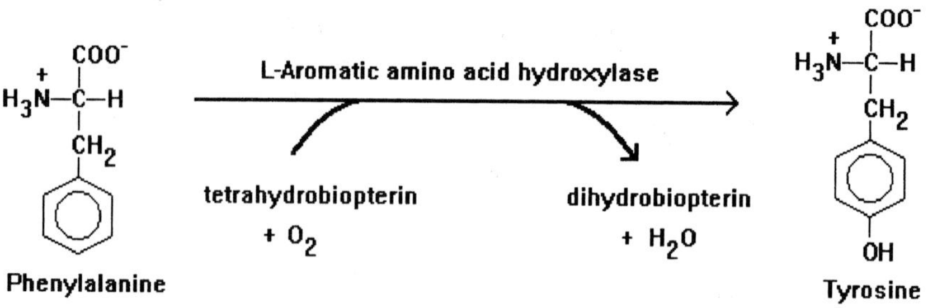

Fig. 3.7   Conversion of phenylalanine to tyrosine.

tyrosinaemia results from a deficiency of tyrosine transaminase (tyrosine aminotransferase), the enzyme that catalyses the transamination reaction that results in the removal of the α-amino group from tyrosine. It is associated with mental retardation, and ocular and dermatological lesions, hence the alternative name of oculocutaneous tyrosinaemia. **Type I tyrosinaemia** results from a deficiency of the enzyme responsible for the final step in the degradation of tyrosine to fumarate and acetoacetate, and is particularly associated with hepatorenal lesions.

The degradation of **histidine** commences with the removal of the α-amino group by means of the reaction shown in Figure 3.8, catalysed by **histidase**, also known as histidine ammonia lyase or histidine α-deaminase, to form urocanate. It can be seen from Figure 3.8 that the final reaction in the degradation to glutamate involves tetrahydrofolate, which is a reduced derivative of the vitamin folic acid. In the recessively inherited inborn error of histidine metabolism known as **histidinaemia** there is a deficiency of histidase, leading to elevated levels of histidine in the blood and urine. The level of histidase can be assayed by measuring the sweat concentration of urocanate. This disorder is usually associated with mental retardation.

Following the removal of the α-amino group, the main pathway of degradation of the **lysine** carbon skeleton takes place via a number of steps to result in the formation of acetoacetyl-CoA. A disorder of lysine metabolism known as **hyperlysinaemia** has been identified, although its cause is not yet clear. It is sometimes, but not always, associated with mental retardation, and lysine administration can result in hyperammonaemia.

The cerebral tripeptide **glutathione** has been mentioned above. Its structure is shown in Figure 3.9, from which it can be seen that it is derived from glutamate, cysteine and glycine. Hence its alternative name of γ-glutamylcysteinylglycine. Its biosynthesis from these three amino acids occurs as part of the **γ-glutamyl cycle**, a simplified diagram of which is

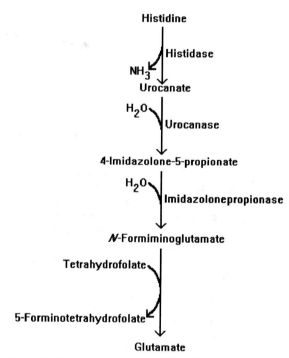

Fig. 3.8   Degradation pathway of histidine to glutamate.

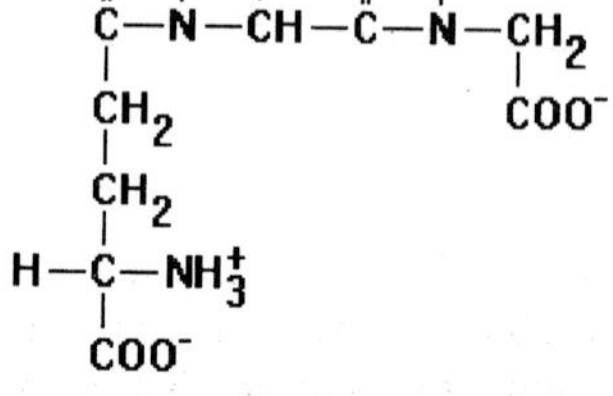

Fig. 3.9   Glutathione.

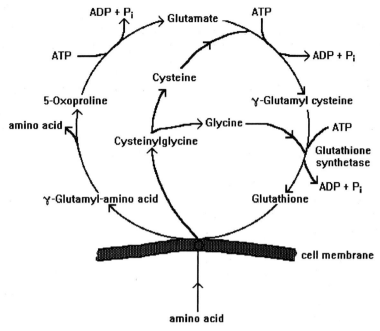

Fig. 3.10   The γ-glutamyl cycle.

shown in Figure 3.10. This cycle is utilized in the transfer of amino acids through the cell membrane, while glutathione and its derivatives have a number of functions including the maintenance of erythrocyte structure, the reduction of methaemoglobin to haemoglobin, and the detoxification of substances such as free radicals and peroxide. Erythrocyte NADPH derived from the pentose phosphate pathway is used to reduce glutathione from its oxidized form to its reduced form, with the production of NADP⁺. The results of deficiency of the enzyme glutathione synthetase have been briefly noted above.

It is convenient to describe one more important amino acid disorder in this section. **Hartnup's disease** is caused by a defect in a permease enzyme that leads to defective gut absorption and renal resorption of neutral amino acids, particularly **tryptophan**. This disorder is inherited as an autosomal recessive and is often associated with mental retardation. It is also associated with cerebellar ataxia and photosensitivity. Pharmacotherapy is with nicotinamide.

LIPIDS

Lipids are esters of fatty acids and alcohols that are relatively insoluble in water. The brain is much richer in lipids than any other mammalian organ, with lipids accounting for approximately one-half of the dry mass of the brain, compared with no more than one-fifth for other organs. In this section the main subjects considered are cerebral lipids and neuronal membranes.

Cerebral lipids

Glycerides

The glycerides comprise a large and important group of cerebral lipids that are found abundantly, particularly in neuronal membranes. Since simple glycerides and free fatty acids are relatively rare in the brain they are not discussed further in this section. Those lipids that contain phosphorus are sometimes known collectively as phospholipids or, if glycerides, glycerophospholipids. The common molecular structure of the glycerophospholipids is shown in Figure 3.11, in which R_1 and R_2 represent fatty acids, and L represents an

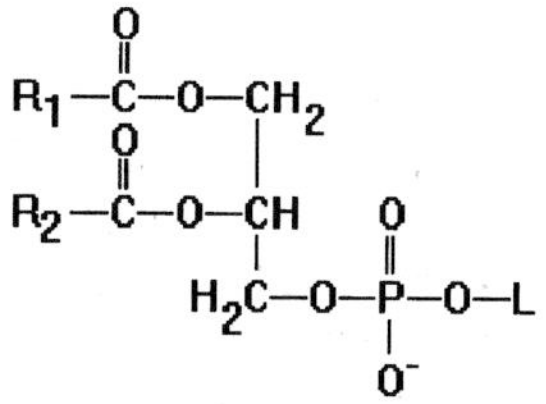

$$
\begin{array}{l}
\quad\quad\quad\; O \\
\quad\quad\quad\; \| \\
R_1-C-O-CH_2 \\
\quad\quad\quad\; O \\
\quad\quad\quad\; \| \\
R_2-C-O-CH \quad\quad O \\
\quad\quad\quad\quad\quad\quad\;\; \| \\
\quad\quad\quad H_2C-O-P-O-L \\
\quad\quad\quad\quad\quad\quad\quad\; | \\
\quad\quad\quad\quad\quad\quad\quad\; O^-
\end{array}
$$

Fig. 3.11   Structure of glycerophospholipids (see text).

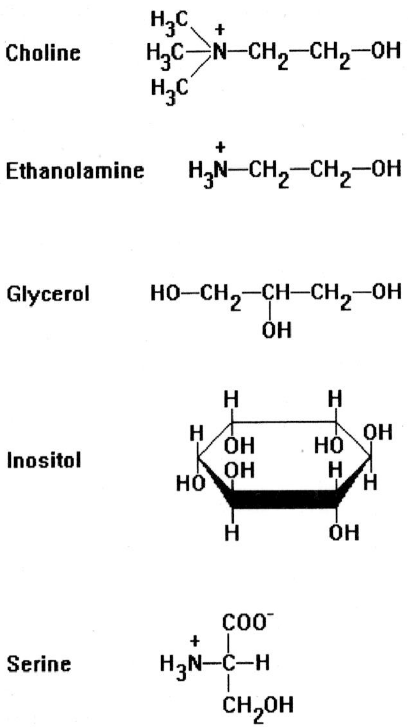

Choline

$$H_3C-\overset{+}{\underset{H_3C}{\overset{H_3C}{N}}}-CH_2-CH_2-OH$$

Ethanolamine $H_3\overset{+}{N}-CH_2-CH_2-OH$

Glycerol $HO-CH_2-\underset{OH}{CH}-CH_2-OH$

Inositol

Serine $H_3\overset{+}{N}-\underset{CH_2OH}{\overset{COO^-}{C}}-H$

Fig. 3.12 Structures of the main alcohols found in phospholipids.

alcohol that is esterified to a phosphatidic acid. Figure 3.12 shows the structures of the more common important alcohols.

Lecithins make up over 5% of the dry mass of the brain. The esterified alcohol in the lecithin molecules is choline (see Fig. 3.12), hence the alternative name phosphatidylcholine. Their fatty acid compositions vary according to the type of membrane structure in which they are found. For example, lecithin in myelin is relatively rich in the fatty acids stearic acid and oleic acid, while mitochondrial and microsomal lecithin is relatively rich in palmitic acid (Fig. 3.13). An important biosynthetic pathway from choline to lecithin, involving the enzymes choline kinase, phosphorylcholine cytidyl transferase and choline phosphotransferase, is shown in Figure 3.14. It can be seen from

(A) $CH_3-(CH_2)_{16}-COOH$

(B) $CH_3-(CH_2)_7-CH=CH-(CH_2)_7-COOH$

(C) $CH_3-(CH_2)_{14}-COOH$

Fig. 3.13 Structures of three of the fatty acids found in lecithins: (**A**) stearic acid; (**B**) oleic acid; (**C**) palmitic acid.

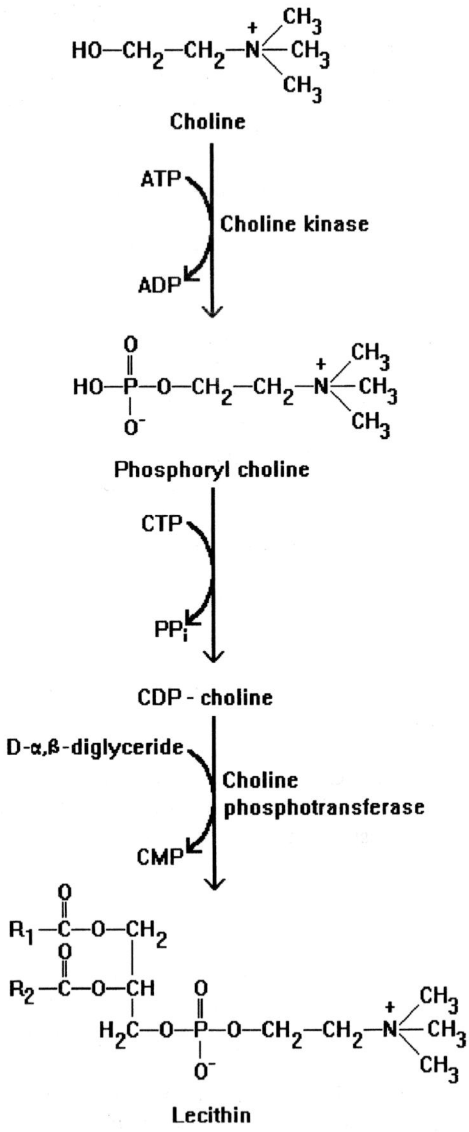

Choline

ATP ⟶ Choline kinase
ADP ⟵

Phosphoryl choline

CTP ⟶
PPᵢ ⟵

CDP - choline

D-α,β-diglyceride ⟶ Choline phosphotransferase
CMP ⟵

Lecithin

Fig. 3.14 Biosynthesis of lecithin from choline.

Figures 3.11 and 3.14 that each molecule of lecithin has two fatty acid groups, labelled R_1 and R_2. **Lysolecithin** has only one such group per molecule, and is involved in the biosynthesis of lecithin from acyl coenzyme A (acyl-CoA) in a reaction catalysed by lysolecithin acyltransferase (see Fig. 3.15). Lysolecithin is also a

lysolecithin + acyl–CoA → lecithin + CoA

Fig. 3.15 Reaction catalyzed by lysolecithin acyltransferase.

product of the biodegradation of lecithin in a hydrolysis reaction catalyzed by phospholipases A_1 and A_2 (see Fig. 3.16). Lysolecithin itself can be hydrolyzed to its corresponding fatty acid and glycerylphosphorylcholine via the action of the enzyme phospholipase B.

$$\text{lecithin} \longrightarrow \text{lysolecithin} + \text{fatty acid}$$

Fig. 3.16 Hydrolysis of lecithin by phospholipases A_1 and A_2.

Initially the phospholipids were classified according to whether, following separation, they were ethanol soluble or insoluble. The former were lecithins and the latter **kephalins**. The kephalins are now known to consist of a number of types of glyceride, including phosphatidylethanolamine, phosphatidylserine, kephalin B and plasmalogens.

The **phospho-inositides** are biosynthesized from the hexahydroxy alcohol inositol (see Fig. 3.12), which is itself found both in the brain and in relatively high concentrations in the cerebrospinal fluid. There are three types of this glyceride known: triphospho-inositide, which can be acted on by phospho-inositide phosphomonoesterase to form diphospho-inositide, which in turn can be acted on by phospho-inositide phosphomonoesterase to form monophospho-inositide. The phospho-inositides are believed to be involved in neurotransmission and cellular calcium ion mobilization.

Sphingolipids

There are four main groups of sphingolipids, most of which contain the alcohol sphingosine (see Fig. 3.17), namely: sphingomyelin, cerebrosides, sulphatides (sulfatides) and gangliosides.

Sphingomyelin molecules contain phosphorus and can therefore also be classed as cerebral phospholipids, of which they comprise approximately 10%. Sphingomyelin is biosynthesized from choline and sphingosine, the latter sometimes being in an acylated form known as **ceramide** (see Figs. 3.18 and 3.19). There are two main biosynthetic reactions, catalysed respectively by

acyl-CoA: sphingosine *N*-acyl transferase and phosphorylcholine ceramide transferase. Three types of sphingolipids are known, and they are an important component of myelin sheaths and cell membranes. The most important disorder of sphingomyelin metabolism is that occurring in **Niemann–Pick disease**, the neuropathological features of which are described in Chapter 7.

The **cerebrosides** contain a carbohydrate group, usually galactose or glucose (hence galactocerebrosides and glucocerebrosides, respectively), in their molecules. The typical structure of a galactocerebroside is shown in Figure 3.20, in which R represents a fatty acid. The most important biosynthetic pathway for the cerebrosides is believed to be one involving the acylation of sphingosine galactoside (also known as psychosine). In a situation analagous to that for the biosynthesis of sphingomyelin, there is a second possible biosynthetic pathway involving a reaction between a galactose compound and ceramide catalysed by UDP-galactose ceramide galactosyl transferase. Whereas galactocerebrosides are found mainly in the central nervous system, the relatively smaller amounts of glucocerebrosides are found mainly in non-nervous tissue outside the central nervous system. Two groups of enzymes that are important in the respective degradation of galactocerebrosides and glucocerebrosides

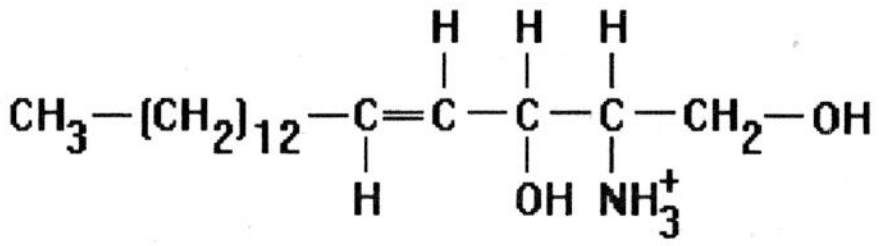

Fig. 3.17 Sphingosine.

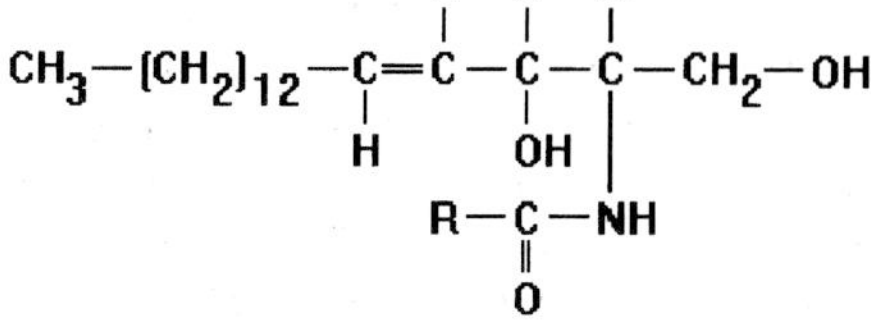

Fig. 3.18 A ceramide.

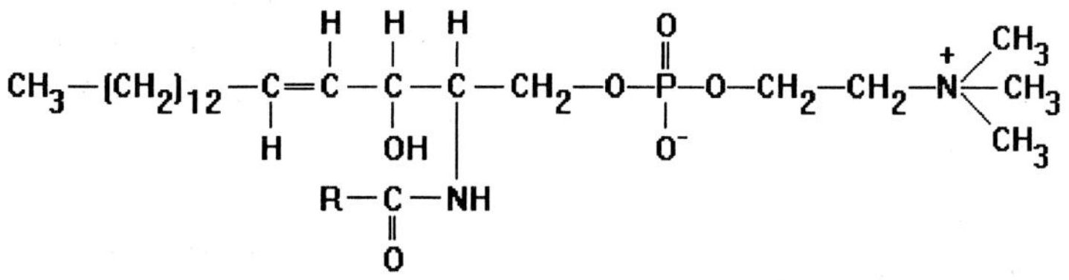

Fig. 3.19 Sphingomyelin.

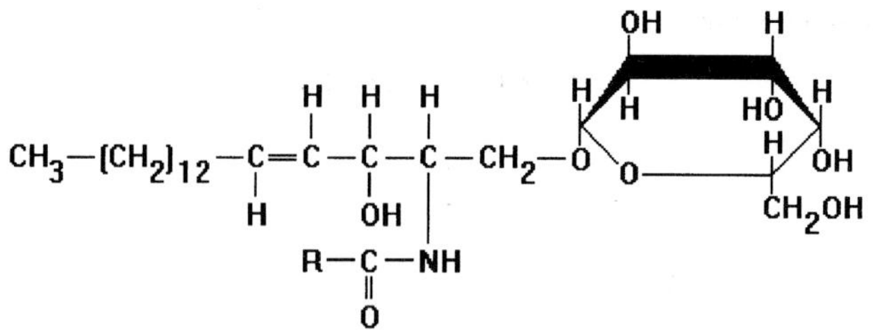

$$CH_3-[CH_2]_{12}-\overset{H}{\underset{H}{C}}=\overset{H}{\underset{}{C}}-\overset{H}{\underset{OH}{C}}-\overset{H}{\underset{R-\overset{}{\underset{O}{C}}-NH}{C}}-CH_2-O$$

Fig. 3.20   Structure of galactocerebrosides.

are the β-galactosidases and β-glucosidases. Defects in both groups of enzymes can lead to neurological disorders. For example, galactocerebroside β-galactosidase deficiency is associated with **Krabbe's disease**, while **Gaucher's disease** is associated with β-glucocerebrosidase deficiency. Neuropathological features are described in Chapter 7.

The cerebral **sulphatides** (spelt sulfatides in North America) have a similar molecular structure to the galactocerebrosides, an important difference being that the carbohydrate group (galactose) has an esterified sulphate group (see Fig. 3.21, in which R_1, is a ceramide residue). This can occur in vivo by means of a reaction catalysed by cerebroside sulphotransferase.

Gangliosides have the most complex and hydrophilic molecules of the sphingolipids. Over 20 types of ganglioside have been identified, the molecules of which have the following moieties in common: galactose, N-acetylgalactosamine, N-acetylneuraminic acid, glucose, and ceramide (see Fig. 3.18, in which R represents a fatty acid). N-acetylneuraminic acid, an acetylated derivative of the amino-containing neuraminic acid, is also known as sialic acid, and is sometimes abbreviated as NANA. The two most important types of ganglioside are known as G_{M1} and G_{M2} (Fig. 3.22). The functions of the gangliosides are thought to include involvement in calcium ion transport, acting as receptors for certain toxins and viruses, membrane excitability, neurotransmission, and possibly providing cell surface recognition features. At the time of writing, details of the biosynthetic and breakdown pathways of the gangliosides and details of pathways for their interconversion are far from complete. However, they are of

clinical importance because of the neurological disorders that are known to result from defects in ganglioside metabolism. Neuropathological features of some types of **gangliosidosis** are considered in the chapter on neuropathology. In **generalized G_{M1} gangliosidosis**, G_{M1} accumulates because of β-galactosidase deficiency. An example of a G_{M2} gangliosidosis, caused by deficiency of the enzyme N-acetyl-β-hexosaminidase A, is **Tay–Sachs' disease**.

Cholesterol

After water, cholesterol and its derivatives are the most common compounds in the brain, making up approximately a tenth of the dry brain mass, with most being present in the free form. Indeed it is estimated that the adult human central nervous system contains about a quarter of the total body mass of cholesterol, with at least half of this being found in myelin. The structure of cholesterol is shown in Figure 3.23. It is a planar molecule consisting of four fused carbon rings. The biosynthesis of cholesterol from acetyl-CoA, via acetoacetyl-CoA, 3-hydroxy-3-methylglutaryl-CoA (HMG-CoA), mevalonic acid, squalene and lanosterol (and a number of other intermediate steps), is similar to that which occurs in other tissues. Most of the cholesterol in the central nervous system is synthesized de novo there; unlike many other tissues, central nervous tissue cholesterol content is not particularly prone to dietary influences. In addition to cholesterol being a component of myelin, cholesterol esters have been found to be important in the developing fetal brain. Outside the central nervous system cholesterol is the precursor molecule of a number of steroid hormones including oestrogen, progesterone and testosterone. Following degradation via reactions such as esterification, cholesterol end products are excreted as bile acids. These bile acids have an important dietary role in aiding the intestinal absorption of the fat soluble vitamins A, D, E and K and of lipids such as fatty acids, monoacylglycerols, and cholesterol itself. This dietary role occurs via emulsification and micelle

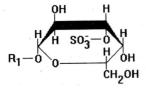

Fig. 3.21   Structure of sulphatides.

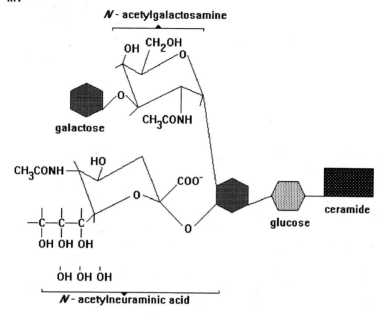

Fig. 3.22   Structure of the GM$_1$ and GM$_2$ gangliosides.

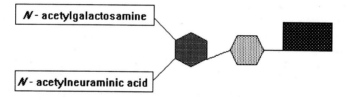

Fig. 3.23   Cholesterol.

formation. Micelles are highly soluble complexes of bile acids with monoacylgycerols, fatty acids, and phospholipids, which have a diameter of between three and ten nanometres and enter intestinal mucosal cells.

Prostaglandins

The prostaglandins are 20-carbon cyclopentane acidic lipids which derive their name from the fact that they were found to be particularly abundant in semen. They are found in nearly every tissue of the body. The most common type in the human brain is prostaglandin F$_{2\alpha}$, the structure of which is shown in Figure 3.24. Its precursor is, as elsewhere in the body, arachidonic acid, an essential fatty acid that may be derived, in the brain, from the action of phospholipase on phospholipids. Among the reported actions of prostaglandins on the central nervous system are changes in arousal (usually sedation), body temperature, mood (subjects can become euphoric) and eating. Prostaglandins appear to modulate neurotransmission, with those of the E series in particular having been shown to act as potent

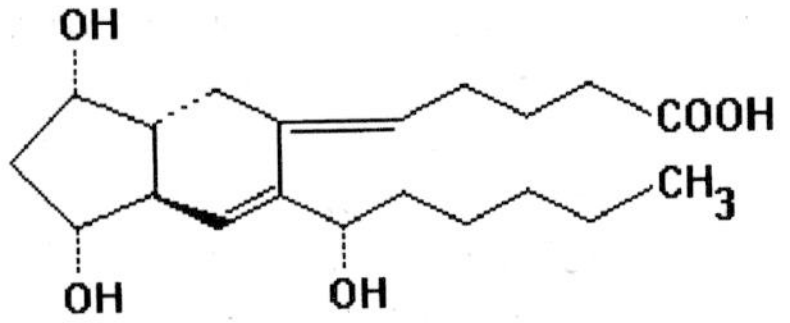

Fig. 3.24   Prostaglandin $F_{2\alpha}$.

inhibitors of feedback neurones involved in catecholamine release. Seizures lead to an increase in the cerebral levels of many of the prostaglandin types, particularly $F_{2\alpha}$, G_2, H_2 and thromboxane B_2.

Neuronal membranes

Myelin

It will be recalled from Chapter 1 that the white matter of the central nervous system consists mainly of myelinated fibres. Therefore, one method used historically to determine the probable lipid composition of myelin was to compare the lipid content of adult mammalian brain white matter with that of its grey matter. When such a study is carried out it is found that white matter is much richer than grey matter in most lipids but particularly in the following types: plasmalogens, sphingomyelin, cerebrosides, sulphatides and cholesterol. These are indeed important components of myelin, with cholesterol accounting for about one-third of the total dry mass of myelin. In all, lipids make up approximately three-quarters of the dry mass, with the remaining one quarter consisting of protein.

Because of the way myelin is laid down, by oligodendrocytes or oligodendroglia in the central nervous system and by Schwann cells in the peripheral nervous system (see Ch. 1), in cross section myelin sheaths have an approximately spiral arrangement.

Fluid mosaic model

As is the case with most biological membranes, the structure of myelin appears to conform to the fluid mosaic model of Singer and Nicolson (1972) in which there is a lipid bilayer matrix in which globular proteins exist both in the bilayer and on the surface. The arrangement of glyceride, sphingolipid and cholesterol molecules in the bilayer is such that the polar heads are at the surfaces, and so able to be in contact with aqueous surroundings, while the hydrophobic tails are sandwiched inside the bilayer, away from polar molecules. As implied by the name of this model, there is a degree of fluidity for both the lipid and protein molecules, with the latter being able to change their

orientation, interchange with nearby proteins, and interact with surrounding substances. Transfer of substances including drugs across the bimolecular membrane sheet by passive diffusion occurs more readily when they are un-ionized. Transfer can also occur by active transport mechanisms, in which the membrane proteins play an important role. The transfer of psychotropic drugs across membranes is considered further in the next chapter.

Synaptic vesicles

Synaptic vesicles are revealed by electron microscopy as existing characteristically at the boutons that form part of the synapses between neurones. They are roughly spherical or ovoid vesicular organelles with a typical diameter of approximately half a micrometer (0.5 µm). The membrane of synaptic vesicles contains both lipids and proteins, with gangliosides, tubulin and actin-like proteins being relatively common.

An important function of the synaptic vesicles is in the process of **neurotransmission**. Neurotransmitters are stored within the vesicles and on electronmicroscopy the density of the synaptic vesicles varies according to the type of stored neurotransmitter. For example, excitatory neurones containing catecholamines tend to be electron dense. Neurotransmission is considered further later in this chapter.

ENERGY METABOLISM

Unlike other organs which can utilize other substances, the brain is dependent on **glucose** supplied by blood for its energy requirements. This is partly a function of the fact that while glucose can readily enter the brain from the bloodstream via an active transport mechanism, other energy-rich substances are restricted by the blood–brain barrier. Furthermore, the brain is not endowed with large stores of energy in the form of lipid and glycogen stores, unlike many other organs. Hence the brain is rich in the enzymes that are involved in glucose metabolism. The currency of energy from the breakdown of glucose is the nucleotide **adenosine triphosphate**, or ATP. This molecule, the structure of which is shown in Figure 3.25, contains relatively large amounts of energy stored in its high-energy phosphoanhydride bonds. Energy is liberated by the hydrolysis of one of these bonds in the conversion of ATP to ADP (adenine diphosphate), and again in the conversion of ADP to AMP (adenine monophosphate). Similar reactions can occur with another set of nucleotides, GTP, GDP and GMP, in which the purine base guanine occurs instead of the purine base

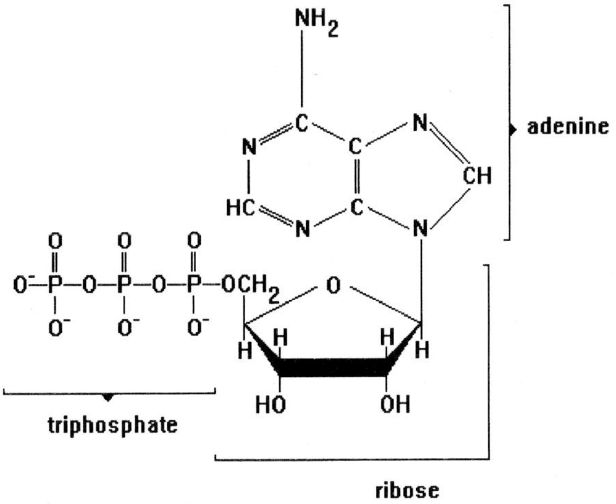

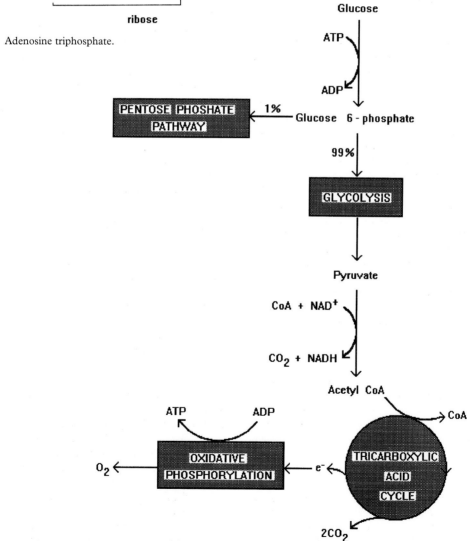

Fig. 3.25 Adenosine triphosphate.

adenine of ATP, ADP and AMP. Similarly, with CTP (contains the pyrimidine base cytosine) and UTP (contains the pyrimidine base uracil). The relatively high energy requirements of the brain, which comprises about 2% of the total body mass, is indicated by the fact that, at rest, the brain extracts about one-tenth of the glucose in its blood supply, and accounts for about a quarter of the total body oxygen requirement.

An overview of the way in which energy is extracted from glucose in the brain is shown in Figure 3.26. About 99% of the glucose is metabolized in the cytoplasm to pyruvate via **glycolysis** (also known as the Embden–Meyerhof pathway), while 1% enters the

Fig. 3.26 Summary of cerebral glucose metabolism.

pentose phosphate pathway. The pyruvate then undergoes oxidative decarboxylation to form acetyl-CoA, which then enters the mitochondrial **tricarboxylic acid cycle** (also known as the citric acid cycle or Krebs cycle), where the acetyl units are oxidized to carbon dioxide and the major electron acceptors NAD^+ and FAD are reduced to NADH and $FADH_2$, respectively, by the transfer of electrons. These electrons are finally transferred by oxygen during **oxidative phosphorylation**. ATP is generated from ADP partly during glycolysis, but mainly during oxi-dative phosphorylation. In all, the complete meta-bolism to carbon dioxide and water of one glucose molecule via glycolysis, the tricarboxylic acid cycle, and oxidative phosphorylation provides the energy for the generation of 36 molecules of ATP (from ADP and orthophosphate P_i).

Glycolysis

Figure 3.27 shows the stages in glycolysis. Since each molecule of glucose gives rise to two molecules of

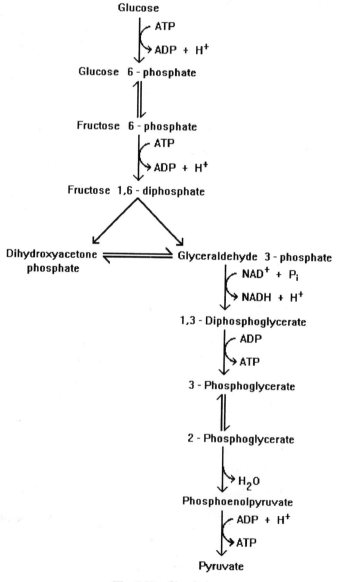

Fig. 3.27 Glycolysis.

glyceraldehyde 3-phosphate, it can be seen that, overall, as a result of glycolysis each molecule of glucose yields two molecules of pyruvate, and the energy for the generation of two molecules of ATP (from two molecules each of ADP and orthophosphate P_i). Two molecules of NAD^+ are also reduced to NADH during this process.

Tricarboxylic acid cycle

Before entering the tricarboxylic acid cycle of the mitochondrial matrix, pyruvate undergoes the following oxidative decarboxylation reaction, catalyzed by the multienzyme pyruvate dehydrogenase complex:

pyruvate + CoA + NAD^+ →
acetyl-CoA + NADH + carbon dioxide

The tricarboxylic acid cycle is shown in Figure 3.28. The overall reaction is as follows:

acetyl-CoA + $3NAD^+$ + FAD + GDP + $2H_2O$ + P_i→
$2CO_2$ + CoA + 3NADH + $FADH_2$ + GTP + $2H^+$

The oxidative decarboxylation of pyruvate to acetyl-CoA and the entry of the latter into the tricarboxylic acid cycle are both inhibited when the cell is rich in ATP.

Also shown in Figure 3.30 are pathways from α-ketoglutarate to the formation of the amino acids glutamate and glutamine, and the neurotransmitter γ-aminobutyric acid (GABA). Also the amino acid aspartate can be formed from oxaloacetate.

A number of cofactors are involved in the above reactions. In particular, **thiamine pyrophosphate** (TPP), derived from vitamin B_1, acts as a cofactor in the oxidative decarboxylation of pyruvate to acetyl-CoA and, within the tricarboxylic acid cycle, the oxidative decarboxylation of α-ketoglutarate to succinyl-CoA (catalyzed by α-ketoglutarate dehydrogenase complex). Thus in vitamin B_1 or thiamine deficiency, which can lead to Wernicke's disease and beriberi, there is a reduction in the activities of the pyruvate dehydrogenase complex and the α-ketoglutarate dehydrogenase complex with a concomitant elevation of the plasma concentrations of pyruvate and α-ketoglutarate. The pentose phosphate pathway is also affected (see below).

Oxidative phosphorylation

Oxidative phosphorylation is a process which couples the oxidation of NADH and $FADH_2$ to the phosphorylation of ADP to ATP. During oxidative phosphorylation the synthesis of ATP from ADP and P_i is coupled to the return of protons from the inner

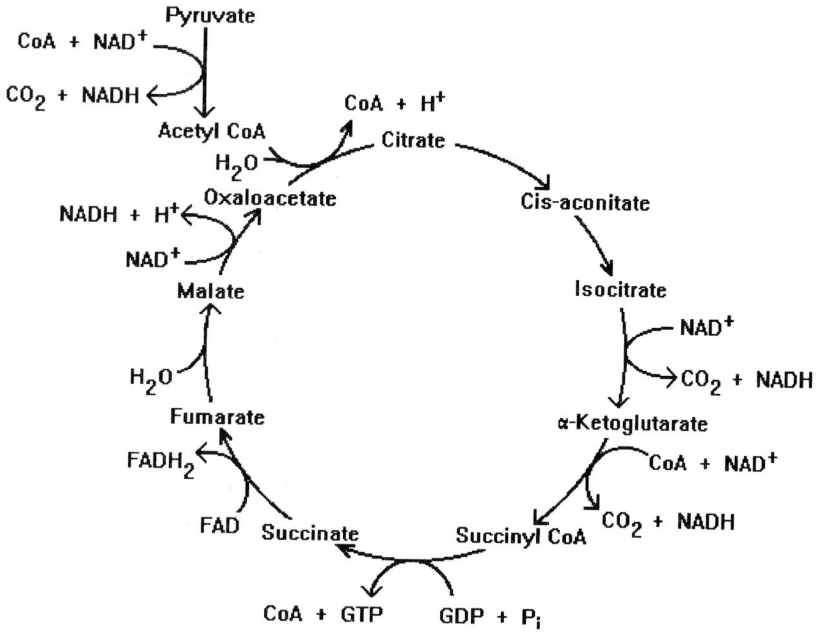

Fig. 3.28 Tricarboxylic acid cycle (citric acid cycle).

mitochondrial membrane to the mitochondrial matrix and the flow of electrons to oxygen by means of a number of carriers of the inner mitochondrial membrane.

The rate of oxidative phosphorylation rises as the concentration of ADP, and therefore the need for ATP, rises, and vice versa.

For each molecule of $FADH_2$, oxidative phosphorylation yields two molecules of ATP, while for each molecule of NADH the corresponding number of molecules of ATP is three. Hence, taking into account the net results of glycolysis and the tricarboxylic acid cycle, the total oxidation of one molecule of glucose to carbon dioxide and water results in a total of 36 molecules of ATP being generated from ADP and P_i.

Pentose phosphate pathway

There are a number of possible pathways in the pentose phosphate pathway. The main oxidative branch proceeds as shown in Figure 3.29. It commences with the glucose 6-phosphate formed from glucose in the first stage of glycolysis (see Fig. 3.27). The result is the generation from each molecule of glucose 6-phosphate of two molecules of the reduced electron and hydrogen donor NADPH (from $NADP^+$) which can be used in reductive biosynthetic pathways, and one molecule of ribose 5-phosphate which can be used in

the biosynthesis of DNA, RNA and nucleotide coenzymes.

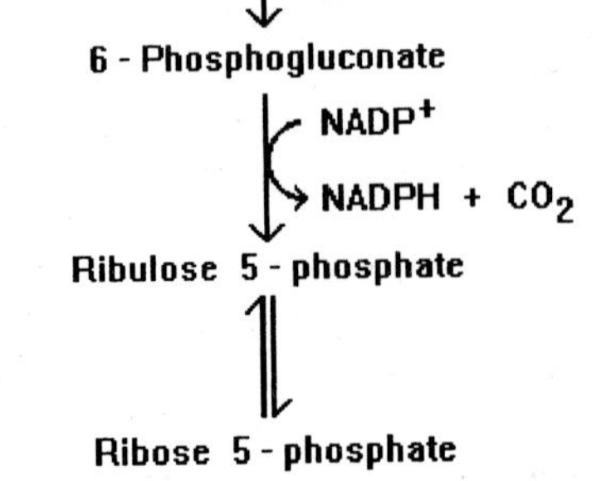

Other pathways allow further connections with glycolysis. When there is a relatively greater cellular requirement for NADPH compared with ribose 5-phosphate, the latter is converted through a number of stages involving the enzymes **transketolase** and **transaldolase** into fructose 6-phosphate and glyceraldehyde 3-phosphate. This can be summarized as:

3(ribose 5-phosphate) →
2(fructose 6-phosphate) + glyceraldehyde 3-phosphate

The last two products, fructose 6-phosphate and glyceraldehyde 3-phosphate, occur in the glycolytic pathway, so that transketolase and transaldolase provide reversible connections between glycolysis and the pentose phosphate pathway. Overall, when one molecule of glucose 6-phosphate follows this pathway the generation of 12 molecules of NADPH (from $NADP^+$) takes place:

glucose 6-phosphate + $12NADP^+$ + $7 H_2O$ →
$6CO_2$ + 12NADPH + P_i + $12H^+$

When there is a relatively greater cellular requirement for ribose 5-phosphate compared with NADPH, glucose 6-phosphate follows a metabolic pathway that does not generate further NADPH. The glucose 6-phosphate is converted, via glycolysis, into fructose 6-phosphate and glyceraldehyde 3-phosphate. The following conversion, the reverse of that previously noted, then occurs:

2(fructose 6-phosphate) + glyceraldehyde 3-phosphate → 3(ribose 5-phosphate)

This is again catalysed by transketolase and transaldolase. Overall, this is part of the following:

5(glucose 6-phosphate) + ATP →
6(ribose 5-phosphate) + ADP + H^+

Thiamine pyrophosphate, already noted as a cofactor in the oxidative decarboxylation of pyruvate and α-ketoglutarate to acetyl-CoA and succinyl-CoA, respectively, needs to be bound to transketolase in order for the latter to function as above in the pentose phosphate pathway. In addition to the changes noted above, thiamine deficiency can therefore also lead to a reduction in the activity of transketolase; erythrocytic transketolase can conveniently be assayed clinically.

GABA shunt

In contrast to the operation of the tricarboxylic acid cycle outside the central nervous system, the stages of

Fig. 3.29 Pentose phosphate pathway: oxidative branch.

this cycle between α-ketoglutarate and succinate are bypassed through the formation of γ-aminobutyric acid (GABA) for approximately one-tenth of the cerebral glucose metabolized. This can occur by virtue of the presence in the brain of a number of forms of the enzyme **glutamic acid decarboxylase (GAD)** which, in conjunction with the presence of the vitamin B_6 derived cofactor **pyridoxal phosphate**, allows the conversion of glutamate to GABA to take place, as shown in Figure 3.30. Pyridoxal phosphate also acts as a cofactor for other enzymes including the transaminases. Vitamin B_6 (pyridoxine) deficiency can lead to a reduction in the activity of such enzymes, and the development of seizures and peripheral neuropathy. An important cause of functional deficiency of this vitamin is from therapy with drugs such as isoniazid, hydralazine, and penicillamine.

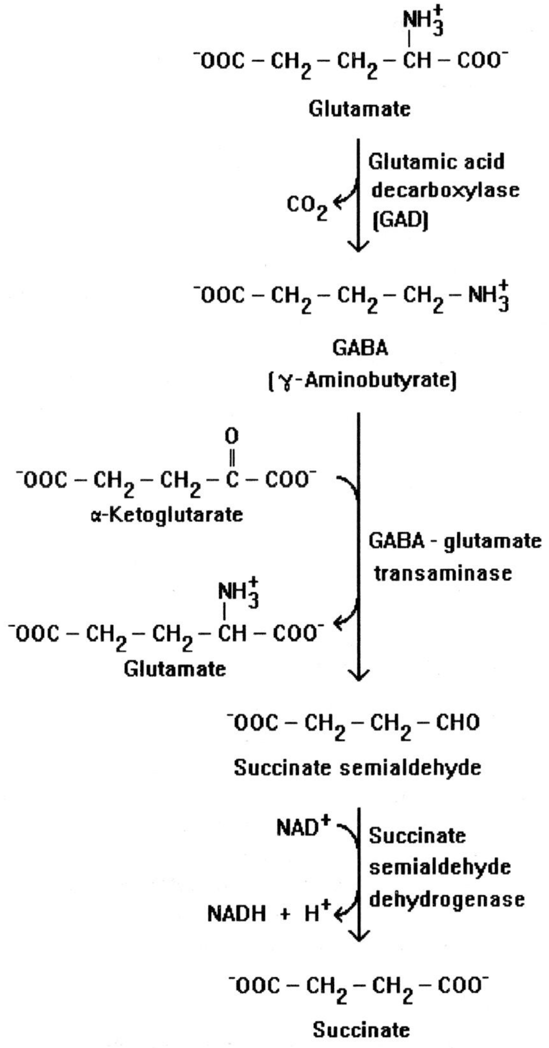

NEUROTRANSMISSION

In the rest of this chapter chemical neurotransmission at synapses is considered. As mentioned above, neurotransmitters are stored within the synaptic vesicles, the characteristics of the vesicles varying with the type of neurotransmitter stored. An increasing number of substances with neurotransmitter functions has been identified in recent years. Furthermore, some of the traditional criteria and characteristics that substances had to fulfil in order to be classed as neurotransmitters have been found not to apply to many of those more recently identified. Thus, for example, Henry Dale's law that implied the existence of only one neurotransmitter per neurone is clearly breached by the phenomenon of coexistence of neuropeptides with classical neurotransmitters mentioned in the last chapter. Moreover, again as was evident in the section on neuro-endocrinology in the last chapter, the distinction between neurotransmitters and neurohormones has become rather blurred in many instances. Individual neurotransmitters and their receptors are discussed in the rest of this chapter, following a brief description of some aspects of receptors in general not previously considered in this book.

The **receptors** to which neurotransmitters bind consist in general of proteins located on the external surface of cell membranes. In recent years there has occurred a large increase in the number and types of receptor characterized. In general there appear to be two main methods by which a receptor responds to neurotransmitter binding in order to cause the specific effect of that neurotransmitter.

In one group of receptors neurotransmitter binding leads to the opening of a transmembrane channel that allows the passage of ions. There is evidence that this is the case for the GABA-A-benzodiazepine receptor, binding of GABA to which leads to the opening of a transmembrane channel for chloride ions. Braestrup and Nielsen (1982) have proposed a model for this receptor complex or ionophore.

The second group of receptors which includes, for example, the adrenergic receptors (apart from α_1), incorporates a receptor–neurotransmitter activated cyclic nucleotide system in which cyclic AMP (cAMP) acts as a secondary messenger; a cyclic AMP cascade allows amplification to take place (see Lefkowitz et al 1984, Sulser 1984). On the other hand, this system is slower than the first type of receptor incorporating

Fig. 3.30 GABA shunt.

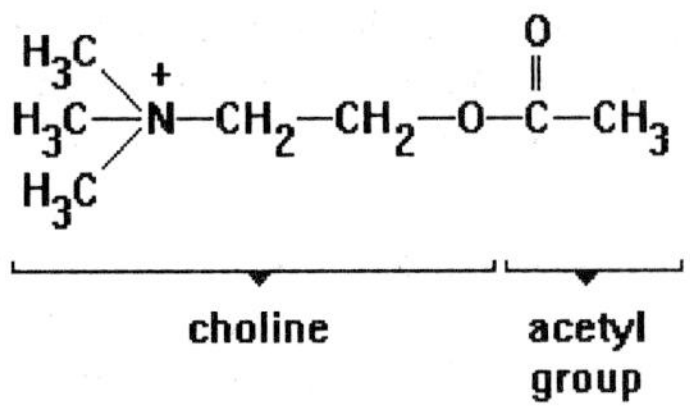

Fig. 3.31   Acetylcholine.

transmembrane ion channels that open in response to neurotransmitter binding.

Acetylcholine

Biosynthesis

The neuronal biosynthesis of acetylcholine from its precursors acetyl-CoA and choline occurs by means of the following reaction, catalyzed by **choline acetyltransferase**:

acetyl CoA + choline → acetylcholine + CoA

The structure of acetylcholine is shown in Figure 3.31. The acetyl-CoA is derived from pyruvate resulting from glycolysis, while the choline is extracellular, being obtained from the plasma. There exist both a low-affinity and a high-affinity transport system for choline, and it is the latter that appears to play the most important role in the regulation of the cellular acetylcholine concentration.

Storage and release

Acetylcholine is stored in synaptic vesicles. Their fusion with the presynaptic membrane, leading to the release of acetylcholine, occurs in response to the entry of calcium ions into the nerve terminal. In the mammalian superior cervical ganglion it has been shown that freshly synthesized acetylcholine tends to be released before that which has been stored for some time.

Metabolism

Rather than being subject to neuronal re-uptake as is the case for many neurotransmitters, following its release from synaptic vesicles into the synaptic cleft acetylcholine is hydrolyzed by **cholinesterase** into choline and ethanoic or acetic acid:

acetylcholine + H_2O → choline + CH_3COOH

There are a number of cholinesterases in the body,

which can be considered to form two main groups. The first, known as **acetylcholinesterase** or true cholinesterase, is generally found in nervous tissue and erythrocytes and is the important class of enzyme so far as the hydrolytic breakdown of acetylcholine in the central nervous system is concerned. The second group is known variously as butylcholinesterase, pseudo-cholinesterase, propionylcholinesterase, or non-specific cholinesterase. It is found mainly in non-nervous tissue and exists in plasma (but not in erythrocytes). Its action on acetylcholine is slower than that of acetylcholinesterase.

The high-affinity choline transport system mentioned above is mainly responsible for the re-uptake of the choline liberated by the action of cholinesterase on acetylcholine. Thus the choline moiety can be efficiently recycled by cholinergic neurones.

Distribution

As mentioned in the previous chapter, acetylcholine is the neurotransmitter for all preganglionic neurones and parasympathetic postganglionic neurones in the autonomic nervous system (see Fig. 2.8). In addition, acetylcholine is the main neurotransmitter used by motoneurones in the spinal cord.

Acetylcholine is found in many parts of the central nervous system, having particularly high concentrations in the caudate nucleus and cerebral cortex. A number of central ascending cholinergic pathways have been identified which project to the limbic system, hypothalamus, thalamus, cerebellum and cerebral cortex. Ascending reticular pathways of the hippocampus have also been found to be cholinergic, as have some afferent neurones of the auditory and visual pathways.

Receptors

There are two groups of cholinergic receptors, namely **muscarinic** and **nicotinic**, which are stimulated respectively by muscarinic and nicotinic agents. Atropine acts as an antagonist at muscarinic receptors, which are subdivided into M_1 and M_2 receptors according to ligand-binding affinity and location. Nicotinic receptors respond faster than muscarinic receptors. Antagonists include d-tubocurarine and dihydro-β-erythroidine.

Central neuropharmacological aspects

Many pharmacological agents that affect non-central cholinergic actions do not cross the blood-brain barrier and are not considered in this section.

In **parkinsonism** there is believed to be a relative excess of central acetylcholine action compared with the action of dopamine. This may result from a deficiency of dopamine, for example in Parkinson's disease. It may also be caused by the use of antidopaminergic agents such as antipsychotic drugs like chlorpromazine, hence their parkinsonian side effects. **Antimuscarinic drugs**, often simply but less correctly referred to as anticholinergic drugs, have been found to have an antiparkinsonian action, and are therefore used in Parkinson's disease and to treat the extrapyramidal side effects of antipsychotics. Those used in Britain include benzhexol, orphenadrine, benztropine, procyclidine, biperiden and methixene. In addition, many antidepressants, such as the tricyclic amitriptyline, as well as antipsychotics have antimuscarinic actions (see Ch. 5) which can result in a number of characteristic side effects. These include dry mouth, urinary retention, constipation, and blurred vision. Antimuscarinic drugs such as scopolamine and hyoscine are also used as anti-emetics, an action they achieve as a result of both their central actions and their antimuscarinic actions on the gastrointestinal tract, where decreased secretions result.

Relatively high doses of these antimuscarinic agents leading to a relatively high level of central antimuscarinic action can result in **delirium**.

Psychiatric aspects

According to the **cholinergic hypothesis of memory dysfunction**, the memory impairment of old age and dementia is associated with changes in central cholinergic activity. In the 1970s Alzheimer's disease was shown to be associated with both a deficiency of central choline acetyltransferase (Bowen et al 1976, Perry et al 1977) and selective loss of central cholinergic neurones (Davies & Maloney 1976). Furthermore, Perry et al (1978) showed that cholinergic abnormalities are positively correlated with neuritic plaques and mental test scores in this condition. In spite of the fact that functional interactions have been shown to exist between cholinergic neuronal activity and a number of other neurotransmitter systems, acetylcholine remains the neurotransmitter that appears most strongly associated with Alzheimer's disease. Further evidence in favour of the hypothesis that cholinergic activity plays an important role in memory comes from studies which have demonstrated that antimuscarinic drugs such as benzhexol can produce memory impairment (see, for example, Potamianos & Kellett 1982, Crawshaw & Mullen 1984, Kopelman 1985). A number of psychopharmacological strategies can be devised, based on the cholinergic hypothesis, for treating Alzheimer's disease. If the hypothesis is correct treatment should increase central functional acetylcholine levels. This could be carried out by increasing the plasma levels of choline available to the brain by increasing the dietary intake. In studies pursuing this strategy the usual method has been to administer relatively high doses of lecithin (phosphatidyl choline), which occurs naturally in many foods. Another strategy could be to reduce the rate of metabolic breakdown of acetylcholine by administering a cholinesterase antagonist that crosses the blood–brain barrier, such as physostigmine.

Yet another method involves the administration of acetylcholine agonists such as bethanechol; this has the practical disadvantage that effective administration has been found to be via the intraventricular route. Again, the release of acetylcholine can be enhanced by using 4-aminopyridine. Trials involving all these methods have been carried out, and in general none has proved clinically useful (Byrne & Arie 1990, Rubin 1990). A number of trials is being carried out with tacrine (9-amino-1, 2, 3, 4-tetrahydroaminoacridine), commonly abbreviated to THA, which is a cholinesterase antagonist and also inhibits cellular potassium ion movement and monoamine oxidase; at the time of writing it is not possible to give a definitive statement as to whether or not this drug may prove useful in the treatment of Alzheimer's disease (Levy 1990).

It has been suggested that cholinergic abnormalities may be associated with **movement disorders** such as Parkinson's disease and Huntington's chorea. As mentioned above, antimuscarinic agents have been shown to be therapeutically beneficial in Parkinson's disease, and this is believed to be because they help reverse the hypothesized imbalance between cholinergic and dopaminergic systems in this condition. Furthermore, in a situation somewhat analagous to the cholinergic hypothesis discussed above, there is evidence that in some cases of this condition there is a reduction in the level of choline acetyltransferase (Ruberg et al 1982, Perry et al 1985). This may be associated with the cognitive dysfunction reported in this condition. Similarly, a reduction in the level of choline acetyltransferase has also been reported in Huntington's chorea, particularly in striatal cholinergic neurones. Since increased dopaminergic activity has been found in the corpus striatum (Spokes 1980), it has been suggested that the choreic movements may result from relative dopaminergic overactivity as compared with cholinergic activity (Marsden 1982).

Tandon and Greden (1989) have proposed a model of cholinergic and dopaminergic interactions in **schizophrenia** in which negative symptoms of schizophrenia are attributed to increased cholinergic activity. This is based on similarities between hypercholinergic states and some of the behavioural negative features of schizophrenia, as well as some similarities in the patterns of hormonal responses and sleep in the two conditions. Indeed there is evidence that acetylcholine is involved in **sleep disorders** (Sitaram et al 1979).

Increased cholinergic activity has also been implicated in **depression** (see, for example, Overstreet 1986).

Amino acids

Some of the amino acids have been found to act as central neurotransmitters. Indeed there is evidence that amino acids act as neurotransmitters at far more central synapses than do all the other types of neurotransmitter put together. The **inhibitory** amino acids, that is, those which from iontophoretic studies have been found to lead to hyperpolarization in neurones, include GABA, glycine, alanine, cystathionine, serine and taurine. Of these, GABA and glycine are the most potent and appear to be the most important. The **excitatory** amino acids, that is, those causing depolarization of neurones, include glutamic acid, aspartic acid, cysteic acid and homocysteic acid. Of these, glutamic acid and aspartic acid appear to be the most important.

GABA

GABA, or γ-aminobutyric acid, the principal inhibitory amino acid neurotransmitter of the brain, is derived from glutamic acid via the action of glutamic acid decarboxylase (GAD) (see above). Its metabolic breakdown to glutamic acid and succinic semialdehyde takes place via the action of GABA transaminase (GABA-T), an enzyme largely associated with mitochondrial cell fractions. The GABA shunt has been described earlier in this chapter.

The distribution of GAD in the central nervous system is closely associated with the distribution of GABA. This is not the case with GABA-T, which is much more widely distributed. The main long GABAergic tracts identified are the projection from the corpus striatum to the substantia nigra, and the Purkinje cells of the cerebellum. GABA is also found in relatively high concentrations in the spinal cord, where it acts as a neurotransmitter for inhibitory interneurones, and in the hippocampus and hypothalamus.

Two classes of GABA receptors have been identified. As mentioned above, $GABA_A$ receptors form complexes with benzodiazepine binding sites and chloride ion channels. Two classes of **sedative and antiepileptic agents**, benzodiazepines and barbiturates, act via $GABA_A$ receptors. The $GABA_A$ antagonist bicuculline, on the other hand, is epileptogenic. Antiepileptic agents can also function by inhibiting the action of GABA-T; this is the mode of action of sodium valproate. In addition to epilepsy, the association of $GABA_A$ receptors with benzodiazepine binding sites has led to the hypothesis that reduced GABAergic activity may be associated with **anxiety disorders** such as panic disorder. The antispasticity agent baclofen is a selective agonist at $GABA_B$ receptors. Although there is some evidence from mammalian experiments that the administration of antidepressants and the equivalent of electroconvulsive therapy is associated with an increase in $GABA_B$ receptors in the cerebral cortex (Lloyd et al 1985), studies testing the hypothetical model in which there is an abnormality in $GABA_B$ receptors in depression have generally failed to support this model. For example, Cross et al (1988) found no significant differences overall between the brain $GABA_B$ receptors of depressed suicide victims and matched controls.

GABA has been shown to be reduced in the brains of patients with **Huntington's chorea** (Perry et al 1973), particularly in the corpus striatum, and it has been suggested that the choreic movements may be caused by an imbalance between dopaminergic activity, which is increased in the corpus striatum (Spokes 1980), and the reduced GABAergic activity. Reduced GABA levels have also been found in the brains of patients with **Alzheimer's disease** (Rossor et al 1982, Hardy et al 1987).

Glycine

Of all the amino acids, the inhibitory amino acid neurotransmitter glycine has the simplest structure (see Fig. 3.1). It is found mainly in the spinal cord, the more rostral parts of the brain stem, and the retina. In addition to its dietary source, glycine can also be biosynthesized from another inhibitory amino acid neurotransmitter, serine, via the action of serine hydroxymethylase. Glycine receptors are blocked by strychnine which, as a result of its inhibition of inhibitory glycinergic interneurones, can lead to muscle spasms. The release of glycine at the same neurones is caused by tetanus toxin, hence the similarity in the clinical actions of the two substances.

Glutamic acid and aspartic acid

Glutamic acid and aspartic acid are probably the main central excitatory neurotransmitters and they are found throughout the central nervous system. There appear to be three types of glutamic acid receptor. Three potent agonists at these three respective receptors have been found to be *N*-methyl-D-aspartic acid, commonly abbreviated to NMDA, quisqualic acid, derived from seeds, and kainic acid, derived from seaweed. Hence these receptors are respectively known as NMDA, quisqualate and kainate receptors.

NMDA receptor functioning, which exhibits voltage dependency and is influenced by magnesium ions, is associated with tonic–clonic epileptic seizure activity and neurodegeneration. Hence NMDA antagonists may be clinically useful both in epilepsy and in neuro-degeneration resulting from anoxia or ischaemia, such as occurs following a cerebrovascular accident (see Meldrum 1985). This is an area that is the focus of a great deal of research effort at the time of writing (see also Cotman et al 1988).

Catecholamines

A catechol nucleus consists of a benzene ring with two hydroxyl (-OH) groups attached. A catecholamine is simply a catechol nucleus with an amine group attached. The catecholamine neurotransmitters are **dopamine**, **noradrenaline**, also known as norepine-phrine in North America, and **adrenaline** or epinephrine.

Biosynthesis

The primary biosynthetic pathway of the three catechol-amines from **tyrosine** is shown in Figure 3.32. Note that as mentioned previously the enzyme aromatic amino acid decarboxylase is also known as dopa decarboxylase. The non-essential amino acid precursor tryosine is normally present in the bloodstream from which it can enter the brain by an active transport mechanism. It can also be biosynthesized from phenyl-alanine, also derived from the diet, by means of the action of phenylalanine hydroxylase. The conversion of tyrosine to 3,4-dihydroxyphenylalanine (DOPA), via the action of **tyrosine hydroxylase**, is the rate limit-ing step in this biosynthetic pathway. The action of tyrosine hydroxylase is stimulated, in a process involv-ing increased calcium ion entry into the cell, when the neurone is repeatedly stimulated and therefore losing its catecholamine store from storage vesicles. Con-versely, the action of tyrosine hydroxylase is inhibited by large catecholamine concentrations. This negative feedback control inhibits further catecholamine bio-synthesis when the catecholamine levels are greater than that which can be stored.

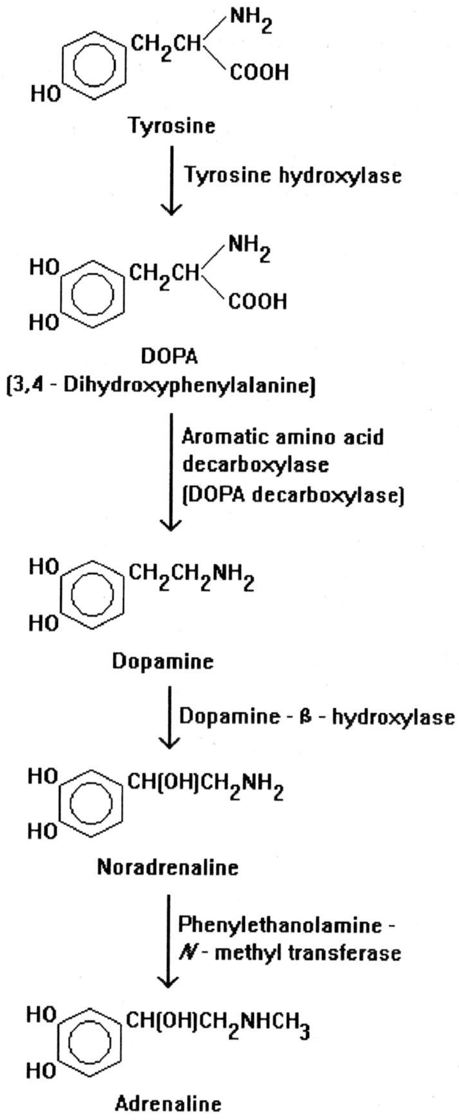

Fig. 3.32   The main biosynthetic pathway of the catecholamines.

Metabolism

There are two main enzymes responsible for the metabolic degradation of noradrenaline and dop-amine: monoamine oxidase (MAO) and catechol-*O*-methyltransferase (COMT).

Monoamine oxidase is found mainly in the exter-nal mitochondrial membrane and deaminates nor-adrenaline and dopamine to their corresponding aldehyde intermediates. It occurs both in nervous and non-nervous tissue. Indeed its presence in the liver and gastrointestinal tract allows dietary sympathomimetic agents such as tyramine to be metabolized, thus preventing the body from being subjected to their actions. Two types of monoamine oxidase have been identified in the central nervous system: MAO_A and MAO_B. MAO_A acts on noradrenaline, serotonin, dopamine, and tyramine, while MAO_B acts on dopamine, tyramine and phenylethylamine.

Catechol-*O*-methyltransferase is also found in most tissues, having relatively high hepatic and renal concentrations. It acts on catecholamines reaching beyond the synaptic cleft in a reaction that requires the methyl donor *S*-adenosylmethionine (SAM) and magnesium ions. The transfer of a methyl group from *S*-adenosylmethionine to the *meta*-hydroxyl group of both catecholamines and other catechols is catalyzed by catechol-*O*-methyltransferase.

The metabolic breakdown pathways, via the actions of these enzymes, of noradrenaline and dopamine are shown in Figures 3.33 and 3.34 respectively. It can be seen that **noradrenaline** is catabolized to either 3-methoxy-4-hydroxyphenylglycol (MHPG) or 3-methoxy-4-hydroxymandelic acid, also known as vanillyl mandelic acid (VMA). The glycol **MHPG** is the major end product from brain metabolism, and some is further conjugated with a sulphate group in what is probably a detoxication reaction that allows the resulting MHPG-sulphate to be actively transported from the brain. Its cerebrospinal fluid and urinary concentrations have been used in studies as indices of brain noradrenaline turnover. However, the urinary concentration of MHPG provides only a crude measure of central noradrenaline metabolism, as a proportion is derived from peripheral metabolism. The major product of noradrenaline catabolism in the peripheral nervous system is VMA; the amount of VMA in the brain is almost insignificant.

Dopamine is catabolized mainly to **homovanillic acid**, conventionally abbreviated to **HVA**, and **dihydroxyphenylacetic acid** or **DOPAC**, and the latter can either be conjugated with sulphate or be further metabolized, via the transfer to it of a methyl group through the action of COMT, to HVA (see Fig. 3.34), which in turn can also be conjugated with

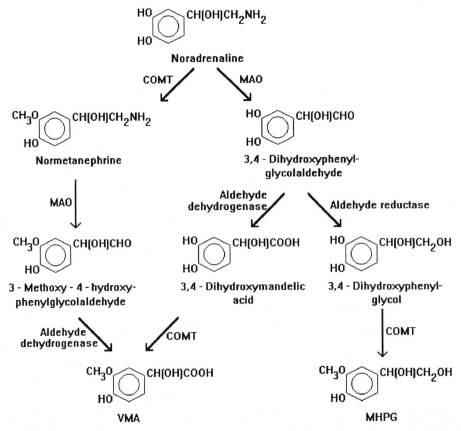

Fig. 3.33 Catabolic pathways for noradrenaline.

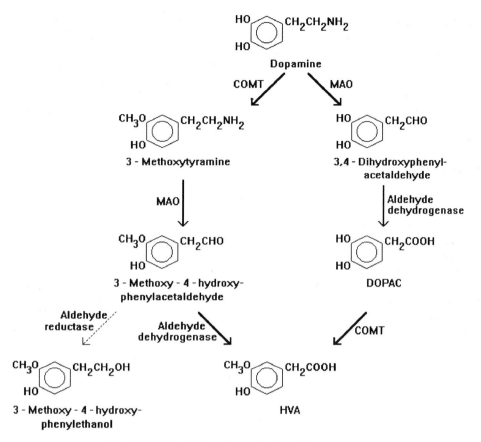

Fig. 3.34 Catabolic pathways for dopamine.

sulphate. Both HVA and DOPAC are central metabolites of dopamine and therefore their brain or cerebrospinal fluid levels can be used as indices of central dopamine turnover. In practice, in human studies it is easier to measure the level of HVA, as DOPAC is a minor metabolite; in infraprimate mammalian species in which DOPAC is a more important dopamine metabolite, however, the measurement of DOPAC levels can be much easier.

Re-uptake

Following their release into the synaptic cleft, the main method of inactivation of dopamine and noradrenaline is via re-uptake by presynaptic neurones. Secondary amine tricyclic antidepressants act by inhibiting such central noradrenaline re-uptake (see below).

Distribution

In the autonomic nervous system **noradrenaline** is the neurotransmitter for sympathetic postganglionic neurones. In the central nervous system there are two important sites of noradrenaline cell bodies: in the locus coeruleus and the brainstem.

The pontine **locus coeruleus** has the greatest central density of noradrenergic neurones and from it arise at least five major noradrenergic tracts. Three of these ascend in the medial forebrain bundle to supply mainly the ipsilateral cerebral cortex, thalamus, hypothalamus, limbic system and olfactory bulb. The fourth noradrenergic tract supplies the cerebellar cortex by way of the superior cerebellar peduncle. The fifth noradrenergic tract descends in the mesencephalon and spinal cord.

The **brainstem** contains a looser collection of noradrenergic neurones. Their precise fibre destinations are not known although the fibres are often found in close association with those from the locus coeruleus.

From a psychiatric viewpoint there are three important **dopaminergic systems**: the mesolimbic–mesocortical system, the nigrostriatal system, and the tuberoinfundibular system.

The **mesolimbic–mesocortical system** originates in the ventral tegmental area, lying medial to the substantia nigra. It can be thought of as being composed of two subsystems. The mesolimbic system projects to the limbic system while the mesocortical system innervates the cingulate, entorhinal and medial prefrontal cortices. The mesolimbic system is closely associated with the noradrenergic innervation of the limbic system mentioned above.

The **nigrostriatal system** originates in the pars compacta of the substantia nigra and projects to the corpus striatum. It is concerned with sensorimotor coordination.

The **tuberoinfundibular system** projects from the arcuate and periventricular nuclei of the hypothalamus to parts of both the adenohypophysis and neurohypophysis. As discussed in the previous chapter, dopamine acts to inhibit the release of the adenohypophyseal hormone prolactin.

Other dopaminergic pathways are found within the olfactory bulb, hypothalamus, retina and medulla oblongata. Other areas relatively rich in dopaminergic neurones include the nucleus tractus solitarius and the vagal dorsal motor nucleus.

Adrenergic systems in the central nervous system are relatively rare and tend to be found in association with noradrenergic systems. Peripherally, as mentioned in the previous chapter, stimulation of the medulla of the suprarenal or adrenal gland causes the release of relatively large quantities of adrenaline and noradrenaline into the bloodstream. This results in similar effects on the body as those of direct sympathetic stimulation, with the exception that these effects last much longer in the case of the bloodborne hormones.

Receptors

So far as the actions of noradrenaline and adrenaline in the peripheral and central nervous systems are concerned, four types of **adrenoceptors** are known: α_1, α_2, β_1 and β_2.

The α_1-adrenoceptors are postsynaptic and, in addition to adrenaline and noradrenaline, phenylephrine and methoxamine are agonists while prazosin, phenoxybenzamine, and phentolamine act as antagonists. The α_2-adrenoceptors, in contrast, are mainly presynaptic and, in addition to adrenaline and noradrenaline, clonidine is an agonist while antagonists include idazoxan and yohimbine.

The β-adrenoceptors tend to be mostly postsynaptic and, unlike α-adrenoceptors, are linked to adenyl cyclase. For both types of β-adrenoceptor isoprenaline is a more potent agonist than either adrenaline or noradrenaline; salbutamol and terbutaline are also agonists at β_2-adrenoceptors. The β_1 antagonists include metoprolol, atenolol and practolol. Propranolol is an antagonist at both β_1 and β_2-adrenoceptors. An important central postsynaptic location for β-adrenoceptors is in the locus coeruleus. Other central locations include the cerebral cortex and corpus striatum, which are mainly of the β_1 type, and the cerebellum, which mainly contains the β_2 type.

At least two types of **dopamine receptors**, D_1 and D_2, are known to exist, but at the time of writing it is not known what the total number of types is.

D_1 receptors stimulate adenylate cyclase and are found in the corpus striatum, retina and olfactory tubercle pyramidal cells. They are also found in the parathyroid gland, where the action of dopamine leads to parathormone release. Unlike D_2 receptors, they are not found in the pituitary gland. Antagonists at D_1 receptors include the butyrophenones (such as benperidol, droperidol, haloperidol and trifluperidol) and thioxanthines (such as flupenthixol and zuclopenthixol).

D_2 receptors, unlike D_1 receptors, probably inhibit adenylate cyclase and are found in the pituitary gland, where they have a hormone release inhibitory action. D_2 receptors are also found in association with the nigrostriatal and mesolimbic (though not the mesocortical) dopaminergic subsystems. Antagonists at D_2 receptors include domperidone, a drug used clinically in the treatment of nausea and vomiting, the substituted benzamide sulpiride, and the thioxanthines.

Central neuropharmacological aspects

Tricyclic antidepressants act as monoamine re-uptake inhibitors. That is, they inhibit the re-uptake of the monoamines noradrenaline and serotonin (see below) by presynaptic noradrenergic and serotonergic neurones, respectively, with the secondary amine tricyclics (see Ch. 5) such as desipramine, nortriptyline and protriptyline having a greater effect on noradrenergic neurones. This action results in an increase in the availability of noradrenaline and serotonin in the synaptic cleft. Although this has been suggested as the mode of antidepressant action of these drugs, this action on monoamines takes place relatively rapidly compared with the clinical antidepressant action which may take at least a fortnight to occur (see Ch. 5).

Another class of antidepressants that acts by increasing the availability of monoamines is the **monoamine oxidase inhibitors**, often abbreviated to MAOIs. As implied by their name, monoamine oxidase inhibitors act by inhibiting the metabolic degradation of monoamines by monoamine oxidase. As with tricyclic antidepressants, the action on monoamine oxidase takes place relatively rapidly compared with the clinical antidepressant action. Their inhibition of the peripheral catabolism of pressor amines, particularly dietary tyramine, can lead to a **hypertensive crisis**, sometimes referred to as the cheese reaction, in patients being treated with monoamine oxidase inhibitors who eat foodstuffs rich in tyramine, for example, cheese (apart from cottage cheese and cream cheese).

It should be noted that both tricyclic antidepressants and monoamine oxidase inhibitors are used in the treatment of conditions other than depression; both types of drug are discussed further in the next two chapters. Furthermore, not all drugs that increase the availability of monoamines are antidepressants. For example, cocaine is a monoamine re-uptake inhibitor.

In addition to its use as a centrally acting antihypertensive agent, the α_2-adrenoceptor agonist **clonidine** is occasionally used clinically in the prophylaxis of migraine, vascular headaches and menopausal flushing. It has also been used in Gilles de la Tourette syndrome and in the treatment of addictions.

In clinical practice β-**adrenoceptor antagonists** such as propranolol may be used in the treatment of hypertension, anxiety, and da Costa's syndrome (also known as neurocirculatory asthenia or cardiac neurosis). They are also used in the prophylaxis of migraine and to treat the tremor associated with conditions such as hyperthyroidism.

Neuroleptics, which include phenothiazines such as chlorpromazine, butyrophenones such as haloperidol, thioxanthenes such as flupenthixol, diphenylbutylpiperidines such as pimozide, and sulpiride, are believed to exert their antipsychotic actions by means of dopamine receptor blockade. In particular, it is believed that it is antagonism of dopaminergic activity in the mesolimbic–mesocortical system that is crucial to this therapeutic action.

The **extrapyramidal side-effects** of the neuroleptics are believed to result from blockade of dopaminergic D_2 receptor neurotransmission in the nigrostriatal system.

Antagonism of dopaminergic activity in the tuberoinfundibular system inhibits the dopaminergic inhibition on the release of the adenohypophyseal hormone prolactin. This accounts for the consequences of **hyperprolactinaemia**, such as galactorrhoea, gynaecomastia and menstrual disturbances, seen as side effects of neuroleptic treatment.

Another side effect seen with long-term neuroleptic treatment is the development of **tardive dyskinesia**. This is believed to result possibly from the development of dopamine postsynaptic receptor supersensitivity following chronic dopaminergic blockade by neuroleptics. Consistent with this model is the observation that tardive dyskinesia is worsened by acute withdrawal of neuroleptic treatment or the administration

of the dopamine precursor L-dopa, but is acutely though transiently improved if there is an increase in the dose of neuroleptic administered. On the other hand, in a situation somewhat analogous to that encountered above with respect to the effects of tricyclic antidepressants and monoamine oxidase inhibitors, the development of dopamine receptor supersensitivity occurs relatively rapidly compared with the time scale for the development of tardive dyskinesia.

Psychiatric aspects

The **monoamine hypothesis of mood disorders** has undergone a number of changes from the original catecholamine or noradrenaline hypothesis of mood (affective) disorders which stated that depression is associated with a depletion of central functional noradrenaline while mania is associated with an excess of central functional noradrenaline (Bunney & Davis 1965, Schildkraut 1965). Currently this hypothesis is stated in terms of the monoamines noradrenaline and serotonin rather than just noradrenaline. Although the monoamine hypothesis is sometimes split into a noradrenaline hypothesis and a serotonin hypothesis, it seems unlikely that these two neurotransmitters act entirely independently of each other in the central nervous system. Maas (1975), for example, divided depression into that caused by noradrenergic dysfunction and that caused by serotinergic dysfunction.

There are a number of findings in favour of the monoamine hypothesis. As mentioned above, both tricyclic antidepressants, which act as monoamine reuptake inhibitors, and monoamine oxidase inhibitors, cause an increase in the availability of noradrenaline and serotonin, and act as antidepressants. Moreover, the use of tricyclic antidepressants in bipolar patients can precipitate episodes of mania. Similarly, amphetamine, which is structurally similar to the catecholamines and releases catecholamines from neurones, is a central nervous system stimulant that lifts mood. It can also lead to a schizophrenia-like clinical picture (see below). Conversely, reserpine, an antihypertensive drug derived from the Indian plant Rauwolfia, depletes central monoaminergic neuronal stores of catecholamines and serotonin and would be expected, under this model, to cause depression. This is indeed the case, with severe depression and suicide having been reported. Other evidence that supports the monoamine hypothesis includes cerebrospinal fluid data, with the serotonin metabolite 5-hydroxyindolacetic acid (5-HIAA) often reported as being reduced in depressed patients (e.g. Agren 1980, Åsberg et al 1984). The cerebrospinal fluid levels of the noradrenaline metabolite MHPG have in general been found not to differ between depressed or manic patients and normal subjects. However studies of depressed patients occasionally do demonstrate lower levels compared with controls (e.g. Post et al 1984). Other supportive evidence includes post mortem and platelet data for serotonin, and studies of urinary noradrenaline metabolites.

Notwithstanding the fact that not all the types of study just mentioned render results that support the monoamine hypothesis of mood disorders, there are other findings that argue against. Two of the most important of these have already been mentioned above. Firstly, there is the delay between the relatively rapid onset of biochemical action of tricyclic antidepressants and monoamine oxidase inhibitors compared with the delay of two to four weeks before the onset of a clinical therapeutic response in depressed patients. Secondly, not all drugs that act as monoamine re-uptake inhibitors have a therapeutic antidepressant action; an example mentioned above is that of cocaine. Conversely, not all antidepressant treatments act by increasing the functional availability of monoamines.

It has also been suggested that **dopaminergic mechanisms** may be involved in **mood disorders**, with functional dopaminergic activity being decreased in depression and increased in mania. There are a number of findings which are consistent with this hypothesis. The mood elevating effect of amphetamine and related compounds has already been mentioned. The dopamine precursor L-dopa has also been found to cause mood elevation, particularly in patients with psychomotor retardation (e.g. Butcher & Engel 1969). Similarly, the dopamine agonist bromocriptine has also been shown to elevate mood, the effect being greater in depression that was part of a bipolar rather than unipolar disorder (Silverstone 1984). Conversely, the depressive action on mood of neuroleptic (dopamine antagonist) pharmacotherapy is a well known clinical observation and is utilized in the treatment of mania. Further support for this hypothesis has been obtained from studies of the cerebrospinal fluid level of the main dopamine metabolite homovanillic acid (HVA). Controlled studies in general demonstrate decreased cerebrospinal fluid HVA concentrations in depression (e.g. Åsberg et al 1984) and increased concentrations in mania (e.g. Swann et al 1983).

According to the **dopamine hypothesis of schizophrenia** the clinical features of this condition are the result of central dopaminergic hyperactivity. This hyperactivity is usually believed to occur within the mesolimbic-mesocortical dopaminergic system mentioned earlier. The dopamine hypothesis arose from psychopharmacological findings. For example, as mentioned above, neuroleptics that are therapeutically useful in the treatment of schizophrenia are dopamine blocking agents. Moreover, there is a positive correlation between the ability of such drugs to block dopamine receptors and their antipsychotic effectiveness (e.g. Snyder et al 1974). Again, the administration of drugs such as amphetamine, which lead to an increase in the functional central activity of catecholamines, including dopamine, by causing their release from catecholaminergic neurones, can induce the development of a psychotic picture that is very similar to that seen in some types of schizophrenia; amphetamine has also been found to worsen the psychotic symptomatology in schizophrenia. A similar effect can occur following the administration of levodopa (L-dopa), the amino acid precursor of dopamine. Johnstone et al (1978) compared the actions of two structural isomers of flupenthixol, one with and one without dopamine receptor blocking activity, and a placebo in a double blind trial on patients with acute schizophrenia. Only those patients receiving the α isomer of flupenthixol, which is a dopamine antagonist, were found to show improvement in their psychotic symptomatology; this was not found with either the placebo or the β isomer of flupenthixol, which did not cause dopamine receptor blocking.

A number of essentially non-psychopharmacological findings also lend support to the dopamine hypothesis, a few of which are now briefly discussed. Post mortem studies of neuroleptic binding, which is an index of dopamine binding, have shown increased binding in brains of schizophrenic patients compared with control brains (e.g. Owen et al 1978, Lee & Seeman 1980). Studies involving animal models of dopamine function have also been carried out in which the administration of dopamine agonists has produced a behavioural picture said to be similar to human psychosis which is reversed by dopamine antagonists (Anden 1975). Although studies of the cerebrospinal fluid levels of the dopamine metabolite homovanillic acid (HVA) in schizophrenic patients compared with controls have in general proved negative, a higher HVA level has been found in those schizophrenic patients with a positive family history of the condition (Sedvall & Wode-Helgodt 1980) or with a history of poor premorbid sexual adjustment (Leckman et al 1981). Plasma dopamine concentrations have occasionally been found to be increased in schizophrenia (Bondy et al 1984) and it has been suggested that the plasma HVA concentration may be a useful index of central dopaminergic activity (Pickar et al 1988).

The fact that studies of the cerebrospinal fluid HVA levels in schizophrenic patients compared with controls have in general proved negative (e.g. Bowers 1973, Berger et al 1980) has been put forward as an argument against the dopamine hypothesis. A number of other arguments against this hypothesis also exist. Under the dopamine hypothesis one might expect there to be a decrease in the baseline plasma prolactin level, as a result of the increased central dopaminergic activity in the tuberoinfundibular system leading to a greater inhibition of the release of prolactin from the adenohypophysis; in general studies have given negative results. Similarly, according to the hypothesis it could be argued that dopamine hyperactivity in the nigrostriatal system should mean that Parkinson's disease, caused by reduced dopaminergic activity in this system, should not occur in a patient with schizophrenia; in practice the two conditions can coexist. Again the argument concerning time lags between biochemical and clinical effects, rehearsed above with respect to antidepressants, can also be applied to the dopamine hypothesis. It has been pointed out that pharmacotherapy with neuroleptics may take at least a fortnight to cause antipsychotic effects in schizophrenic patients, a time interval that is much greater than the biochemical action as shown by the fact that plasma prolactin levels may increase much more rapidly (see Fig. 3.35), and extrapyramidal side effects may begin less than two days after the commencement of the drug. It is also the case that, on the one hand, dopamine receptor antagonists are clinically effective not only in schizophrenia but also in agitation and psychosis not associated with schizophrenia, and, on the other hand, not all cases of acute schizophrenia respond to dopamine antagonists.

The above arguments for and against the dopamine hypothesis have necessarily had to be brief. In spite of the arguments against the hypothesis, which in some cases have been negated by a further consideration of issues such as dopamine receptor distribution and the development of tolerance, the dopamine hypothesis remains very useful. The interested reader is referred to the review by Carlsson (1988) for a more detailed analysis of the status of this hypothesis.

The **transmethylation hypothesis of schizophrenia** is based on the observation that a number of hallucinogenic substances have a similar structure to methylated derivatives of the catecholamines and serotonin. For example, mescaline is an hallucinogen that is a methylated derivative of dopamine. According to this hypothesis schizophrenia is the result of the brain synthesizing endogenous hallucinogens from monoamines via transmethylation. This hypothesis has found support in the finding that administration of the methyl donor methionine, especially in combination with a monoamine oxidase inhibitor, exacerabates schizophrenic symptomatology in chronic schizophrenia (Pollin et al 1961). It has been pointed out, however, that in such studies it was difficult to differentiate between actual exacerabation of schizophrenia and the development of an acute organic brain syndrome superimposed on the chronic schizophrenic picture (Pollin et al 1961, Wyatt et al 1971). Furthermore, overall there is no evidence that the administration of methyl donors such as nicotinic acid are clinically effective in schizophrenia. Another approach adopted in harmony with the transmethylation hypothesis has been the bid to discover the endogenous methylated hallucinogen or hallucinogens supposedly responsible for the development of schizophrenia. Again, in general

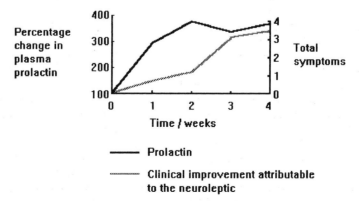

Fig. 3.35 Comparison of the time course of the change in plasma prolactin levels in patients treated with α-flupenthixol with the time course of the clinical improvement attributable to the α-flupenthixol (after Cotes et al 1978).

such studies have proved negative; two of the most studied candidate psychotoxic methlyated substances have been 3, 4-dimethoxyphenylethylamine (see Green & Costain 1981) and the serotonin derivative dimethyltryptamine or DMT (see Murray et al 1979).

As mentioned in the previous chapter, increased saccadic abnormalities have been found in those schizophrenic patients with **tardive dyskinesia** when compared with schizophrenic patients without tardive dyskinesia, the latter in turn having increased abnormalities compared with controls (Thaker et al 1989). It has been proposed that this result is caused by dysfunction of subcortical GABA circuits involved with the frontal eye fields. The result is consistent with a model of schizophrenia put forward by Early et al (1989) in which the condition is associated with dominant hemisphere **unilateral dopaminergic hypoactivity** leading to hyperactivity of the left striatopallidal structures. This is said to affect three parallel cortical–striatal–pallidal–thalamic–cortical limbic, cognitive and motor circuits, which in turn is said to account for a number of features of schizophrenia. A useful aspect of this model is that it provides a framework for further experimental testing.

Reduced functional dopaminergic activity in the nigrostriatal system is associated with **Parkinson's disease** as indicated above, with extrapyramidal side-effects from dopamine antagonists such as neuroleptics being caused by antagonist activity at postsynaptic dopamine receptors in this system. Catecholamine dysfunction has also been noted in a number of other conditions. For example, there is a reduction in the cerebral cortical level of noradrenaline in **Alzheimer's disease** (Adolfsson et al 1979).

Histamine

Because histamine crosses the blood–brain barrier only very poorly, most brain histamine is derived from the amino acid histidine in a reaction catalyzed by histidine decarboxylase. The part of the brain with the highest concentration of histamine is the hypothalamus, from which histaminergic pathways appear to project to the cerebral cortex, thalamus, and parts of the limbic system. As in the periphery, mammalian brains have been found to possess both H_1 and H_2 receptors, both of which are linked to adenyl cyclase. The central effects of drowsiness and hunger associated with antihistamine pharmacotherapy are believed to occur via antagonist activity at central H_1 receptors. Similar side effects of neuroleptics may perhaps also occur in the same way. In general, histamine appears to be able to alter vegetative functions such as thermoregulation and the intake of food and water.

Serotonin

Serotonin is an indolamine and is also known as 5-hydroxytryptamine, conventionally abbreviated to 5-HT. Like dopamine and noradrenaline, serotonin is also a monoamine. Most of the serotonin of the body is found outside the central nervous system, for example in mast cells and platelets.

Biosynthesis

Because serotonin is unable to cross the blood–brain barrier it is biosynthesized in the brain from the essential amino acid precursor **tryptophan** by means of the pathways shown in Figure 3.36. The first reaction,

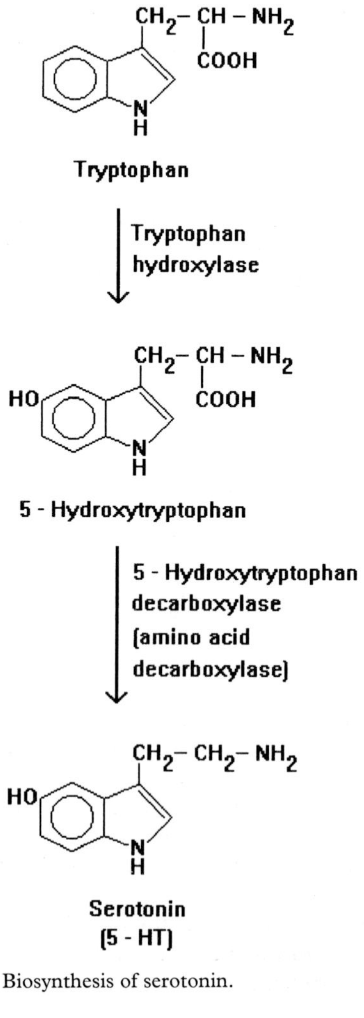

Fig. 3.36   Biosynthesis of serotonin.

catalyzed by **tryptophan hydroxylase**, is the rate-limiting step in this pathway. This enzyme is normally unsaturated, with the result that increased plasma levels of tryptophan result in increased central biosynthesis of serotonin, and vice versa. Hence the sedative serotonergic action that occurs after the intake of foodstuffs such as milk that are relatively rich in this amino acid. The active transport of tryptophan into the brain occurs by means of the 'L system' for large amino acids, mentioned earlier in this chapter, in competition with other amino acids (histidine, methionine, leucine, isoleucine, valine, phenylalanine and tyrosine).

Storage and re-uptake

Serotonin is stored in synaptic vesicles and, following its release into the synaptic cleft, the main method of inactivation is via re-uptake by presynaptic neurones. Such central serotonergic re-uptake is inhibited by tertiary tricyclic antidepressants and the newer more specific 5-HT re-uptake inhibitor antidepressants such as fluvoxamine and fluoxetine.

Metabolism

The catabolism of serotonin occurs via the action of **monoamine oxidase**, in particular MAO_A, to **5-hydroxyindoleacetic acid**, conventionally abbreviated to 5-HIAA, the structure of which is shown in Figure 3.37.

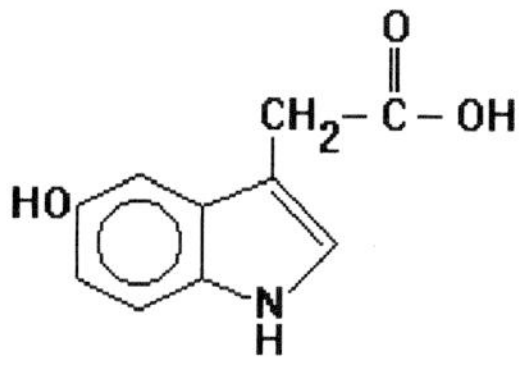

Fig. 3.37 5-Hydroxyindoleacetic acid.

Distribution

The main central nuclei containing serotonin are the midline brain stem **raphe nuclei**. They can be divided variously into three groups of nuclei: the medullary, mesencephalic and pontine raphe groups, whose constituent nuclei are shown in Table 3.1. Major serotonergic pathways from the raphe nuclei are shown in Table 3.2.

A relatively high concentration of serotonin is also found in the **pineal gland**, in which there exist enzymes for the biosynthesis of serotonin, and the biosynthesis

Table 3.1 Serotonergic raphe nuclear groups

Raphe group	Constituent nuclei
Medullary	Nucleus raphe pallidus
	Nucleus raphe obscurus
	Nucleus raphe magnus
Mesencephalic	Nucleus raphe dorsalis
Pontine	Nucleus raphe pontis
	Nucleus centralis superior

Table 3.2 Serotonergic tracts

Tract	Raphe nuclear group origin	Major destinations
Ascending	Mesencephalic	Cerebral cortex
	Pontine	Corpus striatum
		Thalamus
		Lateral hypothalamus
		Limbic system
		Mesencephalic nuclei
		Substantia nigra
Descending bulbospinal	Medullary	Spinal cord
Pontocerebellar	Pontine	Cerebellum
Propriobulbar	Medullary	Inferior olivary complex
	Pontine	Locus coeruleus
		Reticular formation

of **melatonin** (5-methoxy-*N*-acetyltryptamine) from serotonin.

Receptors

As in the case of dopamine receptors, the number of subtypes of serotonin receptors is not known with certainty, but has increased over the last decade or so (see Tricklebank 1987). At the time of writing the following subtypes have been characterized: 5-HT_{1A} (or S_{1A}), 5-HT_{1B}, 5-HT_{1C}, 5-HT_{1D}, 5-HT_2, and 5-HT_3. In addition, there is some evidence for a further subtype, 5-HT_4, and for a division of the 5-HT_2 subtype into 5-HT_{2A} and 5-HT_{2B}.

Central neuropharmacological aspects

As mentioned above in the subsection on the catecholamines, **tricyclic antidepressants** act as monoamine re-uptake inhibitors. It has also been mentioned that antidepressants exist which are more specific **sero-**

tonin re-uptake inhibitors; they include fluoxetine and fluvoxamine, and are discussed further in the next two chapters. Both serotonin re-uptake inhibitors and the tricyclic semi-specific serotonin re-uptake inhibitor clomipramine have therapeutic effects in obsessive compulsive disorder. Again as mentioned above, another class of antidepressants that acts by increasing the availability of monoamines is the **monoamine oxidase inhibitors**, which act by inhibiting the metabolic degradation of monoamines by monoamine oxidase.

For many years L-tryptophan, the amino acid precursor of serotonin, has been used therapeutically as an antidepressant. However, as mentioned in Chapter 5, at the time of writing it has been withdrawn from general clinical use.

High concentrations of serotonin act in the hypothalamic satiety centre to diminish the desire for carbohydrate-rich food intake. This action is utilized by the drug **fenfluramine**, a highly specific serotonin agonist, which may be prescribed as an adjunct to continued dietary treatment of severe obesity in patients not responding to dietary control alone.

At the time of writing, research is taking place on the possible therapeutic value of substances acting at various serotonin receptor subtypes in the treatment of aggression, addiction, suicidal behaviour, schizophrenia and cognitive impairment, including dementia.

Psychiatric aspects

The **monoamine hypothesis of mood disorders** has been considered above in the subsection on the catecholamines, and the psychopharmacological evidence for the involvement of monoamines, including serotonin, has been discussed there. The postulation of an association of serotonergic abnormalities with depression is sometimes known as the **serotonin hypothesis of depression** and is usually taken to imply that depression is associated with decreased serotonergic activity. Evidence in favour of this hypothesis, other than that already mentioned above, is now briefly described.

Changes in serotonergic activity have been found to be associated with many clinical features of depression, including changes in appetite, mood, cognition, sleep, libido and circadian rhythms (see the review by Meltzer & Lowy 1987). Cerebrospinal fluid levels of the main serotonin metabolite 5-HIAA have been found to be either decreased or normal in depressed patients (e.g. van Praag & Korf 1971, Åsberg et al 1984, Meltzer & Lowy 1987). Furthermore, low concentrations of both serotonin and 5-HIAA have been found in post-mortem brain-stem specimens from suicides (reviewed by Mann et al 1989b); the cortical levels have generally been found to be within the normal range in such studies. There is also evidence that depression is associated with reduced levels of plasma tryptophan, and therefore presumably of central serotonin biosynthesis (e.g. Coppen & Wood 1978, DeMyer et al 1981). Since serotonin receptors are found on platelets, which are relatively easy to study directly, platelet imipramine binding sites have been used as an index of central serotonin re-uptake sites. In general, a reduction is found of platelet serotonin uptake sites in depression (e.g. Coppen et al 1978c; Maj et al 1988). Further evidence in support of the serotonin hypothesis has come from studies in which depression of mood follows the reduction of the biosynthesis of serotonin or alteration of serotonin storage by administering reserpine (Goodwin & Bunney 1971), administering parachlorphenylalanine (Shopsin et al 1976), or reducing dietary intake of the amino acid precursor tryptophan (e.g. Young et al 1986, Heninger et al 1989).

The therapeutic actions of the serotonin re-uptake inhibitors and the tricyclic semi-specific serotonin re-uptake inhibitor clomipramine in **obsessive compulsive disorder** have been mentioned above, and provide evidence that this disorder is associated with serotonergic dysfunction. Thorén et al (1980) found that the therapeutic action of clomipramine in obsessive compulsive disorder was such that clinical improvement was significantly correlated with the degree of reduction of cerebrospinal fluid 5-HIAA levels.

The therapeutic use of fenfluramine in the control of food intake has been mentioned above and is one member of a new class of drugs known as **serotonergic anorectics** which enhance serotonergic neurotransmission and have their anorectic effects antagonized by serotonin antagonists (Garattini et al 1989). The serotonergic anorectics appear not to be have the disadvantage of amphetamine-like drugs of causing a stimulating effect leading potentially to dependence; indeed, fenfluramine tends to be sedative.

Animal and human studies have provided evidence that central serotonergic dysfunction is associated with **impulsive aggression** and is reviewed by Coccaro (1989). At the time of writing research is taking place into possible pharmacotherapeutic implications.

Similarly, there is some evidence that central serotonergic dysfunction may be associated with **schizophrenia**. For example, the hallucinogenic drug lysergic acid diethylamide (LSD) appears to act at serotonin receptors. Also, the neuroleptic clozapine acts on the serotonergic system, in addition to acting at dopamine receptors, and appears to be therapeutically effective in a proportion of otherwise resistant schizophrenics with negative symptoms (Kane et al 1988). This latter finding provides some support for a postulated dysfunction occurring in the interaction of dopaminergic and serotonergic systems in schizophrenia. Further support for such an interaction is provided by the finding that the 5-HT antagonist ritanserin activates midbrain dopamine neurones by blocking serotonergic inhibition, implying that normally the dopamine neurones are inhibited by serotonergic systems (Ugedo et al 1989).

Other neurotransmitters

For the sake of completion it should be noted that, in addition to the neurotransmitters discussed already in this section, two further important classes of neurotransmitters are the **peptide neurotransmitters** and the **prostaglandins**. Both have already been discussed earlier in this book.

FURTHER READING

Campbell P N, Smith A D 1988 Biochemistry illustrated. 2nd edn. Chruchill Livingstone, Edinburgh

Cooper J R, Bloom F E, Roth R H 1991 The biochemical basis of neuropharmacology. 6th edn. Oxford University Press, New York

Kruk Z L, Pycock C J 1979 Neurotransmitters and drugs. Croom Helm, Beckenham

McIlwain H, Bachelard H S 1985 Biochemistry of the central nervous system. 5th edn. Churchill Livingstone, Edinburgh

Ryall R W 1989 Mechanisms of drug action on the nervous system. 2nd edn. Cambridge University Press, Cambridge

Stryer L 1988 Biochemistry. 3rd edn. Freeman, New York

4. Neuropharmacokinetics

Neurotransmission has been considered in the last chapter. In this chapter pharmacokinetics, which includes the processes of absorption, distribution, biotransformation and excretion, is considered, followed by a study of basic pharmacokinetic terms and equations. The chapter ends by looking at the pharmacokinetics of drugs commonly used in psychiatry. Other aspects of the pharmacology of these drugs are dealt with in the next chapter.

ABSORPTION

A drug may enter the circulation by being absorbed from the gastrointestinal tract. This is known as **enteral** administration, and includes oral, buccal or sublingual, and rectal administration. **Parenteral** routes include intramuscular, intravenous, subcutaneous, inhalational and topical administration. Other types of administration used less commonly in psychiatric practice include the intranasal and intrathecal routes.

Rate of absorption

The rate of absorption of a drug from it site of administration depends on its form and solubility, and on the rate of flow of blood through the tissue in which it is sited.

The **disintegration** of a drug depends on its pharmaceutical formulation. For example, for orally administered drugs enteric coating slows down the rate of disintegration.

Factors that influence the **solubility** of a drug include its pK_a, the particle size and the ambient pH. The **pK_a** of a drug can be derived from the Henderson–Hasselbalch equation.

$$pH = pK_a + \log_{10}\{[\text{base}]/[\text{acid}]\}$$

Since $\log_{10}1$ is zero, it follows from the equation that a drug is 50% ionized when the pH is equal to the pK_a of the drug.

Oral administration

The major mechanisms for absorption of drugs from the alimentary tract are: passive diffusion, pore filtration and active transport. Most drugs are absorbed by **passive diffusion** through the bimolecular lipid sheet. Such transfer by passive diffusion occurs more readily when the drug is in the un-ionized form. Most drugs are too large to pass across by pore filtration, and most do not fullfil the appropriate specific structural criteria to be carried by active transport systems.

The small intestine is the primary site of absorption of orally administered drugs. Gastric absorption of any importance only occurs with ethanol and weak acids. Enteric coating is used either to protect a drug from gastric acid or to protect the stomach from harmful drug side-effects.

Factors which influence absorption from the gastrointestinal tract include: gastric emptying, gastric pH, intestinal motility, the presence or absence of food in the alimentary canal, intestinal microflora, area of absorption and blood flow. Intestinal motility may be increased in conditions such as anxiety states, leading to an increased rate of absorption of drugs such as diazepam. Also, although in general increased absorption occurs on an empty stomach, the presence of food can lead to increased absorption for some drugs, such as diazepam, metoprolol, propranolol, hydralazine and hydrochlorothiazide.

Rectal administration

The rectal administration of drugs for systemic effects has several advantages over the oral route. It overcomes the problem of patients who cannot swallow, for example because of vomiting. The stomach is bypassed and therefore so too are factors such as the gastric pH and gastric emptying; drugs that are irritant to the stomach may also be given rectally. First pass metabolism (see the section on biotransformation below) is reduced since the portal circulation is avoided to some extent. The rectal route may also be useful in uncooperative patients. In the care of the mentally handicapped diazepam may be administered rectally during an epileptic seizure.

An important disadvantage of rectal administration is an emotional one, this route proving embarrassing to many. Furthermore, since the rectum may contain a variable amount of faecal matter, the rate of absorption may be unpredictable. Frequent use of this route may lead to local inflammation.

Intramuscular injection

The intramuscular administration of psychotropic medication is often used in clinical psychiatric practice for the relief of acute symptoms in disturbed patients. It is also used for administering long-acting depot injections of antipsychotic drugs for maintenance therapy. For example, in the case of fluphenazine decanoate a fatty ester is administered which slowly hydrolyses releasing the free drug. Antimuscarinic drugs such as procyclidine and benztropine may be given by intramuscular or intravenous injection for the emergency treatment of acute drug-induced dystonic reactions. This route may also be useful for giving irritant drugs that cannot be administered subcutaneously.

Lipid-soluble drugs are rapidly absorbed by the intramuscular route, as are low-molecular-weight water-soluble, but lipid-insoluble, drugs. Conversely, high-molecular-weight lipid-insoluble drugs are absorbed only slowly. The rate of absorption of a drug is increased following exercise since this increases muscle blood flow; conditions such as cardiac failure which reduce muscle blood flow lead to a reduction in the rate of absorption.

There are several disadvantages of the intramuscular route. It is likely to be painful and is usually unacceptable as a route for self-administration. There is the danger of damaging structures such as nerves. Thus in gluteal injections the outer upper quadrant is used in order to avoid sciatic nerve damage. Certain drugs such as paraldehyde may cause sterile abscesses. As mentioned above, in conditions of reduced muscle blood flow there is a lowered rate of absorption. For some drugs such as chlordiazepoxide, diazepam and phenytoin, tissue binding or precipitation from solution following intramuscular injection may reduce the rate and extent of absorption as compared with oral administration. Concurrent anticoagulant therapy is a contraindication to the use of the intramuscular route. It should also be noted that following intramuscular injection creatine phosphokinase is released,

which may interfere with diagnostic cardiac enzyme assays.

Intravenous injection

The intravenous route has several advantages. By allowing a drug to enter the systemic circulation rapidly an intravenously administered drug can act rapidly, hence making this route useful in emergencies. Moreover, the drug dosage can be titrated against the response of the patient, especially in the case of slow intravenous infusions, as for example during an abreaction. This route can allow large volumes to be administered slowly, and also allows the administration of drugs that cannot be absorbed by other routes. A further advantage is that this route allows drugs to avoid first-pass metabolism (see below).

Amongst the disadvantages of this route is the fact that adverse reactions may occur rapidly. Furthermore, rapid intravenous administration by bolus injection may lead to dangerously high blood levels. Once administered it is difficult to recall the drug; in comparison stomach washouts and emetics can be used following oral overdosage. Intravenous injections carry the risk of sepsis, thrombosis and air embolism. There is also the risk that the drug may accidentally be injected into the tissues surrounding the vein, or indeed into an artery, leading to necrosis or spasm respectively. This route cannot be used with insoluble drugs.

Intranasal administration

Whilst orally administered peptides are generally ineffective, the intranasal route can be a suitable alternative. Thus, for example, intranasal vasopressin or antidiuretic hormone and its analogues are used in the treatment of pituitary diabetes inspidus. One of the analogues, desmopressin, is also available as a metered-dose nasal spray for the treatment of primary nocturnal enuresis.

It has been mentioned in Chapter 2 that many neuropeptides have a role in psychiatric disorders. If a therapeutic application were to be found for such substances in the future then it is likely that the intranasal route may become increasingly important in psychiatry.

Other routes

Other sites of absorption are rarely or never used with psychotropic drugs, for example, absorption through the skin and lungs, and subcutaneous and intraperitoneal injections. In animal studies, however, subcutaneously implanted osmotic pumps have been found to be effective in delivering continuous controlled amounts of drugs such as dopamine and haloperidol (see, for example, Moore 1979, and Login et al 1979). Research is currently taking place on ways in which to deliver peptides and hormone to the central nervous system across the blood–brain barrier. For example Postmes et al (1983) have used liposomes to carry thyrotropin releasing factor to the central nervous system in laboratory rabbits.

DISTRIBUTION

This refers to the distribution of a given drug between the lipid, protein and water components of the body. The rate and degree of distribution of a drug is influenced hy haemodynamic factors, permeability factors, the blood–brain barrier and the placenta. A discussion of these factors follows. The quantification of drug distribution is considered later in this Chapter.

Haemodynamic factors

It is the most highly perfused organs that receive the highest concentrations of a drug soon after absorption. There follows a gradual redistribution to tissues and organs with lower blood flow rates such as resting skeletal muscle and adipose tissue.

Drugs circulate around the body partly bound to plasma proteins and partly free in the water phase of plasma. Such **plasma protein binding** acts as a reservoir for a drug since it is only the free or unbound fraction that is pharmacologically active. The free and bound fractions are in equilibrium as shown in Figure 4.1. The main plasma binding protein for acidic drugs is **albumin**. Basic drugs, which include many psychotropic drugs such as chlorpromazine and imipramine, can also bind to other plasma proteins including **lipoprotein** and α_1-**acid glycoprotein**.

The extent of plasma protein binding can be changed by a number of factors. It varies with the **plasma drug concentration** usually in the manner shown in Figure 4.2. Another factor influencing plasma protein binding

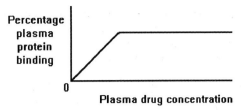

Fig. 4.2   The usual relationship between plasma protein binding and plasma drug concentration.

is the concentration of the plasma protein itself. **Hypoalbuminaemia** can result from diseases of the liver and kidneys, cardiac failure, malnutrition and carcinoma. It can also be the result of surgery and burns. As an **acute-phase reactant**, the plasma level of α_1-acid glycoprotein increases in states of physiological stress such as myocardial infarction and inflammation, for example Crohn's disease. **Hypo-α_1-acid glycoproteinaemia** occurs in the nephrotic syndrome, hepatic disease and malnutrition. The commonest **hyperlipoproteinaemias** are types IIa, IIb and IV. Type IIa may be primary or secondary to a variety of disorders such as hypothyroidism, the nephrotic syndrome, myeloma and hepatic diseases. Type IV may be primary or secondary to diabetes mellitus and obesity. Plasma protein binding can also be influenced by **drug interactions** in which one drug displaces another from protein binding sites, or causes changes in the plasma protein tertiary structure. Such displacements and structural changes may also be caused by changes in the **concentration of physiological substances** such as urea, bilirubin and free fatty acids.

Other reservoirs for drugs include muscle tissue and, especially in the obese, adipose tissue.

Permeability factors

The capillary endothelium is highly permeable except in the brain, where the blood–brain barrier operates (see below). In general, highly lipid-soluble drugs enter cells rapidly, while highly water-soluble drugs enter cells slowly if at all.

Blood–brain barrier and distribution to the brain

The endothelium of the cerebral capillaries lacks the intercellular fenestrations that generally exist in the circulation of the rest of he body. Instead there are **tight junctions** between the adjacent endothelial cells. Moreover, some astrocytic processes end as perivascular feet on the capillaries, forming a **gliovascular membrane**. Together with the capillary **basement**

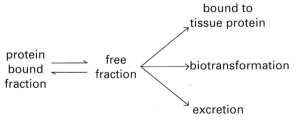

Fig. 4.1   The equilibrium between free and protein bound drug fractions.

membrane, the tight junctions and gliovascular membrane help form the blood–brain barrier. The blood–brain barrier is responsible for preventing the rapid equilibration of some drugs between the blood on the one hand, and the brain and cerebrospinal fluid on the other, in spite of the rich blood supply of the brain.

Since the brain is a highly lipid organ, there is a high rate of penetration for **non-polar highly lipid-soluble** drugs, and for inhalational anaesthetics. However, for highly polar water-soluble drugs and quaternary ammonium salts the rate of penetration, if any, is low. Thus, for example, at the same plasma concentration there is a lower brain concentration of the water-soluble β-adrenoceptor antagonist atenolol than of the lipid-soluble propranolol, leading to fewer central nervous system side-effects with the former. As might be expected, most psychotropic drugs are highly lipid-soluble. Again, the stronger an acidic (low pK_a) or basic (high pK_a) drug, the slower the rate of penetration, because of the higher degree of ionization.

The normal operation of the blood–brain barrier may alter in the presence of **infection**. Thus whereas normally the penicillins, which are water-soluble, do not enter the brain and cerebrospinal fluid to any appreciable extent, they can gain ready access in acute bacterial meningitis.

Owing to the existence of specific **receptors** in the brain for many psychotropic drugs, as mentioned in the previous chapters, psychotropic drug protein binding can take place in the brain, thus forming a central nervous system reservoir. This does not occur in the cerebrospinal fluid, however, as the latter has a very low protein concentration. Hence, following equilibration, the cerebrospinal fluid concentration of a highly protein bound drug will be lower than its total plasma concentration, but may be similar to the free plasma water concentration.

Certain drugs and physiological substances cross the blood–brain barrier using **active transport** mechanisms, for example levodopa.

Small molecules, such as lithium ions, are able to **diffuse** readily into the brain and cerebrospinal fluid from the cerebral ciculation.

Placenta

Drugs may be transferred from the maternal circulation to the fetal circulation across the placenta by passive diffusion, active transport, and pinocytosis.

Drugs taken by the mother may cause teratogenesis during the first trimester, with the third to the eleventh weeks being the time of highest risk. Therefore, if at all possible drugs should be avoided during the first trimester. Any drugs that have to be prescribed during this period should be given in the smallest effective dosage, and well-established drugs which appear to be safe in pregnancy should be used in preference to newer or previously untried drugs.

During the last two trimesters drugs may affect fetal growth and functional development and may lead to toxic side-effects on fetal tissues. Drugs may also affect the newborn child if given during labour.

BIOTRANSFORMATION (METABOLISM)

Some highly water-soluble drugs such as lithium can be excreted unchanged from the body via the kidneys. Others, including most highly lipid-soluble psychotropic drugs, must first undergo biotransformation to reduce their lipid solubility and make them more water-soluble. Such biotransformation, while often converting a pharmacologically active drug to a pharmacologically inactive substance, in some cases produce active metabolites. For example, the tricyclic antidepressant amitriptyline is converted to nortriptyline, which also has antidepressant activity.

Biotransformation occurs mainly in the **liver**. Other sites of biotransformation include the kidneys, the suprarenal or adrenal cortex, the gastrointestinal tract, the lungs, the placenta, the skin and lymphocytes. Hepatic biotransformation may be considered to occur in two broad phases: phase I and phase II, these are summarized in Figure 4.3.

Phase I biotransformation

Phase I biotransformation or metabolism leads to a change in the molecular structure of a drug by means of the **non-synthetic reactions** oxidation, reduction and hydrolysis. Of these **oxidation** is by far the most common, and the most important type is that carried out by **microsomal mixed-function oxidases**, microsomes being partial artefacts of homogenization of

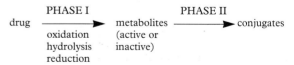

Fig. 4.3 Phases I and II of drug biotransformation.

smooth endoplasmic reticulum. The cytochrome P_{450} isoenzymes are involved in these reactions.

Oxidation reactions taking place in the smooth endoplasmic reticulum include aromatic hydroxylation, aliphatic hydroxylation, O-dealkylation, N-dealkylation, N-oxidation and S-oxidation. This results in metabolites that can take part in the conjugation reactions of phase II biotransformation. Thus, for example, the side-chains of barbiturates undergo aliphatic hydroxylation leading to inactive metabolites. Imipramine is converted by N-dealkylation to desipramine and amitriptyline by the same reaction to nortriptyline; in the case of these antidepressants the metabolites are also pharmacologically active. Again, chlorpromazine, a phenothiazine with an aliphatic side chain, is converted by S-oxidation to the pharmacologically inactive chlorpromazine sulphoxide.

Reduction and hydrolysis reactions taking place in the smooth endoplasmic reticulum are uncommon. An example is pethidine, which is hydrolyzed to the active metabolite meperidinic acid.

Oxidation reactions not taking place in the smooth endoplasmic reticulum are important for certain drugs. Thus, ethanol is oxidized to acetaldehyde by cytosol alcohol dehydrogenase. The acetaldehyde is then further oxidized to ethanoic or acetic acid by aldehyde dehydrogenase. The membrane bound mitochondrial enzymes monoamine oxidase and diamine oxidase convert primary amines to aldehydes or ketones by oxidative deamination. Among the substrates of monoamine oxidase, which is widely distributed in organs such as the brain, liver, intestine and kidney, are the catecholamines, phenylephrine, 5-hydroxytryptamine, tryptamine and tryramine (see chapter 3).

Phase II biotransformation

Phase II biotransformation or metabolism is a synthetic reaction in which **conjugation** takes place between, on the one hand, a parent drug, drug metabolite (a product of phase I biotransformation), or endogenous substance (such as oestrogen), and, on the other hand, a polar (hence water-soluble) endogenous molecule or group such as glucuronic acid, sulphate, acetate, glutathione, glycine and glutamine. The result of phase II biotransformation is a water-soluble conjugate which can be excreted readily by the kidney if its relative molecular mass is less than about 300. Conjugates with a higher relative molecular mass can be excreted in the bile. Following phase II biotransformation drugs and their metabolites almost always lose any pharmacological activity they possessed.

Enzyme induction

As a result of exposure to certain substances, including drugs, there may be an increase in the synthesis or a decrease in the breakdown of an enzyme. This leads to an increase in the activity of the enzyme and is known as **enzyme induction**. It may be accompanied by hypertrophy of hepatic endoplasmic reticulum, an increase in cytochrome P_{450} content, and an increase in hepatic blood flow. Therefore when given drugs induce enzymes involved in their biotransformation, there is an increase in their biotransformation leading to an apparent tolerance to those drugs. Moreover, the induced enzymes may also cause an increase in the biotransformation of other drugs. Thus enzyme induction is a cause of drug interactions. Furthermore, after the inducing drug or drugs are withdrawn, the plasma concentrations of the previously affected drugs will increase, leading to the possibility of toxicity.

The **barbiturates** are enzyme inducers that lead to the increased biotransformation of a relatively large number of drugs, including barbiturates, chlorpromazine, the oral contraceptive pill, phenytoin, tricyclic antidepressants, and vitamin D_3. Similarly, **phenytoin** increases the biotransformation of the oral contraceptive pill, tricyclic antidepressants and vitamin D_3 amongst others. Substances whose biotransformation is increased by **ethanol** (alcohol) include ethanol, barbiturates, bilirubin, meprobamate and phenytoin. Other drugs that induce their own biotransformation include the **phenothiazines, carbamazepine, primidone, chloral hydrate** and **meprobamate. Cigarette smoking** and **caffeine** also cause some enzyme induction but to a lesser degree than most other drugs.

Hence in clinical practice enzyme induction may cause the failure of the oral contraceptive pill, reduced efficacy of drug treatment for epilepsy, and reduced plasma levels of drugs such as chlorpromazine and the tricyclic antidepressants. Smokers may have lower levels of chlorpromazine and benzodiazepines than non-smokers. By increasing the biotransformation of vitamin D_3, antiepileptics and barbiturates can cause osteomalacia.

Enzyme inhibition

The inhibition of enzymes involved in biotransformation of drugs can lead to increased plasma drug levels and toxicity.

Drugs given concurrently may interact by competitive or non-competitive action at enzyme sites. An example of the former is the interaction of diazepam and phenytoin which both compete for microsomal hydroxylation leading to the possibility of phenytoin toxicity when the two drugs are administered together. An example of non-competitive inhibition of microsomal enzymes is that of the antibiotic chloramphenicol with phenytoin, the biotransformation of the latter being inhibited.

Disulfiram inhibits the enzyme aldehyde dehydrogenase. This leads to the accumulation of acetaldehyde after the ingestion of even small quantities of ethanol, causing unpleasant systemic effects such as facial flushing, throbbing headache, palpitations, tachycardia, nausea and vomiting. Hence disulfiram may be used as an adjunct treatment in the treatment of chronic alcohol dependence. Arrhythmias, hypotension, collapse and death may occur in patients taking disulfiram following the ingestion of large doses of ethanol.

Another important interaction is the inhibition of the biotransformation of tricyclic antidepressants by **phenothiazines** and **haloperidol**, resulting in increased side-effects of the tricyclic antidepressants and a reduction in the effective therapeutic dose.

The **monoamine oxidase inhibitors** depend for their antidepressant action on their inhibition of monoamine oxidase, particularly that occuring in brain mitochondria. This is considered further later in this Chapter.

Hepatic impairment tends to affect phase I biotransformation more than phase II. Drugs whose biotransformation is reduced include the barbiturates, the benzodiazepines and the opioid analgesics. Thus these drugs should be used with caution in hepatic impairment, and given in lower dosage if at all.

First-pass effect

The first-pass effect is also known as **first-pass metabolism** and **pre-systemic elimination**, and refers to the biotransformation that an orally absorbed drug undergoes during its passage from the hepatic portal system through the liver before entering the systemic circulation. The first-pass effect can significantly reduce the bioavailability of a drug. For example, only 30–80% of orally administered imipramine and nortriptyline enter the systemic circulation intact, while the figure for orally administered fluphenazine is less than 10%. There is a wide range of inter-individual variability in the extent of the first-pass effect. Thus in some individuals the extent of biotransformation of orally administered phenothiazines before entry to the general systemic circulation is almost 100%; in these patients it is clearly useful to administer the drug by a route, such as intramuscular depot injection, that bypasses the first-pass effect.

Food may reduce the extent of the first-pass effect by increasing hepatic blood flow. The concurrent administration of a drug such as hydralazine that increases hepatic blood flow will have a similar effect. The extent of the first-pass effect may also be reduced in hepatic impairment (caused, for example, by cirrhosis).

EXCRETION

The most important organ for the excretion of drugs and drug metabolites is the **kidney**. Excretion can also take place via the bile and faeces, lungs, saliva, sweat, sebum and milk. In general, with the exception of the lungs, water-soluble polar drugs are excreted more readily by excretory organs than highly lipid-soluble non-polar drugs. Thus, as mentioned above, since most psychotropic drugs are in the latter category they will not be excreted readily until they have undergone biotransformation.

Renal excretion

The amount of drug entering the glomerular filtrate by **filtration** depends on the glomerular filtration rate, the plasma water drug concentration, and the relative molecular mass of the drug. In the proximal renal tubule active **secretion** of charged molecules can occur from the plasma to the glomerular filtrate. There are two systems: one for organic acids such as drug metabolites like glucuronides, drugs such as penicillin, and endogenous substances such as uric acid; the other for organic bases such as amphetamine and histamine.

Initially the glomerular filtrate contains drugs at approximately the same concentration as the plasma water drug concentration. However, as it passes along the nephron the glomerular filtrate becomes progressively more concentrated, leading to the formation of a gradient between the glomerular filtrate and the plasma water of the blood circulation of the nephron. Moreover, the tubular epithelium acts as a lipid membrane. Thus passive tubular **re-absorption** can take place from the glomerular filtrate to the blood. The extent of this re-absorption depends on the lipid solubility of the drug or other substance, which in turn depends on its pK_a and on the pH of the tubular urine. Thus, in alkaline tubular urine acidic drugs tend to undergo less re-absorption, and therefore greater excretion, because they undergo greater ionization and become less lipid-soluble. Under similar conditions basic drugs become un-ionized and undergo greater re-absorption. The converse occurs in acidic tubular urine. Thus forced alkaline diuresis may be used in cases of overdose of acidic drugs such as salicylates and phenobarbitone.

Conversely, in cases of overdose of alkaline drugs such as amphetamine the urine can be made more acidic to encourage excretion of the drug.

In renal impairment the failure to excrete a drug or its metabolites may lead to toxicity. This is particularly important for drugs such as lithium that are highly dependent on renal excretion.

Conditions in which renal clearance increases, such as occurs towards the end of pregnancy, can lead to a decrease of the drug plasma concentration.

Biliary and faecal excretion

Drug metabolites may be excreted by the liver into the intestinal tract in the **bile**. This route is particularly important for the glucuronides which, together with other organic anions, enter the bile by means of active transport. Similar carrier mechanisms exist for organic cations. Once in the intestinal tract the drug metabolites may either be re-absorbed, a process known as the **enterohepatic cycle**, or, less commonly, be excreted in the faeces. Also excreted in the faeces are unabsorbed orally administered drugs.

Breast milk

Excretion of drugs in maternal breast milk, although quantitatively unimportant as a form of elimination, is important clinically in breast-feeding. A drug entering the milk in pharmacologically significant concentrations can lead to toxicity in infants. At lower doses drugs may cause hypersensitivity in the infant. Since most psychotropic drugs are lipid-soluble they are able, in their non-ionized form, to diffuse through the epithelium of glands and enter the milk. Because milk is more acidic than plasma, substances with a high pK_a are likely to be concentrated in breast milk, while those with a low pK_a will have a concentration in the milk that is lower than the plasma water concentration. Not all psychotropic drugs, however, need to be avoided during breast feeding. For example, the tricyclic antidepressants generally enter the milk in amounts too small to be harmful.

Cycling processes

Cycling processes refer to cycles in which many drugs are excreted and then re-absorbed into the bloodstream. One example, the **enterohepatic cycle**, has been mentioned above. The effect of a cycling process on an affected drug is to increase its half-life (see below) in the body. Another effect in the case of the enterohepatic cycle is to increase the pharmacological effect of a drug after eating, when large amounts of bile usually enter the intestinal tract.

Other examples of cycling processes include **gastric excretion**, in which drugs such as pethidine are re-absorbed from the intestinal tract following excretion into the stomach, and **salivary excretion** in which drugs excreted into the saliva are similarly re-absorbed. Salivary excretion occurs actively for some drugs such as lithium and may offer a means of therapeutic monitoring of drug levels although the relationship between the salivary concentration and the plasma concentration is not always constant.

PHARMACOKINETIC TERMS AND EQUATIONS

In this section the concepts of volume of distribution and half-life are first considered, followed by a discussion of the formulae and equations used in calculations involving bio-availability, first-order and zero-order elimination, intravenous administration and multiple dosing.

Volume of distribution

The volume of distribution is a theoretical concept that relates the mass of a drug in the body to the blood or plasma concentration. It is conventionally given the symbol V_d, and can be defined by the following formula:

$$V_d = D/C$$

where D is the mass of a drug in the body at a given time, and C is concentration of the drug at that time in the blood or the plasma, depending on the fluid being measured.

The volume of distribution is the size of the conceptual compartment for a drug and is the volume of fluid that would be required in theory to account for the total mass of the drug in the body. It does not have an anatomical meaning, and usually does not equal an anatomical volume. Likewise, it does not have a physiological meaning.

The volume of distribution of a drug can exceed the total body volume if there is a heavy concentration of the drug in a particular type of tissue. In general, the volume of distribution tends to be lower in drugs that are highly protein-bound, and higher in drugs that have a higher lipid solubility. In the latter case such drugs are clearly better able to enter adipose tissue, for example. In adults the volume of distribution may have a value of well over 300 litres for certain drugs.

Strictly speaking, the volume of distribution has units of volume. However, because its value can vary with body mass, the volume of distribution of a drug is often given in units of volume per kilogram body mass.

Half-life

There are two types of half-life in common use. The **distribution half-life** is half the time taken for a drug to reach its volume of distribution. A drug with a short distribution half-life usually exerts its therapeutic and toxic effects more quickly. The **elimination half-life** of a drug is the time taken for its plasma concentration or its total mass in the body to decrease by one-half. It is conventionally represented by the symbol $t_{1/2}$. For example, suppose a given drug is being administered to the body by a constant intravenous infusion such that it achieves a constant plasma concentration of C_0. If at time zero the infusion were to be suddenly stopped, and if, further, the drug had a constant half-life, $t_{1/2}$, then the plasma concentration of the drug would decline after time zero in the fashion shown in Table 4.1.

Clearance

The clearance of a drug is the volume of biological fluid cleared of the drug in unit time. It is conventionally given the symbol Cl and it can be shown that:

$$Cl = kV_d$$

where k is the first-order (see below) elimination rate constant. The clearance can also be related to the half-life of the drug by the following equation:

$$Cl = (V_d \ln 2)/t_{1/2}$$

When the volume of distribution is measured in units of volume, then the clearance is measured in units of volume per time, for example ml min^{-1}. When the volume of distribution is measured in units of volume per mass, then the clearance is measured in units of volume per time per mass, for example ml min^{-1} kg^{-1}.

The total body clearance is the sum of the individual clearances by individual eliminating organs. That is:

$$Cl_{total} = Cl_{renal} + Cl_{hepatic} + \dots$$

Bio-availability

The bio-availability of a drug is the fraction of the dose administered that actually reaches the systemic circulation. By convention it is symbolized by the letter F. Clearly the maximum value that F can have is 1, and this would occur in the situation when the whole of a drug reached the systemic circulation following intravenous administration. Thus:

$$0 \leq F \leq 1$$

It can be shown that in a graph of the drug plasma concentration versus time for a given drug administered by a given route, the area under the curve, AUC, is directly proportional to the amount of the drug that enters the systemic circulation. Thus following oral administration of a drug, assuming complete absorption from the gastrointestinal tract, the bio-availability of the drug can be calculated from the following expression:

$$F = \frac{\text{AUC after oral administration of a drug}}{\text{AUC after intravenous administration of the same dose of the drug}}$$

Similar calculations can be carried out for drugs administered by other routes.

First-order elimination (linear kinetics)

In first-order elimination the rate of elimination of a drug is **directly proportional** to the plasma concentration of the drug at a given time. In other words, the greater the plasma concentration the greater the rate of transfer of a drug across a cell membrane or the greater the rate of biotransformation. This is also known as linear kinetics, and can be represented mathematically as

$$dC/dt = -kC$$

where C is the plasma concentration of the drug, k is a constant, and t represents time. Therefore it follows by integration that:

$$C = C_0 e^{-kt}$$

where C_0 is the value of the plasma concentration at

Table 4.1 The plasma concentration of a drug as a function of its elimination half-life (see text)

Time	Plasma concentration
Zero	C_0
$t_{1/2}$	$C_0/2$
$2t_{1/2}$	$C_0/4$
$3t_{1/2}$	$C_0/8$
$4t_{1/2}$	$C_0/16$
$5t_{1/2}$	$C_0/32$
...	
$nt_{1/2}$	$C_0/2^n$

time zero. Taking the natural logarithm of each side of the equation yields:

$$\ln C = \ln C_0 - kt$$

Since the elimination half-life, $t_{1/2}$ is the time taken for a given value of C to fall to $\frac{1}{2}$C, it follows that C can have the value of $C_0/2$ at time $t_{1/2}$ in the last equation:

$$\ln (C_0/2) = \ln C_0 - kt_{1/2}$$

Therefore,

$$t_{1/2} = (\ln 2)/k$$

where ln 2 is the natural logarithm of 2 and has a fixed value of 0.6931 to four significant figures. Since k is also a constant, it follows that the elimination half-life is **constant** in first-order elimination.

Zero-order elimination (saturation kinetics)

In zero-order elimination the rate of elimination is **constant**. For example, when the total amount of a drug in the body rises above a certain threshold, various processes involved in its elimination can become saturated and the rate of elimination will no longer be proportional to the plasma drug concentration but instead will be constant. This may occur in the case of enzyme mediated biotransformation reactions which may become saturated because of the finite amount of enzyme available. This type of kinetics is also known as a rate-limited process or saturation kinetics.

An example of a drug which can show zero-order elimination is ethanol. Below a plasma concentration of approximately 0.1 mg ml^{-1} the elimination of ethanol is a first-order process. However, above this concentration there is saturation of the enzyme ethanol dehydrogenase with the result that elimination is now zero-order, and remains so until the plasma concentration falls below this threshold. Phenytoin is an example of a drug for which the threshold for changing between first-order and zero-order elimination occurs within the therapeutic dosage scale.

The half-life in zero-order elimination is no longer constant but in fact increases as the plasma concentration of the drug increases, and vice versa.

Intravenous injection

In the simple **one-compartment model** of kinetics it is assumed that following intravenous injection a drug simply dissolves in the whole of one compartment immediately. This compartment has no anatomical or physiological meaning, but is assumed to be the plasma. If the drug were then to be cleared from that

compartment by first-order elimination, there should theoretically be an exponential relationship between the plasma drug concentration and time. This results in a straight line profile for the plot of the logarithm of the plasma concentration versus time, of the form:

$$\ln C = C_0 - kt$$

where C_0 is the plasma concentration of the drug immediately following intravenous injection. This is shown graphically in Figure 4.4; the value of k can be calculated from the gradient of this line.

For the majority of psychotropic drugs a **two-compartment model** applies following intravenous injection. In this model there is one **central compartment** and a **peripheral compartment** that is larger and more slowly equilibrating, as shown schematically in Figure 4.5. Figure 4.6 shows the plot of the logarithm of the plasma concentration versus time for this model. It can be seen that there is a biphasic fall in the plasma concentration following intravenous injection. Mathematically it can be shown that the value of the plasma concentration in the central compartment C_1, is given by:

$$C_1 = Ae^{-\alpha t} + Be^{-\beta t}$$

where A, B, α and β are constants whose values can be calculated from a plot such as that in Figure 4.6.

Constant intravenous infusion

Since $V_d = D/C$ (see above), then at steady state (ss), for a drug obeying linear kinetics and being

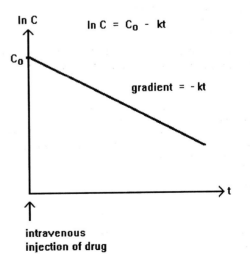

Fig. 4.4 The logarithm of the plasma concentration versus time in a one-compartment model for a drug administered by intravenous injection and obeying linear kinetics.

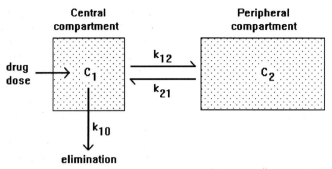

Fig. 4.5 Diagrammatic representation of a two-compartment model. C_1 and C_2 are the drug concentrations in the central and peripheral compartments, respectively; k_{12} and k_{21} are the first-order rate constants for drug transfer between compartments; k_{10} is the first-order elimination constant.

administered by constant intravenous infusion, when the rate of elimination of the drug is equal to the rate of intravenous infusion, it follows that:

$$D_{ss} = C_{ss}V_d$$

where C_{ss} is the plasma concentration at steady state. It can also be shown mathematically that theoretically the plasma concentration of the drug, C, at any time, t, is:

$$C = C_{ss}(1 - e^{-kt})$$

where $k = (\ln 2)/t_{\frac{1}{2}}$. Therefore the fraction of the steady state attained at any given time, C/C_{ss}, is given by:

$$1 - e^{-kt}$$

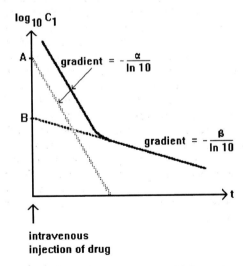

Fig. 4.6 The logarithm of the plasma concentration versus time in a two-compartment model for a drug administered by intravenous injection and obeying linear kinetics.

It is clear from this expression that the rate of attainment of the steady state plasma concentration, C_{ss}, depends only on the **half-life** of the drug, as shown in Table 4.2.

Multiple dosing

For a drug obeying linear kinetics and being administered at a regular dose interval, it can be shown that theoretically for a one-compartment model the mean steady state concentration at steady state, C_m, is given by:

$$C_m = \frac{Dt_{1/2}F}{V_dT.\ln 2}$$

where T is the dose interval and F is the bioavailability factor (see above). Since $1/\ln 2 = 1.44$ to three significant figures, it follows that:

$$C_m = \frac{1.44Dt_{1/2}F}{V_dT}$$

For the same situation, it can also be shown that the ratio of the maximum plasma concentration achieved, C_{max}, to the minimum plasma concentration, C_{min}, is

Table 4.2 The percentage of the steady state concentration reached for a drug obeying linear kinetics and being administered by constant intravenous infusion

Percentage of C_{ss} reached	Time taken
50	$t_{1/2}$
75	$2t_{1/2}$
88	$3t_{1/2}$
94	$4t_{1/2}$
97	$5t_{1/2}$

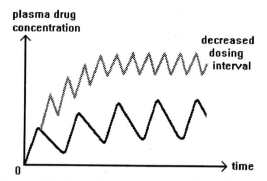

Fig. 4.7 The effect of changing the dosing interval on the plasma concentration versus time profile of a drug being administered at regular intervals.

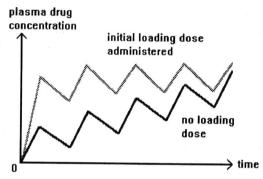

Fig. 4.8 The effect of using an initial loading dose on the plasma concentration versus time profile of a drug being administered at regular intervals.

twice the ratio of the regular dosing interval to the half-life of the drug. That is:

$$C_{max}/C_{min} = 2T/t_{1/2}$$

The graph in Figure 4.7 shows the effect on the plasma concentration of altering the dosing interval, while Figure 4.8 shows the effect of using an initial loading dose.

PHARMACOKINETICS AT THE EXTREMES OF AGE

The special pharmacokinetic considerations at the extremes of age are considered briefly in this section.

It is important when considering the pharmacokinetics of drugs in **children** not to think of the child's body as simply a smaller version of an adult body. For example, in the neonate the total **body water** and the extracellular body water are a much higher proportion of the total body mass than in the adult. The proportion of body mass that is **adipose tissue** in the neonate is lower than in the adult, thus causing the volume of distribution of lipid-soluble drugs, such as most psychotropic drugs, to be lower. On the other hand, the volume of distribution of highly **plasma protein**-bound drugs, such as phenytoin, is higher in neonates owing to their reduced albumin concentration and different plasma protein-binding capacity when compared with adults. Furthermore, it is not until the child has reached three to five months that the adult **glomerular filtration rate** is reached. Moreover, since it is not until the age of approximately two months that the **hepatic microsomal enzymes** involved in biotransformation are fully active, it follows that most drugs undergo biotransformation only slowly up to this age. There is also reduced **gastric acidity** and an increased **gastric emptying time** in the neonate. The **blood–brain barrier** is also more permeable in neonates than in older children and adults, making neonates very sensitive to psychotropic drugs.

In the **elderly** there tends to be a reduction in the **total body mass**, and in the proportion of body mass that is composed of **water** and **muscle tissue**. At the same time, there is an increase in the relative proportion of **adipose tissue**, resulting in an increase in the volume of distribution for many lipid-soluble drugs, such as the benzodiazepines. There may be changes in the concentrations of **plasma proteins**, often as a result of illness rather than ageing. Typically the plasma albumin level is reduced and the plasma γ-globulin level is increased; the former can lead to reduced binding of drugs such as phenytoin, warfarin and pethidine. Reduced **hepatic biotransformation** in the elderly can lead to a reduction in the hepatic clearance of drugs such as diazepam, propranolol, and some of the tricyclic antidepressants. Mainly as a result of the reduction in the glomerular filtration rate with age, there is a reduction in the **renal clearance** of many drugs. This is particularly important in the case of lithium, which should therefore be used in lower doses in the elderly, and should be maintained at the lower end of the adult plasma therapeutic range in maintenance therapy.

PHARMACOKINETICS OF PSYCHOTROPIC DRUGS

In this section the pharmacokinetics of major groups of psychotropic drugs are considered. Other aspects of their pharmacology are considered in Chapter 5.

Benzodiazepines

Benzodiazepines are normally absorbed well when administered orally. It has been found that antacids reduce the rate of absorption of orally administered chlordiazepoxide. Absorption following intramuscular administration is less rapid than that following oral administration and usually leads to lower peak plasma concentrations. Intravenous administration, on the other hand, gives high peak plasma concentrations.

The plasma protein binding of the benzodiazepines is high, being approximately 95% for diazepam, and 87%–88% for chlordiazepoxide.

The liver is the main site of benzodiazepine biotransformation. Hepatic oxidation is involved in the biotransformation of long-acting benzodiazepines such as diazepam whilst non-oxidative pathways are important for shorter-acting benzodiazepines. As shown in Figure 4.9, the benzodiazepine metabolites are interrelated, with desmethyldiazepam, also known as nordiazepam, having a central position. It is also evident from Figure 4.9 that some benzodiazepine metabolites, such as oxazepam, are used therapeutically.

Most of the benzodiazepines have long elimination half-lives, making it possible to reach therapeutic steady-state plasma concentrations with once-daily dosages. Some of the important metabolites also have a long half-life; that for desmethyldiazepam, for example, is 96 hours. There is in general an increase in the half-life of the benzodiazepines with increasing age.

β-Adrenoceptor antagonists

Following oral administration there is good absorption with peak plasma concentration being achieved one to three hours later.

The degree of plasma protein binding is high for propranolol but low for most of the other β-adrenoceptor antagonists.

The degree to which the β-adrenoceptor antagonists undergo biotransformation varies. Those that undergo

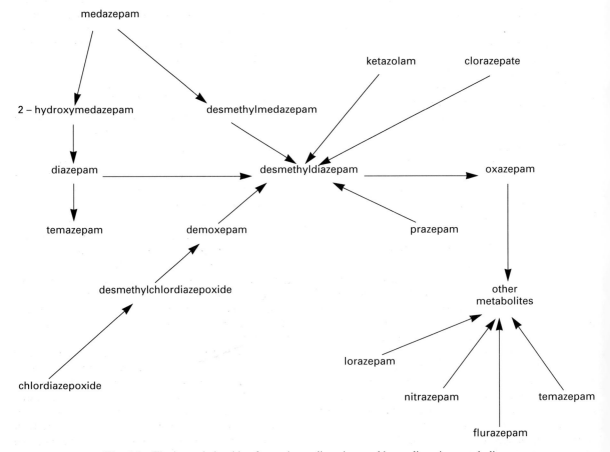

Fig. 4.9   The interrelationship of some benzodiazepines and benzodiazepine metabolites.

almost total biotransformation include alprenolol, oxprenolol and propranolol. Those with an intermediate level of biotransformation of 40–60% include pindolol and sotalol. Atenolol, nadolol and practolol are examples of members of this group of drugs that undergo almost no biotransformation at all. Some metabolites also act as β-adrenoceptor antagonists as, for example, in the case of the propranolol metabolite 4-hydroxypropranolol.

The elimination half-lives of the different β-adrenoceptor antagonists have a wide range of values. For example that for propranolol is less than four hours, whilst nadolol has a half-life more than 16 hours. With chronic administration there tends to be a reduction in the first-pass effect, which in turn results in an increase in the half-life.

Other anxiolytics and hypnotics

The **barbiturates** are derivatives of barbituric acid that are rapidly absorbed following oral administration and undergo biotransformation mainly in the liver. They may be divided on the basis of the elimination half-life into: a short-acting group used in the induction of anaesthesia (for example, thiopentone); a medium-acting group, including amylobarbitone and quinalbarbitone, used in the treatment of patients already taking them who are suffering from severe intractable insomnia; and a long-acting group used as anticonvulsants (for example, phenobarbitone and primidone). Phenobarbitone is considered further in the section on anticonvulsants below.

Buspirone is a non-benzodiazepine anxiolytic which is a member of the azaspirodecanedione group of compounds. It is well absorbed orally, peak plasma concentrations occurring after 60–90 minutes. At therapeutic levels the plasma concentration is linearly related to the oral dose. It is approximately 95% plasma protein bound. The liver is the main site of biotransformation and the elimination half-life is between two and 11 hours.

Chloral hydrate is a chlorinated alcohol which is well absorbed when given orally. It undergoes rapid reduction, mainly in the liver and blood, to the less polar active metabolite trichloroethanol via the enzyme alcohol dehydrogenase. This metabolite has an elimination half-life of approximately eight hours. It is excreted either following glucuronide conjugation as urochloralic acid, or after conversion to trichloroacetic acid. Since the latter metabolite has an elimination half-life of the order of four days, it is clear that it can accumulate following multiple dosing. In addition to both these further metabolites being excreted by the kidneys, some of the urochloralic acid is also excreted in the bile. The chloral derivative **paraldehyde** is a polymer of acetaldehyde and is usually administered by deep intramuscular injection or per rectum. It may also be given, in specialist centres only, by slow intravenous injection diluted with 0.9% sodium chloride solution. As well as being excreted renally following hepatic biotransformation, paraldehyde is also excreted by the lungs.

Chlormethiazole has a structure related to that of part of the vitamin B$_1$ molecule. It is absorbed rapidly following oral administration, peak plasma concentrations being achieved by one hour. The drug undergoes an extensive hepatic first-pass effect when given orally. Its elimination half-life is less than an hour.

Meprobamate is a propanediol and is absorbed well when given orally, peak plasma concentrations being achieved one to two hours later. Its main metabolite, the hydroxylated derivative, is not pharmacologically active. It causes an important degree of hepatic induction and its elimination half-life is approximately ten hours.

Zopiclone is a recently introduced non-benzodiazepine hypnotic which is a member of the cyclopyrrolone group of compounds. It is well absorbed orally, inducing sleep within 30 minutes. Its main site of biotransformation is the liver, and it has a relatively short elimination half-life of 3.5–6 hours.

Tricyclic antidepressants and related drugs

The tricyclic antidepressants are in general lipid-soluble and therefore readily absorbed from the gastrointestinal tract. This is related to the large volume of distribution that the tricyclic antidepressants have; for imipramine values have been found to be in the range 28 to 61 lkg^{-1} body mass. The anticholinergic effect that occurs following absorption causes decreased gastric emptying, which in turn leads to delayed oral absorption.

In general, the plasma protein binding of the tricyclic antidepressants is high. For example, at therapeutic plasma concentrations approximately 90% of clomipramine is bound, which the figure for imipramine is approximately 85%. This high protein binding combined with the large volume of distribution results in poor dialysability for these antidepressants following overdosage.

There is evidence that some of the secondary amine tricyclic antidepressants, such as nortriptyline, may have a **therapeutic window** whereby plasma concentrations that are too low or too high are not associated with an adequate clinical response (see, for example, Åsberg et al 1971). However, studies on the tertiary amine tricyclic antidepressants such as amitriptyline have failed to show consistent evidence of a similar therapeutic window. For example, no such effect was found for amitriptyline in the study by Coppen et al (1978a).

There is extensive biotransformation prior to excretion. The main metabolite of imipramine is desipramine, which is active. Nortriptyline is an active metabolite of amitriptyline. Both these antidepressant metabolites are monomethyl derivatives that result from demethylation of the side chain of the parent tricyclic antidepressants. Ring hydroxylation also occurs in the biotransformation of some members of this group.

Orally administered tricyclic antidepressants undergo an extensive first-pass effect; for example, that for imipramine is over 50%. Their lipid-solubility leads to a large volume of distribution, which in turn means that the plasma concentrations achieved with given oral doses are relatively low. The long elimination half-life of members of this group allows therapeutic steady-state plasma concentrations to be achieved with oral administration once per day, usually at night.

Maprotiline has a similar structure to that of the tricyclic antidepressant nortriptyline with the addition of an ethylene bridge across the central ring structure, making it tetracyclic. It is well absorbed orally, and has an elimination half-life a little longer than that of secondary amine tricyclic antidepressants.

Mianserin is a piperazinoazepine tetracyclic antidepressant which is well absorbed orally, peak plasma concentrations occurring within two to three hours. It is highly plasma protein bound and has an elimination half-life that varies between approximately eight and 19 hours.

Trazodone is a triazolopyridine. It is better absorbed orally when taken with food. Its elimination half-life is approximately four hours.

Viloxazine is a bicyclic tetrahydroxazine which is well absorbed orally, peak plasma concentrations occurring after approximately two hours. It has no known active metabolites and its elimination half-life is approximately five hours.

Selective 5-HT re-uptake inhibitors

Some of the above drugs, for example **trazodone** and the tricyclic antidepressant **clomipramine**, are potent serotonin (5-HT) re-uptake inhibitors. However, they are not selective in this action; for example the principal metabolite of clomipramine, desmethylclomipramine, potently inhibits the re-uptake of noradrenaline as well as serotonin. The first selective 5-HT re-uptake inhibitor used in psychiatry was the bicyclic antidepressant **zimelidine**; it was withdrawn from use in Britain owing to toxic effects. Currently there are four more recently introduced selective 5-HT re-uptake inhibitors available: fluvoxamine, fluoxetine, sertraline and paroxetine.

Fluvoxamine is a unicyclic antidepressant which is absorbed rapidly following oral administration. Hepatic biotransformation leads to the formation of metabolites which are not pharmacologically active and which are excreted by the kidneys. The elimination half-life of fluvoxamine is approximately 15 hours. Steady-state plasma concentrations are reached within a fortnight of treatment with a constant daily dose.

Fluoxetine is a substituted phenylpropylamine. It is absorbed slowly following oral administration with peak plasma concentrations being reached after six to eight hours. Its elimination half-life is one to three days, whilst its major metabolite, N-demethylfluoxetine, has been found to remain in the plasma for up to two weeks or more.

Paroxetine is a piperidine derivative which is well absorbed following oral administration. The metabolites resulting from first-pass metabolism are not known to be clinically active. The elimination half-life of paroxetine is variable but generally about 24 hours. Steady-state plasma concentrations are reached within a fortnight of treatment with a constant daily dose.

L-Tryptophan

L-Tryptophan is an essential amino acid serotonin precursor that is readily absorbed following oral administration and readily crosses the blood–brain barrier, unlike serotonin itself (see Ch. 3). As mentioned in the previous chapter, at the time of writing this drug has been withdrawn from general use, following reports of the occurrence of an eosinophilia–myalgia syndrome associated with its use. In addition to being administered on its own, L-tryptophan has in the past sometimes been used clinically to potentiate the therapeutic effect of tricyclic antidepressants and monoamine oxidase inhibitors. Studies to date have shown that neither the free nor the total plasma concentrations of L-tryptophan predict the clinical response to either L-tryptophan on its own or in combination with tricyclic antidepressants.

Monoamine oxidase inhibitors

The monoamine oxidase inhibitors can be divided into the **hydrazine** type, such as iproniazid, isocarboxazid and phenelzine, and the **non-hydrazine** type, of which tranylcypromine is the only example currently in use in British psychiatry. They are in general lipid-soluble and therefore well absorbed from the gastrointestinal tract and readily cross the blood–brain barrier.

The major metabolic pathway for the hydrazine mono-amine oxidase inhibitors iproniazid and isocarboxazid appears to be acetylation. Thus it is possible that toxic levels may be reached in slow acetylators. The acetylator status would not appear to be of such importance in the case of phenelzine for which oxidation, rather than acetylation, is probably the major degradative route (Robinson et al 1985).

The elimination half-life of the monoamine oxidase inhibitors has been found to be very short; in one study that for tranylcypromine was less than two hours (Simpson et al 1985). However, after stopping therapy with this medication, the inhibition of monoamine oxidase usually continues even when it is not possible to detect any remaining amount of the drug in the body. Therefore the dietary and drug restrictions (see Ch. 5) should be continued for at least two weeks following discontinuation of monoamine oxidase inhibitor therapy.

New monoamine oxidase inhibitors such as moclobe-mide are being introduced that show reversible inhibition of monoamine oxidase.

Phenothiazines

As discussed in greater detail in Chapter 5, the phenothiazines can be divided into three groups on the basis of the nature of a side-chain substitution: those with **aliphatic** side chains such as chlorpromazine; the **piperazines** such as trifluoperazine; and the **piperidines** such as thioridazine. Chlorpromazine is mainly considered in this section as it is the phenothiazine whose pharmacokinetics have been studied in greatest detail.

Chlorpromazine is highly lipid-soluble but when administered orally it is incompletely absorbed, with a bio-availability of approximately 30%. Peak plasma concentrations are reached less than three hours after oral administration, and there is a wide inter-individual variation in plasma concentrations for a given oral dose. There is extensive degradation of the oral drug by not only the liver, but also intestinal mucosal enzymes (a prehepatic first-pass effect). Furthermore, chlorpromazine causes the auto-induction of hepatic and intestinal mucosal enzymes. Moreover, the high lipid-solubility leads to a high volume of distribution which in turns also contributes to a relatively low plasma concentration following oral administration. On the other hand, intramuscular administration bypasses the gastrointestinal and hepatic first-pass effect, leading to increased bio-availability. Indeed, the peak plasma concentration of chlorpromazine achieved following intramuscular injection is three to five times that achieved following oral administration.

There is a high degree of plasma protein binding for chlorpromazine, of 90–95%. It is concentrated in some tissues such as the brain.

In addition to the liver, the biotransformation of chlorpromazine, which has 168 potential metabolites, also takes place in the brain, kidneys, gastrointestinal tract and lungs. The more important routes include: demethylation, forming nor_1-chlorpromazine and nor_2-chlorpromazine; hydroxylation, forming 3-hydroxychlorpromazine and 7-hydroxychlorpromazine; oxidation, forming N-oxides and sulphoxides; and conjugation with glucuronate and sulphate. Many of the metabolites are pharmacologically active; for example 7-hydroxychlorpromazine has a dopamine receptor blocking effect which is approximately the same as that of chlorpromazine itself.

The elimination half-life of chlorpromazine has been found to vary widely between individuals, having values between two hours and one day.

The pharmacokinetics of **thioridazine** have been found to be similar to those of chlorpromazine. Its absorption from the gastrointestinal tract can be modified by its strong anticholinergic action. A diurnal variation in the rate of biotransformation has been described, with a slower rate of excretion occurring during sleep. One of the most important metabolites, mesoridazine, has a dopamine receptor blocking effect which is approximately the same as that of thioridazine itself. Since this metabolite has been found to have a free plasma concentration approximately 50 times that of the parent drug, it has been hypothesized that meso-ridazine is responsible for most of the therapeutic action of thioridazine. However, a study by Smith et al (1984) found no relationship between clinical response in patients with a diagnosis of schizophrenia or schizo-affective disorder and the steady-state plasma or red blood cell concentration of either thioridazine or mesoridazine. Another metabolite which may be of importance is sulphoridazine, which is more antidop-aminergic than either thioridazine or mesoridazine (Kilts et al 1984). The elimination half-life of thioridazine varies between ten and 36 hours, being greater in old age.

Butyrophenones

The butyrophenones include benperidol, droperidol, haloperidol and trifluperidol (see Ch. 5). Of these haloperidol has been studied the most and is therefore considered in this section.

Haloperidol is well absorbed following oral administration, with a bioavailability of approximately 60%. A linear relationship exists between the dose and steady state plasma concentrations. The major route for biotransformation is via hepatic oxidative dealkylation. The metabolites are thought not to be pharmacologically active and enzyme autoinduction has not been demonstrated. As with thioridazine there may be a diurnal rhythm to the rate of biotransformation, it being slowed during sleep. Haloperidol has an elimination half-life of between 12 and 38 hours.

Other neuroleptics

The **diphenylbutylpiperidines**, which include fluspirilene and pimozide, are less sedating than chlorpromazine and have a prolonged action. For example, the elimination half-life of pimozide is approximately 18 hours. Pimozide is administered orally and fluspirilene by intramuscular injection (see Ch. 5).

The **thioxanthenes** include flupenthixol and zuclopenthixol. Flupenthixol, which is less sedating than chlorpromazine and has greater extrapyramidal side effects, is thought to possess some antidepressant activity. The absorption and biotransformation of the thioxanthenes are similar to that of the phenothiazines.

Sulpiride is a substituted benzamide which may have fewer extrapyramidal side effects than chlorpromazine and which may be less likely to cause tardive dyskinesia. It is administered orally and has an elimination half-life of approximately eight hours.

Loxapine is a recently introduced neuroleptic which is well absorbed following oral administration. It undergoes biotransformation mainly by hydroxylation, demethylation, and N-oxidation. Its major route of excretion is via the kidneys.

There are relatively few pharmacokinetic data on the other neuroleptics.

The **depot neuroleptics** are considered in detail in Chapter 5. From a pharmacokinetic viewpoint it is clear that depot preparations deliver a relatively greater concentration of the parent drug to the brain by bypassing the gastrointestinal and hepatic first-pass effect.

Lithium

The lithium salts used in clinical psychiatry are the carbonate and citrate. They are administered orally following which they are rapidly absorbed, with the peak plasma concentration being reached between three and five hours later. It should be noted that the bio-availability of different preparations vary widely. Controlled release preparations are also available.

There is no plasma or tissue protein binding. The total body lithium can be predicted from the plasma concentration (Bergner 1977). It is feasible to use salivary levels as an alternative to blood analysis to monitor plasma concentrations. Details of plasma lithium levels and patient monitoring are considered in Chapter 5. Being an element, lithium has no metabolic products.

The mean elimination half-life in adults is approximately 13 hours, and has been found to increase with age. Over 95% of lithium is excreted by the kidneys. Normally the renal clearance does not depend on the plasma concentration (Thomsen & Schou 1968). Thus, normally the plasma concentration is approximately directly proportional to the maintenance dose.

Lithium ions readily pass into the glomerular filtrate with 70–80% being re-absorbed in the proximal tubules. None is re-absorbed in the distal tubules, and the re-absorption is not affected by diuretics that act only on the distal tubules. Lithium and sodium cations compete for proximal tubular re-absorption. Therefore lithium retention can be caused by sodium deficiency and by sodium diuresis. Further details of the causes, effects and treatment of lithium toxicity appear in Chapter 5.

Anticonvulsants

The pharmacokinetics of the more commonly used anticonvulsants are discussed in this section. Although it also has a therapeutic role in bipolar mood disorder, as discussed in Chapter 5, carbamazepine is conveniently considered in this section with other anticonvulsants.

The absorption of **carbamazepine** following oral administration is generally slow but overall it is good. There is some evidence that the drug may affect gastric motility. At therapeutic concentrations carbamazepine is 75% plasma protein-bound. In principle it is possible to monitor the free plasma concentration using salivary concentrations in place of taking blood samples. Hepatic biotransformation is important with autoinduction of enzymes probably taking place. One clinically important product which is itself antiepileptic is carbamazepine-10,11-epoxide. Following a single dose the elimination half-life is approximately 25–60 hours, while after chronic administration it decreases to 10 hours. This is probably caused by enzyme autoinduction. Therefore this drug needs to be taken approximately 12 hourly.

Ethosuximide is absorbed well following oral administration and has a very low level of plasma protein binding. Two major pharmacologically inactive metabolites are formed by hepatic biotransformation.

In adults the elimination half-life is approximately 70 hours; it is approximately 30 hours in children. Ethosuximide can be administered once daily.

Following oral or intramuscular administration **phenobarbitone** is absorbed well but the peak plasma concentration may occur only after six hours. At therapeutic concentrations it is 50% plasma protein bound. Biotransformation results in about half of the parent drug being changed to pharmacological inactive products, mainly by hepatic parahydroxylation. The amount of phenobarbitone excreted unchanged in the urine varies from approximately 10–40 %. It is an acidic drug (pK$_a$ 7.4) and the urinary excretion depends on the pH of the urine, as well as its volume. Forced alkaline diuresis is used in cases of poisoning. The approximate elimination half-life in adults is 100 hours and 40 hours in children, and the drug can be administered once daily.

The absorption of **phenytoin** following oral administration is generally good but there is a wide inter-individual variation. When calcium is administered with it the bio-availability of the phenytoin is reduced.

Following intramuscular administration phenytoin tends to crystallize out at the injection site and the rate of absorption and peak plasma concentration are lower than following oral administration. At therapeutic concentrations phenytoin is 90% plasma protein-bound. It is displaced from plasma protein binding by a number of other drugs. Biotransformation takes place mainly by hepatic parahydroxylation, with the main metabolite being 5-parahydroxyphenyl-5-phenyl-hydantoin. Saturation of this enzymatic reaction takes place at phenytoin concentrations that lie within the therapeutic range. Less than 5% of the parent drug is excreted unchanged. The elimination half-life is generally between 12 and 120 hours, but it is not possible to calculate a single half-life owing to the occurrence of zero-order kinetics (see above).

Sodium valproate is well absorbed following oral administration and is 90% plasma protein-bound. The elimination half-life is between seven and ten hours. Because there is a wide inter-individual variation in the free plasma concentrations it is useful to monitor these directly.

FURTHER READING

Goodman-Gilman A, Rall T W, Nies A S, Taylor P (ed) 1990 Goodman and Gilman's: The pharmacological basis of therapeutics. 8th edn. Pergamon, New York

Meltzer H Y (ed) 1987 Psychopharmacology: the third generation of progress. Raven Press, New York

Rang H P, Dale M M 1991 Pharmacology. 2nd edn. Churchill Livingstone, Edinburgh

Silverstone T, Turner P 1988 Drug treatment in psychiatry. 4th edn. Routledge, London

Spiegel R 1989 Psychopharmacology: an introduction. 2nd edn. Wiley, Chichester

Tyrer P J (ed) 1982 Drugs in psychiatric practice. Butterworths, London

5. Clinical Psychopharmacology

In this chapter the main drugs used in psychiatry will be described. Because some of the principles underlying their use have been described in the previous two chapters this will not be repeated again apart from relevant accounts of their mechanisms of action. In each section the main benefits and risks of the drugs will be described as it is only by understanding and judging these that the proper use of drugs is achieved.

TERMINOLOGY

The terminology of drugs used in psychiatry is constantly changing and varies between descriptions based on the main use of drugs (e.g. anti-depressants), simple chemical descriptions (e.g. lithium salts) and description by main pharmacological action (mono-amine oxidase inhibitors). Because these can change considerably it is wise to be aware of all three types of description (Table 5.1).

ANTIPSYCHOTIC DRUGS

The main use of antipsychotic drugs is in the treatment of schizophrenia, and for this reason they are sometimes known as antischizophrenic drugs. As they are also used in the treatment of the acute symptoms of mania, in organic psychoses (including those induced by drugs) it is more appropriate to describe them as antipsychotic drugs. In the treatment of these disorders these drugs are given in relatively high dosage but many are also used in lower dosage for their mild sedative actions. These drugs are also described as 'major tranquillizers', but this description is to some extent undermined as the same drugs in lower dosage can equally well be described as 'minor tranquillizers'.

In high dosage antipsychotics act by post-synaptic blockade of dopamine receptors in the central nervous system. Their antipsychotic effects are shown primarily by dopamine blockade in the mesolimbic system, but

Table 5.1 Terminology of psychotropic drugs

Clinical description	Chemical names	Pharmacological type
Antipsychotics	Phenothiazines Butyrophenones Thioxanthenes Diphenylbutyl-piperidines Substituted benzamides	Neuroleptics and dopamine-blocking drugs
Drugs for movement disorders	Levodopa Orphenedrine Procyclidine Amantadine Selegiline drugs Bromocriptine Tetrabenazine Lysuride	Dopomine precursors Anticholinergic drugs Dopamine agonists Dopamine-depleting (tetrabenazine)
Anti-anxiety drugs (including sedatives and hypnotics)	Benzodiazepines Cyclopyrrolones Barbiturates Propranolol Chlormethiazole Buspirone	GABA-facilitating drugs Beta-blocking drugs 5-HT partial agonists
Anti-depressants	Tricyclic anti-depressants Tetracyclic anti-depressants Monoamine oxidase inhibitors Oxazines New anti-depressants	Noradrenaline re-uptake inhibitors 5-HT re-uptake inhibitors Monoamine oxidase inhibitors
Drugs for bipolar affective psychosis	Lithium salts Carbamazepine	
Psycho-stimulants	Amphetamines Pemoline	Dopamine potentiating drugs
Gonadal hormones	Oestrogens Progestogens Androgens Antiandrogens	Oestrogens Progestogens Androgens Anti-androgens

Table 5.2 Comparison of antipsychotic and tranquillizing dosages of the main antipsychotic drugs

Main drug class	Drug name	Dosage	
		Antipsychotic (High dose mg day^{-1})	Tranquillizing (Low dose mg day^{-1})
Phenothiazines	Chlorpromazine	> 200	< 50
	Thioridazine	> 150	< 50
	Perphenazine	> 8	< 2
	Oxypertine	> 200	< 80
	Prochlorperazine	> 75	< 25
Butyrophenones	Haloperidol	> 8	< 2
	Droperidol	> 30	< 10
Thioxanthenes	Flupenthixol	> 6	< 2
	Zuclopenthixol	> 20	< 6
Diphenylbutylpiperidine	Pimozide	> 6	< 2
Substituted bezamide	Sulpiride	> 200	Not used
Dibenzoxazepine	Loxapine	> 20	Not used
Dibenzodiazepine	Clozapine	> 150	Not used

similar dopamine blockade in the nigro-striatal system can lead to parkinsonism and other drug-induced movement disorders, and in the tubero-infundibular system to hormonal changes. In lower dosage these drugs do not block dopamine receptors but exert their sedative actions through a combination of antimuscarinic (anticholinergic) and antihistaminic effects. It is therefore important for the practitioner to realize at what dose-levels dopamine blockade is likely to occur, and a list of the most commonly used drugs and their dosage ranges (Table 5.2) summarizes this. However, it is important not to be too slavish in following this table as there are considerable inter-individual differences in the pharmacokinetics and metabolism of these drugs. In all instances the antipsychotic dose blocks dopamine receptors and the tranquillizing dose does not.

Antipsychotic dosage

All drugs used as antipsychotics block post-synaptic dopamine receptors (Creese et al 1976) and their effects cannot be explained by antimuscarinic, anti-histaminic and anti-adrenergic effects (Johnstone et al 1978). Although the sedative effects are immediate, the antipsychotic ones take up to three weeks to become evident in the treatment of schizophrenia, but only after a few days in the treatment of mania. Symptomatic improvement in organic psychoses also takes place more quickly. This more rapid improvement in these conditions may be related to simple sedation being of greater therapeutic value than in schizophrenia. The sedative effect can be made more rapid by intravenous of intramuscular use; this is made particular use of with droperidol, which, when given between 5 and 15 mg intravenously up to six times daily, is effective in controlling disturbed behaviour rapidly.

Because all these drugs block dopamine receptors they are liable to produce movement disorders. There continues to be debate over the prophylactic use of anticholinergic (antimuscarinic) drugs when antipsychotic drugs are being used. Approximately 60% of patients have no parkinsonian or dyskinetic symptoms after taking antipsychotic drugs (Mandel & Oliver 1961) and so prophylactic treatment is unnecessary in this proportion. Movement disorders are more common when higher dosage is being used and when drugs known to have greater propensity for producing extrapyramidal effects are chosen (e.g., perphenazine, fluphenazine, trifluoperazine, prochlorperazine and haloperidol). The piperidine phenothiazines (e.g., thioridazine, pericyazine), loxapine and clozapine are much less likely to produce extrapyramidal unwanted effects. In deciding whether or not to use anti-parkinsonian drugs prophylactically the previous response of patients when exposed to antipsychotic drugs can often be more valuable than any other information. Movement disorders are most likely to occur within 48 hours of starting therapy or within three days of receiving a depot injection of an antipsychotic drug.

Low-dose antipsychotic drug treatment

In low dosage these drugs have long been used for the treatment of mild agitation and anxiety and also as hypnotics. Although they are far from ideal for this purpose they do have the one major advantage of not producing pharmacological dependence and for this reason can be chosen selectively in patients who are clinically prone to addictive behaviour. The sedative effect can be immediate but is sometimes delayed for up to three days. Some antipsychotic drugs have energizing as well as sedative effects. One of the best known of these is flupenthixol which has been shown to have both antidepressant and antianxiety effects in low dosage (Young et al 1976).

The antipsychotic drugs are divided into several chemical groups which are all of roughly similar efficacy in the treatment of psychotic disorders. However, there are some differences between them, mainly with regard to unwanted effects, and so their benefits, indications and risks are best discussed separately.

Benefits

There is little doubt that the antipsychotic drugs have been a major advance in the treatment of psychotic disorders despite their considerable handicaps. Just over 10% of patients fail to make any clinical response to antipsychotic drugs in any form and unfortunately these cannot be identified in advance. The new dibenzodiazepine compound, clozapine, has been found to be effective in resistant schizophrenia and may have a special place in this condition. Another 25% of patients receive some benefit from antipsychotic drug treatment but still have persistent symptoms and handicaps no matter what dose of the drug is chosen.

Although there are differences in response to different drugs there is little evidence of consistent superiority of one compound over another. It is usually wise to continue treatment with one drug for several weeks before deciding to abandon it in favour of another. In studying response to drugs it is important not to be too rigid. In addition to great variations in the individual pharmacokinetics there are similar individual differences in pharmacodynamics. This means that an individual may respond badly to one antipsychotic drug but improve dramatically when treated with another that appears to be essentially similar.

Because antipsychotic drugs only suppress symptoms they are required to be given for the natural history of the disorder. Because most of them are used for the treatment of schizophrenia, which is frequently a chronic disease, it is usually necessary to continue treatment in

maintenance dosage subsequently, certainly for months and sometimes for years. Maintenance dosage is about 60% of that required to resolve symptoms in the initial phase. Irrespective of duration of maintenance treatment there is clear evidence from double-blind trials that patients who change from active drug to placebo relapse significantly more frequently than those remaining on medication (Hirsch et al 1973, Hogarty et al. 1974) and the same applies when patients are cut to dosages below levels sufficient to produce dopamine receptor blockade, sometimes called micro-dose maintenance treatment (Kane et al 1983).

Because maintenance therapy often needs to be long-term, depot injections are often used for new patients who are not compliant with oral therapy. The only real purpose of giving depot injections is to ensure compliance and, although patients with schizophrenia are often unreliable in taking medication regularly, it should not be assumed that all patients requiring maintenance treatment should preferably be taking injectable drugs. Depot injections consist of the drug in a form that allows for slow release. The most common form of depot preparation is the antipsychotic drug prepared as an ester, most frequently as a decanoate (flupenthixol, fluphenazine, zuclopenthixol, haloperidol) but the depot phenothiazine, pipothiazine, is given as the palmitate salt, and fluspirilene is given in aqueous injection at weekly intervals.

All other depot preparations are given at intervals between two and eight weeks, and more frequent administration cannot be pharmacologically justified despite its frequent use in some hospitals. Dosage should not be rigid, and the best type of prescription involves giving the drug within a flexible dosage range (Johnson 1975), reducing at settled periods in the patient's life and increasing again at times of stress, particularly when high 'expressed emotion' is present. This reduces the total drug dosage given and so reduces the incidence of tardive dyskinesia.

Risks

The main risks of antipsychotic drugs are summarized in Table 5.3. Most of these follow directly from post-synaptic dopamine blockade in the nigrostriatal and tubero-infundibular systems, and the corpus striatum, or from the known antimuscarinic effects of the drugs. The neuroleptic malignant syndrome is the most serious of the unwanted effects of antipsychotic drugs. It describes a toxic delirious state with rigidity and apparent catatonia associated with fever, which can progress to coma and death. Its exact cause remains to be determined; some claim it is more common when antipsychotic drugs are combined with antiparkinsonian

Table 5.3 Major unwanted effects of antipsychotic drugs

Unwanted Effects	Main features	Likely cause
Acute dystonias	Muscular spasms in back (opisthotonos), neck (torticollis), jaws (trismus), eyes (oculogyric crisis), and eyelids (blepharospasm)	Dopamine blockade in nigro-striatal system
Parkinsonian symptoms (also called pseudo-parkinsonism)	Rigidity, coarse tremor, postural flexion, festinating gait, mask-like face and hypersalivation	ditto
	Akathisia (motor restlessness)	Dopamine blockade in corpus striatum and mesocortical system
Tardive dyskinesia	Syndrome of choreiform or athetoid movements	Dopamine receptor supersensitivity
Anticholinergic effects	Dry mouth, difficulty in micturition, stuffy nose, constipation and blurring of vision	Anti-muscarinic effects
Hormonal effects	Impotence, galactorrhoea, gynaecomastia and amenorrhoea	Hyper-prolactinaemia following dopamine receptor blockade
Cardiac effects	Tachycardia, postural hypotension and ECG irregularities	Antiadrenergic effects
Neuroleptic malignant syndrome	Fever, excitement extrapyramidal symptoms (may be fatal)	Direct toxic effect
Other effects	Weight gain, hypothermia, blood dyscrasias, obstructive jaundice, skin rashes and epileptic seizures	

ones (Westlake & Rastegar 1973). Tardive dyskinesia and the other dyskinesias are discussed in more detail below.

Phenothiazines

The phenothiazines are the oldest antipsychotic drugs; and the introduction of the first of them, chlorpromazine, heralded the beginning of what has been called the 'psychopharmacological revolution' in 1950 when it was first introduced for the treatment of schizophrenia and other psychotic disorders. There are many phenothiazines available and these are separated into three main groups on the basis of their important subsidiary effects (Table 5.4). The main members of the **aliphatic** group are chlorpromazine itself, methotrimeprazine and promazine. The **piperidines** include thioridazine, pericyazine and the depot preparation, pipothiazine palmitate; and the **piperazines** include perphenazine, fluphenazine, trifluoperazine and prochlorperazine. The aliphatic phenothiazines are used mainly when sedation is required as well as antipsychotic effects. Chlorpromazine is the most commonly used and has had much the longest period of use. Obstructive jaundice and skin rashes are sometimes noted as idiosyncratic effects, and blood dyscrasias such as aplastic anaemia have been reported. Because these sedative effects are found in low-dosage, chlorpromazine is popular as an anti-anxiety drug treatment. This group has pronounced antihistaminic effects and can be used to control nausea and vomiting. Promazine has been particularly used for the control of agitation and restlessness in the elderly.

The **piperidines** are usually chosen when there is concern about extrapyramidal side-effects. If such side-effects are common in acute treatment it may be appropriate to choose the depot preparation, pipothiazine palmitate, for maintenance therapy. Thioridazine is the most commonly used but care has to be taken in long-term use. A daily dosage of 500 mg or more for longer than six months is liable to lead to retinitis pigmentosa and blindness.

The **piperazines** are sometimes described as energizing phenothiazines but this is a misnomer as they merely have fewer sedative effects. Their main advantage is in producing fewer autonomic effects and this may be of importance in patients who are concerned by

Table 5.4 Classification of phenothiazines and their subsidiary effects[1]

Type	Subsidiary effects		
	Sedation	Extrapyramidal	Autonomic
Aliphatic compounds	+++	++	++
Piperidines	++	+	+++
Piperazines	+	+++	+

[1]There is no evidence that these groups differ in antipsychotic efficacy.

symptoms such as increased heart rate and postural hypotension.

Butyrophenones, thioxanthenes and diphenylbutylpiperidines

These three groups are similar to piperazine phenothiazines in having relatively few sedative and autonomic effects but pronounced extrapyramidal ones. This group includes the depot preparations flupenthixol, haloperidol, fluspirilene and droperidol, all except droperidol and fluspirilene being given as the decanoate ester. Droperidol is used mainly as acute sedation for psychotic patients, where it is often given by intramuscular or intravenous injection. It produces fewer autonomic effects than chlorpromazine and has a rapid onset of action. It is sometimes used for the acute sedation of children.

Haloperidol is also given frequently in oral form (both tablets and liquid) and by intramuscular injection. It is frequently used in the treatment of mania. In lower doses it has been found to be effective in attenuating the rapid mood and behaviour disturbance of impulsive and borderline personality disorders and for relieving anxiety (Soloff et al 1986). The drug oxypertine is primarily used in the treatment of anxiety and has energizing properties as well. It is related to the phenothiazines and can be used in chronic schizophrenia with many negative symptoms. Fluspirilene, which is given weekly by intramuscular injection (2–8 mg) is reserved for patients who require closer monitoring or who cannot tolerate the oily esters that constitute most of the other depot preparations. Triperidol is a butyrophenone that is sometimes used as an alternative to haloperidol in tablet form.

Substituted benzamides

These compounds, of which sulpiride is far the best known, specifically block D_2 dopamine receptors, whereas other antipsychotic drugs tend to block both D_1 and D_2 receptors. There is some evidence that this selectivity may be associated with reduced risk of developing tardive dyskinesia and possibly also of other unwanted effects. However, this is difficult to establish clearly and at present the evidence does not allow definite conclusions to be drawn.

Loxapine and clozapine

Loxapine is a dibenzoxazepine which is similar to other classes of antipsychotic drugs but which may have a lower incidence of extrapyramidal effects in a clinically recommended dose of 60–100 mg daily. Clozapine is different from other antipsychotic drugs in blocking 5-HT, alpha-adrenergic and histamine receptors with much less dopamine receptor-blocking activity. It is certainly an effective antipsychotic drug and may be of particular interest because it is effective in patients with schizophrenia who are resistant to other compounds. Unfortunately it carries the major risk of blood dyscrasias, and for this reason all patients being treated with the drug have to have their white cell counts monitored weekly for the first 18 weeks of treatment and every 2 weeks subsequently. It is recommended that the drug is withdrawn if the white count falls below 3000 mm^{-3}. Because of these risks it is suggested that the drug is given initially in a dose of 25–50 mg daily before reaching full therapeutic doses of 200–450 mg daily.

Tardive dyskinesia

Although this syndrome is thought to be a consequence of the long-term use of antipsychotic drugs, most likely caused by dopamine receptor supersensitivity (Clow et al 1978), this theory does not fit all the facts. In particular there is some evidence that chronic antidepressant drug treatment may precipitate the syndrome also. Tardive dyskinesia is characterized by rotatory movements of the tongue, lips and facial muscles (originally described as the bucco-linguo-masticatory syndrome) (Brandon et al 1971), but also including similar athetoid movements of the hands and feet and sometimes of the trunk.

Support for the view that supersensitivity of dopamine receptors could be at least partly responsible for the syndrome comes from evidence that increasing the dose of the original dopamine-blocking drug reduces its features. As this increases receptor blockade it will also block at least some of the (presumed supersensitive) receptors. Treatment with anti-parkinsonian drugs also tends to aggravate tardive dyskinesia which is in keeping with the supersensitivity hypothesis.

There is good evidence for this theory from animal studies (Clow et al 1978) but unfortunately this knowledge has not helped treatment. Treatment by dopamine depletion with tetrabenazine and reserpine, cholinergic drugs such as physotigmine and choline, and GABA-facilitating drugs such as sodium valproate and benzodiazepines have all been tried without convincing evidence of efficacy (Mackay & Sheppard 1979). L-Dopa, lecithin, baclofen, verapamil and vitamin E have also been tried without obvious success. In about a third of patients the syndrome has a limited time-span and improves spontaneously (which again supports the dopamine supersensitivity theory). Unfortunately, however, in most patients there is no improvement after stopping the antipsychotic drug, and in some cases the syndrome is made worse. It is possible for the movement disorders to be either independent of, or associated with, the schizophrenic process so that it is sometimes necessary to continue the antipsychotic drug even in the presence of tardive dyskinesia to avoid relapse of schizophrenic symptoms. A satisfactory treatment of tardive dyskinesia remains one of the main challenges of psychopharmacology.

Among these conditions should also be included other late-developing movement disorders, including tardive dystonia (Burke & Kang 1988) and tardive akathisia (Barnes & Braude 1985) which are similar to the acute equivalents but become manifest during chronic therapy. The dystonias include spasm of the larynx, torticollis and blepharospasm; a form of axial dystonia causing the trunk to list to one side is predictably called the Pisa syndrome (Yassa 1985).

Drugs for movement disorders

The main use of these drugs in psychiatry is in the treatment of drug-induced parkinsonism following use of antipsychotic drugs. The main compounds used are procyclidine (also available by intramuscular or intravenous injection), orphenadrine, benztropine and benzhexol (the oldest of the group). All these are anticholinergic drugs which improve tremor and stiffness but can produce antimuscarinic effects of their own in higher dosage. As discussed earlier they are usually given at the time of onset of dystonia or parkinsonian symptoms although they can also be used prophylactically. Levodopa (L-Dopa) is given for the primary treatment of parkinsonism as it is converted into dopamine in the brain. Because it undergoes decarboxylation to dopamine in the systemic circulation it is normally given with a peripheral decarboxylase inhibitor such as benserazide or carbidopa.

L-Dopa is not recommended for the treatment of parkinsonian symptoms and has many side-effects, including nausea, vomiting, hypotension and induction of psychoses. It is used for the treatment of primary parkinsonism, sometimes in combination with other drugs including amantadine, selegiline (a selective inhibitor of monoamine oxidase-B (MAO-B), bromocriptine, or lysuride. Tetrabenazine, in addition to being sometimes for the treatment of tardive dyskinesia, is used in the treatment of the athetoid movements of Huntington's chorea.

The use of intramuscular or intravenous anticholinergic drugs such as procyclidine and benztropine is necessary in the treatment of acute dystonic reactions such as torticollis, opisthotonos, trismus and oculogyric crises. These dystonic reactions are sudden and are sometimes misdiagnosed as 'hysterical' reactions. Patients are often very frightened during them and there is sometimes a danger of serious consequences, as for example when the tongue becomes fixed in the back of the throat and obstructs the airway. The intravenous administration of an anti-parkinsonian drug (e.g., procyclidine 10 mg) normally relieves the symptoms within a few minutes.

There is no satisfactory drug treatment for akathisia, the syndrome of motor restlessness, independent of anxiety, that is thought to be caused by blockade of dopamine receptors in the mesocortical system (Marsden & Jenner, 1980).

ANTI-ANXIETY DRUGS

The main drugs used for treating anxiety and inducing sleep are shown in Table 5.5. Although many of these are marketed specifically for treating anxiety or insomnia their similarities are too great for useful separation to be made for this clinical purpose. Put simply, an anti-anxiety drug in somewhat higher dosage is an effective hypnotic, and a hypnotic drug in somewhat lower dosage is effective in anxiety. Whereas in the past only the drugs that had sedative or hypnotic effects were considered in this context it has been appreciated in recent years that many other drugs, not specifically marketed for treating anxiety, are also extremely effective and in some cases supplant the use of conventional anti-anxiety drugs.

Mechanism of action

The likely mechanisms of action of the main anti-anxiety drugs are listed in Table 5.5. Most of the common drugs for treating anxiety have immediate sedative actions. Although many of these act on different receptors in the central nervous system, the final common pathway to sedation is through facilitation of γ-amino-butyric acid (GABA) transmission. GABA is the most widespread inhibitory neurotransmitter in the central nervous system and, when stimulated, most activity is reduced. Benzodiazepines constitute the major member of this drug group. These act directly by binding with benzodiazepine receptors — which are themselves linked to GABA receptors in an ionophore or complex involving GABA and benzodiazepine receptors and a chloride channel. Barbiturates, propanediols and chloral hydrate act directly on GABA receptors, and the cyclopyrrolones act on a

Table 5.5 Classification of anti-anxiety drugs

Type	Main examples of individual drugs	Main mechanism of action
Sedative/hypnotic benzodiazepines	Diazepam Temazepam Triazolam Lorazepam	
barbiturates	Amylobarbitone sodium	GABA-facilitation
cyclopyrrolones	Zopiclone	
propanediols	Meprobamate Chlormezanone Chlormethiazole Chloral hydrate	
Azospirodecanediones	Buspirone Gepirone	5-HT$_{1A}$ partial agonist
Beta-blocking drugs	Propranolol Atenolol Oxprenolol	Peripheral beta-blockade
Antihistamines	Promethazine Cyclizine Chlorpheniramine	Histamine-receptor blockade
Anti-depressant drugs	See next section	See next section
Antipsychotic drugs	See previous section	See previous section

different site close to benzodiazepine receptors in the GABA–benzodiazepine complex.

The azospirodecanediones have primary action on 5-hydroxytryptamine (5-HT$_{1A}$) receptors. This has led to a great deal of interest in the possible importance of 5-HT receptors in both anxiety and depression, and there are now several models that regard pathological anxiety and depression as associated with an excess and deficiency of 5-HT activity respectively (e.g., Wise et al 1972, Eison 1990).

There are also many other potential anti-anxiety compounds that appear to act primarily on 5-HT receptors. These include ritanserin, which blocks 5-HT$_2$ receptors, and odansetron, which antagonizes 5-HT$_3$ receptors, which in both animal and human studies have been shown to have anti-anxiety activity.

Beta-blocking drugs are thought to act primarily by peripheral blockade of beta-adrenergic receptors in cardiac and skeletal muscle in particular. This evidence is based on the following experimental findings: intra-arterial propranolol injections can reduce tremor, so there is no possible central mediation (Marsden et al 1967); beta-blocking drugs that do not cross the blood–brain barrier are just as effective in

reducing anxiety as are compounds that do cross the barrier (Bonn et al 1972); and propranolol, in the dosages used to treat anxiety, has little or no effect on other central nervous system functions, including the electroencephalogram (Tyrer & Lader 1974). Although peripheral blockade therefore appears to be the main site of action of these drugs the possibility of central blockade having some relevance cannot be discounted entirely.

Antihistamines

Antihistamines have long been known to have sedative effects and this follows from their blockade of central histamine (H$_1$) receptors. Antipsychotic drugs, as has been noted in the previous section, also possess antihistaminic properties and this contributes a significant part of the sedative action of these compounds also.

Other drugs

Although antidepressants are not discussed in any detail in this section it is important to be aware that both the tricyclic and newer antidepressants and the monoamine oxidase inhibitors have significant anti-anxiety effects as well as antidepressant ones. These anti-anxiety effects are independent of the sedative anticholinergic properties of these drugs as they occur simultaneously with the antidepressant effects between three and four weeks of treatment, whereas the anticholinergic effects are immediate. A distinction should therefore be made between the immediate sedative calming effects of antidepressants and the energizing anti-anxiety effects that take place after several weeks. The best documentation of the anti-anxiety effects of antidepressants has been shown with the relatively new diagnosis of panic disorder.

Benzodiazepines

These drugs remain the most commonly prescribed psychotropic drugs although their use has declined steadily over the past ten years because of concerns about dependence. Most of the benzodiazepines in general use are agonists but there are many other drugs in the series, including antagonists and inverse agonists that can have anxiogenic properties. One of these drugs, flumazenil, a competitive antagonist, is used in anaesthesia to reverse the anaesthetic effects of benzodiazepines.

Although the main use of benzodiazepines is for the treatment of anxiety and sleep disturbance these drugs also have anti-aggressive, muscle relaxant and anticonvulsant properties. They are also used for premedication in anaesthesia and midazolam, available for instramuscular and intravenous use, is mainly used for this purpose. All the benzodiazepines have a rapid onset of action, with effects beginning within an hour of administration by mouth and within seconds of intravenous use.

Benefits

In short-term use benzodiazepines are the most effective drugs for the treatment of anxiety and are also superior to psychological treatments (Quality Assurance Project 1985). After two or three weeks of regular treatment, however, both psychological and other drug treatments, particularly anti-depressants, become more effective (e.g., Kahn et al 1986; Tyrer et al 1988). This loss of superiority is not only because other forms of treatment have a delayed onset of action; it is also because tolerance develops to all the effects of benzodiazepines at a variable rate after repeated treatment (File 1985). Benzodiazepines are therefore best used for the emergency treatment of anxiety and insomnia rather than planned long-term use. Despite their speed of action, however, they are still not sufficiently rapid in their onset of clinical effects to treat attacks of panic which reach a peak within a few minutes of symptoms first becoming manifest. Some newer compounds (e.g., Ro 16–6028) (Katschnig et al 1988) have been tested and may offer a solution to this problem as they act rapidly.

Benzodiazepines are also effective in the treatment of some forms of epilepsy, particularly involving the temporal lobe, and two compounds, clobazam and clonazepam, are used primarily for this purpose. The anti-aggressive properties of the drug are useful in the emergency treatment of disturbed behaviour, when lorazepam given by intramuscular injection, or diazepam given intravenously, are effective rapidly.

Risks

In acute dosage there are few risks with benzodiazepines. The group of drugs as a whole is remarkably safe, although there can be additive effects with alcohol and other depressant drugs, particularly other drugs of the sedative–hypnotic type. A major indication of excessive dosage is drowsiness. Anterograde amnesia (i.e., amnesia from the time of onset of drug action) may also be a problem in higher dosage, although it may be a beneficial effect in some conditions (e.g., premedication for dental phobia).

All benzodiazepines have the propensity for psychomotor impairment. This affects psychological and motor functioning so tasks requiring coordination and vigilance (e.g., driving, monitoring machinery) may be impaired by treatment. Because anxiety fluctuates considerably from hour to hour, unwanted effects such as psychomotor impairment may be noted after acute anxiety has passed but the effects of the drug persist. This is more likely with long acting benzodiazepines (see below) and are most marked in the demonstration of 'hangover' effects of hypnotic benzodiazepines. One benzodiazepine, clobazam, may lead to less psychomotor impairment than others (Hindmarch 1979).

The risks with benzodiazepines become much more prominent once regular prescription is continued for 4 weeks or longer. This is mainly because of the risk of dependence on the drug. Dependence on benzodiazepines is shown mainly by the exhibition of a withdrawal syndrome after reduction or discontinuation of the drug (Tyrer 1984a). The other major features associated with dependence are: craving, drug-seeking behaviour, escalation of dosage and marked tolerance, are not nearly so marked, although it is now becoming appreciated that tolerance is a much greater problem than was once thought. There is also some evidence that the more potent benzodiazepines, particularly those with a short elimination half-life such as triazolam and lorazepam, may carry greater risks of dependence than other benzodiazepines (Tyrer & Murphy 1987).

The withdrawal syndrome is characterized by symptoms that can be grouped into three categories: (a) those that are typical of anxiety such as palpitations, trembling, panic, dizziness, nausea and other bodily symptoms, together with depressed mood; (b) symptoms of perceptual disturbance, including depersonalization and derealization, hypersensitivity to all sensory modalities, particularly to auditory stimuli, and distorted perception of height and space, tinnitus, formication (itching sensations in the skin like ants crawling over it), a peculiar taste in the mouth, and influenza-like symptoms; (c) epileptic seizures, confusional states or paranoid psychotic episodes. The third of these is unequivocal evidence of a withdrawal syndrome, the second is likely evidence when two or symptoms are present, and the first is not in itself indicative of a withdrawal reaction. However, if any of the symptoms show a clear onset of effect after withdrawal which then reaches a peak and then slowly resolves it is likely that such symptoms are those of a withdrawal reaction.

Dependence can begin after as little as four weeks in regular dosage and there have been claims for withdrawal symptomatology following a single dose of the short-acting drug, triazolam (Morgan & Oswald 1982), in which symptoms the day following a nightly hypnotic dose can be regarded as those of withdrawal. This is not generally agreed but after stopping regular treatment after four to six weeks there is unequivocal symptomatic change indicative of a withdrawal reaction (Fontaine et al 1984, Power et al 1985). Symptoms of withdrawal normally begin within 24 hours of stopping short-acting benzodiazepines and up to six days after stopping a long-acting one. The term 'rebound insomnia' and 'rebound anxiety' are sometimes used in the context of withdrawal reactions (e.g., Kales et al 1978). However, there is no fundamental qualitative difference between the symptoms of rebound and those of withdrawal, it is just more common to use the term withdrawal when drugs have been taken for a longer period when the symptoms are greater after stopping the drug. The syndrome lasts for up to five weeks after stopping the drug and there is some evidence for a 'post-withdrawal syndrome', with symptoms persisting for up to two years after all traces of the drug have been eliminated from the body (Tyrer 1991).

Clinical use of benzodiazepines

Because the efficacy of benzodiazepines decreases after regular use and is also accompanied by the risk of dependence it is normally recommended that these drugs are given for up to four weeks but no longer in regular dosage (Priest & Montgomery 1988). There is also some concern about the possibility that the amnesiac effects of benzodiazepines may promote the dynamic mechanism of denial and repression after an emotionally significant stress (e.g., bereavement of a close relative). For this reason some clinicians are wary of prescribing benzodiazepines for treatment of stressful events that require personal adjustment but hard data for this view are lacking.

The major uses for benzodiazepines are in the treatment of epilepsy (where the drugs may be considered for long-term treatment), the detoxification of alcohol dependence (where the anti-convulsant effects of benzodiazepines may be particularly valuable), and for the short-term treatment of anxiety and insomnia. In most cases it is wise for the clinician to predict the duration of treatment before the first prescription is given in order to avoid the possibility of long-term dependence.

However, there continue to be clinicians who can argue persuasively for long-term benzodiazepine use, particularly in elderly patients who tend to become confused when benzodiazepines are withdrawn and do not require the same levels of vigilance and fine coordination that many younger people do. There are also those who are concerned that withdrawal of benzodiazepines may lead to more serious addictions, particularly alcohol dependence, and therefore justify long-term prescription of benzodiazepines as the lesser of two evils. Again, however, it is hard to find data that give adequate support for this view.

Use of non-benzodiazepine sedatives

Because all these drugs act primarily by facilitating GABA transmission they have the same risks as benzodiazepines as well as some of their advantages. All carry the major risk of pharmacological dependence and it is likely that they will not differ significantly from the benzodiazepines in this respect. These risks have been established for the propanediols such as meprobamate and for chlormethiazole (Tyrer 1982a) but not for chlormezanone, which, despite its apparent efficacy, has not been popular as an anxiolytic. The cyclopyrrolones such as zopiclone (used as an hypnotic) and suriclone (used for anxiety) have shown fewer sedative effects of these drugs with insufficient evidence to suggest that they are less prone to dependence (Goa & Heel 1986).

Azaspirodecanediones

These comprise several members, of which the best known is buspirone, which have anti-anxiety (and possibly antidepressant) properties. They differ from the sedative hypnotics in having no muscle relaxant or anticonvulsant effects, no obvious risk of dependence, and a delayed onset of clinical action (Lader 1988).

For reasons that are not fully understood, patients who have previously taken benzodiazepines tend to do badly on buspirone. This may be because buspirone differs from benzodiazepines in producing a dysphoric rather than a euphoric effect after initial administration though there may be other factors also, including the possibility that benzodiazepines have a long-term influence on benzodiazepine receptors that continues long after the drug has been eliminated from the body (Tyrer 1991). It is therefore not surprising that buspirone tends to be ineffective for treating benzodiazepine withdrawal symptoms (Ashton et al 1990).

Ipsapirone and gepirone are azaspirodecanediones which are likely to be marketed in the near future for the treatment of anxiety or depressive disorders. The main use of these compounds is in the treatment of anxiety in patients who are at risk of dependence (Table 5.6). Patients dependent on alcohol are a prime example.

Beta-blocking drugs, antihistamines and other anti-anxiety drugs

Beta-blocking drugs have been in use for the treatment of some forms of anxiety for the past 25 years. Since the first study demonstrating their possible value (Granville-Grossman & Turner 1966) it has been established that these drugs have a relatively small place in the treatment of anxiety in clinical psychiatric practice even though they are demonstrably more effective than placebo (Tyrer 1988). It is likely that their major effects are peripheral and symptoms mediated through beta-receptors are most likely to be helped. These include awareness of fast heart beat, flushing, palpitations and tremor. The most obvious use of beta-blockers is in the treatment of performance anxiety in acute stress situations, such as speaking in public and playing a musical instrument. If avoidance of tremor is particularly important (e.g. playing the violin) beta-blocking drugs may be of particular help (James et al 1977). The main advantage of beta-blockers is that they have no sedative effects or sensorimotor impairment and have no risk of dependence.

There are other major uses in the treatment of anxiety associated with somatic orientation, sometimes called somatosthenic anxiety (Tyrer 1982b) in which the patient gives a special emphasis to somatic symptoms. Most patients in this group are not seen by psychiatrists but by general practitioners or hospital physicians.

Antihistamines are well established drugs with a long history of successful use in the treatment of mild anxiety and insomnia from childhood onwards. The sedative effects are rapid in onset but drowsiness is common in doses needed to reduce anxiety. Although the dependence risk of these drugs is low there is still some potential for abuse, with both cyclizine and diphenhydramine being reported as addictive. However, problems of withdrawal following long-term low dose treatment have not been reported. The main effects of anti-anxiety drugs are summarized in Table 5.6. Because of the dependence risks of the sedative/hypnotic group it is wise to restrict treatment to several weeks as a maximum or to use the drugs in intermittent dosage to avoid dependence from developing. If longer-term treatment is required the anti-depressants (see below), buspirone or beta-blocking drugs (for the special forms of anxiety indicated above) are preferable.

ANTI-DEPRESSANTS

Although the general term 'anti-depressants' is appropriate for this group of compounds in that they are all used in the treatment of depressive illness, it has become realized in recent years that they have a spectrum of activity which is much greater. They are effective in the treatment of phobias and obsessional disorders, all forms of anxiety, and hypochondriacal and somatoform conditions. In reviewing their use all these conditions will be taken into account, although this inevitably leads to some overlap with other sections in this chapter.

Mechanism of action

The exact mechanism of action of anti-depressants is unknown. However, it is clear from studies carried out over the last 40 years that central monoamines, both

Table 5.6 Comparison of the main clinical properties of anti-anxiety drugs

Drug group	Speed of action	Sedation and sensorimotor impairment	Risk of dependence	Efficacy	Main indications
Sedative/ hypnotics	Fast	+++	+++	+++	Short-term anxiety
Dibenzobicyclo-octadienes	Slow	+	+	+	Unclear
Antihistamines	Fast	++	+	+	Mild anxiety and insomnia
Azaspirodec-anediones	Slow	±	−	++	Anxiety in dependence-prone patients
Beta-blocking drugs	Fast	−	−	+	Performance anxiety

noradrenaline and 5-hydroxytryptamine (5-HT), are closely involved with the action of these drugs and are almost certainly related in some way to their anti-depressant effects. Because it was shown soon after the introduction of monoamine oxidase inhibitors (MAOIs) to clinical practice that these drugs acted by inhibiting the oxidation of amines (Costa & Brodie 1964), and that the tricyclic anti-depressants inhibited the re-uptake of these amines (Sulser et al 1962) (so that they are sometimes abbreviated to MARIs (monoamine re-uptake inhibitors)), it was clear that amines had a part to play in depressive disturbance. When it was found that the hypotensive drug, reserpine, sometimes led to severe depression in patients treated for hypotension and that this drug also depleted the central monoamines (Quetsch et al 1959), the nature of the relationship appeared to be clear. It led to the so-called monoamine hypothesis of affective disorders (Schildkraut 1965) which suggested that in depressive illness there was a relative deficiency of noradrenaline and 5-HT, and, conversely, that in mania there is a relative excess of these amines.

This hypothesis was deceptively simple and is now known to be wrong in this form. The main problem is that a single dose of an anti-depressant drug will either inhibit monoamine oxidases significantly or prevent the re-uptake of noradrenaline or 5-HT, yet the anti-depressant effects are delayed for up to four weeks after starting treatment. It is therefore necessary to explain this delay whilst still retaining the central importance of monoamine function.

When it was found that chronic administration of anti-depressants leads to changes that are not found with acute treatment, the most noted of which is a decrease in beta-adrenoceptor numbers and function (commonly known as 'down-regulation') (Sulser et al 1978), then it appeared that an explanation of the delay in anti-depressant effect was forthcoming. The amine hypothesis of depression could therefore be modified as the 'supersensitive amine receptor hypothesis'. According to this theory there was a central deficiency of cerebral amines that only led to depression when beta-adrenoreceptors became supersensitive in an attempt to compensate for this deficiency (Tyrer & Marsden 1985). When anti-depressant treatment was given there was a rapid rise in amine levels but no change in depression until the sensitivity of the beta-adrenoreceptors had become reduced or down-regulated. As this change occurs between two and four weeks after starting anti-depressant therapy it seemed reasonable to conclude that the change in sensitivity of beta-receptors preceded and was responsible for the change in mood. In the late 1970s most of the anti-depressants available for clinical use were shown to down-regulate beta-adrenoreceptors after long-term treatment and this appeared to give a common explanation of their anti-depressant effect. True, there were differences between anti-depressants and their ability to inhibit the reuptake of different amines but although this allowed the separate description of selective noradrenaline and 5HT-re-uptake inhibitors this did not contradict the main hypothesis.

A fly in the ointment is that there are several compounds with unequivocal antidepressant activity that do not seem to down-regulate beta-receptors. They include mianserin, paroxetine, and most of the other new selective 5-HT reuptake inhibitors apart from sertraline (Johnson 1990).

Although there have been many new anti-depressants introduced to clinical practice since the original prototypes, iproniazid (the first monoamine oxidase inhibitor) and imipramine (the first tricyclic antidepressant), it is surprising how little has changed over the years. In simple terms, monoamine oxidase inhibitors and tricyclic anti-depressants have been made 'cleaner' in that they produce fewer side-effects and less toxicity while their basic efficacy is unchanged. The supposed mechanism of action of all these compounds is summarized in Table 5.7.

Tricyclic antidepressants

This group includes some of the oldest anti-depressants in clinical practice which have enjoyed continuous use for 30 years and some newer ones that differ mainly in producing fewer unwanted effects. There is little doubt that these drugs are effective in the treatment of depression (Morris & Beck 1974), and in recent years they have also been shown to be effective in the treatment of generalized anxiety (Kahn et al 1986), panic (Sheehan et al 1980) and obsessive–compulsive disorder (Marks et al 1980). In all studies there are initial effects of the drugs that are unrelated to their main clinical action. Many of these are unwanted effects (Table 5.8) but some can be desired, particularly sedation in the agitated depressed patient. These effects improve somewhat over time despite maintaining the same dosage because of the development of tolerance. Between 10 days and four weeks after starting treatment the main anti-depressant (or anti-anxiety, antipanic or anti-obsessional) effect is shown. This often occurs suddenly and is associated with greater self-confidence, energy and self-esteem. This is the change which is thought to be related to down-regulation of beta-adrenoceptors (see above).

Once clinical improvement has taken place it is

Table 5.7 Likely mechanisms of action of anti-depressants

Group	Typical member	Mechanism of action
Tricyclic anti-depressants	Amitriptyline Imipramine Dothiepin Lofepramine	Mixed NA and 5-HT re-uptake inhibition leading to down-regulation of adrenergic beta-receptors
Selective NA* re-uptake inhibitors	Maprotiline	Ditto, except that NA re-uptake inhibition is paramount
Selective 5-HT re-uptake inhibitors	Fluvoxamine Fluoxetine Paroxetine Sertraline	Ditto, except that 5-HT re-uptake inhibition is paramount and down-regulation of beta-receptors often absent
Hydrazine monoamine oxidase inhibitors (MAOIs)	Phenelzine Isocarboxazid	Complete (suicide) inhibition of monoamine oxidase (MAO)
Reversible monoamine oxidase inhibitors (MAOIs)	Tranylcypromine Moclobemide	Reversible inhibition of monoamine oxidase (MAO)
Amine precursors	L-tryptophan L-hydroxy-tryptophan	Increased synthesis of 5-HT
Other anti-depressants	Mianserin	See text

*NA = noradrenaline

common to reduce anti-depressants to maintenance dosage and continue them for at least another six months although a great deal will depend on circumstances of the individual case. It is usual to reduce the dose of the anti-depressant to approximately two-thirds of that necessary to produce clinical benefit initially and this is retained as the maintenance dose. For many tricyclic anti-depressants the standard acute treatment dose is of the order of 150 mg daily and the maintenance dose about 100 mg (Table 5.9).

GENERAL ISSUES CONCERNING ANTI-DEPRESSANT THERAPY

Benefits

Despite occasional studies throwing doubt about the anti-depressant efficacy of these compounds there is overwhelming evidence from the majority of studies that these drugs are effective in the treatment of depressive illness and many of the anxiety disorders (Table 5.9). In all instances there is a delay between the initial administration of the drug and onset of the main therapeutic effect of the order of one week at minimum (found with the monoamine oxidase inhibitor, tranylcypromine) and six weeks. The therapeutic effects are sometimes noted suddenly (e.g., a patient wakes up one morning having lost virtually all depressive symptoms) but more commonly the main improvement is shown over a period of a few days.

In general it is believed that all the main anti-depressants are equally efficacious and that no significant progress has been made since the introduction of imipramine in producing a more powerful anti-

Table 5.8 Adverse effects of tricyclic anti-depressants and monoamine oxidase inhibitors

Drug	Sedation	Anticholinergic effects (e.g. dry mouth, constipation)	Anti-adrenergic effects (e.g. postural hypotension)	Weight gain	Toxicity in overdose
Imipramine	+	++	++	++	++
Desipramine	++	++	+++	++	+++
Clomipramine	+	++	++	++	+
Trimipramine	+++	++	++	++	+++
Lofepramine	±	+	+	+	−
Amitriptyline	+++	++	++	++	+++
Protriptyline	+	++	++	++	+
Nortriptyline	++	++	++	++	++
Dothiepin	++	++	++	++	+++
Phenelzine	±	++	++	++	+++
Isocarboxazid	−	+	+	+	+
Tranylcypromine	−	++	++	++	+++

Table 5.9 Comparison of efficacy of anti-depressants in depression and neurotic disorders[1]

Main drug group	Level of efficacy compared with placebo control					
	Severe depression	Moderate depression	'Atypical depression'	Panic disorder	Obsessive–compulsive disorder	GAD
Tricyclic anti-depressants (e.g. imipramine 75–200 mg day^{-1})	++ (Morris & Beck 1974)	++ (Paykel 1989)	++ (Paykel 1989)	++ (Liebowitz et al 1988)	++ (Marks 1983)	++ (Kahn et al 1986)
Monoamine oxidase inhibitors (e.g. phenelzine 30–90 mg/day^{-1})	+ (Paykel 1979)	++ (Paykel 1989)	++ (Tyrer & Shawcross 1988)	++ (Liebowitz 1989)	++ (Jenike et al 1983)	++ (Tyrer 1989b)
Selective 5-HT-re-uptake inhibitors (e.g. fluoxetine 20–60 mg/day^{-1})	+/++ (van Praag et al 1987)	++ (Montgomery 1988a)	++ (Reimherr et al 1984)	++ (Sheehan et al 1988)	++ (Goodman et al 1989)	?
Selective anti-depressants: maprotiline, trazodone, viloxazine, mianserin	+/++ (Paykel 1989)	++ (Pinder 1988)	?	?	?	?

[1]The terminology used is that of the ICD-10 classification. Atypical depression includes mixed anxiety–depressive disorders and depressive conditions associated with somatization symptoms. GAD = generalized anxiety disorder.

depressant. However, there is dispute about this issue. Unfortunately the newer anti-depressants cannot be tested in severely depressed populations using placebo comparisons because of ethical considerations and so most trials that use anti-depressants are carried out in patients with moderate depression. However, there remains some doubt that the newer compounds are as effective as the original tricyclic anti-depressants (van Praag et al 1987). On the other hand, there are also suggestions that some of the newer compounds are more effective in resistant depressed patients (Reimherr et al 1984, Barker et al 1987) although there are no controlled studies supporting this. There are also suggestions that monoamine oxidase inhibitors have a slightly different spectrum of therapeutic activity, being more effective in anxiety disorders than tricyclic anti-depressants although less effective in severe depression (Tyrer & Shawcross 1988). There is also some circumstantial evidence suggesting that the newer 5-HT reuptake inhibitors may have a similar profile to monoamine oxidase inhibitors in this respect.

There is also evidence with all these compounds that treatment needs to be continued beyond the time of initial therapeutic benefit for varying periods which have yet to be determined satisfactorily. The drugs can also be effective as prophylactics in the treatment of recurrent depression. The best evidence for this has been shown with amitriptyline and imipramine (Mindham et al 1973, Coppen et al 1978b), and fluoxetine (Montgomery et al 1988). All anti-depressants can trigger manic episodes in predisposed patients, so it is normally advised to use lithium in those who have bipolar affective psychosis and reserve other anti-depressants for unipolar illness.

Risks

There are many more differences between the risks of anti-depressants than between their benefits (Table 5.10). Almost all anti-depressants in therapeutic dosage have potential unwanted effects early in treatment. Although some of these may be relatively minor, because they are not accompanied by obvious therapeutic improvement, they can often discourage patients from continuing therapy. This is one of the reasons why compliance with anti-depressants is so poor, particularly in general practice (Johnson 1974). Sometimes apparently unwanted effects can be desired, as, for example, sedation with a tricyclic anti-depressant in agitated depression, but more often they are unwanted.

Table 5.10 Comparison of risks of main anti-depressant drugs

Main drug group	Anti-muscarinic	Anti-adrenergic	Anti-histaminic	Epilepto-genic	Weight gain	'Serotonin syndrome'	Gastro-intestinal symptoms
Tricyclic antidepressants	+++	++	++	+	++	−	−
MAOIs	++	+++	+	+	++	-	±
5-HT selective re-uptake inhibitors	−	−	-	±	−	+	++
Other anti-depressants:							
Maprotiline	+	±	−	++	+	−	−
Trazodone	−	−	−	?[1]	−	−	−
Mianserin	±	±	−	±	−	−	−
Viloxazine	±	−	±	+	−	−	+

[1]? = insufficient information.

Many of the new anti-depressants have been introduced because of their lower incidence of unwanted effects in acute dosage although, as yet, there is no good evidence that compliance with these compounds is superior to the original tricyclic anti-depressants.

Another major risk is that of toxicity in overdose. Depressed patients are more likely to attempt suicide than most other psychiatric groups, and although many methods can be chosen it is important not to add to the list unnecessarily. Studies of the incidence of fatalities due to overdoses of anti-depressant drugs and adverse drug reactions have demonstrated that the original tricyclic anti-depressants have a much higher incidence of fatality in overdose than the newer compounds (Cassidy & Henry 1987, Farmer & Pinder 1989). The figures vary between 80 deaths per million prescriptions for desipramine to less than six deaths per million prescription for mianserin.

The third major risk area is with idiosyncratic adverse drug reactions that cannot be predicted because they are so rare but which can be extremely serious. For reasons that are far from clear, many of the newer anti-depressants have been associated with severe idiosyncratic reactions of this nature that have sometimes led to the withdrawal of the drug from clinical prescription. These include zimelidine, which was withdrawn because of evidence that it precipitated ascending myelitis (Guillain–Barré syndrome) and nomifensine, which precipitated haemolytic anaemia. There has also been concern about the propensity of mianserin to cause bone marrow suppression, rarely leading to agranulo-cytosis, the rare incidence of priapism with trazodone and epileptic seizures with maprotiline.

Individual anti-depressant drugs

Tricyclic anti-depressants

Imipramine and amitriptyline, the two original tricyclic anti-depressants, remain popular in clinical practice for the treatment of depression and anxiety. The development of panic disorder as a clinical entity largely followed from the work of Klein (1964), who used imipramine in certain types of anxiety disorder and found that panic was improved in particular. The use of this drug (and other anti-depressants) in anxiety is now widespread, with a somewhat wider dosage range than in depression (e.g., 25–300 mg day^{-1}). Imipramine is also effective in generalized anxiety (Kahn et al 1986).

Other similar compounds include dothiepin, desipramine, trimipramine, protriptyline, nortriptyline and butriptyline, which differ only in their propensity for antimuscarinic and antihistaminic effects (see Table 5.8). A more recent introduction, lofepramine, is much safer in overdosage than the other compounds, particularly its main metabolite, desipramine (Cassidy & Henry 1987). The reason for this is far from clear but it appears that there is decreased bio-availability of lofepramine and antagonism between its effects and those of desipramine (Leonard 1987).

Clomipramine is also a tricyclic anti-depressant closely related to imipramine. It is unusual in that it is a selective re-uptake-inhibitor of 5-HT, although not as selective as some of the more recently introduced compounds. It has been used more for the treatment of obsessive–compulsive and phobic disorders than in depressive ones. It is undoubtedly effective in obsessional disorders although its effects may only be marked in the first few weeks of therapy (Marks et al 1980). There is dispute over whether clomipramine is effective because it is specifically anti-obsessional (Montgomery 1980) or only effective when there is coincidental depression (Mark & O'Sullivan 1989).

Monoamine oxidase inhibitors (MAOIs)

These drugs preceded the tricyclic anti-depressants by 2–3 years, and were first discovered to have anti-depressant properties when iproniazid was being tested for its efficacy as an anti-tuberculosis drug. Although they were found to be efficacious in depression in some studies in others they were no better than placebo. Examination of the reasons for this showed that good response was only achieved when the drugs were given in higher dosage or for longer than four weeks in regular dosage (Paykel 1979, Tyrer 1979).

Subsequently it has been established that the hydrazine MAOIs such as iproniazid, phenelzine and isocarboxazid, are less rapidly acting than the non-hydrazine drugs such as tranylcypromine.

The MAOIs fell into disfavour both because they appeared to be less effective than the tricyclic anti-depressants, most influentially in a trial organized by the Medical Research Council (1965), but also because of their interaction with foods containing tyramine (usually called the 'cheese reaction' (Blackwell 1963)) and other drugs including indirectly acting sympathomimetic amines such as phenylephrine and ephedrine, opiates and oral hypoglycaemic drugs. These interactions were sometimes fatal. Sudden hypertensive crises can arise when tyramine, which is normally deaminated by monoamine oxidase in the gut wall, enters the systemic circulation, and similar crises can be precipitated by many amines in the systemic circulation.

The main foods containing tyramine are cheese, seasoned game, pickled herring, yeast extracts (e.g. Bovril), chianti, fortified wines and beer. Noradrenaline and adrenaline, which are metabolised by other pathways, are safe when taken with an MAOI. Hypotensive crises can occur with opiate analgesics such as pethidine when taken with MAOI although the combination is safe if careful titration of dosage is carried out first.

Similar fears were generated by the combination of tricyclic anti-depressants and MAOIs, but the combination is now realized to be safer then originally thought (Schuckit et al 1971), and sometimes effective in resistant depression when all other treatments have failed (Tyrer & Murphy 1990). However, when monoamine oxidase inhibitors and tricyclic anti-depressants are given together as first-line treatment in depressive disorders they show no advantages over the individual drugs given alone (Young et al 1979, Razani et al 1983, Shawcross & Tyrer 1987). If the combination is to be used the two drugs should either be given together or the MAOI added to the treatment regime of a patient already on a tricyclic anti-depressant. Combinations with selective 5-HT reuptake inhibitors are dangerous and should be avoided; the safest combination is with amitriptyline as this may confer protection against the cheese reaction (Pare et al 1982).

In addition to food and drug interactions, MAOIs also have antimuscarinic effects, stimulate appetite and weight gain (see Table 5.8), are sometimes hepatotoxic, and may be asssociated with some risk of dependence. This last hazard is most often described with tranylcypromine although it is difficult to separate relapse from withdrawal effects in any drug with a delayed onset of action (Tyrer 1984b) and a definite verdict cannot be delivered.

Recently, newer MAOIs have been introduced which have much less propensity to food and drug interactions. These include L-deprenyl, or selegiline, which is primarily used for the treatment of parkinsonism, but which also has some limited anti-depressant efficacy. Monoamine oxidase exists in two enzyme forms, MAO-A and MAO-B, the inhibition of MAO-A being associated with most of the adverse effects of these compounds. L-Deprenyl is a selective inhibitor of MAO-B but in doses that are anti-depressant it also inhibits MAO-A to some extent (Mann et al 1989a) and is not entirely free of the danger of food interactions.

Most of the standard MAOIs are irreversible inhibitors of MAO-A; the confusing term 'suicide inhibitors' is often applied to them as they 'kill' all the MAO in the body and the enzyme has to be synthesized again. This explains why the forbidden foods and drugs should not be taken again for at least three weeks after MAOIs are stopped. New reversible inhibitors such as moclobemide and brofaromine are likely to become available on prescription soon. They are less likely to produce the 'tyramine reaction' and have short half-lives, so it is hoped that they will become a suitable alternative treatment for depressive illnesses. If they have the same clinical profile as the original MAOIs they are likely to be of value in mixed-anxiety depressive states, sometimes called atypical depression, and other anxiety disorders (Tyrer & Harrison-Read 1990).

Selective 5-HT re-uptake inhibitors

These drugs have been introduced to clinical practice at the rate of about two each year since 1989. They include fluvoxamine, fluoxetine, sertraline, paroxetine and citalopram, and have been used primarily for depressive illness but also in panic, phobic and obsessional disorders. They differ from other anti-depressants mainly in their side-effect profiles, summarized in Table 5.10. Thus, they rarely produce significant sedation, dry mouth, weight gain or postural hypotension, but instead have other effects not commonly found with tricyclic anti-depressants, including nausea and vomiting, and agitation more than sedation. In extreme form the agitation can be part of a 'serotonin syndrome' in which there is overactivity, fever and confusion, probably due to excessive serotonin levels (Feighner & Boyer 1990). There have also been reports of akathisia and extrapyramidal effects with fluoxetine (Lipinski et al 1989) thought to be due to dopamine inhibition by excessive 5-HT activity. There has also been concern about impulsive behaviour with these compounds, and although they are not associated with significant toxicity in overdose, there have been reports of increased suicidal behaviour with fluoxetine that have attracted considerable media attention in the United States but which conflict with other evidence suggesting a lower tendency to suicidal behaviour (Montgomery 1988b). These drugs have also not been tested fully for their dependence potential.

The place of selective 5-HT re-uptake inhibitors in psychiatry is still far from clear and there appear to be too many compounds available. They could have a place in the treatment of resistant depressive illness (Reimherr et al 1984), in bulimia and other eating disorders, in phobic and obsessional conditions (Goodman et al 1989), in panic and in mixed anxiety–depressive states, but none of these indications is unequivocal at present.

Other anti-depressants

L-Tryptophan, a dietary precursor of 5-HT, has been recommended for the treatment of resistant depressive illness for many years, usually as an adjunct to an existing anti-depressant, but this drug has now been withdrawn following reports of hepatotoxicity. Amoxapine is a metabolite of the antipsychotic drug loxapine, and is a dopamine receptor antagonist. It has

anti-depressant properties but its predisposition to extrapyramidal side effects is a handicap.

Mianserin is different from all other anti-depressants. It has a nucleus of four rings and is often termed a tetracyclic anti-depressant. It does not down-regulate beta-adrenoceptors or inhibit re-uptake of noradrenaline and is only a weak inhibitor of 5-HT. It is an antagonist of presynaptic alpha-receptors and this may offer a clue to its mode of action. It has become a popular drug in the treatment of depression in the elderly because of its low incidence of postural hypotension and antimuscarinic effects such as dry mouth and constipation.

Trazodone is an antagonist of 5-HT which has no antimuscarinic effects whatsoever and can therefore be given to patients, for example, with glaucoma and prostatic symptoms. It is effective as an anti-depressant but somewhat less so than other anti-depressants. Viloxazine and maprotiline are selective inhibitors of noradrenaline reuptake (although not advertized as such in the way that their 5-HT equivalents are). They are also effective anti-depressants but their use has been inhibited to some extent by their epileptogenic potential (although this is only a little greater than other anti-depressants and antipsychotic drugs) and gastrointestinal symptoms (mainly nausea).

It is worth emphasizing that electroconvulsive therapy (ECT), although decreasing in use, remains an important option in the treatment of severe depression, particularly when the illness is life-threatening or has failed to respond to adequate courses of drug treatment (Kendell 1981). Bilateral treatment is somewhat more effective then unilateral treatment (Gregory et al 1985) although it creates more memory disturbance.

Lithium salts

It is curious that the drug with the simplest chemical structure known to psychopharmacology is also the most complex in its effects. The precise mechanism of action of lithium is unknown but it is undoubtedly effective in any one of its basic salts (carbonate, citrate, acetate, sulphate or monoglutamate) in preventing attacks of recurrent depression and mania, and, to a lesser extent, in treating them. The main problem with lithium is its low therapeutic index (i.e., ratio of toxic to therapeutic doses), which is of the order of 3–4 only. It is therefore necessary to keep a regular check on serum (or plasma) lithium levels in patients receiving treatment. Lithium is excreted mainly by the kidney, and any impairment of renal function virtually excludes treatment in view of the risk of toxicity. Before deciding on treatment with lithium it is therefore necessary to determine renal function, either by full renal function studies in those patients suspected of poor function, or simple tests (e.g. serum creatinine, electrolytes, urea), in those who are physically well. Lithium can produce a form of nephrogenic diabetes insipidus, with excessive thirst and micturition, because of its effects on the kidney, but these effects usually disappear when the drug is reduced or stopped.

The most serious toxic effect is an encephalopathy, with cerebellar ataxia, vomiting, confusion, drowsiness and dysarthria. The first sign of this is often gastrointestinal disturbance with diarrhoea as a prominent symptom. Serum lithium levels may occasionally be within the normal range even when toxic symptoms are present.

Thyroid and cardiac function are also required to be satisfactory before starting lithium. This is because lithium impairs the uptake of iodine by the thyroid gland and can induce primary hypothyroidism, and can also be dangerous in those with impaired cardiac function. The drug should also be avoided during the first trimester of pregnancy because it may create cardiac abnormalities (primarily affecting the tricuspid valve).

The primary use for lithium is in the prophylaxis of bipolar affective psychosis. Serum levels are estimated between 8 and 12 hours after the last dose and aim for lithium levels between 0.5 and 1.0 mmol l^{-1} for prophylactic purposes, with lower levels in the elderly (Jefferson 1983). The normal minimum requirements for prophylactic treatment are one significant episode of depression and mania within two years. Serum estimations are frequent at first, up to twice weekly, but may be decreased to once every three months in established therapy. Thyroid function needs checking at approximately yearly intervals and renal function tests may also need repeating.

Lithium is also used for the treatment of mania, when somewhat higher serum levels may be necessary to produce benefit but should be kept below 1.5 mmol l^{-1} to guard against toxicity (Tyrer 1985). Toxic interactions have been reported between lithium and antipsychotic drugs such as haloperidol and thioridazine, but there is controversy over when the incidence of such adverse effects is really higher than when either drug is used singly. Lithium is also anti-aggressive and may be of value in the treatment of persistent antisocial or disturbed behaviour (e.g. Sheard et al 1976).

In case of resistant mania or depression carbamazepine is sometimes used as an alternative to, or sometimes in combination with, lithium. Although satisfactory controlled studies are hard to carry out in this population, the evidence suggests that carbamazepine is both effective as a prophylactic and a treatment (Post et al 1987, S P Tyrer 1989). Because carbamazepine may depress the white cell count regular checks on this are necessary, particularly early in treatment. Calcium channel blockers such as verapamil and nifedipine have also been used to treat mania; they may act by blocking 5-HT receptors.

Other psychotropic drugs

Disulfiram blocks the action of acetaldehyde dehydrogenase; thus, the metabolism of alcohol is interrupted and acetaldehyde accumulates in the body, leading to severe headaches, marked hypotension and cardiovascular collapse in extreme instances. This may help some patients who are dependent, or potentially dependent, on alcohol to avoid drinking again after successful detoxification. Much depends on the patient's motivation; sometimes disulfiram is given by intramuscular implant to ensure compliance, but as absorption from these implants is erratic they are not generally recommended. Disulfiram may occasionally cause rashes, peripheral neuropathy and impotence.

The hyperkinetic syndrome in children is sometimes treated with psychostimulants such as methylphenidate and D-amphetamine. This represents one of the few indications for these drugs but even this is controversial as well as difficult to explain pharmacologically. Treatment should be short-term wherever possible because these drugs stunt growth and have a strong propensity to dependence.

There are several hormones that are used in psychiatry and other drugs that are used to correct hormonal imbalance. Although they can not be regarded as strictly primary psychotropic drugs they deserve some mention. They include oestrogens, used for the treatment of premenstrual symptoms (without satisfactory evidence of efficacy), male trans-sexuals wishing to become females, and hormone replacement therapy after the menopause; testosterone and other androgens, used for increasing libido in women (not effective in men apart from those with Klinefelter's syndrome); bromocriptine, used to treat impotence which follows from hyperprolactinaemia (e.g. from antipsychotic drugs); and antilibidinal drugs such as benperidol (a butyrophenone related to haloperidol), goserelin and cyproterone acetate, primarily used in forensic psychiatry to prevent sexual offences such as rape and exhibitionism when it is felt that abnormal sexual drive is a major cause of offences.

FURTHER READING

Goodman-Gilman A, Rall T W, Nies A S, Taylor P (ed) 1990 Goodman and Gilman's: The pharmacological basis of therapeutics. 8th edn. Pergamon, New York

Meltzer H Y (ed) 1987 Psychopharmacology: the third generation of progress. Raven Press, New York

Rang H P, Dale M M 1991 Pharmacology. 2nd edn. Churchill Livingstone, Edinburgh

Silverstone T, Turner P 1988 Drug treatment in psychiatry. 4th edn. Routledge, London

Spiegel R 1989 Psychopharmacology: an introduction. 2nd edn. Wiley, Chichester

Tyrer P J (ed) 1982 Drugs in psychiatric practice. Butterworths, London

6. Genetics

This chapter first outlines the structure and functions of the nucleic acids deoxyribonucleic acid (DNA) and ribonucleic acid (RNA), before discussing chromosomes and cell division. This is followed by details of genetic disorders including examples of relevance to psychiatry. Finally there is a section that deals with the principles of genetic studies in psychiatry, again with appropriate examples being given.

NUCLEIC ACIDS

In this section the structure and functions of deoxyribonucleic acid (DNA), a molecule by means of which hereditary information is passed from one generation to the next, and ribonucleic acid (RNA), of which there are several types, are described.

Nucleotides

A nucleotide is made up of three components: a nitrogenous **purine** or **pyrimidine** base; a pentose sugar; and a phosphate moiety. The first two components linked together without the phosphate moiety form a molecule known as a **nucleoside**.

The two most important purine bases are **adenine** and **guanine**, and their structures are shown in Figure 6.1. Figure 6.2 shows the structures of the three most important pyrimidine bases, namely, **cytosine**, **thymine** and **uracil**.

When the pentose sugar is D-ribose the nucleoside is termed a **ribonucleoside**. When the pentose sugar is

D-2-deoxyribose the nucleoside is known as a **deoxyribonucleoside**. The structures of these two pentose sugars are shown in Figure 6.3.

A nucleotide is a nucleoside with a phosphate moiety attached, usually esterified at the 5^1 hydroxyl group of the pentose sugar. The phosphate moiety can be a single phosphate, a diphosphate, or a triphosphate. The names of the ribonucleosides and ribonucleotides found in RNA are shown in Table 6.1, while the names of the deoxyribonucleosides and deoxyribonucleotides found in DNA are shown in Table 6.2. Taking the base adenine as an example, the ribonucleotide with a monophophate group attached to the ribose is known as adenosine 5^1-phosphate or adenylate, and is conventionally abbreviated to AMP, as mentioned in Chapter 3. The corresponding diphosphate is ADP, and the triphosphate is ATP. The corresponding deoxyribonucleotides are prefixed with the letter d: dAMP, dADP and dATP. The nomenclature of ribonucleotides and deoxyribonucleotides with other bases is similar with the A (adenine) replaced by G for guanine, C for cytosine, and either T for thymine in deoxyribonucleotides (in DNA) or U for uracil in ribonucleotides (in RNA).

In addition to their roles in DNA and RNA, nucleotides have other important functions, some of which have been mentioned in Chapter 3. For example, ATP and GTP act as energy-rich molecules in many important cellular reactions, while cyclic AMP (cAMP) has a role in the second messenger systems of many hormones. Nucleotides based on adenine are also found in

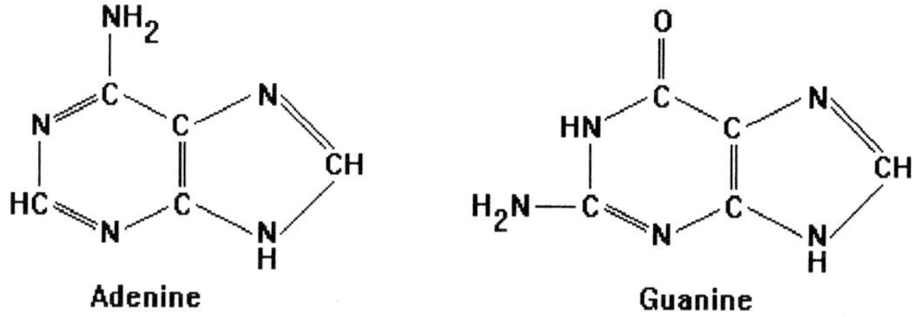

Adenine   Guanine

Fig. 6.1   The purine bases adenine and guanine.

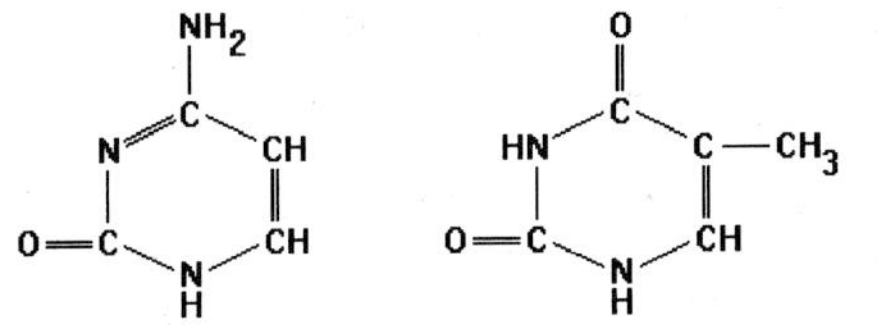

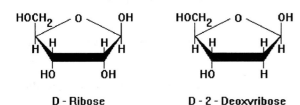

Fig. 6.2   The pyrimidine bases cytosine, thymine and uracil.

D - Ribose   D - 2 - Deoxyribose

Fig. 6.3   Structures of D-ribose and D-2-deoxyribose.

Table 6.1   Ribonucleosides and ribonucleotides found in RNA

Base	Ribonucleoside	Ribonucleotide
Adenine (A)	Adenosine	Adenylate (AMP)
Guanine (G)	Guanosine	Guanylate (GMP)
Cytosine (C)	Cytidine	Cytidylate (CMP)
Uracil (U)	Uridine	Uridylate (UMP)

Table 6.2   Deoxyribonucleosides and deoxyribonucleotides found in DNA

Base	Deoxyribonucleoside	Deoxyribonucleotide
Adenine (A)	Deoxyadenosine	Deoxyadenylate (dAMP)
Guanine (G)	Deoxyguanosine	Deoxyguanylate (dGMP)
Cytosine (C)	Deoxycytidine	Deoxycytidylate (dCMP)
Thymine (T)	Deoxythymidine	Deoxythymidylate (dTMP)

coenzyme A (CoA), FAD, $FADH_2$, NAD^+ and NADH.

Deoxyribonucleic acid

Structure

The structure of DNA as a macromolecule consisting of a double stranded helix was demonstrated by Watson and Crick (1953). Each strand consists of a linear polymer of deoxyribonucleotides linked by phosphodiester bonds. The phosphodiester bond between any two deoxyribonucleotides in DNA is formed by a condensation reaction in which the 3' carbon atom of the deoxyribose of one deoxyribonucleotide is linked by an ester bond to the monophosphate group of the other deoxyribonucleotide, with the elimination of a water molecule formed from the 3' hydroxyl group of the deoxyribose and a hydrogen ion from the phosphate. Thus each DNA strand in effect has a sugar–phosphate backbone as shown in Figure 6.4.

The deoxyribonucleotides that make up DNA can each contain any one of four bases. These are the purines, adenine (A) and guanine (G), and the pyrimidines, cytosine (C) and thymine (T). These four bases

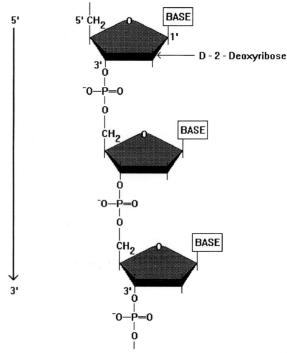

Fig. 6.4   Part of a single strand of DNA showing the sugar-phosphate backbone.

can be thought of as constituting a genetic alphabet, with the order of bases being the means of storing information in the molecule. It should be noted that although, strictly speaking, the letters A, G, C and T stand for the bases themselves, in practice these letters are often used a shorthand for the whole deoxyribonucleotides, so that a given sequence of DNA deoxyribonucleotides can be represented simply as a sequence of the four letters. By convention this sequence is written such that the direction for the DNA strand from the 5^1 end of the deoxyribose to the 3^1 end corresponds to the direction left to right for the written letters.

The bases of one of the strands of DNA are linked to the bases of the other strand of the double helix in a precisely determined way by means mainly of hydrogen bonds. In order for there to be a precise fit between the two strands of the double helix, devoid of distortions, an adenine base of one strand must pair with a thymine base of the other strand by means of two hydrogen bonds, and a guanine base of one strand must pair with a cytosine base of the other strand by means of three hydrogen bonds. These two possible arrangements are shown in Figure 6.5. Other forces such as van der Waals forces and hydrophobic forces also play a role in maintaining the structure of the double helix. Owing to steric considerations, naturally occurring human DNA is usually in the form of a right-handed double helix, although the left-handed form, known as Z DNA, may occasionally occur. Some DNA double helices are in the form of closed circular molecules which may themselves be supercoiled.

Since there is a one-to-one mapping between the

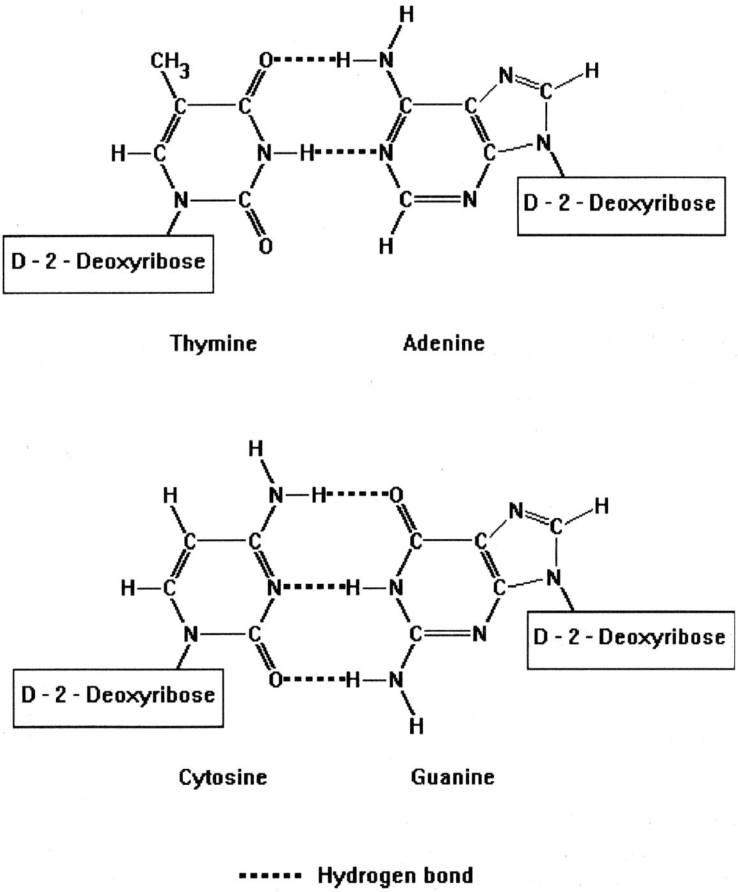

Fig. 6.5 Hydrogen bonding between purine and pyrimidine bases found in DNA.

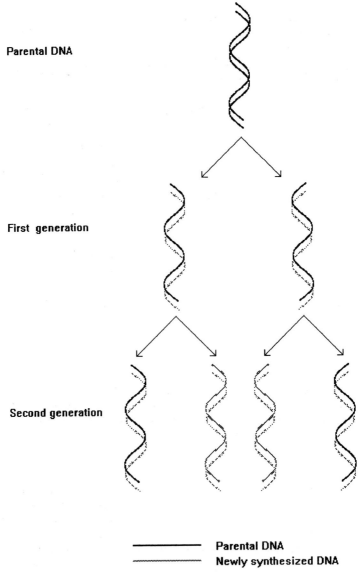

Parental DNA

First generation

Second generation

——————— **Parental DNA**
〰〰〰〰〰 **Newly synthesized DNA**

Fig. 6.6   Semiconservative replication of DNA.

bases of one strand of a DNA double helix and the bases of the other strand, with A paired with T, and C paired with G, it follows that the two strands are complimentary and each can act as a template for the other in the replication of DNA. Thus, during the replication of a DNA double helix, as the two strands are separated, a process known as the **denaturing** of DNA, a new strand of DNA can form around each original strand such that the new strand around a given original strand has the same structure as the original complimentary strand of the original strand, as shown

in Figure 6.6. Hence DNA replication is said to be **semiconservative**.

Functions

DNA is the molecule that carries information related to characteristics that are inherited from one generation to the next. This information is passed to new somatic cells by means of the accurate replication of DNA in the process of **mitosis** described below. The genetic information carried in the gametes (spermatozoa

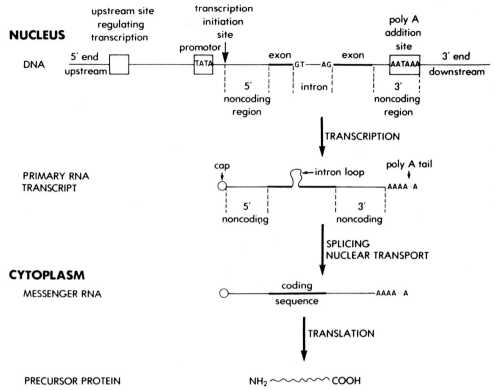

Fig. 6.7 Eukaryotic gene structure and the steps involved in eukaryotic gene expression. (Reproduced with permission from Emery A E H, Rimoin D L 1983 Principles and practice of medical genetics vol 1. 1st edn. Churchill Livingstone, Edinburgh.)

and ova) is not a duplication of that found in somatic cell nuclei but is derived by means of the process of **meiosis**, also described below. The total genetic information found in the DNA of somatic cells (including neurones and neuroglia) is known as the **genome**, while the biological units of heredity have classically been referred to as **genes**.

As indicated above, the genetic information is carried in the form of the order of the nucleotide base pairs in the DNA molecule. This nucleotide sequence encodes the information required for protein biosynthesis. If the individual nucleotides are considered to be letters of the genetic code, then there are four letters: A, C, G and T. A group of such letters can be considered to form a word in the genetic code. Now, since proteins are made up of 20 different types of amino acids (see Ch. 3), the minimum length of a word of the genetic code that would enable all the amino acids to be encoded is three letters. With four nucleotide possibilities for each of these letters, this implies that a sequence of three nucleotides can give rise to as many as 64 different possible words (4^3 =

64). It has been shown that such words, known as a **codons**, do indeed each consist of a sequence of three nucleotide bases. Not all the possible codons code for amino acids; three codons represent the command 'stop', and one can represent the command 'start' in addition to coding for the amino acid methionine. The **genetic code** of the DNA is read or **transcribed** and then decoded or **translated** by RNA (see below).

A typical gene does not consist merely of a continuous sequence of codons. Instead, the codons tend to be grouped into regions known as **exons**, with intervening sequences of nucleotides known as **introns** that do not code for amino acids. Specific nucleotide sequences are usually also found at the beginning and end of genes which have functions such as enabling transcription to occur accurately. The structure of a typical gene is shown in Figure 6.7. The introns and the parts of the DNA molecule between exons and transcription regulation regions do not have any known function and are sometimes referred to as **junk DNA**.

Ribonucleic acid

Structure

The two most important ways in which the structure of ribonucleic acid (RNA) differs from that of DNA are, firstly, that the pentose sugar found in RNA is D-ribose and not D-2-deoxyribose (see Fig. 6.3), and secondly, that the pyrimidine base **uracil** occurs in RNA in place of the pyrimidine base thymine of DNA.

Like DNA, RNA can have a double helix conformation. However, RNA is also found in a number of other types of structural forms, and is usually single stranded.

Functions

Three types of RNA exist in cells: messenger RNA (mRNA), transfer RNA (tRNA), and ribosomal RNA (rRNA).

The function of **mRNA** is to carry the genetic information from the DNA in the chromosomes of the cell nucleus to the cytoplasm where protein biosynthesis can take place. The **transcription** of genetic information from the relevant part of the DNA molecule onto mRNA occurs first by small portions of the relevant part of the DNA double helix being successively separated through the action of **RNA polymerase** thereby enabling a single strand of the DNA to be bound to by nucleotides with comple-

mentary base pairs (A with U, and so on). The new molecule which is formed, for which the nucleotide sequence of the DNA acts as a template, is **pre-mRNA**. The latter is then further processed, in order to remove or splice out complementary nucleotide sequences from transcribed introns, and the final product is mRNA. Thus the mRNA carries the genetic information required for the biosynthesis of a particular protein in the form of the nucleotide base triplets of the genetic code (which is complementary to that of the original DNA) shown in Table 6.3.

Having carried the required genetic information from the nucleus to the cytoplasm, mRNA acts as a template in its own turn in the process of **translation**, in which the genetic code is, in effect, deciphered to allow the formation of a peptide chain, and thence proteins such as enzymes and those with structural roles. This deciphering is carried out by **tRNA** molecules, each of which possesses at one end a three nucleotide **anticodon**. Some anticodons contain modified nucleotides (other than A, C, G and T) such as dihydro-uridine and pseudo-uridine. At the opposite end of the tRNA molecule an amino acid can be attached via the action of the enzyme **aminoacyl tRNA synthetase**. The amino acid attached corresponds to the anticodon and since there are 20

Table 6.3   The genetic code (from mRNA to amino acids)

Position number 1 (5' end)		Position number 2			Position number 3 (3' end)
	U	C	A	G	
U	Phe	Ser	Tyr	Cys	U
	Phe	Ser	Tyr	Cys	C
	Leu	Ser	STOP	STOP	A
	Leu	Ser	STOP	Trp	G
C	Leu	Pro	His	Arg	U
	Leu	Pro	His	Arg	C
	Leu	Pro	Gln	Arg	A
	Leu	Pro	Gln	Arg	G
A	Ile	Thr	Asn	Ser	U
	Ile	Thr	Asn	Ser	C
	Ile	Thr	Lys	Arg	A
	Met (START)	Thr	Lys	Arg	G
G	Val	Ala	Asp	Gly	U
	Val	Ala	Asp	Gly	C
	Val	Ala	Glu	Gly	A
	Val (Met)	Ala	Glu	Gly	G

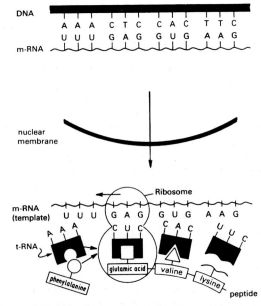

Fig. 6.8   Diagrammatic summary of transcription and translation. (Reproduced with permission from Emery A E H, Mueller R F 1988 Elements of medical genetics (student notes). 7th edn. Churchill Livingstone, Edinburgh.)

common amino acids there must be at least 20 specific aminoacyl tRNA synthetases. Figure 6.8 is a diagrammatic summary of the processes of transcription and translation. It can be seen that the growing peptide chain is attached to the tRNA at a site distant to that of the anticodon. In order for the next amino acid to be attached to this peptide chain, a tRNA with an anticodon complementary to the next mRNA codon to be translated recognizes this codon, and the chain attaches to the latest amino acid addition.

The process of translation just described takes place on **ribosomes**, which consist of proteins combined with **rRNA**. By bringing all the components for translation together at ribosomes, the process of protein biosynthesis occurs much faster than if translation were to occur in solution in the cytoplasm.

DNA mutations

It is clear from the foregoing discussion of transcription and translation that mutations in the sequence of nucleotides of DNA can lead to changes in the structure of polypeptide chains. Mutations may be caused by environmental factors, such as chemicals and ultraviolet and ionizing radiation, or they may be inherited.

Substitution

The commonest type of DNA mutation is the substitution of one or more nucleotide base pairs for one or more others, as shown in Figure 6.9A. Substitutions affecting just one nucleotide base pair are known as **point mutations**. Because of the redundancy in the genetic code (see Table 6.3), with more than one codon nucleotide triplet sequence often coding for a given amino acid, substitution may not lead to a change in the amino acid structure of the polypeptide chain formed during translation. For point mutations, the substitution of one pyrimidine for another (C for T or vice versa), or of one purine for another (A for G or vice versa) is known as a **transition**, whereas a **transversion** entails the substitution of a pyrimidine (C or T) by a purine (A or G), or vice versa.

Insertion

Figure 6.9B illustrates a typical mutation involving the insertion of one or more nucleotide base pairs in the genome. Because the genetic code is in the form of nucleotide triplets which are read as triplets during translation, if the number of nucleotide base pairs

inserted into the DNA is not a multiple of three, then the insertion leads to a shift in the **reading frame** during translation, so that a large number of incorrect amino acids are attached to the polypeptide chain. On the other hand, if the insertion is of number of nucleotide base pairs that is a multiple of three, then during the process of translation the nucleotide sequence distal to the insertion may be read as triplets correctly with no shift in the reading frame.

Deletion

A deletion is shown in Figure 6.9C. So far as the reading frame is concerned, a similar consideration to that for insertions applies, with the deletion of a number of nucleotide base pairs that is not a multiple of three leading to a shift in this frame.

Translocation

A translocation involves the breakage of a part of one chromosome and the addition of this part to another chromosome of a different chromosome pair. Translocations are considered further below in the subsection on chromosomal abnormalities.

CHROMOSOMES AND CELL DIVISION

The way the genome is arranged into chromosomes and passed on in cell division is considered in this section.

Chromosomes

The mammalian genome is arranged in the cell nucleus on chromosomes. If the genome is likened to an encyclopaedia of genetic information, then the chromosomes can be thought of as being the volumes of the encyclopaedia, with each species having a specific number of such volumes, or chromosomes, per somatic cell nucleus. In normal humans there are 46 chromosomes per somatic cell nucleus. These consist of a pair of **sex chromosomes** and 44 **autosomes**, the latter consisting of 22 pairs of chromosomes. The sex chromosomes are known as the X chromosome and the Y chromosome; normal females carry XX, while normal males carry XY.

In practice it is only during certain stages of the process of mitotic or, for gonadal cells, meiotic cell division, described below, that it is feasible to study the chromosomes of a cell nucleus. During this process the chromosomes, which are normally very thin and long-drawn out, and therefore difficult to identify under the

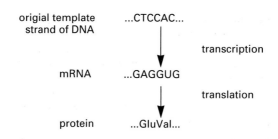

(a) Substitution

DNA ...CTCAAC...

mRNA ...GAGUUG...

protein ...GluLeu...

DNA ...CTCCAC...

mRNA ...GAGUUC...

protein ...GluVal...

there is no change in the
amino acid in this case owing
to the redundancy in the
genetic code

(b) Insertion

DNA ...CTCGGCAC...

mRNA ...GAGCCGUG...

protein ...GluPro...

shift in reading frame

the number of inserted
nucleotides is a multiple of 3

DNA ...CTCTTCCAC...

mRNA ...GAGAAGGUG...

protein ...GluLysVal...

the nucleotide sequence distal
to the insertion is read correctly

(c) Deletion

DNA ...CCCAC... (deletion of T)

mRNA ...GGGUG...

protein ...Gly...

a shift in the reading frame has occurred

Fig. 6.9 DNA mutations: (A) substitution; (B) insertion; (C) deletion.

microscope, become shortened and thickened. It is possible to stimulate somatic cell division, arrest this at a suitable stage, and then disperse the chromosomes so that they can be fixed, stained, photographed, identified, and arranged to show the chromosomal make-up of the cells. Such an arrangement is known as the **karyotype**. Since there are 46 chromosomes in the normal human somatic cell nucleus, a shorthand for the normal human female karyotype is 46, XX, while that for the normal human male karyotype is 46, XY.

The process of karyotyping reveals that in the normal human the autosomes consist, as mentioned above, of 22 pairs of chromosomes. The two chromosomes in each pair are similar to each other and are sometimes referred to as being **homologous**. However, the two chromosomes of each homologous pair are not identical to each other in the genetic information they carry, with one chromosome in each pair having originally been inherited from the subject's father and the other chromosome from the subject's mother.

Karyotyping is not necessary to determine the number of X chromosomes in each somatic cell nucleus. It has been found that the number of **Barr bodies** — also known as chromatin bodies — seen microscopically in the cell nucleus, is related to the number of X chromosomes in that nucleus. The relationship is such that the number of Barr bodies is equal to the number of X chromosomes minus one. Thus normal females (XX) and normal males (XY) have respectively one Barr body and no Barr bodies per somatic cell nucleus. On the other hand, the corresponding numbers of Barr bodies for individuals with a female appearance or **phenotype**, but a genetic make-up or **genotype** of XO (Turner's syndrome) and XXX (superfemale) are zero and two, respectively.

Each chromosome has a somewhat constricted region known as the **centromere** that is particularly evident during the processes of mitosis and meiosis described below. The position of the centromere is such that a chromosome is usually divided into two arms of differing lengths. The short arm is conventionally denoted by the letter **p**, and the long arm by the letter **q**. Chromosomes with a centrally or almost centrally positioned centromere are known as **metacentric**, while the term **acrocentric** is used for those chromosomes in which the centromere is very near to one end.

Various stains have been found to render a transverse banding pattern to chromosomal arms. These patterns have been found to have characteristic features for each chromosomal arm,thereby allowing individual chromosomes to be identified other than by chromosomal length during karyotyping. In an internationally agreed classification system, numbers have been attached to the bands produced under specific staining conditions. The result is that a **chromosomal map** has been produced which, rather like a library index system, allows locations in the genome to be classified. In this system, the number (1 to 22) or letter (X or Y) of the chromosome is first given, followed by a p (for short arm) or q (long arm), and then a digit corresponding to a stretch of the chromosome lying between two relatively distinct morphological landmarks (such as the centromere and the end of a chromosomal arm) and known as a **region**, and fourthly a digit corresponding to a **band** derived from staining properties. This system was agreed at the International Paris Conference in 1971. For example, the Fragile X syndrome, which is responsible for a relatively large proportion of cases of mental retardation in males, is associated with a fragile site on the long arm (q) of the X chromosome at a position in region 2 and band 7. Therefore the location of this fragile site can be more conveniently described as being at Xq27. Where necessary, this system can be find tuned further by specifying subbands written using a decimal point after the band digit, for example Xq27.1.

A given gene is located at the same position on the same chromosome and can exist as one of a number of alternative forms known as **alleles**. In the somatic cell nucleus each gene exists in an allelic pair with each allele having been inherited from the gamete of each parent.

Cell division

Mitosis

Mitosis is the process of nuclear division by means of which many somatic cells (although not including differentiated neurones) undergo cell division throughout the lifetime of an individual. The stages of mitosis, from the resting interphase, through prophase, metaphase, anaphase and telophase, are shown diagrammatically in Figure 6.10. It can be seen from Figure 6.10 that before entering prophase each chromosome replicates longitudinally into two identical sister **chromatids** which initially remain attached to each other at the single centromere. After the chromosomes have become aligned at the nuclear equator during metaphase, the centromere is also duplicated and each pair of sister chromatids separates to form two identical chromosomes, each with its own centromere that is attached to the mitotic spindle during anaphase. During this phase each sister of the new daughter chromosomes appears to be pulled by its centromere along the spindle to opposite poles of the nucleus. The result is that when the nucleus divides during telophase, each daughter nucleus has a full and identical complement of chromosomes to the original nucleus. In other words, a replication of the genome has taken place, and subsequently each new daughter somatic cell inherits the same genetic information as that possessed by the original cell.

Meiosis

Meiosis is a process, involving two stages of cell division, that occurs in gamete formation, or **gametogenesis**. During this process chromosomal division occurs once, and not twice, as shown in Figure 6.10.

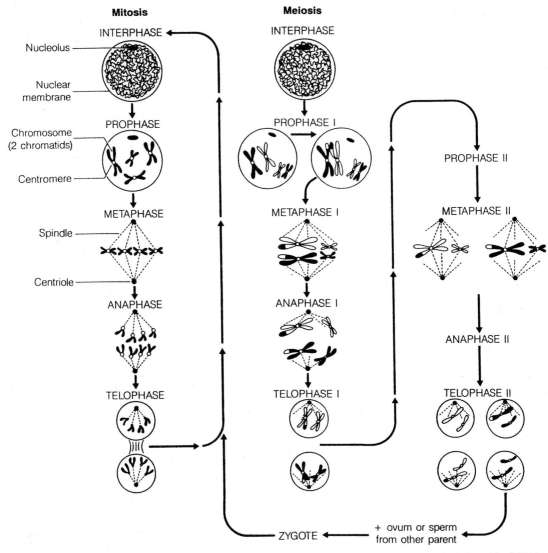

Fig. 6.10   The stages of mitosis and meiosis. (Reproduced with permission from Macleod J, Edwards C, Bouchier I 1991 Davidson's principles and practice of medicine. 10th edn. Churchill Livingstone, Edinburgh.)

The result is that each gamete inherits one of each of the 22 original pairs of homologous chromosomes. Each gamete also inherits one of the two original sex chromosomes. In a normal female this means that her gametes, or ova, each contain one X chromosome (from the original XX genotype). This also implies that in a normal male half the gametes, or spermatozoa, each contain an X chromosome, and half contain a Y chromosome (from the original XY genotype). Hence during reproduction it is the sex chromosome of the fertilizing spermatozoon that determines whether the zygote is genetically male (XY) or female (XX). It can

also be seen that each gamete must contain a total of 23 chromosomes, or just one-half of the original somatic cell nucleus chromosome count of 46 (22 pairs of homologous chromosomes plus either XX or XY); hence the gametes are said to be **haploid**. On fertilization of a haploid ovum by a haploid spermatozoon, it can be seen that the normal chromosomal number of 46 (22 pairs of homologous chromosomes plus either XX or XY) is restored in the somatic cell nuclei of the zygote.

During the prophase of the first meiotic division, there is an alignment and contact of homologous chromosome pairs which allows genetic information to **crossover** between adjacent chromatids (see Figure 6.10). Since the location of a given gene is the same on the chromatids of homologous dividing chromosome pairs, this process, also known as **recombination**, leads to a change in the alleles carried by the chromatids at the end of the first meiotic division. The number of types of crossover that can occur is exceedingly large and leads to the extremely large variety of gamete genotypes that each individual produces. The inheritance of different alleles is considered in the next section.

GENETIC DISORDERS

In this section genetic disorders are considered under the following classification: Mendelian inheritance, chromosome disorders, and non-Mendelian inheritance.

Mendelian inheritance

Single gene defects are inherited according to Mendelian patterns, named after the nineteenth century Austrian monk and botanist Gregor Mendel whose experiments with the breeding of the garden pea gave rise to **Mendel's laws**. These single gene defects include deletions, insertions, and inversions.

When a subject possesses two identical alleles at the same locus for a given trait he or she is said to be **homozygous** for that trait. If the subject possesses two different alleles at that locus the term **heterozygous** is used. Although more than two different alleles may exist in the human population, clearly in any individual no more than two alleles can normally be present because the chromosomes exist as 22 homologous pairs of autosomes together with one pair of sex chromosomes.

In the following discussion of the law of uniformity and Mendel's first and second laws, we shall suppose that at a given locus the two alleles R and r may exist for a certain trait. Clearly there are two possible homozygous genotypes for that trait, namely RR and rr. In the one possible heterozygous genotype, Rr (which is the same as rR), if the phenotypic expression is that seen in RR individuals rather than in rr individuals, then R is said to be **dominant** over r; R is the dominant allele, and r is the **recessive** allele. As an illustration, the implications of an early genetic model for eye colour, in which blue eyes were held to be recessive and brown eyes to be dominant, will be examined briefly. According to this model, if we were to consider only brown-eyed and blue-eyed indi-

viduals, then R would be the allele for brown eyes and r the allele for blue eyes. Individuals with the genotypes RR and Rr will have the phenotypic expression of brown eyes (R is dominant), while a phenotypic expression of blue eyes will result only from the homozygous genotype rr (r is recessive). In practice, this model has been negated by numerous cases of brown-eyed individuals being born to two blue-eyed parents. A more appropriate model for eye colour is probably one involving polygenic inheritance, which is discussed below under non-Mendelian inheritance; the interested reader is referred to the review by Rufer et al (1970).

Law of uniformity

Consider the mating of two homozygous parents with the respective genotypes RR and rr. Since their gametes contain the alleles R and r, respectively, then the next generation, conventionally known as the F1 generation, can only have the genotype Rr. This is shown in Figure 6.11.

Mendel's first law

Mendel's first law is also known as the **law of segregation** and is illustrated by the example in Figure 6.12. As mentioned above, if R is the dominant allele and r is a recessive allele, then the genotype of the heterozygote, Rr, contains both alleles. However, it is only the dominant allele, R, that is important so far as the phenotype is concerned. When two heterozygote individuals (Rr) mate, it can be seen from Figure 6.12 that, on average, the genotypes of the offspring will be RR, Rr, and rr, in the ratio 1:2:1. This is because segregation of the alleles occurs. With R being dominant, this means that the phenotype ratio of the offspring on average is (3 dominant):(1 recessive),

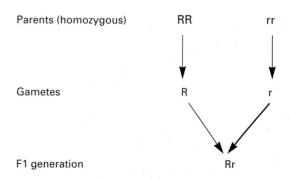

Parents (homozygous)   RR   rr

Gametes   R   r

F1 generation   Rr

Fig. 6.11   The law of uniformity.

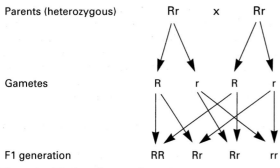

Fig. 6.12   The law of segregation.

since the genotype Rr has the same phenotype as RR. On the other hand, if the genotype Rr were to lead to a phenotype that is intermediate between those from RR and rr, then the phenotype ratio of the offspring on average is (1 dominant):(2 intermediate):(1 recessive).

Mendel's second law

Mendel's second law is also known as the **law of independent assortment** and is illustrated by the example in Figure 6.13. In this example two loci are now considered, with alleles Rr and Ss. Figure 6.13

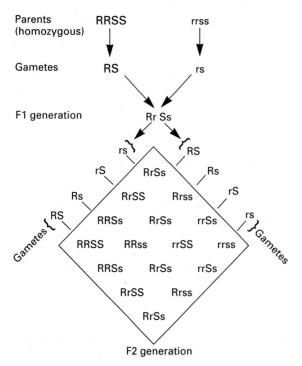

Fig. 6.13   The law of independent assortment.

shows what takes place when two homozygous individuals, with genotypes RRSS and rrss, respectively, mate. Members of the F1 generation will always have the genotype RrSs. If now two individuals with the genotype RrSs mate, there are 16 genotypes possible in the F2 generation, as shown in the diagram. This is the result of the independent assortment of different alleles.

Autosomal dominant disorders

Autosomal dominant disorders result from the presence of an abnormal allele which is dominant and causes the individual carrying it to manifest the abnormal phenotypic trait. There are a number of features of their transmission which allow autosomal dominant disorders to be identified. The phenotypic trait is present in all individuals carrying the dominant allele and in successive generations of a family without missing generations. Hence autosomal dominant disorders are said to show **vertical** transmission from one generation to the next. Both male and female individuals are affected. Male to male transmission can occur. The occurrence of the phenotypic trait in offspring is not solely dependent upon parental consanguineous matings. If a normal individual (genotype rr, say) mates with a heterozygote individual (genotype Rr, where R is the abnormal dominant allele), then on average one half of the offspring (with genotype Rr) will manifest the abnormal phenotypic trait. Following the much rarer event of two heterozygote individuals (Rr) mating, on average three quarters of the offspring (with genotypes RR and Rr) will manifest the abnormal phenotypic trait. Mating between two heterozygote individuals is much more common in consanguineous matings than in non-consanguineous matings. In the event that one of the parents is homozygous (genotype RR), all the offspring will manifest the abnormal phenotypic trait.

Family pedigrees can be drawn using conventional symbolic notation. Some of the symbols used are shown in Figure 6.14. Figure 6.15 shows a typical family pedigree for an autosomal dominant disorder such as Huntington's chorea.

When the autosomal dominant disorder reduces the fertility of affected individuals, and most cases observed are sporadic, then new cases are more likely to be the result of new dominant **mutations**. This is particularly likely to be so if the average parental age is raised. An example of a condition in which the majority of new cases is caused by new mutations is acrocephalosyndactyly type I, more commonly known as the typical Alpert syndrome or Alpert's syndrome. In this condition, which causes skull malformation,

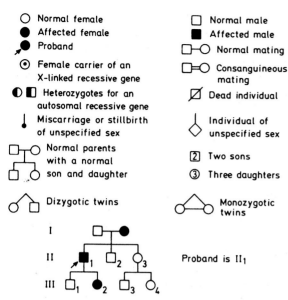

Fig. 6.14 Common symbols used in drawing family pedigrees. (Reproduced with permission from Emery A E H, Rimoin D L 1983 Principles and practice of medical genetics vol 1. 1st edn. Churchill Livingstone, Edinburgh.)

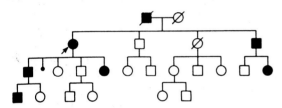

Fig. 6.15 Characteristic pedigree of an autosomal dominant trait. (Reproduced with permission from Emery A E H, Rimoin D L 1983 Principles and practice of medical genetics vol 1. 1st edn. Churchill Livingstone, Edinburgh.)

syndactyly, and, usually, mental retardation, there is a raised mean parental age.

Occasionally a pedigree may show a pattern of affected individuals that appears to be consistent with vertical transmission caused by an autosomal dominant type of inheritance when in fact the disorder is caused by an infectious agent. An example of such a condition is Creutzfeldt–Jakob disease, which is discussed in the next chapter. This can be a problem when the cause of a disorder of unknown or uncertain aetiology is being investigated.

Huntington's chorea, also known as Huntington's disease, has already been mentioned as an example of a disorder seen in psychiatric clinical practice which is inherited as an autosomal dominant trait. Another well-established example is acute intermittent porphyria, also known as porphobilinogen deaminase deficiency

or the Swedish type of porphyria (owing to its comparatively high prevalence of the order of 1 in 1 000 in Northern Sweden). There is also strong evidence of a dominantly inherited defect in an autosomal gene being responsible for some cases of Alzheimer's disease (Nee et al 1983, Foncin et al 1986) and Pick's disease, also known as lobar atrophy of the brain (Schenk 1959).

The phacomatoses consist of four disorders which in many cases are probably caused by autosomal dominant inheritance or new dominant mutations. The first phacomatosis is tuberous sclerosis, for which tuberose sclerosis, adenoma sebaceum and epiloia are synonyms, and which is sometimes associated with psychotic symptomatology and mental retardation. Mental retardation is also sometimes a feature of the second and third disorders, namely, neurofibromatosis or von Recklinghausen's disease, and von Hippel–Lindau syndrome. The fourth phacomatosis, Sturge–Weber syndrome, is also occasionally associated with mental retardation. However, compared with the other three phacomatoses, the evidence for autosomal dominant inheritance in the Sturge–Weber syndrome is poor, although father to son transmission has been described in this condition (Debicka & Adamczak 1979).

A number of other conditions that can result in mental retardation are sometimes caused by an autosomal dominant allele, either inherited or resulting from a new mutation. In addition to acrocephalosyndactyly type I, mentioned above, they include the following conditions: the acrocallosal syndrome, in which absence of the corpus callosum, macrocephaly, and severe mental retardation occur (Schinzel & Schmid 1980); acrodysostosis (Jones et al 1975); a syndrome referred to by its constituent abnormalities as alopecia, psychomotor epilepsy, pyorrhoea and mental subnormality (Shokeir 1977); aortic arch anomaly with peculiar facies and mental retardation (Strong 1968); convulsive disorder and mental retardation (Juberg & Hellman 1971); the De Barsey syndrome of cutis laxia, corneal clouding, and mental retardation (Hoefnagel et al 1971); dominant mental retardation (Dekaban & Klein 1968); the periodic paralyses (Pearson & Kalyanaraman 1972); and progressive ophthalmoplegia with scrotal tongue and mental deficiency (Levic et al 1975).

Autosomal recessive disorders

Autosomal recessive disorders result from the presence of two abnormal recessive alleles (rr, say) in the genotype which cause the individual (male or female) carrying them to manifest the abnormal phenotypic trait. There are a number of features of their transmission which allow autosomal recessive disorders to be identified. In general individuals who are heterozygotes (carrying one abnormal recessive allele, r, and a normal dominant allele, R, say) are carriers of the disorder who do not usually manifest the abnormal phenotypic trait. The most likely mating of an affected individual (genotype rr) is with a normal homozygous individual (RR) and will result in all the offspring being unaffected heterozygous carriers (Rr). If such a heterozygous carrier (Rr) mates with another heterozygous carrier (Rr), the average ratio of affected offspring (rr) to unaffected offspring (Rr or RR) will be 1:3, that is, a quarter of the offspring will on average be affected. In the much rarer event of two affected individuals (rr) mating, a situation more common in consanguineous matings, all the offspring will also be affected homozygotes (rr).

It can be seen, therefore, that an autosomal recessive disorder tends to miss generations. When it does occur, if the number of siblings is large then usually on average a quarter of them are affected, whereas the parents (F1 generation) and the children (F3 generation) of the index cases are likely not to be affected. Thus, unlike the vertical transmission of autosomal dominant disorders, autosomal recessive disorders tend to show a **horizontal** pattern (see Fig. 6.16).

If the abnormal phenotypic trait is secondary to an enzyme deficiency, then a heterozygous carrier may show a **partial deficiency** of that enzyme on testing. For example, the autosomal recessive disorder homocystinuria, which is often associated with mental retardation, is secondary to a deficiency of the enzyme cystathionine synthetase. Deficiency of this enzyme leads to an increased level of homocystine in the blood and urine, and an elevated blood methionine level. In heterozygous carriers a partial deficiency of cyst-

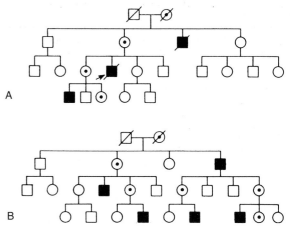

Fig. 6.17 Characteristic pedigree of an X-linked recessive trait in which affected males do not reproduce (**A**) and in which affected males may reproduce (**B**).

athionine synthetase may be present and can be tested for by administering a loading dose of L-methionine. Similarly, there exist loading tests for detecting heterozygous carriers of other autosomal recessive disorders such as phenylketonuria.

Both homocystinuria and phenylketonuria are disorders of protein metabolism. There are many other such disorders which are probably also autosomal recessive conditions which may be seen in psychiatric clinical practice. Some of these disorders have been previously described in Chapter 3. They are summarized in Table 6.4 with references being given where required, for example, for those conditions not previously described that are exceedingly rare.

Similarly, there is a large number of autosomal recessive disorders of carbohydrate and lipid metabolism. Some of these, including storage diseases such as mucolipidoses, mucopolysaccharidoses and sphingolipidoses, are given in Table 6.5. It should be noted that for some of these disorders, for example, Gaucher's disease, more than one type exist.

There are many other autosomal recessive disorders that are relevant to psychiatry; some of these are given in Table 6.6. Note that Wilson's disease, also known as hepatolenticular degeneration, is discussed in the next chapter. There is some evidence that a transketolase defect inherited as an autosomal recessive may give rise to a genetic component in the Wernicke–Korsakoff syndrome (Blass & Gibson 1979).

There are also some autosomal recessive disorders that are of importance with respect to psychopharmacology. For example, an inherited arene oxide detoxification defect can lead to toxicity following

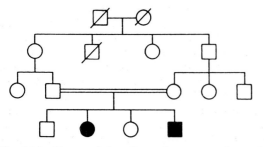

Fig. 6.16 Characteristic pedigree of an autosomal recessive trait. (Reproduced with permission from Emery A E H, Rimoin D L 1983 Principles and practice of medical genetics vol 1. 1st edn. Churchill Livingstone, Edinburgh.)

Table 6.4 Autosomal recessive disorders of protein metabolism

Aminoaciduria with mental deficiency, dwarfism, muscular dystrophy, osteoporosis and acidosis (Stransky *et al* 1962)
Aminoadipicaciduria (Fischer & Brown 1980)
Argininaemia or arginase deficiency
Beta-methylcrotonylglycinuria (Eldjarn *et al* 1970)
Cystathioninuria
Cysteine peptiduria (Ben-Ami *et al* 1973)
Cystinuria
Dicarboxylic aminoaciduria (Melancon *et al* 1977)
Formiminotransferase deficiency or formiminoglutamicaciduria (Arakawa *et al* 1968)
Hartnup's disease
Histidinaemia or histidase deficiency
Homocarnosinosis (Sjaastad *et al* 1976)
Homocystinuria
Hydroxyprolinaemia or 4-hydroxy-L-proline deficiency
Hyperlysinaemia or lysine-ketoglutarate reductase deficiency
Hyperlysinuria with hyperammonaemia (Brown *et al* 1972)
Hyperprolinaemia types I and II
Indolylacroyl glycinuria (Mellman *et al* 1963)
Isovalericacidaemia (Budd *et al* 1967)
Lysine malabsorption syndrome (Omura *et al* 1976)
Maple syrup urine disease, also known as branched-chain ketoaciduria and ketoacid decarboxylase deficiency
Mercaptolactate-cysteine disulphiduria (Ampola *et al* 1969)
Methionine malabsorption syndrome, also known as Smith–Strang disease and oasthouse urine disease (Smith & Strang 1958)
3-Methylglutaconicaciduria (Greter *et al* 1978)
Methylmalonicaciduria
Non-ketotic hyperglycinaemia
Ornithinaemia
Phenylketonuria
Renal histidinuria (Kamoun *et al* 1981)
Saccharopinuria (Carson *et al* 1968)
Stimmler's syndrome or alaninuria with microcephaly, dwarfism, enamel hypoplasia, and diabetes mellitus (Stimmler *et al* 1970)
Type II tyrosinaemia, also known as oculocutaneous tyrosinaemia, tyrosine transaminase (or aminotransferase) deficiency, and the Richner–Hanhart syndrome
Urea cycle disorders mentioned in Chapter 3 (hyperammonaemia, citrullinaemia, and arginosuccinic aciduria) together with, in general, other rarer disorders of the urea cycle such as the hyperornithinaemia-hyperammonaemia-homocitrullinuria syndrome (Shih *et al* 1969)
Urocanase deficiency (Yoshida *et al* 1971)
Valinaemia

Table 6.5 Autosomal recessive disorders of carbohydrate and lipid metabolism

Cephalin lipidosis (Baar & Hickmans 1956)
Fucosidosis or alpha-L-fucosidase deficiency (van Hoof & Hers 1968, Patel *et al* 1972)
galactosaemia
Gaucher's disease
Generalized gangliosidosis (particularly types I and II)
Hereditary fructose intolerance or fructosaemia
Hurler's syndrome or mucopolysaccharidosis type I
Krabbe's disease or globoid cell leukodystrophy
Lactosylceramidosis or neutral beta-galactosidase deficiency (Burton *et al* 1977)
Mannosidosis or lysosomal alpha-D-mannosidase deficiency (Ockerman 1967)
Metachromatic leucodystrophy
Morquio's syndrome
The mucolipidoses
Niemann–Pick disease
Pompe's disease, also known as cardiomegalia glycogenica diffusa, acid maltase deficiency and glycogen storage disease II
Sanfilippo's syndrome types A, B, C and D, also known respectively as mucopolysaccharidosis types IIIA, IIIB, IIIC and IIID
Scheie's syndrome or mucopolysaccharidosis type I
Sea-blue histiocyte disease or syndrome (Silverstein & Ellefson 1972)
Sulphatidosis or multiple sulphatase deficiency (Basner *et al* 1979)
Tay-Sachs disease or amaurotic idiocy
Tricarboxylic acid cycle defect (Blass *et al* 1972)
Triosephosphate isomerase deficiency (Mohrenweiser 1981)
Von Gierke's disease, also known as hepatorenal glycogenosis and glycogen storage disease I

probably inherited as an autosomal recessive (Turkington 1972).

X-linked disorders

X-linked, or sex-linked, disorders result from the presence of an abnormal allele on the X chromosome. As with autosomal disorders, X-linked disorders can be either recessive (see Fig. 6.17) or dominant.

Normal males (XY) have just one X chromosome per somatic cell nucleus. Therefore, in the case of **X-linked recessive disorders**, all male offspring inheriting the abnormal recessive allele will manifest the phenotypic trait since there is no other X-linked normal allele with which the abnormal allele can pair in the somatic cell nucleus; such males are termed **hemizygous** for the X-linked mutant gene. On the other hand, because normal females (XX) have two X chromosomes per somatic cell nucleus, female offspring inheriting one abnormal X-linked recessive allele (X_r, say) from one parent and one corresponding

pharmacotherapy with phenytoin, including the fetal hydantoin syndrome, since arene oxides are intermediate metabolites in the catabolism of this anticonvulsant (Hanson et al 1976, Phelan et al 1981). Similarly, whereas normally treatment with phenothiazines such as chlorpromazine leads to hyperprolactinaemia, this does not occur in the syndrome of isolated prolactin deficiency, a disorder that is

Table 6.6   Other autosomal recessive disorders

Achalasia-microcephaly syndrome (Williams *et al* 1978)
Acrodysostosis with mental retardation and nasal hypoplasia
  (Niikawa *et al* 1978)
Agenesis of cerebral white matter (Waggoner *et al* 1942)
Agenesis of the corpus callosum (Naiman & Fraser 1955)
Alexander's disease (Alexander 1949)
The alopecia-epilepsy-oligophrenia syndrome of Moynahan
  (Moynahan 1962)
A syndrome of aniridia, cerebellar ataxia, and mental
  retardation (Gillespie 1965)
A syndrome of partial aniridia, unilateral renal agenesis, and
  psychomotor retardation (Sommer *et al* 1974)
Ataxia with myoclonus epilepsy and presenile dementia (Skre
  & Loken 1970)
A syndrome of ataxia, deafness and mental retardation
  (Berman *et al* 1973)
Athyreotic cretinism (Ainger & Kelley 1955)
Bardet-Biedl syndrome (Schachat & Maumenee 1982)
Bird-headed dwarfism, also known as nanocephaly, and
  Seckel's syndrome (Sauk *et al* 1973)
Bird-headed dwarfism, Montreal type (Fitch *et al* 1970)
The camak syndrome of cataracts, microcephaly,
  arthrogryposis, and kyphosis (Lowry *et al* 1978)
The camfak syndrome of cataracts, microcephaly, failure to
  thrive, and kyphoscoliosis (Scott-Emuakpor *et al* 1977)
Cerebellar hypoplasia (Norman & Urich 1958)
The cerebelloparenchymal disorders
Choroid plexus calcification and mental retardation (Lott *et al*
  1979)
Coat's disease of deafness, muscle weakness, and mental
  retardation (Small 1968)
Cockayne's syndrome (Paddison *et al* 1963, Proops *et al*
  1981)
Cohen's syndrome (Kousseff 1981)
Craniodiaphyseal dysplasia (Joseph *et al* 1958)
Craniofacial dyssynostosis (Neuhauser *et al* 1976)
Crome's syndrome (Crome *et al* 1963)
The Dubowitz syndrome (Parrish & Wilroy 1980)
Dwarfism associated with mental retardation (Mollica *et al*
  1972)
The Dyggve-Melchior-Clausen syndrome (Bonafede and
  Beighton 1978)
The dysequilibrium syndrome, also known as non-progressive
  cerebellar disorder with mental retardation (Schurig *et al*
  1981)
Ectodermal dysplasia associated with mental retardation
  (Bowen & Armstrong 1976)

Table 6.6   (continued)

Familial photogenic epilepsy with mental retardation (Daly *et*
  al 1959)
Friedrich's ataxia (Davies 1949)
Hallervorden–Spatz disease (Dooling *et al* 1974)
The Hall–Riggs mental retardation syndrome (Hall & Riggs
  1975)
The happy puppet syndrome, also known as Angelman's
  syndrome (Williams & Frias 1982)
Ichthyosiform erythroderma associated with mental
  retardation (Tay 1970)
A syndrome of ichthyosis, mental retardation, dwarfism, and
  renal impairment (Passwell *et al* 1975)
Kifafa seizure disorder (Jilek-Aall *et al* 1979)
The Laurence–Moon syndrome (Schachat & Maumenee
  1982)
Leukomelanoderma associated with mental retardation
  (Berlin 1961)
Macrocephaly or megalencephaly (de Myer 1972)
The Marinesco-Sjogren syndrome (Sjogren 1950)
The Mast syndrome (Cross & McKusick 1967)
The megalocornea-mental retardation syndrome or MMR
  syndrome (Neuhauser *et al* 1975)
Mental retardation syndrome, Mietens–Weber type (Mietens
  & Weber 1966
Mental retardation syndrome, Buenos Aires type (Mutchinick
  1972)
A syndrome of metaphyseal dysostosis, conductive hearing
  loss, and mental retardation (Rimoin & McAlister 1971)
Microcephaly (primary) (Kloepfer *et al* 1964)
Muscle–eye–brain disease or MEB disease (Raitta *et al* 1978)
The Lafora type of myoclonus epilepsy or Lafora's disease
  (Janeway *et al* 1967)
Martsolf's syndrome (Martsolf *et al* 1978)
A syndrome of mental retardation, epilepsy, palpebral
  conjunctival telangiectasias, and IgA deficiency (Aguilar *et*
  al 1978)
The N syndrome (Hess *et al* 1974)
The oculocerebral syndrome with hypopigmentation (Cross *et*
  al 1967)
The oculoosteocutaneous syndrome (Tuomaala & Haapanen
  1968)
The oculorenocerebellar syndrome (Hunter *et al* 1982)
Rubinstein's syndrome or the Rubinstein–Taybi syndrome
  (Rubinstein & Taybi 1963, Johnson 1966)
Salla disease (Aula & Autio 1980)
Turcot's syndrome (Turcot *et al* 1959)
Wilson's disease or hepatolenticular degeneration

X-linked normal allele (X_R, say) from the other parent will not normally develop the abnormal phenotype but instead will be carriers (X_RX_r). It is only in the extremely rare event that a daughter inherits the abnormal X-linked recessive mutant genes from both parents that she will develop the phenotypic trait (X_rX_r). It follows that if an unaffected female carrier (X_RX_r) were to mate with a normal male, then on average half of their sons would be affected (X_r) and half their sons would be unaffected (X_R), while half of

their daughters would be carriers (X_RX_r) and half their daughters would be neither affected nor carriers (X_RX_R). X-linked disorders often reduce the ability of affected males to reproduce, for example because of early death. In what is therefore the less likely situation of an affected male (X_rY) mating with an unaffected female (X_RX_R) it can be seen that none of the sons (X_RY) will manifest the abnormal phenotype and that all the daughters (X_RX_r) will be carriers (see Fig. 6.17).

It follows from the above discussion that in X-linked

recessive inheritance male to male transmission does not occur and female heterozygotes are carriers. In attempting to identify X-linked recessive inheritance, additional evidence may be available in the form of enzyme activity. In a given family, affected males may demonstrate enzyme deficiency on testing, while carrier females and the mother may show only partial deficiency; the father (unless he is affected) will have normal enzyme activity.

In the case of **X-linked dominant inheritance** when an affected male mates with an unaffected female, then clearly all the daughters and none of the sons will be affected, since male to male transmission cannot occur (a son inherits a Y sex chromosome and not an X sex chromosome from his father). On the other hand, when an affected heterozygous female mates with an unaffected male, on average half the sons and half the daughters will be affected.

Examples of disorders of relevance to psychiatry which in general show X-linked recessive inheritance are given in Table 6.7.

A smaller number of disorders of relevance to psychiatry show X-linked dominant inheritance. They may include: the Aicardi syndrome (Bertoni et al 1979); the Coffin–Lowry syndrome (Lowry et al 1971); and the Rett syndrome (Haas 1988).

Chromosomal abnormalities

Chromosomal abnormalities can arise during both meiosis and mitosis. Those occurring in meiosis may lead to abnormal fertilizing gametes and in turn to abnormalities in the resulting offspring. Chromosomal abnormalities arising in somatic cell mitosis can lead to a cell line which may give rise to a relatively small number of abnormal cells which the affected individual will possess in addition to his or her normal cells; such an individual is known as a **mosaic**. The first type of chromosomal abnormality is of importance in clinical psychiatry and is considered further in this section. It should be noted that fetal chromosomal abnormalities are often lethal and lead to spontaneous abortions early in the pregnancy.

Chromosomal abnormalities can be classified into those affecting the sex chromosomes and those affecting autosomes.

Types

Both autosomal abnormalities and sex chromosome abnormalities can take the form of numerical and structural disorders.

In **numerical** disorders, or **aneuploidy**, there is an

Table 6.7 X-linked recessive disorders

Borjeson's syndrome (Borjeson et al 1962)
Cerebellar ataxia (Shokeir 1970, Malamud & Cohen 1958)
Charcot–Marie–Tooth peroneal muscular dystrophy combined with Friedreich's ataxia (van Bogaert & Moreau 1939–1941)
Partial agenesis of the corpus callosum (Menkes et al 1964)
Cutis verticis gyrata, thyroid aplasia, and mental deficiency (Akesson 1965)
Dwarfism, cerebral atrophy, and generalized keratosis follicularis (Cantu et al 1974)
The FG syndrome (Keller et al 1976, Riccardi et al 1977)
The fragile X syndrome
Hunter's syndrome or mucopolysaccharidosis type II
Hyperglycerolaemia
Hyperuricaemia
Intrauterine growth retardation, mental retardation, and microcephaly (Warkany et al 1961)
The Lesch–Nyhan syndrome
Microphthalmia
Nerve deafness, optic nerve atrophy, and dementia (Jensen 1981)
Ornithine transcarbamylase deficiency
The oculocerebrorenal syndrome of Lowe
The Paine syndrome (Paine 1960, Seemanova et al 1973)
Pelizaeus-Merzbacher disease (Tyler 1958)
The Plott syndrome of congenital laryngeal abductor paralysis (Opitz et al 1978)
Diffuse Scholz type cerebral sclerosis (Ford 1966)
Spastic athetotic paraplegia (Bundey & Griffiths 1977)
Testicular feminization syndrome
The van den Bosch syndrome (van den Bosch 1959)
X-linked cutis laxa or Ehlers-Danlos syndrome type IX (Lazoff et al 1975)
X-linked (congenital) hydrocephalus or hydrocephalus secondary to congenital stenosis of the aqueduct of Sylvius (Burton 1979)
A number of types of X-linked mental retardation not associated with the fragile X site, for example, the Renpenning type (Fox et al 1980)
X-linked parkinsonism (Johnston & McKusick 1961)
X-linked spastic paraplegia (Johnston & McKusick 1962)
The W syndrome (Pallister et al 1974)

increase or decrease in the number of chromosomes leading to an abnormal total number. Most possible forms are lethal. However, as is discussed below, an increase in the chromosome set by one autosome, leading to three autosomes instead of just a pair, and termed **trisomy**, is not always lethal and is seen, for example, in Down's syndrome. On the other hand, a reduction in the chromosome set by one autosome, known as **monosomy**, is not seen in live babies. In the case of sex chromosome abnormalities, both an increase by one (or more) sex chromosome, for example in Klinefelter's syndrome, and the decrease by one sex chromosome that occurs in Turner's syndrome, are compatible with fetal survival.

The **structural** disorders can incude: a **deletion** of part of a chromosome; a **translocation** of a part of one chromosome to another chromosome of a different pair; an **inversion** of a translocated chromosomal portion so that it comes to be oriented in the opposite direction to its original one; a **reciprocal translocation** in which parts of two chromosomes, each belonging to a different pair, are swapped; a **ring** chromosome formed by two broken ends of a chromosome fusing with each other; and an **isochromosome** resulting from separation of dividing sister chromatids at the centromere along an axis horizontal to that of the chromosome instead of longitudinally so that the chromosome consists of two short arms (pp) or two long arms (qq).

Nomenclature

It will be recalled that the normal male and female somatic genotypes can be represented respectively as 46, XY and 46, XX. That is, the total number of chromosomes is written first, followed by the sex chromosomes present per somatic cell nucleus.

Autosomal abnormalities resulting from trisomy of a chromosome N are represented by +N. For example, the genotype of a male baby with Patau's syndrome caused by trisomy 13 can be written as 47, XY, +13 and in general this genotype can be represented as 47, +13.

Sex chromosome abnormalities caused by a numerical chromosomal disorder are simply written as the total number of chromosomes followed by the sex chromosomes. For example, the genotype of Turner's syndrome is usually 45, XO or just 45, X.

The symbol '−' is used to denote either that a chromosome is missing, in which case the '−' precedes the appropriate chromosome number, or it can be used to represent a deletion. If the deletion is from the short arm of a given chromosome N, say, then this is written as Np−. Similarly, '+' can be used to represent the presence of additional chromosomal substance. For example, the presence of additional substance on the long arm of chromosome N is written as Nq+.

Autosomal abnormalities

Autosomal abnormalities usually result in mental retardation and usually affect more than one organ system. The commonest such causes of mental retardation are now briefly discussed.

The commonest autosomal abnormality is **Down's syndrome**, originally termed Mongolism, which has an incidence of between one in 600 and one in 700 live briths and is characterized by bradycephaly, widely spaced eyes with epicanthic folds and oblique palpebral fissures, Brushfield spots, a small nose and mouth, a horizontally furrowed tongue, a high arched palate, malformed ears, broadening and shortening of the neck and hands, a single transverse palmar crease, curvature of the fifth finger, an increased range of joint movements, and hypotonia. Down's syndrome is associated with an increased incidence of cataracts, congenital cardiac disease, umbilical herniae, respiratory infections, and acute leukaemia. In approximately 85% of cases the IQ is less than 50. Children with Down's syndrome usually have a pleasant personality and often enjoy music. Compared with the previously high infant mortality, many individuals with Down's syndrome now survive to their fourth decade and beyond. Most of the older individuals develop the neuropathology seen in Alzheimer's disease (Oliver & Holland 1986) and the clinical features of dementia.

Approximately 95% of cases of Down's syndrome are the result of **trisomy 21** (47, +21) following **non-disjunction** during meiosis. In such cases the homologous pair of chromosomes 21 fails to separate during one of the anaphase stages of meiosis and both enter the same gamete. Fertilization of such an abnormal gamete gives rise to a zygote with trisomy 21. The risk of non-disjunction occurring increases with increasing maternal age and as a result an older mother is more likely than a younger one to give birth to a baby with trisomy 21, trisomy 13 (Patau's syndrome), or trisomy 18 (Edwards's syndrome).

Approximately 4% of Down's syndrome cases are caused by **translocation** involving **chromosome 21**. This can involve the exchange of chromosomal substance between chromosome 21 and chromosomes 13, 14, 15, 21 or 22.

The remainder of cases of Down's syndrome are **mosaics** resulting from non-disjunction in a cell division at or after the first cell division following fertilization. One cell line gives rise to cells with nuclei having a normal chromosome set, while in the other cell line the somatic cell nuclei each contain a third **chromosome 21**.

The association of Down's syndrome with Alzheimer's disease noted above, together with the finding that there may be an increased incidence of Down's syndrome in the relatives of patients with Alzheimer's disease (Heston et al 1981), provided support for the hypothesis that an important Alzheimer's disease genetic locus exists on chromosome 21. There is now strong evidence that this is indeed the case as is discussed later in this chapter.

Edward's syndrome is caused by **trisomy 18** (47, +18) and has an incidence of between one in 3 500 and one in 6 700 live births. It is characterized by severe mental retardation and other neurological disorders, micrognathia, and abnormalities in the shapes of the skull, chest wall, fingers and toes. There is an increased incidence of congenital cardiac disease in this condition, and the majority of patients die within the first year of life.

Less common is **Patau's syndrome**, caused by **trisomy 13** (47, +13), which has an incidence of approximately one in 6 700 live births. In addition to severe mental retardation, Patau's syndrome is characterized by neurological disoreders, microcephaly, microophthalmia, cleft palate and hare lip, and polydactyly. There is an increased incidence of cataracts, congenital cardiac disease, and deafness. Approximately 95% of patients die within the first three years of life.

The **cri du chat syndrome** is caused by a partial **deletion of the short arm of chromosome 5**, that is, 5p–. It has an incidence of approximately one in 50 000 live births and derives its name from a characteristic kitten-like cry occurring in patients as a result of abnormalities of the superior larynx. Other characteristics include severe mental retardation, low body mass at birth, microcephaly and hypertelorism.

Other causes of mental retardation include trisomy 8, trisomy 22, 4p–, 13q–, 18p–, and 18q–.

Sex chromosome abnormalities

Klinefelter's syndrome is a sex chromosome abnormality in which phenotypic males possess more than one X chromosome per somatic cell nucleus. Usually there is one extra X chromosome, so that the genotype is usually **47, XXY**. However, the following genotypic variations can occur: 48, XXXY, 49, XXXXY, and 48, XXYY. The incidence of Klinefelter's syndrome is estimated to be between one in 400 and one in 600 live male births, and as with Down's syndrome, it has an increasing incidence with inceasing maternal age. Many of the characteristic clinical features are seen following puberty and include hypogonadism, gynaecomastia, reduced or absent postpubertal facial hair, a female distribution of pubic hair, and body proportions that are eunuchoid. Although the libido is sometimes normal, patients with Klinefelter's syndrome are usually sterile owing to azoospermia. Mild mental retardation may occur in 47, XXY individuals but many have an IQ in the normal range. Severe mental retardation tends to be found in individuals with the other rarer genotypic variations mentioned above.

As implied by its name, the genotype of the **XYY syndrome** is 47, XYY. This sex chromosome abnormality affects phenotypic males and has an incidence of approximately one in 700 live male births. Individuals tend to be taller than average. They also tend to score somewhat lower than average on tests of IQ and this may be associated with the finding that they are more likely than males with a normal genotype to be found in prison and maximum security hospitals (Jacobs et al 1965). The latter finding was previously thought to be evidence that XYY individuals are more likely to show criminal behaviour and to be aggressive. However, this now seems less likely especially in view of the finding that this genotype occurs more commonly in the general population than was previously thought (see Witkin et al 1976).

The genotype of phenotypic females with the **triple X syndrome** is 47, XXX. It has an incidence of approximately one in 1 000 live female births and is usually associated with a normal physical appearance. However, menstrual abnormalities, mental retardation, and psychotic symptomatology may occur. Females with more than two X chromosomes are sometimes termed 'super females' and with increasing numbers of X chromosomes, for example 48, XXXX, such individuals tend to suffer from increasing degrees of severe mental retardation.

Phenotypic females with **Turner's syndrome** usually have the genotype **45,X** (also sometimes written as 45, XO). Turner's syndrome is much rarer than the above sex chromosome abnormalities, having an estimated incidence of approximately one in 10 000 live female births. Theoretically this incidence might be expected to be much higher, given that 45, X is the complementary genotype of both 47, XXY and 47, XXX following meiotic non-disjunction, and the latter two sex chromosome abnormalities each has a much higher incidence. One possible explanation is that at fertilization 45, X may occur as commonly as would theoretically be expected, but subsequently the majority of fetuses with this genotype are spontaneously aborted. Indeed, the genotype 45, X is relatively common in abortuses. Besides 45, X, other genotypes that can give rise to Turner's syndrome include X chromosomal deletions, isochromosomes, and rings. Characteristic features of Turner's syndrome include short stature, neck webbing, low posterior hairline margin, a shield-shaped chest wall, widely spaced nipples, cubitus valgus, and ovarian dysgenesis leading to primary amenorrhoea and absent secondary sexual characteristics. There is an increased incidence of coarctation of the aorta and renal disorders. Individuals with Turner's syndrome tend on average to score lower than females with a normal genotype on IQ tests, particularly on tests of the performance IQ; nevertheless, many individuals are of normal intelligence.

Non-mendelian inheritance

With the possible exception of mental retardation, the genetic components of the majority of psychiatric disorders are associated with neither chromosomal abnormalities nor pure Mendelian inheritance. Some aspects of non-Mendelian inheritance are now described. Rather than using the term phenotype, in discussing non-Mendelian inheritance it is useful to distinguish between the **endophenotype** which refers to the more endogenous characteristics of an organism such as metabolic products, in contrast to the more clearly manifest **exophenotype** such as hair colour.

Incomplete penetrance

Even when a disorder is caused by a single dominant allele, an individual inheriting that gene may not necessarily develop the disorder because the gene is said to have incomplete penetrance. The penetrance can be defined as the proportion of individuals inheriting a particular genotype who manifest the exophenotype that this genotype can cause. For example, if in a condition caused by an autosomal dominant gene family pedigrees show that the exophenotype manifests itself according to the appropriate Mendelian pattern of inheritance discussed earlier, then the gene has a penetrance of 100%. On the other hand, for a disorder such as acute intermittent porphyria which is caused by a dominantly inherited defect in the activity of the biosynthetic enzyme δ-ALA synthetase, not all individuals inheriting the defect actually manifest the exophenotypic disorder. Thus the inheritance of acute intermittent porphyria is not strictly Mendelian. Rather, its inheritance can be said to be by an autosomal dominant gene with incomplete penetrance. If in an individual an autosomal dominant gene has a penetrance of zero then the disorder may appear to skip a generation.

Variable expressivity

If a dominant mutant gene causes the manifestation of exophenotypes that vary in scope and degree of severity between different individuals inheriting that gene, with such variation not being secondary to other factors such as the environment, then the gene is said to have variable expressivity.

Polygenic inheritance

Most inherited characteristics show a continuous population distribution and are said to be inherited on a polygenic basis if they are the result of many genes acting independently. Each gene may have an individual effect, but overall the characteristic is the result of the cumulative independent gene effects. If none of the relevant genes is dominant and the effect of each gene is relatively small, then a plot of the frequency of occurrence against the characteristic is assumed to give a Normal distribution curve (see Chapter 8) in the general population.

Multifactorial inheritance

Most inherited characteristics for which there is a polygenic component as just described are also influenced by environmental factors. Examples include height, body mass, skin colour and intelligence. Such characteristics which are a function of the combined effect of a polygenic component and an environmental component are said to be the result of multifactorial inheritance.

It seems probable that many disorders are also the result of multifactorial inheritance. In the **multifactorial-threshold model** of the development of disorders which have a multifactorial genetic predisposition, it is supposed that there is an underlying continuously graded **liability** to developing the disorder. This liability is a function of both the polygenic predisposition and environmental factors. It is usually further supposed that a plot of the frequency of the disorder against its liability yields a Normal or Gaussian distribution curve (see Chapter 8) for the general population, as shown in Figure 6.18. According to this simple model, there is a reduced threshold liability in relatives of affected individuals for the development of a multifactorially inherited disorder. This is shown in Figure 6.18 in which the frequency–liability curve for relatives is shifted to the right compared with the frequency–liability curve for the general population. In this way, this model accounts for the observation that for many disorders the familial incidence is greater than the population incidence. It should be noted that a number of more sophisticated computer models, for example with multiple thresholds, exist that provide a better simulation of reality.

Heritability

The heritability of a trait is the proportion of the total phenotypic variance of the trait that is contributed by the genetic component. Variance is a measure of dispersion about the mean and is described in Chapter 8. Heritability is conventionally given the notation h^2 and often expressed as a percentage. It allows an estimate

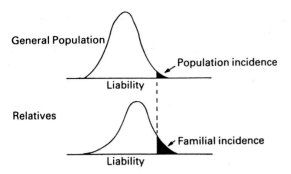

Fig. 6.18 Hypothetical frequency-liability curves showing the reduced threshold liability in relatives of affected individuals for the development of a multifactorially inherited disorder under the multifactorial-threshold model. (Reproduced with permission from Emery A E H, Mueller R F 1988 Elements of medical genetics (student notes). 7th edn. Churchill Livingstone, Edinburgh.)

to be made of the relative proportions of the genetic and environmental components of a multifactorial trait. If a trait is wholly genetic in its aetiology, with no environmental component, then h^2 has a value of one, or 100%.

If the total phenotypic variance is given by V_p, the variance contributed by additive genetic factors by V_G, and the variance contributed by environmental factors by V_E, then

$$V_P = V_G + V_E$$
and $h^2 = V_G/V_P$

GENETIC STUDIES IN PSYCHIATRY

In civilized societies many of the methods used in genetic studies of infrahuman species, for example selective breeding, are clearly not available for use in psychiatric genetic studies. Nevertheless, it has been possible to accumulate a vast corpus of information relating to the genetic aspects of psychiatric disorders by using classical methods based essentially on phenotypic studies. Such methods allow valuable data to be derived from naturally occurring situations, both biological and social, in a non-invasive way. They include family studies, twin studies, and adoption studies. For disorders with a multifactorial inheritance the data so derived can then be compared with data pertaining to the general population by means of biometric analysis, for example the determination of the heritability, h^2, mentioned above. Although census data could in theory also provide valuable data, particularly in relatively isolated communities, in practice such data are not usually helpful in the study

of most psychiatric disorders.

More recently, with the advent of the so-called new genetics, it has become increasingly possible to apply advances in molecular genetics to genotypic and endophenotypic studies. At the time of writing, the data from the application of these new techniques to psychiatric disorders is accumulating very rapidly. It is to be hoped that genes associated with psychiatric disorders will increasingly be characterized.

Family studies

Methodology

The principle of family studies involves comparison of the rates of illness in the first and second degree relatives of affected individuals, termed index cases or **probands**, with the corresponding rates in the general population. The first degree relatives include the parents, siblings, and children of the proband, and on average have 50% of the genome in common with the proband. The second degree relatives include grandparents, uncles, aunts, nephews, nieces and grandchildren, and on average have 25% of the genome in common with the proband. If a psychiatric disorder has a genetic component then one would expect family clustering of the disorder to be observed, with a higher rate of illness being found in the biological relatives of the proband than in the general population. Furthermore, within the group of relatives, a higher rate of illness would be expected in first degree relatives than in second degree relatives.

Difficulties

Determining whether or not the close relatives of the proband are also affected by a psychiatric disorder at a cross section in time will not usually suffice. This is because such disorders need to be considered in a longitudinal way with respect to time; they do not begin at the same age in all affected individuals, and once present they may remit and relapse. A solution to this problem is to use lifetime **expectancy rates** or **morbid risks**. Since some of the relatives may not have reached an age range during which they may be affected by the disorder, an **age correction** such as Weinberg's method may also be used.

Another difficulty with family studies is that they do not allow a good separation of genetic from **environmental factors**, since a shared environment can lead to similarities between relatives. This problem finds its solution in twin studies and adoption studies, which are discussed below.

Clinical examples

Compared with a lifetime expectancy of developing **schizophrenia** in the general population of approximately one per cent or less, with a lifetime risk for the first degree relatives of non-schizophrenic controls estimated at between 0.2 and 0.6% (Kendler 1986), the corresponding figures are significantly higher in the relatives of schizophrenic probands, as shown in Table 6.8 (see Gottesman & Shields 1982; Weissman et al 1986). It can be seen that in general the lifetime expectancy rates are greater in first degree relatives than in second degree relatives.

The lifetime expectancy rates for developing a **mood disorder** have also been found to be raised in the first degree relatives of probands, with figures between six and 40% being calculated in a range of studies (McGuffin & Katz 1986), compared with 1–2% in the general population. When mood disorders are divided into cases of bipolar mood disorder, in which at least one episode of mania has occurred, and unipolar mood disorder, in which only depressive episodes have occurred (Leonhard 1959), then in general family studies demonstrate increased lifetime expectancy rates of both bipolar and unipolar disorder for the first degree relatives of bipolar probands, but increased lifetime expectancy rates of only unipolar disorder in the first degree relatives of unipolar disorder probands. This is illustrated in Table 6.9 which consists of pooled results from family studies of unipolar and bipolar disorder published between 1966 and 1984 (McGuffin & Katz 1986).

Compared with a general population prevalence of approximately 3% in men and 6% in women, family

Table 6.8 Approximate lifetime expectancy rates for the development of schizophrenia in the relatives of schizophrenic probands, given to the nearest percentage point

Relationship	Lifetime expectancy rate to the nearest percentage point
Parents	6
All siblings	10
Siblings (when one parent is also schizophrenic)	17
Children	13
Children (when both parents are schizophrenic)	46
Grandchildren	4
Uncles, aunts, nephews and nieces	3

Table 6.9 Lifetime expectancy rates of unipolar and bipolar mood disorder in first-degree relatives of unipolar and bipolar disorder probands (after McGuffin & Katz 1986)

Proband type	Lifetime expectancy rate to the nearest percentage point	
	Bipolar disorder	Unipolar disorder
Bipolar disorder	8	11
Unipolar disorder	1	9

studies of anxiety neurosis report an increased prevalence of approximately 15% in first degree relatives of probands with **anxiety neurosis** (for example, McInnes 1937; Brown 1942; Noyes et al 1978). Within the anxiety disorders there is evidence that **panic disorder**, but not generalized anxiety disorder, is familial (Crowe et al 1983).

Increased prevalence rates of **obsessional traits** (Carey & Gottesman 1981) and **hysteria** (Ljungberg 1957) have also been found in the respective first degree relatives of probands with those conditions.

There is evidence that there is more likely to be a positive history of familial **alcoholism** in more severe male alcoholics (Latcham 1985).

Theander (1970) reported that there was a 6–10% prevalence of **anorexia nervosa** among the sisters of female probands with this eating disorder, compared with a prevalence of up to 2% in a comparable control group.

Compared with a general population prevalence of 0.02–0.04%, siblings of probands with **childhood autism** or autistic disorder have a risk for this disorder of approximately 2% (Folstein & Rutter 1977).

An increased risk of **Alzheimer's disease** has also been found in the relatives of probands with this form of dementia (Larsson et al 1963, Heston et al 1981).

Twin studies

Methodology

The principle of twin studies involves comparison of the rates of illness in the sibling twins or cotwins of monozygotic (identical) and dizygotic (fraternal or non-identical) probands. Monozygotic twins are uni-ovular, resulting from the division of one fertilized ovum. Hence each monozygotic twin has a genome identical to that of its twin sibling or cotwin. In contrast, dizygotic twins are binovular, resulting from the fertilization at approximately the same time of two different ova, each by a separate spermatozoon. Hence dizygotic twins, like normal siblings, are first degree relatives who share, on average, one half of their genome.

In twin studies of discrete traits, such as the presence of absence of a given psychiatric disorder, pairs of both monozygotic twins and dizygotic twins are identified in whom at least one member, the proband, has the index trait or disorder. The rate of concurrence of the index trait or disorder in the cotwins of the probands is then investigated. This rate of concurrence is known as the **concordance rate**. In twin studies of continuous traits, such as the IQ, a **correlation coefficient** (see Chapter 8) rather than a concordance rate is derived from the data obtained.

In twin studies of psychiatric disorders, if the effect of the environment is assumed to be equal for both kinds of twins, then a higher concordance rate in monozygotic twins than in dizygotic twins, giving a relatively high value to the ratio of the monozygotic twin concordance rate to the dizygotic twin concordance rate, can be considered to be the result of genetic effects; the higher this ratio the greater is assumed to be the genetic component of the disorder. On the other hand, if there is no significant difference between the two concordance rates, so that the value of the ratio of the monozygotic twin concordance rate to the dizygotic twin concordance rate is equal to or approximately equal to one, then it can be considered that there is no significant genetic component. It should be noted that a high concordance rate for monozygotic twins does not in itself imply a genetic component. Rather, this concordance rate has to be compared with that for dizygotic twins, so that it is the value of the ratio of the two concordance rates which is assumed to be a function of the size of the genetic component.

Difficulties

Two types of concordance rate are in use that usually yield differing numerical results. The **pairwise** rate considers pairs of twins and is the number of concordant pairs divided by the total number of pairs, and is often expressed as a percentage. The **probandwise** rate considers each proband and is the number of cotwins (of the probands) in whom the index disorder is concurrent divided by the total number of cotwins, again often expressed as a percentage. Compared with the pairwise rate, the probandwise rate can give a higher result because each affected cotwin may be sampled independently so that in effect each cotwin may become a proband for the other and so be doubly counted. In comparing studies it is clearly important, therefore, to be aware of the method used to ascertain the concordance rate.

In the past the determination of **zygosity** was less accurate than in more recent studies, partly because it was often made by the same observer who made the clinical diagnosis in twin studies, rather than a more objective independent observer. This difficulty has now largely been solved through the use of more objective measures of zygosity such as serological analysis. Furthermore, more detailed modern diagnostic criteria have reduced the degree of diagnostic variability of earlier studies.

A major difficulty with twin studies concerns bias introduced by the method of **sampling** used. For example, some of the earlier twin studies were based on samples drawn from the hospitalized population, thereby increasing the proportion of very ill and chronically ill individuals studied. Clearly, there will be a bias towards the opposite type of individual in studies based on samples drawn from essentially healthy populations such as members of the armed forces. The best solution to this difficulty is probably the use of **twin registers** which keep a record of all the twins born in a particular hospital or preferably, as occurs in Scandinavia, in the whole country.

Another source of error is introduced if organic central nervous system abnormalities contribute to the disorder under study. This is because twins are at greater risk of central nervous system abnormalities secondary to **birth injury** or **congenital abnormalities** than are normal siblings, with monozygotic twins being at greater risk than dizygotic twins. In such cases the results of twin studies may err in either direction.

There is a tendency for marriage or cohabitation to occur between poeple with similar exophenotypic characteristics. For example, tall people tend to choose a tall mate and vice versa. This parental phenomenon, known as **assortative mating**, can lead to a relative increase in the rate of illness in dizygotic twins compared with monozygotic twins.

Error can also be introduced into the results of twin studies through the use of **age correction** techniques such as that of Weinberg mentioned above in connection with family studies. This is because such techniques may make the assumption that given a particular lifetime probability for a given disorder, this can be divided uniformly throughout the at risk age range. In practice, however, concordant twins tend to manifest a disorder within a relatively short period of time of each other. For instance, there exist examples of monozygotic twins living separately who have independently developed acute appendicitis or severe toothache within hours of each other. As a result, the application of age correction techniques can lead to an increase in the estimated concordance rate, even to the extent of giving an impossible value of greater than 100%. It can, therefore, be better to avoid the use of such techniques in twin studies if possible.

It has been mentioned that the application of twin studies makes the assumption that the effect of the **environment** is equal, or nearly so, for both kinds of twins. However, monozygotic twins tend to identify with each other in a psychological sense more than dizgotic twins. Moreover, monozygotic twins are often treated more alike than dizygotic twins by other people such as their parents and teachers. A solution to this difficulty is the use of adoption studies, described below, which allow the genetic and environmental components of a disorder or trait to be better teased apart. Another strategy for investigating the environmental component of a disorder is provided by a comparison of concordant monozygotic twins with discordant dizygotic twins; this strategy has been used in schizophrenia research (Gottesman & Shields 1976, Murray & McGuffin 1988). One finding from adoption studies is that monozygotic twins reared apart sometimes have a greater similarity than monozygotic twins reared together. This may be a result of monozygotic twins reared together each attempting to be different from his or her cotwin, a phenomenon known as the **competing effect**.

Clinical examples

In the case of **schizophrenia**, twin studies have reported a higher concordance rate for monozygotic twins than dizygotic twins. Representative weighted figures for the probandwise rates based on the results of more recent twin studies involving the use of twin registers are 46 per cent for monozygotic twins compared with 14% for dizygotic twins (Gottesman & Shields 1976; Murray & McGuffin 1988). It will be noted that, as would be expected in theory, the concordance rate for dizygotic twins is comparable with the lifetime expectancy of schizophrenia in the siblings of schizophrenic probands reported in family studies.

Twin studies also provide support for the presence of an important genetic component in **mood disorders**. Representative weighted figures for the pairwise rates based on the results of six studies published between 1930 and 1974 which did not differentiate between unipolar and bipolar mood disorder are 69% for monozygotic twins and 13% for dizygotic twins (Gershon et al 1976, McGuffin & Katz 1986). In a large twin study by Bertelsen et al (1977) using the Danish Twin Register (Hauge et al 1968) the probandwise rate for monozygotic twins was 67%,

compared with 20% for dizygotic twins (giving a monozygotic to dizygotic concordance ratio of over 3:1). This study did differentiate between unipolar and bipolar mood disorder and reported a higher probandwise concordance rate for bipolar monozygotic probands, of 79%, compared with 54% for unipolar monozygotic probands; there was little difference between the dizygotic concordance rates. The monozygotic to dizygotic concordance ratio for bipolar mood disorder was approximately 4:1 for bipolar probands, compared with a corresponding ratio of approximately 2:1 for unipolar probands. As with family study results, this supports the hypothesis that there may be a greater genetic contribution in bipolar mood disorder than in the unipolar form.

In harmony with the family study of Crowe et al (1983) mentioned above, well constructed twin studies also suggest that there is a genetic contribution to **panic disorder** but not to generalized anxiety disorder (Murray & McGuffin 1988).

Owing to a paucity of cases, twin studies have not generally been able to provide robust data on **obsessive–compulsive disorder**; those published point to a genetic component (see, for example, Carey & Gottesman 1981). Similarly, few cases of hysteria have been studies in twins; the twin study by Slater (1961) found a zero concordance rate for both 12 monozygotic and 12 dizygotic twins.

In twin studies of **alcoholism**, Kaij (1960) found that monozygotic twins had a significantly higher concordance rate for alcoholism than did dizygotic twins, whereas a study at the Maudsley Hospital in London found no such difference in concordance rates (Murray et al 1983). A large Finnish study of the use of alcohol in adult twins found a difference in alcohol problems between monozygotic and dizygotic twins in younger but not older individuals (Partanen et al 1966).

A twin study of **anorexia nervosa** by Holland et al (1984) found a significantly higher concordance rate for female monozygotic twins than female dizygotic twins.

A twin study of **childhood autism** or autistic disorder by Folstein and Rutter (1977) reported a higher concordance rate in monozygotic twins than in dizygotic twins.

Adoption studies

Methodology

Adoption studies involve the study of individuals who, from an early age, have been brought up by unrelated adoptive parents instead of their biological parents. This allows a separation of the genetic and environmental components of a given trait or disorder. There are a number of types of comparison that can be studied in relation to psychiatric disorders.

In **adoptee studies** one or both of the biological parents of the individuals studies are known to have had the given disorder, whereas the adoptive parents have not. The rate of occurrence of the disorder in these adoptees is compared with that in control adoptees whose biological and adoptive parents have not had the disorder. A genetic contribution to the disorder leads to a higher rate in the former group compared with the control adoptees.

In **adoptee family studies** both the biological and the adoptive parents of adopted individuals who develop a given disorder are studied. The rate of occurrence of the disorder in the biological parents is compared with that in the adoptive parents. The presence of a genetic contribution to the disorder leads to a higher rate in the biological parents than in the adoptive parents.

In **cross fostering studies** two sorts of adoptee are considered. In the first case, one or both adoptive parents have a given disorder while bringing up the adoptee, whereas the biological parents have not had the disorder. In the second, one or both biological parents of the other sort of adoptee has had the disorder whereas the adoptive parents have not. The rates of occurrence of the disorder in the two sorts of adoptee are compared. Although this method provides a very good strategy for the investigation of genetic and environmental components, in practice it is very difficult to find suitable cases for comparison.

A clearer distinction between genetic and environmental components can be obtained when the adoptees involve **monozygotic twins**. Various situations are possible. For example, one monozygotic twin may be brought up by an adoptive parent suffering from a given disorder while his or her cotwin may be brought up by adoptive parents neither of whom has suffered from that disorder.

Difficulties

The number of cases that can be found that fulfil the criteria for one of the above types of adoption study is often very few, particularly if adoption registers are unavailable. Furthermore, even when suitable cases have been found, there is the difficulty of the relatively long period of time required for such studies.

It will have been noted that ideally information on both biological parents is required. However, in many cases information on the biological father may be unavailable, for example in cases of children born out of wedlock where the father has not been identified.

The experience of adoption may have indeterminate psychological sequelae for the adoptees and the process of adoption itself is likely to be nonrandom with efforts often having been made to find adoptive parents who are similar to the adoptees. It is important, therefore, to compare data obtained from an index group of adoptees with control adoptees.

In studies of monozygotic twins adopted separately from an early age, even very soon after birth, it cannot be assumed that the environmental influences on each twin are roughly equivalent. There is evidence that twins can experience significantly different environments in utero, for example with respect to the biochemical environment, and the childbirth experiences may be markedly different.

Clinical examples

Heston (1966) studied 47 adults with a mean age of approximately 36 who had been born to mothers who had a known diagnosis of **schizophrenia**. The 47 individuals had been separated from their biological mothers within three days of birth following the Oregon state laws in force at the time. They were then brought up by non-schizophrenic families unrelated to the biological mother. A comparison with an age and gender matched control group of fostered children born to non-schizophrenic mothers was carried out. In the index group five out of 47, or approximately 11%, were found to be schizophrenic, compared to none of the control group. The higher age-corrected figure for the index group was comparable to the lifetime expectancy in children raised in their biological families where one parent has schizophrenia. It should be noted, however, that the biological fathers were not studied. In the index group there was also an increased level of other conditions, both abnormal such as antisocial personality and neurosis, and supernormal such as artistic creativity, which Heston felt were related to schizophrenia, the so-called **schizophrenia spectrum disorder**.

A larger series of adoption studies on schizophrenia, involving collaboration between Danish and American psychiatrists, was started in 1965 and used the national Danish adoption register. An adoptee study similar to that by Heston gave comparable results (Rosenthal et al 1975). An adoptee family study used an index group of adoptees who had been adopted early and who went on to develop schizophrenia, and a matched control group of adoptees who had no history of schizophrenia. A higher rate of schizophrenia was found in the biological relatives of the index group than in either the non-biologically related adoptive relatives of the index group or in the relatives of the control group (Kety et al 1976; Kendler & Davis 1981). Although these results support the hypothesis that there is an important genetic component to schizophrenia, it could be argued that the higher rate of schizophrenia in the biological relatives of the index group might be a function of similar maternal environments both before birth (the intrauterine environment) and after birth (prior to adoption). Both these maternal factors were controlled for by a study of the paternal half-siblings of the index and control groups. Again, the rate of schizophrenia was found to be higher in the paternal half-siblings of the index group (13%) than in the paternal half-siblings of the control group (less than 2%) (Kety et al 1976). A cross-fostering study by Wender et al (1973) found a higher rate of schizophrenia spectrum disorder in a group of adoptees with normal adoptive parents but a schizophrenic biological parent compared with a group of adoptees with normal biological parents but an adoptive parent who became schizophrenic.

In an adoption study by Mendlewicz and Rainer (1977) on 29 adoptees with a history of bipolar **mood disorder**, 16 of the biological parents (28%) were reported to suffer from a mood disorder (the number was 18, or 31%, when other forms of psychiatric disorder were included) compared with seven of the adoptive parents (12%). By comparison, 26% of the biological parents of 31 bipolar non-adoptees were also found to suffer from a mood disorder. Although this is comparable to the 28% reported for the biological parents of bipolar adoptees, three-quarters of the latter group of biological parents had a history of only unipolar depressive disorder.

In another even smaller adoption study of mood (in the main unipolar) disorder, Cadoret (1978) studied eight adopted away children with a biological parent who had a mood disorder. Their adoptive parents were normal and three of the adoptees (38%) developed a unipolar disorder. This compared with the development of a mood disorder in eight (7%) out of a control group of 118 adoptees whose biological parents did not have a history of having a mood disorder (they either suffered from a psychiatric disorder other than mood disorder of had no psychiatric disorder at all). A confounding difficulty with the results of this study is that it also reported a higher rate of psychiatric problems in the adoptive relatives of the index group of adoptees than in the adoptive relatives of the control group.

In contrast, another adoption study of mood disorder by von Knorring et al (1983) reported an increase in the rate of mood disorder in neither the biological nor the adoptive parents of a group of 56 adoptees, 51 of whom had a history of unipolar mood disorder with the remaining five having a history of bipolar mood disorder, when compared with the biological and adoptive parents of a matched control group of adoptees.

There have been a number of adoption studies of **alcoholism**, beginning with that of Roe (1944). Roe studied the adult adjustment of adopted children with an alcoholic biological parent and found no significant difference between their drinking behaviour in their early twenties and that of a matched sample of control children whose biological parents did not have a history of alcoholism. This study has been critized on the grounds of having a small sample size and of not presenting the diagnostic criteria for alcoholism.

Adoption studies of alcoholism since that of Roe have in general offered support for the hypothesis that there is an important genetic contribution to this addiction, with alcoholism being passed on from alcoholic biological parents to their sons even when the latter are adopted away. In particular, three adoption studies in the 1970s, in Denmark (Goodwin et al 1973), Sweden (Bohman 1978), and the United States (Cadoret & Gath 1978), have reported similar findings. Adopted away sons of alcoholics brought up by non-alcoholic adopted parents were found to be three to four times more likely to develop alcoholism than control groups of adopted away sons of non-alcoholics. Again, adopted away sons of alcoholics were found to have a similar rate of development of alcoholism as their non-adopted brothers, the latter being brought up by their biological parents. The American study reported a higher rate of childhood conduct disorder in the sons of alcoholics but as adults the studies found that when both groups were brought up by non-alcoholic adoptive parents, the sons of alcoholics were no more likely to suffer from non-alcoholic psychiatric disorders than were the controls. The results for the daughters of alcoholics are more difficult to interpret with the proportions of such daughters developing alcoholism (4% in the Danish study, for example) being much lower.

Molecular Genetics

A major difficulty in identifying genes that might be associated with psychiatric disorders is the enormous size of the human genome, with the total number of nucleotide base pairs in the 46 chromosomes estimated to be at least 3×10^9. However, this fact does not make gene identification as daunting as it might otherwise be because of two inherent properties of DNA. Firstly, a large proportion of the DNA of the human genome consists of base pair sequences with no known function. Secondly, the task of identifying genes is simplified by the fact that information is encoded in each strand of DNA in a one dimensional form, that is, in the sequence of nucleotide bases, as mentioned earlier in this chapter. Hence, as portions of the genome are characterized, their nucleotide base pairs can be ordered sequentially.

Gene analysis has become possible because of a number of key technical advances including, in particular, the use of gene probes, the discovery of bacterial restriction endonucleases, and the invention of Southern mapping. These and other techniques are now described. This chapter then ends with a brief discussion of their application to the study of certain psychiatric disorders.

Denaturation

Another inherent property of DNA that is used in gene analysis is the fact that its two strands are complementary. By increasing the temperature beyond the **melting temperature** of the DNA, conventionally denoted as T_m, double-stranded DNA can be **denatured**, or melted, into its two constituent separate strands. Conversely, a reduction of the temperature allows the reverse process of **hybridization** to occur.

Owing to the fact that G–C nucleotide base pairs each contain three hydrogen bonds compared with two for A–T pairs, as mentioned earlier in this chapter, more energy is required to separate the former. Therefore, the value of T_m increases as the proportion of G–C base pairs in the DNA increases.

Another variable that affects the value of T_m is the ionic concentration of the solution in which the DNA is to be denatured. Moreover, the presence of substances that destabilize hydrogen bonds will reduce the T_m. These variables are exploited in the laboratory with the use of appropriate buffers and hydrogen bond destabilizers to enable denaturation to take place at lower temperatures that reduce the risk of damage to the strands.

Gene probes

The construction of gene probes exploits the fact that two complementary strands of DNA hydridize. Gene probes are lengths of DNA constructed such that they have a nucleotide base sequence complementary to that of a given part of the genome. Under suitable conditions the gene probe will hybridize with the index part of the genome. The sequence of a gene probe does not usually need to be exactly complementary to that of the index stretch of DNA; if the gene probe is relatively large and the majority of its nucleotide bases are complementary, then the gene probe will normally still hybridize to the required part of the genomic DNA. For the study of gene mutations, smaller gene probes, known as oligonucleotide probes, can be constructed which are able to detect single-base mutations. Radioactive labelling of gene probes can be used in order to use the latter as hybridization probes that will seek out complementary genomic DNA. Since it is possible for a strand of DNA to hybridize with a strand of RNA, hybridization probes can also be used to find complementary RNA sequences.

One method of making gene probes is from genomic DNA which has been fractionated by using restriction endonucleases (see below). Gene probes can also be constructed from messenger RNA using an enzyme first found in RNA tumour viruses and known as **reverse transcriptase**. As its name implies, reverse transcriptase allows the biosynthesis to occur from a given mRNA molecule of DNA that has a nucleotide base sequence complementary to that of the mRNA, this process being the reverse of transcription. DNA so formed is known as **complementary DNA**, or **cDNA** for short.

Restriction endonucleases

Restriction endonucleases, or restriction enzymes, are a class of enzymes, discovered in bacteria, which cleave DNA only at locations at which specific nucleotide base pair sequences occur. This base pair sequence is the same for a given restriction endonuclease, with different restriction endonucleases having different characteristic base sequences. These target sequences are usually four to six nucleotide base pairs in length.

The identification of restriction endonuclease cleavage sites for a DNA molecule allows a one-dimensional **restriction map** to be constructed for that DNA. Distances between points on this map are measured in base pairs, or bp, with one kilobase (kb) being defined as a thousand base pairs. Following cleavage by one restriction endonuclease, the distance between successive cleavage sites on the resulting restriction map is usually of the order of a few hundred base pairs to several kilobases.

With the exception of monozygotic twins, restriction maps of the genome of different individuals would differ, mainly as a result of random base changes that can delete pre-existing cleavage sites or introduce new ones. For the most part such base changes do not lead to a phenotypic change. The polymorphism at a cleavage site is referred to as a **restriction fragment length polymorphism**, abbreviated to **RFLP**, because it gives rise to fragments of different lengths following digestion with restriction endonucleases. RFLPs are usually inherited in a simple Mendelian manner and so can be used as DNA markers that allow alleles to be followed in the DNA of families. They are particularly useful in the molecular genetic study of psychiatric disorders because they can be used as gene markers in cases where the genes themselves have not been characterized. In addition to RFLPs, there are also inherited regions of DNA with adjacent multiple repeats, a variable number of times, of a given sequence which can also be used as genetic markers; such VNTR (variable number of tandem repeats) loci have found a particular use in forensic human identification. (Somewhat confusingly, some sequences of highly repetitive DNA are also known as satellites, a term also used to refer to small amounts of chromatin attached to the short arms of acrocentric chromosomes.)

Instead of cleaving both strands of a given stretch of a DNA molecule at the same place, in the case of some restriction endonucleases the cleavage is such that small lengths of complementary single strands of DNA protrude from the double stranded parts. These are known as **sticky ends** because of their tendency to hybridize with complementary sticky ends.

Restriction endonucleases are named after the bacteria in which they occur, with Roman numerals being used when more than one restriction endonuclease originates from the same bacteria. For example, the restriction endonuclease *Eco*RI originates from *Escherichia coli* (the R denotes that this is a restriction endonuclease as opposed to non-restriction endonucleases from the same bacteria), and *Hin*dIII originates from *Haemophilus influenzae* serotype d.

Southern blotting

Southern blotting is named after its inventor, Edwin Southern, and is a standard technique used in the analysis of fragments of DNA, often most conveniently extracted from peripheral blood leucocytes, that has been cleaved by restriction endonucleases. The different fragments are separated by size using agrose gel electrophoresis. The gel is then treated with an alkaline solution in order to denature the DNA fragments. Southern blotting is the technique of transferring these DNA fragments to a nylon or nitrocellulose filter by overlaying the gel with the filter and overlaying the latter with paper towels, and blotting a solution through the gel to the paper towels.

Following Southern blotting, in order to identify the fragments of interest on the filter the latter is exposed to a specific radioactively labelled gene probe which can bind to the appropriate DNA fragments with complementary sequences which have already been denatured by the previous treatment with alkali. Excess quantities of the radioactive gene probe that have not bound are removed and the filter is then autoradiographed to expose the position of radiolabelled bound DNA fragments. The size of these fragments can be estimated by running a reference solution of labelled DNA fragments of known size on the original gel or by comparison with the results of using a different restriction endonuclease.

The above procedure gives information about stretches of DNA of interest and about the adjacent stretches of DNA up to the nearest restriction sites. By repeating this procedure with a number of different restriction endonucleases it is possible to build up a more comprehensive restriction map of a particular stretch of the genome. It is also possible to work out the actual nucleotide base sequence of the stretch of DNA of interest either manually, which can be a very long procedure, or more rapidly by automation. Automated techniques also exist for the extration of DNA from cells and, in cases in which only very small quantities of DNA are available, for amplifying the amount of DNA through a process known as the polymerase chain reaction. It is the aim of the ambitious and expensive Human Genome Project to utilize automated technology in order to locate all the genes and sequence the entire genome.

Recombinant DNA

A recombinant DNA molecule has DNA sequences derived from more than one organism, for example human (genomic or cDNA) and bacterial. They are made by inserting the foreign, say human, DNA fragment into the genome of a vector which is a bacterial plasmid, a bacteriophage, or a cosmid.

Bacteria plasmids are extrachromosomal circular bacterial DNA molecules which replicate independently of bacterial chromosomes. As shown in Figure 6.19, a given stretch of human DNA, with sticky ends, cleaved by a given restriction endonuclease, can be inserted into a plasmid by cleaving the plasmid with the same restriction endonuclease and allowing the human (foreign) DNA to be spliced into the plasmid

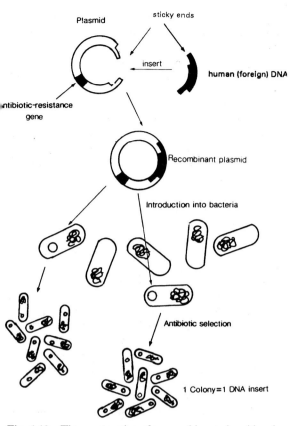

Plasmid

sticky ends

insert

human (foreign) DNA

antibiotic-resistance
gene

Recombinant plasmid

Introduction into bacteria

Antibiotic selection

1 Colony=1 DNA insert

Fig. 6.19 The construction of a recombinant plasmid and molecular cloning. Foreign DNA is inserted into a bacterial plasmid which contains an antibiotic resistance gene. On infection, bacteria which contain the plasmid are able to grow in the presence of antibiotics, whereas those without plasmids cannot. Plating bacteria on antibiotic selection medium therefore produces bacterial colonies each of which is derived from a single bacterium and which contains a pure population of recombinant plasmid. (Reproduced with permission from Reynolds E H, Trimble M R 1988 The bridge between neurology and psychiatry. Churchill Livingstone, Edinburgh.)

DNA, thereby forming a recombinant, hybrid, or chimeric plasmid. A plasmid is used which contains at least one antibiotic resistance gene so that following the re-introduction of the recombinant plasmids into bacteria appropriate antibiotic selective pressure can be used to allow the bacteria containing recombinant plasmids to reproduce. In this way a pure bacterial colony is obtained which contains multiple recoverable copies of the original stretch of human DNA. This technique is known as **molecular cloning**. It is the basis for constructing a **gene library** which is a set of cloned DNA fragments which represents all the genes of an organism or of a given chromosome. Gene probes can be used to identify the genes in a gene library.

Bacteriophages, or phages, are viruses that can infect bacteria, reproduce inside them, and then be released for further cycles of re-infection by causing lysis of the host bacteria. The insertion of foreign DNA into the genome of a bacteriophage, to form a recombinant, hybrid, or chimeric bacteriophage, does not prevent the reproductive cycle just outlined, so that large amounts of the original foreign DNA can be obtained. The size of DNA fragments that can be cloned in bacteriophages is up to approximately 20 kb and is around twice the maximum size that can be cloned in plasmids. Even longer fragments, of up to approximately 45 kb, can be cloned in cosmids, which consist of hybrids constructed of both plasmids and bacteriophages.

Linkage analysis

Since two genes that are in close proximity on the same chromosome are likely to be inherited together they are said to be linked, with the phenomenon of being inherited together being called linkage. Suppose the two alleles occurring at two linked loci in a double heterozygote are A and a, and B and b, respectively. There are clearly only two ways, or **linkage phases**, in which these alleles can occur in the pair of chromosomes, as shown in Figure 6.20; in (i) AB/ab alleles A and B are said to be in **coupling** while in (ii) alleles A and B are in **repulsion**, with coupling and repulsion referring to whether two alleles are on the same or opposite chromosomes of a pair. In the following consideration it is assumed that the linkage phase is known.

The **recombinant fraction** is a measure of how often the alleles at two loci are separated during meiotic recombination, and is conventionally denoted by θ. The minimum value of θ is zero, which is the case for two loci that are completely linked. In the opposite case of two entirely unlinked loci which act as if they are on separate chromosomes, θ has a value of 0.5 (not one because the chromosomes occur in pairs and recombination will occur on average half the time).

Let us consider the results of meiosis for the double heterozygote AB/ab considered above. Since the recombinant fraction is θ, the chromosomal products of recombination, Ab and aB, will occur with a total probability of θ, and therefore each with a probability of $\theta/2$. Similarly, the chromosomal products when recombination does not take place, AB and ab, will occur with a total probability of $(1 - \theta)$, so that each occurs with a probability of $(1 - \theta)/2$.

(i) AB/ab

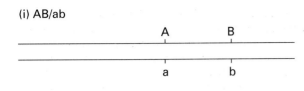

(ii) Ab/aB

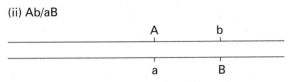

Fig. 6.20   Two possible linkage phases (see text).

It follows that in a two-generation family pedigree, if the genotypes of the parents are AB/ab and ab/ab, then the probability of each of the following four offspring genotypes is as stated:

$Pr(AB/ab) = (1 - \theta)/2$

$Pr(Ab/ab) = \theta/2$

$Pr(aB/ab) = \theta/2$

$Pr(ab/ab) = (1 - \theta)/2$

For loci that are relatively near each other, the value of θ is related closely to the actual so-called map distance between the loci. Map distances are measured in centimorgans (cM), one hundred of which make one morgan (named after the geneticist Thomas Hunt Morgan who carried out some of the earliest studies of linkage). One centimorgan is approximately 1000 kb, and up to about 30 cM the value of the percentage recombinant fraction is the same as the number of centimorgans (that is, 1 cM corresponds to a value of θ of 0.01 or 1%). However, beyond a map distance of approximately 30 cM the relationship between map distance and the recombinant fraction is no longer nearly linear.

Linkage analysis has been made easier by the **lod score method** of Morton (1955) which allows useful information to be gleaned from small pedigrees, including even those consisting of two generations, even in cases in which the linkage phase is not known, as well as using information from families in which the linkage phase is known. Morton used the term lod to mean logarithm of the odds, and the lod score is denoted by z. The lod score for a given value of θ is given by:

lod score, $z_\theta = \log_{10}(P_1/P_2)$

where $P_1 =$
the probability of there being linkage, for a given θ ($0 \leq \theta < 0.5$)

and $P_2 =$
the probablity of there being no measurable linkage ($\theta = 0.5$)

The method is a sequential test in which the value of z is calculated for a series of hypothetical values of θ from zero to 0.5. The value of θ that gives the greatest value for z is known as the **maximum likelihood estimate** of θ, and is the best estimate that can be made for the value of the recombinant fraction from the available data.

Applications to psychiatric disorders

In the case of inherited psychiatric disorders with known genetic loci there is little difficulty in identifying the molecular pathology by applying the above techniques. For those disorders caused by gross chromosomal lesions restriction maps will identify the lesions because of the changes in the size of the appropriate DNA fragments detected by gene probes. To identify the molecular pathology of those disorders caused by point mutations either oligonucleotide probes or nucleotide base sequencing can be used.

The genetic loci are not yet known for the majority of psychiatric disorders with a genetic component. In such cases even the biochemical defects are often not known. It is in these cases that linkage analysis is proving to be a powerful investigative tool. Large pedigrees are used preferentially. This is because large, multigenerational pedigrees are likely to be more useful than nuclear families, and contain more genetic information than the latter. The availability of large sibships reduces the number of families needed to reveal linkage and large families also help to overcome problems of heterogeneity of linkage. Genetic markers, including RFLPs, showing genetic variation or polymorphism are used, and lod scores are calculated between the unknown defective allele(s) and the marker loci. By convention a lod score of three, representing odds of 1000 to one in favour of linkage, is arbitrarily taken as proof of linkage. Negative lod scores below minus two are taken as implying there is no linkage, and lead to the exclusion of a given region of the genome. Since lod scores are logarithmic ratios, results from different pedigrees can be combined by addition. Positive lod scores may be used as the basis of studying adjacent DNA markers in order to increase the probability of detecting linkage. Another related method relies on the use of candidate genes when there is reason to believe that the protein the gene codes for may be defective in the disorder. For example, if cDNAs for subunits of a given neurotransmitter receptor are available, possibly as a result of a form of reverse genetics in which mRNAs expressed in the central nervous system are characterized, and there is psychopharmacological evidence that this neurotransmitter may be involved in a certain psychiatric disorder, then the cDNAs can be used as candidate genes in the investigation of that disorder, with for example RFLPs related to these genes being used as linkage markers. A third method is to concentrate on a particular part of the genome because of information that indicates its relevance to a disorder of unknown genetic locus (or loci). This has been mentioned earlier in this chapter in connection with the link between the neuropathology of Down's syndrome, known to result from trisomy 21, and Alzheimer's disease. Because of this association, effort was concentrated on chromosome 21 in the search for the inherited defect in Alzheimer's disease. Difficulties that occur in the use of linkage strategies are partly a result of the complexity of psychiatric disorders, and include diagnostic difficulties, aetiological heterogeneity, and assortative mating. Using the lod score method, parameters relating to inheritance, such as the penetrance, need to be specified and may not be known. Indeed the true mode of inheritance itself may be uncertain. Sometimes psychiatric disorders show a change in rate of occurrence over time as a result of a cohort effect; it is important that this factor is incorporated in linkage analyses. It may be that these difficulties account in part for the fact that some initially promising findings showing linkage between certain loci and psychiatric disorders, such as schizophrenia and bipolar mood disorder, have not been successfully replicated by other research teams (see below).

RFLPs and linkage analysis can also be used in the prenatal and presymptomatic diagnosis of inherited disorders even if the details of the molecular pathology of the disorders are unknown.

Clinical examples

Following a report of the cosegregation of partial trisomy of chromosome 5 with **schizophrenia** in a family of Chinese origin living in Canada (Bassett et al 1988) several groups of researchers have investigated whether a susceptibility locus for schizophrenia exists in the same region on the long arm of that chromosome. Linkage between a region on the long arm of chromosome 5 and schizophrenia was reported by

Sherrington et al (1988) in Icelandic and British families. However, in the same issue of *Nature* Kennedy et al (1988) reported evidence against linkage of schizophrenia to markers on chromosome 5 following an investigation of a northern Swedish pedigree. Subsequently St Clair et al (1989) also failed to find any such linkage in Scottish families. Other researchers have also argued against there being evidence for such linkage (Diehl et al 1989; Kaufmann et al 1989; Aschauer-Treiber et al 1990) and at the time of writing the results of Sherrington et al (1988) have not been successfully replicated.

Following their investigation of a relatively large American Old Order Amish pedigree, Egeland et al (1987) reported linkage between a locus on the short arm of chromosome 11 and **bipolar mood disorder**. However, as in the case of schizophrenia and chromosome 5, other researchers have failed to replicate this finding (Hodgkinson et al 1987; Detera-Wadleigh et al 1987; Gill et al 1988). Prior to the recent advances in techniques of molecular genetics, there was evidence that an X-linked gene may be involved in the genetic transmission of a subtype of bipolar mood disorder, for example because of the linkage of an X-linked form of colour blindness with this psychiatric disorder and the existence of X-linked pedigrees (Winokur et al 1969; Mendlewicz et al 1972). Evidence which argued against X-linkage included cases of bipolar mood disorder not being linked with the X-linked form of colour blindness, and cases of father to son genetic transmission of this psychiatric disorder. However, this could be explained by postulating the existence of aetiological heterogeneity. Mendlewicz et al (1987) and Baron et al (1987) have reported linkage between bipolar mood disorder and a region on the long arm of the X chromosome close to the locus for the fragile X syndrome.

Prior to the 1980s linkage studies had excluded approximately one fifth of the human genome as the location for the gene for the autosomal dominant progressive neurodegenerative disorder **Huntington's chorea**. A systematic linkage study was carried out by Gusella et al (1983) on the DNA from two large pedigrees, one in the United States, and the other a large affected pedigree in Venezuela the members of which could be traced to one original Huntington's chorea sufferer. It might be expected that such a search would take a very long time but fortunately one of the first gene probes used detected a polymorphic DNA marker linked to the Huntington's chorea locus. The marker was located on chromosome 4. Further linkage analyses showed that this marker was separated from the disease locus by a distance of only about 4 cM (Harper et al 1985). The gene for Huntington's chorea has now been localized to the most distal band of the short arm of chromosome 4 (4p16.3) using loci very tightly linked to the disease locus (Gilliam et al 1987, Wasmuth et al 1988) and mapping (Whaley et al 1988). Although the Huntington's chorea gene has not been characterized at the time of writing, this can be expected to occur soon. These findings can clearly also be used in prenatal and presymptomatic diagnostic investigations; a discussion of the ethical implications of such investigations is beyond the scope of this book.

It has already been mentioned how the associations of Down's syndrome with **Alzheimer's disease** indicated that the Alzheimer's disease locus might be located on chromosome 21. It should be noted, however, that many cases of Alzheimer's disease do not appear to be inherited. As described in the next chapter, a characteristic neuropathological feature of Alzheimer's disease is the occurrence of neuritic plaques which contain an amyloid core. This amyloid protein is known as A4 or beta amyloid. It is now known that this protein is derived from a large precursor protein, known as beta-amyloid precursor protein, or β-APP (Kang et al 1987). Here, the term β-APP gene is used in place of the earlier references to the gene coding for the amyloid A4 or beta protein. When this gene was cloned and localized to the long arm of chromosome 21 (Kang et al 1987, Goldgaber et al 1987; Tanzi et al 1987a) this was hailed by some as being the same as discovery of the Alzheimer's disease gene, a belief reinforced by the report that this gene was duplicated in cases of sporadic Alzheimer's disease (Delabar et al 1987). However, the celebration proved to be premature because further analysis showed that the β-APP gene was not tightly linked to the Alzheimer's disease locus (Tanzi et al 1987b, van Broeckhoven et al 1987). Moreover, the report of duplication of the β-APP by Delabar et al (1987) has since been retracted by some members of the same group (Rahmani et al 1989). Furthermore, partial sequencing of the β–APP cDNAs cloned from the brains of three patients with sporadic Alzheimer's disease did not reveal mutations (Vitek et al 1988). At around the same time as the excitement over the β-APP gene, linkage studies confined to the long arm of chromosome 21 using DNA from four pedigrees affected by an inherited form of Alzheimer's disease reported that the disease locus is located on the proximal part of the long arm of that chromosome (St George-Hyslop et al 1987). This finding has since been replicated by other groups (Goate et al 1989; van Broeckhoven et al 1989). Although two other recent studies (Pericak-Vance et al 1988; Schellenberg et al 1988) appear not to support this finding, a careful analysis of their results shows that most of the families have a late age of onset and therefore non-inherited forms of Alzheimer's diseases may have been included; analysis of the familial early onset cases in these two studies is indeed consistent with a chromosome 21 locus (Hardy 1990). More recently, attention has again been given to the β-APP gene. The possibility has been raised that a mutation in the β-APP gene may be of aetiological significance in some cases of Alzheimer's disease, with the finding by Goate et al (1991) of a mutation in the β-APP gene in two familial Alzheimer's disease pedigrees.

A fragile site on the X chromosome occurs in an inherited form of mental retardation known as the **fragile X syndrome** which principally affects males (Sutherland 1979). After Down's syndrome it is the second most common form of mental retardation and affects almost 0.1% of males. Females carrying fragile X sites may demonstrate varying degrees of the exophenotypic expression seen in males, while cases exist of males with fragile X sites who do not manifest any features. As mentioned earlier in this chapter the fragile site has been localized to Xq27; close linkage has been found between the fragile site and the coagulation factor IX locus in that region of the X chromosome (Camerino et al 1983).

FURTHER READING

Campbell P N Smith A D 1988 Biochemistry illustrated. 2nd edn. Churchill Livingstone, Edinburgh

Davies K E 1986 Human genetic diseases. IRL Press, Oxford

McGuffin P, Murray R M (eds) 1991 The new genetics of mental illness. Butterworth-Heinemann, Oxford

Stryer L 1988 Biochemistry. 3rd edn. Freeman, New York

Weatherall DJ 1990 The new genetics and clinical practice. 3rd edn. Oxford University Press, Oxford

7. Neuropathology

The structure and functioning of the central nervous system have been discussed earlier in this book. In this chapter neuropathological aspects are discussed with particular reference to how these relate to clinical psychiatry. The selection of topics has been governed by the fact that this is neither a textbook of neurology nor of clinical psychiatry. Thus epilepsy is not discussed separately; rather, in those cases in which epilepsy is a consequence of a neuropathological process, this is mentioned in the text dealing with the latter. In this chapter neuropathological topics are considered in the following order: neurocellular reactions to disease; pathophysiological reactions to disease; focal cerebral disorder: clinical features; cerebrovascular disorders; infections; trauma; demyelinating diseases; metabolic disorders; Wernicke–Korsakoff syndrome; Parkinson's disease; dementia; cerebral tumours; and brain imaging.

NEUROCELLULAR REACTIONS TO DISEASE

In this section the basic reactions of neurones, astrocytes, oligodendrocytes, ependymal cells and microglia to central nervous system disease are considered. Since this chapter is concerned with diseases of the central nervous system rather than of the peripheral nervous system, the latter not being directly related to psychiatry, such reactions as Wallerian degeneration and axonal degeneration are not discussed in this section. Anatomical aspects of neurones and neuroglia, including details concerning lysosomes, lipofuscin and melatonin, have been discussed in Chapter 1 and are not repeated in this section.

Neurones

Reactions to ischaemia

It will be recalled from Chapter 3 that unlike other organs the brain is dependent on glucose supplied by blood for its energy requirements and that approximately 99% of this glucose is metabolized in the cyto-plasm by glycolysis, the tricarboxylic acid cycle and oxidative phosphorylation. Hence neurones are particularly susceptible to **hypoglycaemia** such as that which follows ischaemia (which is discussed later in this chapter). Moreover, since the complete metabolism of glucose to carbon dioxide and water is dependent on a supply of oxygen, it follows that neurones are also susceptible to **hypoxia**, which again is usually associated with ischaemia.

The sequence of neuropathological ischaemic neuronal changes begins with **microvacuolation** which occurs in experimental mammalian studies between five and 15 minutes following ischaemia. The microvacuoles consist mainly of swollen mitochondria. Further ischaemia leads to irreverible neuronal damage and the stage of **ischaemic cell change** which is characterized by shrinkage of the neurones and a change in the shape of their nuclei. This is followed by **homogenizing cell change** which is characterized by a further reduction in the size of the nuclei and an increasingly homogeneous cytoplasmic appearance, resulting finally in **ghost cell change** in which the cytoplasm can no longer be recognized in the necrotic neurones.

Phagocytosis

Neuronal phagocytosis occurs particularly in cases of acute neuronal necrosis and is also known as neuronophagia.

Central chromatolysis

Central chromatolysis is characterized by enlargement of the perikaryon (or neurocyte) and nucleolus, dispersion of the Nissl substance from the central perikaryon causing it to disappear centrally (hence the name of this neuropathological change), displacement to the periphery of the nucleus, and increased numbers of lysosomes. Central chromatolysis occurs following injury to the perikaryon or axon hillock. Neuronal recovery or degeneration may follow, depending in part on the type of neurone and the degree of injury.

Neurofibrillary degeneration

This neuropathological change leads to the formation of highly insoluble **neurofibrillary tangles**, made up of thick bundles of neurofibrils, in the perikaryon. They are stained well by silver impregnation. Ultrastructurally they are mainly composed of paired helical filaments of approximately 8 nm diameter with a periodicity of approximately 80 nm. Neurofibrillary tangles are occasionally present in the brains of elderly people. Their abundant presence in the brains of individuals with **Alzheimer's disease**, particularly in the cerebral cortex, is a characteristic neuropathological feature of this dementia. There is a correlation between their density and the degree of cognitive impairment in Alzheimer's disease. Neurofibrillary tangles also occur in adult individuals with Down's syndrome who, as mentioned in the previous chapter, develop the neuropathology seen in Alzheimer's disease. Other conditions associated with neurofibrillary tangles include postencephalic parkinsonism, the parkinsonism-dementia complex of Guam, amyotrophic lateral sclerosis, progressive supranuclear palsy, and the punch-drunk syndrome. Neurofibrillary tangles are also sometimes seen in patients who have received aluminium, for example because of its presence in the water used in renal dialysis, and it has been suggested that aluminium toxicity may play a role in the pathogenesis of Alzheimer's disease. In support of this argument is the fact that in the past many dialysis patients exposed to aluminium developed dementia.

Inclusion bodies

Viral inclusions are intranuclear or intracytoplasmic inclusions that are usually visible under the light microscope and are indicative of viral infection. Viral intranuclear inclusions are particularly associated with infections with herpes simplex virus type I and cytomegalovirus, subacute sclerosing panencephalitis (caused by measles virus), and progressive multifocal leukoencephalopathy (caused by papovavirus). Viral intracytoplasmic inclusions are associated with rabies (caused by a rhabdovirus) and some infections with cytomegalovirus.

In **granulovacuolar degeneration** there occur intracytoplasmic vacuoles of up to 5 μm in diameter which each contain an argyrophilic granule. This neuropathological change occurs mainly in the middle pyramidal layer of the hippocampus. Like neurofibril-

lary tangles it is a neurodegenerative change that occurs in normal individuals with increasing age but is more prevalent in both Alzheimer's disease and in progressive supranuclear palsy; in the latter disorder this neuropathological change occurs in brain stem nuclei.

Another type of intracytoplasmic neuronal inclusion bodies that also occur frequently in Alzheimer's disease are eosinophilic rod-shaped filamentous structures known as **Hirano bodies**. They consist mainly of actin filaments.

A characteristic neuropathological feature of Pick's disease is the presence of argyrophilic intracytoplasmic neuronal inclusion bodies known as **Pick's bodies** which consist of neurofilaments, paired helical filaments and endoplasmic reticulum.

Similarly, a characteristic neuropathological feature of Parkinson's disease is the presence in the substantia nigra and locus coeruleus of eosinophilic intracytoplasmic neuronal inclusion bodies known as **Lewy bodies** which consist of filaments that are often arranged in such a way that the Lewy bodies may appear to have dense cores surrounded by less dense rims.

Metabolic disorders

Characteristic neuropathological changes occur in metabolic disorders that cause neuronal damage. An enzyme deficiency can lead to neuronal storage of intermediate substrates or the products of alternative metabolic pathways. This can lead to enlargement of neurones with changes in the neuronal shape and the position of the nucleus, and the presence of intracytoplasmic inclusion bodies.

Astrocytes

Primary astrocytosis

Primary astrocytosis is the reaction of astrocytes to sublethal astrocytic damage and is seen mainly in hepatolenticular degeneration, or Wilson's disease, and hepatic encephalopathy. The altered astrocytes are known as **Alzheimer type II astrocytes** and contain enlarged vesicular nuclei and sometimes nucleoli. There is no change in the number of astrocytic processes. Alzheimer type II astrocytes are found mainly in grey matter.

Secondary astrocytosis

Secondary astrocytosis is the reaction of astrocytes to damage to other components of the central nervous system. It is an important neurocellular reaction characterized by astrocytic hyperplasia and hypertrophy.

Early in secondary astrocytosis, particularly in white matter, the astrocytes hypertrophy displacing their nuclei to the periphery. Because of their plump appearance they are termed **gemistocytic astrocytes**. If resolution of the process occurs there is loss of cytoplasm in the gemisotocytic astrocytes which now become **fibrillary astrocytes**. The latter then form the **neuroglial scar tissue** mentioned in Chapter 1, which is made up mainly of astrocytic processes.

If resolution does not occur, prolonged secondary astrocytosis causes the formation of eosinophilic **Rosenthal fibres** in astrocytic processes.

Hypoxia

As with other neurocellular components of the central nervous system, prolonged hypoxia is lethal to astrocytes. The latter characteristically react by enlarging and then losing their nuclei, a neuropathological change known as **cloudy swelling**.

Oligodendrocytes

Oligodendrocytic reactions to damage include **acute swelling** and the clustering of oligodendrocytes around ageing and degenerating neurones, a process known as **satellitosis**.

Since oligodendrocytes form the myelin sheaths of nerve fibres of the central nervous system (see Ch. 1), oligodendrocytic damage can result in **demyelination**.

Ependymal cells

Chronic irritation of the ependyma can lead to a focal compensating non-specific subependymal astrocytic proliferation known as **granular ependymitis**.

Microglia

As mentioned in Chapter 1, injury to the central nervous system activates the microglia, which may become elongated **rod cells** with enlarged nuclei, and they migrate to the site of injury where neuronophagia takes place. They may become filled with lipids, derived from necrotic tissue, and are then known as **lipid phagocytes**. If they become laden with haemoglobin breakdown products they are known as **siderophages**.

It is possible that many of the active macrophages seen at sites of central nervous system injury are derived not from microglia but from monocytes emigrating from the circulating bloodstream.

PATHOPHYSIOLOGICAL REACTIONS TO DISEASE

In this section the following pathophysiological reactions are discussed: raised intracranial pressure, cerebral oedema and hydrocephalus. Cerebral oedema and hydrocephalus also in themselves lead to raised intracranial pressure.

Raised intracranial pressure

Aetiology

By increasing the volume within the rigid adult cranium, intracranial expanding or space-occupying lesions lead to raised intracranial pressure. Such lesions include tumours, haemorrhages, abscesses and granulomas. As mentioned above, cerebral oedema and hydrocephalus also cause raised intracranial pressure, as does meningitis.

Stages

The relationship between intracranial pressure and the volume of intracranial contents, which include the brain, blood, cerebrospinal fluid, and any pathological entities such as intracranial expanding lesions, is shown in Figure 7.1. It can be seen that, initially, a small intracranial expanding lesion causes little, if any, change in the intracranial pressure. This is because the increased intracranial volume is compensated for

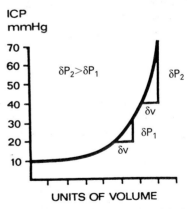

Fig. 7.1 The relationship between intracranial pressure (ICP) and the volume of the intracranial contents. As the volume increases and ICP rises, uniform increments of volume (δv) lead to progressively larger increases in ICP ($\delta P_2 > \delta P_1$). Increases or decreases in volume cause correspondingly greater changes in ICP on the steep part of the curve. (Reproduced with permission from Adams J H, Graham D I 1988 An introduction to neuropathology. Churchill Livingstone, Edinburgh.)

mainly by a reduction in the volume of both the intracranial venous blood and the cerebrospinal fluid of the ventricles and subarachnoid space. This is known as the stage of **spatial compensation**.

From Figure 7.1 it can be seen that as the leeway provided by spatial compensation is used up, the intracranial pressure begins to rise gradually with progressive increases in the volume of intracranial contents. In order to maintain the cerebral perfusion pressure, which would otherwise decrease, during this second stage there is a rise in the systemic arterial blood pressure. This is known as the **Cushing response**.

It is clear from Figure 7.1 that, with further progressive increases in the volume of intracranial contents, a critical third stage is reached in which only a slight increase in the volume of the intracranial contents leads to a rapid rise in the intracranial pressure which is too great to allow a compensatory increase in the systemic arterial blood pressure to maintain the cerebral perfusion pressure. Clinically, the condition of the patient may suddenly deteriorate with the decrease in the cerebral blood flow.

The fourth and final stage is of **cerebral vasomotor paralysis** leading to brain stem death, and follows on from the previous stage very quickly.

Effects

It will be recalled that intracranial compartments are formed by the tough dural folds, with the cerebral hemispheres being largely separated by the falx cerebri, and the cerebellum being similarly separated from the cerebral hemispheres by the tentorium cerebelli.

A supratentorial intracerebral expanding lesion initially causes cerebrospinal fluid displacement and distortion of the shape of the brain so that there is shift in the midline structures and flattening of the gyri of the affected hemisphere against the dura. The displacement of cerebrospinal fluid in the ventricular system and the subarachnoid space makes the lateral ventricle of the affected side smaller than the contralateral lateral ventricle and the brain surface of the affected side drier. With a further increase in the volume of the supratentorial intracerebral expanding lesion displacement of the brain itself into adjacent intracranial compartments occurs. This is known as **herniation** and there are three types which are shown diagrammatically in Figure 7.2: subfalcine, tentorial and tonsillar.

A **subfalcine hernia**, also known as a supracallosal or cingulate hernia, is the result of herniation of the cingulate gyrus under the inferior free edge of the falx cerebri, superior to the corpus callosum. If the anterior

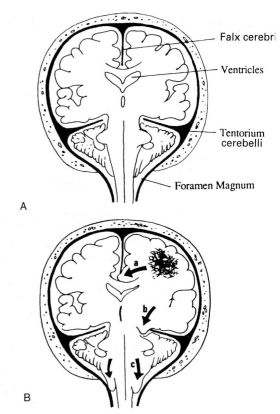

Fig. 7.2 (**A**) The intracranial compartments; (**B**) the effects of an intracranial expanding lesion. a = subfalcine herniation; b = tentorial herniation; c = tonsillar herniation. (Reproduced with permission from Adams J H, Graham D I 1988 An introduction to neuropathology. Churchill Livingstone, Edinburgh.)

cerebral artery is compressed this can lead to infarction in the distal areas supplied by it.

From a clinical viewpoint the most important hernia caused by a supratentorial intracerebral expanding lesion is a **tentorial hernia**, also known as an uncal or transtentorial hernia. It is can be considered to be made up of two components: medial and caudal herniation through the incisura of the tentorium cerebelli of the medial ipsilateral temporal lobe, including the uncus and parahippocampal gyrus; and caudal herniation of the rostral brainstem, hypothalamus, and mammillary bodies through the tentorial incisura. The second component is sometimes referred to as central herniation. The importance of tentorial herniation arises as a result of its effects on other structures. The first component of tentorial herniation leads to displacement and compression of the mesencephalon, which is pressed against the free edge of the contralateral tentorium and may lead to the formation of a mesencephalic surface groove known as the **Kernohan notch** or groove. This is associated with compression of the aqueduct, the contralateral cerebral peduncle (which is pressed against the free edge of the contralateral tentorium), and the ipsilateral oculomotor nerve (between the posterior cerebral artery and the free edge of the tentorium). The oculomotor nerve compression results in the early clinical sign of ipsilateral pupillary dilatation, while the involvement of the posterior cerebral artery can cause infarction in the distal area supplied by it. Caudal brain displacement can lead to rupture of pontine and mesencephalic blood vessels, producing haemorrhages in the midline and laterally, known as **Duret haemorrhages**. These haemorrhages are lethal if vital brain-stem centres are affected.

A **tonsillar hernia**, also known as a cerebellar cone, results from herniation of the cerebellar tonsils through the foramen magnum. Although it can be caused by a supratentorial intracerebral expanding lesion it is more likely to occur because of an infratentorial expanding lesion, when it is also likely to be more severe. The effects of a tonsillar hernia include infarction of the cerebellar tonsils, and obstruction of the flow of cerebrospinal fluid which in turn increases the already raised intracranial pressure and so forms a vicious circle. The most important effect, however, is compression of vital medullary centres, particularly the respiratory centre, as this is lethal unless rapidly treated. Clearly the process of tonsillar herniation or cerebellar coning will be worsened by the removal of cerebrospinal fluid during lumbar puncture; hence the contraindication of this investigation in cases of suspected intracranial expanding lesions.

Infratentorial expanding lesions can also cause rostral herniation of the cerebellum through the incisura of the tentorium cerebelli. This is known as a **reversed tentorial hernia**. Compresison of the superior cerebellar arteries leads to infarction of the superior cerebellum while less often distortion of the parahippocampal gyri also occurs.

Chronic moderately raised intracranial pressure leads to **bone erosion** which may be detectable by skull radiography. Regions most often eroded include the diaphragma sellae (which separates the optic chiasma from the pituitary) and posterior clinoid processes. In children there may be separation of the skull sutures and radiographs may also show a typical beaten-brass appearance caused by skull thinning associated with convolutional markings.

Clinical features

In addition to specific features already mentioned above, the general clinical features of raised intracranial pressure include headache, which may be made worse by lying down or straining and is associated with nausea and vomiting; reduced pulse and raised systolic blood pressure; and a reduced level of consciousness leading to coma and death.

Another result of raised intracranial pressure is **papilloedema**. Papilloedema is swelling of the optic nerve head (optic disc). The disc is pinker than usual in the early stages, with the veins appearing full. This is followed by blurring of the nasal margin of the optic disc which spreads to the temporal margin. At the same time the physiological cup fills with exudate and thus becomes obliterated. The surface of the optic disc may become raised owing to the swelling. Severe papilloedema of acute onset may be accompanied by areas of haemorrhage. The optic disc may become paler in chronic papilloedema as a result of optic atrophy. Raised intracranial pressure leads to a rise in cerebrospinal fluid pressure which enlarges the subarachnoid space around the optic nerve, and so causes retinal venous compression.

Cerebral oedema

Definition

True cerebral oedema is an increased brain volume owing to increased water and sodium ion content. However, brain swelling which is not strictly oedematous may result from congestion caused by vasodilatation, as occurs in cases of subdural haematoma, hypercapnia, hypoxia, and, particularly in children, in acute head injury.

Classification and pathological features

Clinically, the commonest type of cerebral oedema is **vasogenic oedema** in which there is increased capillary permeability leading to leakage of fluid containing protein and electrolytes (including sodium ions) into the extracellular space. It commonly occurs focally around tumours, abscesses, and regions of contusion, haemorrhage and infarction. Vasogenic oedema is also associated with meningitis and lead encephalopathy. It tends to be more evident in white matter which becomes paler microscopically with an accumulation of gemistocytic astrocytes. Prolonged existence of the oedematous fluid in the white matter may lead to myelin breakdown and the presence of lipid phagocytes.

Cytotoxic oedema, also known as cellular oedema, is a form of intracellular oedema that affects grey matter more than white matter, causing neurocellular enlargement. It can be caused by ischaemia and by specific experimental cytotoxins owing to impairment of the

ATP-dependent cell membrane sodium ion pump that normally helps to maintain osmotic equilibrium. It can also occur in other conditions that alter osmotic equilibrium, such as diabetic ketoacidosis and haemodialysis. Vasogenic and cytotoxic oedema may coexist in cases of ischaemia.

The increased periventricular tissue water content that occurs in **interstitial oedema** is seen in hydrocephalus and is discussed in the subsection on hydrocephalus below. This form of cerebral oedema is also known as hydrocephalic oedema.

Hydrostatic oedema is the result of an acute increase in intravascular pressure and occurs in hypertensive encephalopathy and following neurosurgery on patients with raised intracranial pressure.

Clinical features

The rigidity of the skull means that cerebral oedema will raise intracranial pressure and lead to the clinical features mentioned above.

Because it commonly occurs focally around regions of pathology, vasogenic oedema can give rise to focal symptoms and signs which are more pronounced than those caused by the primary pathology and which may diminish if there is resolution of the oedema. One method employed clinically to reduce the degree of vasogenic oedema is pharmacotherapy with corticosteroids such as dexamethasone and betamethasone which have a very high glucocorticoid activity but almost no mineralocorticoid activity. The glucocorticoid steroids do not have such a therapeutic effect with other types of cerebral oedema however.

Unlike vasogenic oedema, cytotoxic oedema causes more generalized effects such as fits and a reduced level of consciousness leading to coma. Clinically it may be possible to compensate for the osmotic disequilibrium with a rapid intravenous infusion of the osmotic diuretic mannitol.

Hydrocephalus

Definition

Hydrocephalus is an increase in the intracranial cerebrospinal fluid volume. It is associated with dilatation of the ventricular system.

Aetiology

In the case of **primary hydrocephalus** an increased volume of cerebrospinal fluid within the cranial cavity can result from increased formation of cerebrospinal fluid, an obstruction to its circulation, or decreased absorption.

Increased formation of cerebrospinal fluid by the choroid plexuses is a rare cause of hydrocephalus and occurs in papillary tumours of the choroid plexus.

The commonest cause of primary hydrocephalus is **obstruction**, in either the ventricular system or subarachnoid space, to the free circulation of cerebrospinal fluid. Ventricular system obstruction can be the result of tumours, fibrotic adhesions caused by meningitis, the Dandy–Walker syndrome, and aqueduct stenosis. Because of their small sizes, lesions causing obstruction at the cerebral aqueduct (of Silvius) and interventricular foramina (of Monro) can themselves be of a small size. Obstruction in the subarachnoid space may result from tumours and adhesions caused by inflammation, trauma, or haemorrhage. Subarachnoid space obstruction may also result from congenital malformations such as the Arnold–Chiari malformation.

It will be recalled that absorption of the cerebrospinal fluid is mainly by the arachnoid villi of the dural venous system. **Decreased absorption** of cerebrospinal fluid may result from increased intracranial venous pressure, which may be caused by an intracranial expanding lesion, or reduced intracranial venous drainage caused by increased intrathoracic pressure. Reduced absorption of cerebrospinal fluid may also occur following obstruction of the arachnoid villi by, for example, neoplastic cells or siderophages (following subarachnoid haemorrhage).

Secondary hydrocephalus is also known as compensatory hydrocephalus or hydrocephalus ex vacuo. There is an increased intracranial volume of cerebrospinal fluid to compensate for the loss of brain tissue occurring, for example, as a result of cerebral atrophy.

Classification of primary hydrocephalus

In **obstructive hydrocephalus** there is an obstruction to the free circulation of cerebrospinal fluid as mentioned above, while there is no such obstruction in **non-obstructive hydrocephalus**. Obstructive hydrocephalus is usually **non-communicating** which implies that cerebrospinal fluid cannot freely pass from the ventricular system to the subarachnoid space. Non-obstructive hydrocephalus, on the other hand, is **communicating**.

Hakim (1964) identified a form of primary hydrocephalus with normal cerebrospinal fluid pressure. This is now most commonly known as **normal pressure hydrocephalus** (see also Hakim & Adams 1965). It differs from the above types in being both obstructive and communicating. It is caused by an obstruction in the subarachnoid space which prevents cerebrospinal fluid from being reabsorbed but allows it to flow into the subarachnoid space from the ventricular system. Monitoring studies have demonstrated that there may be episodes of raised cerebrospinal fluid pressure during sleep and it has therefore been suggested that a better term for this syndrome might be **intermittent hydrocephalus**.

Since the types of primary hydrocephalus are usually associated with raised cerebrospinal fluid pressure (with the notable exception of normal pressure hydrocephalus) they are also sometimes termed **hypertensive hydrocephalus**.

Pathological features

As mentioned above, hydrocephalus is associated with dilatation of the ventricular system. For obstructive hydrocephalus the parts of this system that are enlarged depend on the site of the obstructing lesion. Associated pathological changes include atrophy of periventricular white matter, erosion of structures in the skull such as the sella turcica and clinoid processes, atrophic changes in the olfactory and optic nerves and, if it occurs in infants prior to the closure of the skull sutures, enlargement of the head.

Since there is little or no change in the overall volume of intracranial contents, secondary hydrocephalus does not lead to an increased cerebrospinal fluid pressure.

Clinical features

As would be expected, the clinical features of those types of hydrocephalus associated with raised intracranial pressure are essentially those of raised intracranial pressure described earlier in this section. Impaired attention, concentration and initiative may also occasionally be seen.

In infants there may also be enlargement of the head, as mentioned above, and associated mental retardation and spasticity. Blindness may also occur. It should be noted, however, that there are many cases of such infants not being affected intellectually; some have gone on to study successfully at university level.

In normal pressure hydrocephalus the features of raised intracranial pressure are generally absent. The syndrome mainly occurs in the seventh and eighth decades of life. Varying degrees of cognitive impairment and physical slowness occur. Other features include unsteadiness of gait, urinary incontinence, and nystagmus.

FOCAL CEREBRAL DISORDER: CLINICAL FEATURES

Although not strictly belonging in a chapter on neuropathology, it is convenient to discuss briefly the clinical features of focal cerebral disorder before describing neuropathological conditions, such as infections and tumours, that can give rise to them. It should be noted that the clinical features give an indication of the location of the pathology, but usually do not imply the nature of the pathology itself. For example, frontal lobe syndrome can be caused by a number of different disorders such as a tumour, trauma, Pick's disease and neurosyphilis.

Frontal lobe

The types of change of personality which occur in association with frontal lobe lesions include disinhibition, reduced social and ethical control, sexual indiscretions, financial and personal errors of judgement, elevated mood, lack of concern for the feelings of other people, and irritability. These are primarily related to prefrontal impairment, and in frontal lobe damage are associated with perseveration, utilization behaviour (for example, putting on a pair of spectacles when seen, writing when a pen comes within grasp, and eating and drinking whenever food and water are seen), and pallilalia (repetition of sentences or phrases). All these features demonstrate rigidity of thinking and stereotyped repetition. Other characteristic features include impairment of attention, concentration and initiative. Aspontaneity, slowed psychomotor activity, motor Jacksonian fits, and urinary incontinence may also occur as part of a frontal lobe syndrome.

If the motor cortex or deep projections are affected this may result in a contralateral spastic paresis or aphasia. Posterior dominant frontal lobe lesions may cause apraxia of the face and tongue, primary motor aphasia, or motor agraphia. Anosmia and ipsilateral optic atrophy may result from orbital lesions.

Temporal lobe

Dominant temporal lobe lesions may cause sensory aphasia, alexia and agraphia. Posterior dominant temporal lobe lesions may cause features of the parietal lobe syndrome mentioned below.

Non-dominant temporal lobe lesions may cause hemisomatognosia, prosopagnosia, visuospatial difficulties, and impaired retention and learning of nonverbal patterned stimuli such as music.

Bilateral medial temporal lobe lesions may cause amnesic syndromes.

The personality changes that may occur are similar to those caused by frontal lobe lesions. Other features of temporal lobe lesions include psychotic symptomatology, epilepsy, and a contralateral homonymous upper quadrantic visual field defect.

Parietal lobe

Features of the parietal lobe syndrome include visuospatial difficulties such as constructional apraxia (for example difficulty in buttoning one's coat) and visuospatial agnosia, topographical disorientation, visual inattention, sensory Jacksonian fits, and cortical sensory loss. The last neurological feature results in agraphaesthesia, asterognosis, impaired two point discrimination and sensory extinction.

Dominant parietal lobe lesions may cause primary motor aphasia (caused by anterior lesions), primary sensory aphasia (caused by posterior lesions and leading to agraphia and alexia), motor apraxia, Gerstmann's syndrome (dyscalculia, agraphia, finger agnosia, and right-left disorientation), bilateral tactile agnosia, and visual agnosia (caused by parieto-occipital lesions).

Non-dominant parietal lobe lesions may cause anosognosia, hemisomatognosia, dressing apraxia and prosopagnosia.

Occipital lobe

Features of the occipital lobe syndrome include a contralateral homonymous hemianopia, scotomata and simultanagnosia. Bilateral lesions may result in cortical blindness.

Dominant occipital lobe lesions may cause alexia without agraphia, colour agnosia, and visual object agnosia.

The following are more common with non-dominant lesions: visuospatial agnosia, prosopagnosia, metamorphopsia (image distortion) and complex visual hallucinations.

Corpus callosum

Acute severe intellectual impairment may occur. With extension into other parts of the central nervous system neurological signs of involvement of the frontal lobe, parietal lobe, or diencephalon may result. For a patient in whom the left hemisphere is dominant, loss of contact between the dominant hemisphere speech centres and the non-dominant hemisphere can lead to left-sided apraxia to verbal commands and asterognosis in the left hand.

Diencephalon and brain stem

Characteristic features of midline lesions include amnesia of the Korsakoff type, hypersomnia and akinetic mutism. Intellectual impairment or occasionally a rapidly progressive dementia may also be seen. Personality changes resemble those caused by frontal lobe lesions except that loss of insight is less likely. Features of raised intracranial pressure may occur and have been mentioned above.

Pressure on the optic chiasma may lead to visual field defects.

Thalamic lesions may cause hypalgesia to painful stimuli and sensory disorders similar to those seen in the parietal lobe syndrome.

Hypothalamic lesions may cause polydipsia, polyuria, increased body temperature, obesity, amenorrhoea or impotence, and an altered rate of sexual development in children.

Pituitary lesions may cause various endocrine disorders.

Brain-stem lesions may cause palsies of cranial nerves and disorders of long tract motor and sensory functions.

CEREBROVASCULAR DISORDERS

In this section the cerebrovascular accidents or strokes and extracerebral haemorrhage are considered. The subsection on extracerebral haemorrhage deals mainly with subarachnoid haemorrhage because of its psychiatric sequelae. Multi-infarct dementia is considered later in this chapter in the section on the dementias. The blood supply to the central nervous system has been outlined in Chapter 1 and will not therefore be repeated here.

Cerebrovascular accidents

Definition

Cerebrovascular accidents or strokes are terms used clinically to refer to focal non-convulsive neurological disorders of sudden onset that are believed to be caused by cerebrovascular dysfunction.

Aetiology

Approximately four-fifths of cerebrovascular accidents are caused by cerebral infarction with the remaining one-fifth resulting from both intracerebral haemorrhage and subarachnoid haemorrhage caused by ruptured intracranial aneurysms; subarachnoid haemorrhage is considered below in the subsection on extracerebral haemorrhage.

Cerebral infarction

A cerebral infarct is a localized region of necrosis of cerebral tissue and its parenchymal components secondary to varying degrees of vascular occlusion. No gross neuropathological changes are visible during the first few hours following the occurrence of cerebral infarction. The first such change is a slight softening and change in colour of the infarcted region. This is followed by marked swelling which may give rise to features of intracranial expanding lesions mentioned earlier in this chapter. With resolution there is a reduction in this swelling, and therefore of neurological features caused by it, and the formation of cysts in larger infarcts.

The causes of cerebral infarction include: embolism and thrombosis, hypertension, and cerebral arteritis.

Cerebral **embolism** is the commonest cause of cerebral infarction. The emboli most often arise from atherosclerotic thrombotic occlusions, particularly of the carotid and vertebrobasilar arteries. Prior to causing cerebral infarction, platelet emboli from these arterial occlusions may give rise to sudden focal non-convulsive neurological disorders from which patients rapidly recover completely; these are known as **transient ischaemic attacks**. Emboli can also arise from the heart following myocardial infarction or atrial fibrillation.

In addition to its role in the pathogenesis of atherosclerosis, **hypertension** is associated with small infarcts known as **lacunes** which occur particularly in the basal ganglia, thalamus, and white matter of the cerebral hemispheres. Lacunes may be caused by **hyaline arteriosclerosis** produced by hypertension, or by emboli. Hypertension is also associated with mutiinfarct dementia, hypertensive encephalopathy, intracranial haemorrhage, and Binswanger's disease.

Cerebral **arteritis** may be caused by a variety of microorganisms (as in meningovascular syphilis), collagen diseases, and damage from radiation. The collagen diseases responsible include systemic lupus erythematosus, polyarteritis nodosa, temporal arteritis, and Bechet's disease. **Systemic lupus erythematosus** affecting the central nervous system can cause cranial and peripheral nerve lesions, depression, phobias, epilepsy; disorientation and hallucinations may also occur, but usually secondary to pharmacotherapy with corticosteroids.

Vascular occlusion is also associated with intravenous drug abuse.

Intracerebral haemorrhage

Most episodes of intracerebral haemorrhage result from rupture of microaneurysms in small perforating arteries or rupture of larger vessels. Hypertension is an associated factor. Rarer causes include abscesses, tumours, and clotting disorders. The neuropathological effects of a blood clot are those of intracranial expanding lesions, for example ventricular distortion and raised intracranial pressure.

Cerebral arterial syndromes

The clinical features resulting from occlusion of different cerebral arteries are now briefly outlined.

Middle cerebral artery occlusion can result in: contralateral hemiparesis; cortical sensory loss; contralateral hemianopia; aphasia, if the dominant hemisphere is affected; agnosic syndromes and body image disturbances if the non-dominant hemisphere is affected; and clouding of consciousness.

Anterior cerebral artery occlusion can result in: contralateral hemiparesis, with the leg being affected more severely than the arm; grasp reflex; cortical sensory loss; motor aphasia; clouding of consciousness; personality change of the frontal lobe dysfunction type; changes in mental functioning similar to those seen in dementia; and incontinence.

Occlusion of the **internal carotid artery** may be asymptomatic if the circle of Willis is patent. Otherwise it may cause a middle cerebral infarction-type picture, together with monocular blindness, unilateral loss of the carotid pulse, and an ipsilateral Horner's syndrome.

Posterior cerebral artery occlusion can result in: contralateral hemianopia; visual hallucinations; visual agnosia; spatial disorientation; visual perseveration; alexia without agraphia, if the dominant occipital lobe and the splenium of the corpus callosum are affected; a contralateral thalamic syndrome; cerebellar ataxia; and, in the case of bilateral occipital lobe infarction, cortical blindness.

So far as the **vertebrobasilar system** is concerned, total occlusion of the basilar artery is rapidly fatal. Partial occlusion of the basilar artery may cause: features of brain stem, pyramidal, and ipsilateral cerebellar lesions: ipsilateral cranial nerve palsies; peduncular hallucinosis; the locked-in syndrome; states of bizarre disorientation; and excessive dreaming.

Psychiatric sequelae

The psychiatric sequelae of cerebrovascular accidents include: dementia, as mentioned above; organic personality change, such as irritability and apathy, which may be more the result of widespread atherosclerosis than of focal changes following cerebrovascular accidents; depressed mood; and, rarely, paranoid-hallucinatory syndromes.

Extracerebral haemorrhage

Extradural haemorrhage

Extradural haemorrhages result from traumatic damage to one of the meningeal arteries, usually the middle meningeal. Extradural haematomas form rapidly, owing to the arterial nature of the bleeding, and lead to raised intracranial pressure. Clinically, temporary loss of consciousness at the time of the trauma is typically followed by a lucid interval during which the patient appears to have recovered. However, if untreated then within a week the patient becomes comatose and dies.

Subdural haemorrhage

Subdural haemorrhages can result from relatively minor head injury, such as after a fall, causing traumatic damage to veins that connect the cerebral venous system with the dural venous sinuses. They occur most often in the elderly, in individuals suffering from chronic alcoholism, and in children. Because of the venous nature of the bleeding the formation of subdural haematomas is slower than that of extradural haematomas. Once formed the haematoma will gradually increase in size partly through the absorption of fluid from the cerebrospinal fluid.

Acute subdural haematomas may cause fluctuating levels of consciousness, but with delayed onset.

Chronic subdural haematomas may be overlooked, particularly in alcoholics and the elderly. They may give rise to vague physical complaints, headache, confusion, and, more rarely, a progressive contralateral weakness and hemiplegia. Seizures and features of raised intracranial pressure may also occasionally occur.

Subarachnoid haemorrhage

The most common cause of subarachnoid haemorrhage is rupture of berry aneurysms that have developed at regions of weakness of the medial arterial wall at arterial bifurcations in the circle of Willis. Other types of aneurysm that may rupture are arteriosclerotic and mycotic. Other causes of subarachnoid haemorrhage include rupture of an arteriovenous malformation, traumatic damage, intracerebral haemorrhage extension, and haemorrhage from a tumour. As indicated above, hypertension is an important predisposing factor.

Clinically, the patient may complain of a sudden severe headache at the time of onset of the subarachnoid haemorrhage, and there may sometimes be loss of consciousness. Nausea and vomiting may also occur. Neck stiffness, photophobia, and a positive Kernig's sign are present owing to meningeal irritation. Localizing focal neurological disorders may occur which may include cranial nerve lesions, aphasia, and hemiparesis. With continuing bleeding, signs of raised intracranial pressure may be seen.

Psychiatric sequelae have been found to be relatively common in individuals who survive. In a large study of 261 survivors, Storey (1967, 1970) found cognitive impairment in 40%; personality impairment in 19%, including 'organic moodiness' and frontal lobe syndrome-like changes; occasional improvement in personality following anterior bleeds; and symptoms of anxiety and depression in a quarter of cases. Amnesic syndromes, usually transient and resembling a Korsakoff picture, have been reported in survivors in other studies such as those of Tarachow (1939), Walton (1953), Theander and Granholm (1967), and Logue et al (1968).

INFECTIONS

In this section the following conditions, which may have psychiatric manifestations, are considered: neurosyphilis; virus encephalitis; meningitis; cerebral abscess; Creutzfeldt–Jakob disease; and the acquired immunodeficiency syndrome or AIDS. Other conditions that may have mental manifestations but are very rare in developed countries, for example kuru and trypanosomiasis, are not discussed in this section.

Before considering individual conditions it is useful to remind oneself of the sites of infection. In general, an infection by any organism may affect either the meninges and cerebrospinal fluid, in which case it is known as **meningitis**, or the brain parenchyma, which is known as **encephalitis.** Inflammation of both regions together is known as **meningoencephalitis.** Working from the skull internally, bacterial infections of the bone, extradural space, subdural space, subarachnoid space, cerebral tissue, and ventricles may respectively cause osteitis, an extradural abscess, a subdural abscess, leptomeningitis, a cerebral abscess, and ventriculitis. It should be noted that although infection of the subarachnoid space is strictly leptomeningitis, it is more commonly referred to as meningitis, and this is the terminology that is followed in this chapter. Pachy-meningitis refers to infections of the extradural space and subdural space. Viruses usually cause a meningo-encephalitis; if the spinal cord is also affected this is known as **encephalomyelitis**.

Neurosyphilis

Aetiology

Syphilis is a bacterial infection caused by the spirochaete *Treponema pallidum.* It is usually transmitted through sexual contact but can also occur as a congenital infection following transplacental transmission.

Classification

The clinical classification of syphilis is into four stages. Primary syphilis is the first clinical stage in which a chancre and lymphadenopathy occur. Secondary syphilis may occur more than three months following infection and is characterized by a bacteraemia affecting many sites. The bacteria may enter the central nervous system. Neurological effects in secondary syphilis include headaches and acute meningitis, and more rarely cranial nerve palsies and convulsions. Following a latent period in untreated syphilis which may last many years, tertiary or late syphilis occurs in which localized gummata are present. Owing to effective antibiotic pharmacotherapy this stage now rarely occurs. Following this stage quaternary syphilis manifests itself as either cardiovascular syphilis or neurosyphilis. It should be noted that some authors do not recognize a fourth clinical stage of syphilis and include neurosyphilis as a form of tertiary syphilis.

The cerebrospinal fluid changes in neurosyphilis include increased cells and protein, in addition to positive results to the appropriate antibody tests.

The most important manifestations of neurosyphilis are meningovascular syphilis, tabes dorsalis, and general paralysis of the insane (GPI). Pathological and clinical features of these three groups are given below following an outline of the features of localized central nervous system gummata. A rare manifestation of neurosyphilis is syphilitic amyotrophy which affects the anterior horns of the spinal cord.

Gummata

Syphilitic gummata are granulomatous lesions that occur in the meninges attached either to the dura mater alone or, more commonly, to both the dura mater and the underlying part of the brain. Central necrosis and lymphocytic and plasma cell invasion of the lesions occur.

Gummata typically occur over the cerebral hemispheres and cerebellum. They can give rise to convulsions and, depending on their location, focal neurological features such as aphasia and hemiplegia.

Meningovascular syphilis

Meningovascular syphilis typically occurs within a decade or so of the time of the initial infection. A characteristic granulomatous meningitis occurs in which fibrosis of the meninges may affect the cranial nerves. Cranial nerve palsy may also be the result of endarteritis obliterans, leading to ischaemic necrosis and arterial thrombosis. Other pathological features include hypertrophic pachymeningitis causing thickening of the cervical dura mater, and myelitis.

Clinically, meningovascular syphilis may present with headache, lethargy, irritability and malaise. Neck stiffness, convulsions, delirium and psychotic symptomatology may also occur. An alternative presentation is the occurrence of focal neurological features such as cranial nerve palsy, particularly affecting the second, third, fourth or eighth cranial nerves, and features of cerebrovascular accidents. Meningeal anterior and posterior spinal root thickening may cause muscle wasting and pain, respectively.

Tabes dorsalis

Tabes dorsalis develops some 10 to 35 years after the primary infection, and sometimes occurs with general paralysis of the insane (described below). In tabes dorsalis there is initially spinal cord posterior root sensory neuronal degeneration usually in the lower thoracic and lumbar nerve roots; the rarer cases in which the cervical nerve roots are mainly affected are known as cervical tabes. This is followed by posterior white column atrophy.

The spinal cord posterior involvement leads to the characteristic clinical feature of lightning pains which are severe stabbing pains lasting a few seconds and most often affecting the legs, although other regions of the body may also be affected. Another characteristic feature is the occurrence of sudden severe episodes of abdominal pain and vomiting that may last for days which are known as tabetic crises. Posterior white column involvement leads to loss of proprioception which in turn causes an ataxia with a characteristic gait that is wide-based and stamping; Romberg's sign is positive. Loss of vibration sense in the lower limbs can also be demonstrated clinically. There may also be loss of deep pain sensation leading to neuropathic or Charcot's joints. Severe foot ulceration can result from loss of cutaneous sensation. Areflexia occurs in the legs but in pure tabes dorsalis the plantar reflexes are flexor. A neurogenic bladder is also sometimes present.

Argyll–Robertson pupils are seen mainly in tabes dorsalis and general paralysis of the insane. The pupils are small and irregular. There is no light reflex but the pupils do constrict with accommodation. They are usually caused by interruption of the fibres from the pretectal nucleus to the Edinger–Westphal nucleus (see Ch. 1) by a syphilitic lesion.

General paralysis of the insane

General paralysis of the insane is also known as general paresis and usually develops 10 to 20 years after the initial infection. There is atrophy of the cerebral cortex which is usually severest in the frontal and temporal lobes. Histologically, a subacute encephalitis is present in the regions of cortical atrophy with neuronal degeneration, astrocytosis, and the presence of siderophages, lymphocytes, plasma cells and spirochaetes. A chronic ventriculitis is present and there is meningeal thickening with lymphocytic infiltration.

Presenting clinical features include impairment of memory and concentration, frontal lobe dysfunction-type personality changes, and minor emotional symptoms. In a survey of 91 cases by Dewhurst (1969) a depressive picture was reported to occur in about one quarter of cases, and a grandiose picture in one-tenth of cases. Dementia was found in one-fifth of cases in the same series. Manic elated, neuraesthenic and schizophreniform presentations are also sometimes seen.

Other clinical features of general paralysis of the insane include: epilepsy; Argyll–Robertson pupils; tremor, either cerebellar or extrapyramidal, affecting the tongue, limbs and trunk; dysarthria; and focal neurological features such as aphasia and hemiplegia.

Virus encephalitis

Aetiology

Many viruses have been found to cause encephalitis. In Britain the most common causes are mumps, Echo, Coxsackie, measles, Herpes simplex, Herpes zoster, Epstein–Barr, and adenoviruses. The pandemic encephalitis that began during the First World War which is known as encephalitis lethargica is of unknown origin although most believe it was probably viral. The acquired immune deficiency syndrome or AIDS can also cause an encephalitis; AIDS is discussed later in this section.

Pathological features

There are a number of characteristic histological changes that are seen in virus encephalitis. These include infiltration of the cerebral parenchyma and subarachnoid space by inflammatory cells, and neuronal necrosis that varies in extent and site according to the causative virus. In addition, many of the neurocellular reactions discussed in the first section of this chapter also occur. They include: the presence of viral inclusions; activation of microglia, with the presence of rod cells and lipid phagocytes; and astrocytosis.

Clinical features

Generalized features that occur include headache, nausea and vomiting, drowsiness, and fits. In addition, focal neurological features may also occur. Features of raised intracranial pressure may also be present, as may neck stiffness if there is meningeal involvement. Pyrexia, if present, is usually low-grade. Impaired consciouness, delirium, and more rarely, psychiatric disorders may also be part of the presentation. For example, Misra and Hay (1971) reported three cases who were admitted as psychiatric inpatients with a provisional diagnosis of shizophrenia (see also Crow 1978).

With resolution of the acute episode a number of complications may emerge. These include personality changes, depression, chronic anxiety and dementia. Episodes of virus encephalitis may contribute to childhood behaviour disorders (Greenbaum & Lurie 1948).

Cases of encephalitis lethargica are now very rare. It is also known as epidemic encephalitis and postencephalitic Parkinsonism owing to the occurrence of the features of Parkinson's disease as a complication. Post-encephalitic personality change and psychosis have also been reported (Fairweather 1947, Hall 1929, Davison & Bagley 1969).

Myalgic encephalomyelitis (ME) is also known as the chronic fatigue syndrome or postviral syndrome. Prolonged periods of lethargy occur, often accompanied by headache, myalgia and features of depression. The disorder often affects young people and may occur in epidemics (such as the outbreak at the Royal Free Hospital in London in 1955) and may follow an illness. Some cases may be associated with viral infections, but others are associated with either other infective agents or no known infective cause. In spite of its name, encephalitis is not a neuropathological feature.

Meningitis

Most cases of meningitis are relatively easy to diagnose, with the main presenting features including headache, nausea and vomiting, photophobia, neck stiffness, and irritability. However, in the case of **tuberculous meningitis** the presentation is subacute with a prodrome of general malaise being common. Apathy and personality change may occur during the prodromal stage so that the diagnosis is easily missed. Tuberculous meningitis is now rare in developed countries. It is usually caused by the bacterium *Mycobacterium tuberculosis* which spreads to the meninges via the bloodstream during miliary infection. It gives rise to granulomatous inflammatory lesions which may affect the cranial nerves and give rise to cranial nerve palsies, particularly of the second, third, fourth, sixth, seventh and eight nerves. The granulomatous inflammation may also cause raised intracranial pressure by decreasing the reabsorption of cerebrospinal fluid, and it may cause fits by affecting the cerebral cortex.

Cerebral abscess

Cerebral abscesses are caused by bacteria such as streptococci, staphylococci, pneumococci, and members of the family Enterobacteriaceae. The primary infection is usually in the middle ear, the sinuses, or the lungs. Although the diagnosis is often clear, with a presentation of the primary infection and features caused by an intracranial expanding lesion, cerebral abscesses may form insidiously and may be wrongly diagnosed as psychiatric disorders because of the development of depressive symptoms and change in personality.

Creutzfeldt–Jakob disease

Aetiology

Creutzfeldt–Jakob disease is a rare progressive dementia which is known to be transmissible. It is believed to be caused by infection with either a **prion**, a glycoprotein viral subparticle lacking ribonucleic acid, or a **slow virus.** These are similar to the viral infective agents believed to be responsible for the now extremely rare neurodegenerative human disorder **kuru,** which was reported as occurring in New Guinea following cannibalism, and the animal neurodegenerative disorders of **scrapie** in sheep and **bovine spongiform encephalopathy** (BSE) in cattle. It is now believed that the increased incidence of BSE in Britain in the late 1980s was the result of including bonemeal derived from scrapie-infected sheep in cattle feed. However, at the time of writing there is no evidence that any cases of human Creutzfeldt–Jakob disease have arisen from eating beef, mutton or lamb. On the other hand there is evidence that Creutzfeldt–Jakob disease may be transmitted from infected humans through procedures such as corneal transplantation (Duffy et al 1974), depth electroencephalography with contaminated electrodes (Bernoulli et al 1977), and neurosurgery using contaminated instruments (Will & Masters 1982).

Pathological features

There may be little or no gross atrophy of the cerebral cortex evident in rapidly developing cases. In those who survive longest gross neuropathological changes seen may include selective cerebellar atrophy, generalized cerebral atrophy and ventricular dilatation.

Histologically, there is evidence of neuronal degeneration without inflammation. Astrocytic proliferation occurs particularly in the cerebral cortex, basal ganglia, brain–stem motor nuclei, and spinal cord anterior horn cells. A characteristic feature of the grey matter of the cerebral cortex is the presence of multiple vacuoles. This gives the cerebral cortex a spongy appearance and is known as **status spongiosus.** Degeneration also occurs in spinal cord long descending tracts.

Clinical features

Creutzfeldt–Jakob disease is a rare form of presenile dementia with an equal incidence in men and women. Its clinical features depend on the parts of the brain most affected and include: memory impairment and personality change with slowing, fatigue and depressed mood; features typical of parietal lobe dysfunction, such as Gerstmann's syndrome; epileptic fits; myoclonic jerks; psychotic symptomatology; extrapyramidal features; cerebellar ataxia; dysarthria; and dysphagia. Because of the mental manifestations, and particularly in cases of temporary remission of early neurological features, the presentation may be mistaken for a functional psychiatric disorder.

Acquired immunodeficiency syndrome (AIDS)

Definition

A case of AIDS is a reliably diagnosed disease indicative of a defect in cell-mediated immunity occurring in a person who has serum antibodies to the human immunodeficiency virus (HIV) and no other known cause for diminished cell-mediated immunity. Diseases indicative of diminished cell-mediated immunity include: opportunistic viral, bacterial, fungal, protozoal and helminthic infections, for example pneumonia caused by *Pneumocystis carinii;* and secondary neoplasia such as Kaposi's sarcoma occurring in individuals aged less than 60 years.

Aetiology

The acquired immunodeficiency syndrome, or AIDS, is caused by the retrovirus **human immunodeficiency virus** or **HIV** which is present in the body fluids of affected individuals. It is transmitted through contact with some of these fluids, particularly blood and semen, mainly by sexual intercourse (both homosexual and heterosexual) and intravenous drug abuse using contaminated needles. HIV can also be transmitted through the administration of contaminated blood or blood derivatives. For example, it has been transmitted to haemophiliacs through contaminated clotting factor concentrates. It is rare for HIV to be transmitted through breast milk, and at the time of writing there are no reported cases of transmission through saliva. AIDS can also occur congenitally in the babies of infected mothers either in utero or through contact with maternal blood during parturition.

Pathological features

The virus attaches to T4 helper cell surface receptors causing a reduction of the numbers of these lymphocytes (and the formation of antibodies). This in turn is responsible for many of the clinical features of AIDS owing to a reduction in the body defences against infections generally. Most cases of AIDS involve the central nervous system and the pathological features of this are now described.

Encephalitis, usually subacute, affects approximately one-third of patients overall, and almost all patients who survive for a relatively long time. It is usually caused by HIV but may also result from other opportunistic infections, such as with cytomegalovirus, herpes simplex, herpes zoster and mycobacteria.

Meningitis may be caused not only by the more common pathogens responsible for meningitis in non-AIDS cases, but also by less common pathogens such as the fungus *Cryptococcus neoformans,* other fungi, and amoebae.

Another common neuropathological feature of AIDS is the occurrence of **cerebral abscesses,** usually multiple and caused by *Toxoplasma gondii.*

Other neuropathological features include primary cerebral lymphoma, myelopathy associated with subacute encephalitis, retinitis and peripheral neuropathy.

Clinical features

The clinical development of AIDS in infected individuals is the basis of the classification system of the Centers for Disease Control, Atlanta, Georgia. The first group in this classification is the development of an **acute seroconversion illness** soon after infection in some individuals. The second group refers to the **asymptomatic infection** that occurs in most infected individuals, who are seropositive, for a few months to several years. During this stage HIV can be transmitted to others. The next stage that usually follows (Group III) is of **persistent generalized lymphadenopathy.** Approximately one-third of this group go on to develop AIDS itself within the next five years.

There are a number of clinical manifestations of HIV infection which are brought together in five subgroups (A to E) of Group IV of the above classification. The first subgroup is constitutional disease or the **AIDS-related complex** (ARC). ARC includes generalized features such as decreased body mass, chronic fatigue, night sweats, pyrexia, myalgia, diarrhoea and cutaneous infections. ARC is not strictly part of AIDS as defined above, but the majority of those who survive go on to develop AIDS itself.

The next subgroup consists of HIV-related **neurological disease.** A wide range of neurological and neuropsychiatric complications can occur. Subacute encephalitis can cause insidious cognitive impairment and, with a superimposed opportunistic infection, a confusional state. The clinical picture may gradually progress to one of **AIDS dementia** with loss of cognitive functioning. It should be noted that initially the progressive loss of cognitive function may be mistaken for a depressive illness. Moreover, the latter can occur in its own right in AIDS. Motor and behavioural functioning are also affected and patients may manifest pyramidal signs and ataxia. Features typical of meningitis are also seen in AIDS; the cranial nerves most often affected are the fifth, seventh and eight. The clinical features of cerebral abscesses have already been mentioned. If it occurs, **primary cerebral lymphoma** may present with headache, seizures, and focal neurological deficits, or with a progressive dementia. When present, clinical features of other neuropathological changes, such as peripheral neuropathy, are also seen.

Subgroup C consists of diseases resulting from **opportunistic infections,** such as *Pneumocystis carinii* pneumonia. So far as the central nervous system is concerned, **meningitis** may be caused most often by *Cryptococcus neoformans,* and **encephalitis** by cytomegalovirus, JC virus and SV40. Cytomegalovirus infections are more likely to cause **retinitis** and therefore blindness. As mentioned above, opportunistic infections, particularly with *Toxoplasma gondii,* may given rise to multiple **cerebral abscesses.**

Subgroup D consists of **secondary neoplasia,** the commonest being Kaposi's sarcoma. The presentation of primary cerebral lymphomas has been mentioned above; because of the similarity of presentation, the clinical differentiation of such a secondary neoplasm from multiple abscesses can be difficult. Subgroup E is a collection of other conditions such as thrombocytopaenia.

The psychosocial and neuropsychiatric aspects of HIV infection and AIDS are reviewed by Catalan (1988) and King (1990).

TRAUMA

In this section the neuropathological features of head injury and the punch-drunk syndrome are considered. Some aspects of the clinical mental manifestations are also described briefly; a more comprehensive account is to be found in Lishman (1987).

Non-missile head injury

In peacetime, non-missile head injuries are much more common than head injuries caused by missiles. The commonest causes in Britain are road traffic accidents, with alcohol often being involved, and falls.

Pathological features

The pathological effects of head injury can be classified either into primary and secondary, or into focal and diffuse damage. The first classification is used in this section.

The **primary** effects of head injury are considered first. **Diffuse axonal injury,** particularly in the white matter, results from the effects of shearing and tensile forces; haemorrhagic lesions of the corpus callosum, sometimes with tearing of the interventricular septum, and brain stem are relatively common. **Cerebral contusions** result mainly from damage to focal brain regions in contact with bony protuberances; common sites are the inferior surfaces of the frontal and temporal lobes. Cerebral contusions that take place at the site of impact are known as **coup,** while those occurring at the opposite pole are known as **contrecoup.** The latter is more prominent when the head is moving at the time of impact, as occurs during road traffic accidents and falls. In addition to surface contusions, haemorrhagic foci may occur in parasagittal regions. They are caused by greater gliding of subcortical cerebral tissue than surface cortical tissue at the time of injury, and these lesions are known as **gliding contusions.** Although the presence of a skull **fracture** is usually associated with a strong force of impact, it does not necessarily imply that the head injury is more severe. On the other hand, intracranial haemorrhage is more common when there is a fracture. Moreover, depressed fractures may lead to cortical lacerations. Fractures can also give rise to **otorrhoea** and **rhinorrhoea,** that is, leakage of cerebrospinal fluid from the ear and nose, respectively.

A common **secondary** effect of head injury is the formation of intracranial **haematomas.** As mentioned above, they are more likely to occur in association with skull fractures. The different types of intracranial haemotoma described earlier in this chapter may occur, causing effects mentioned there. Although cerebral oedema may occur as a result of an intracerebral haematoma, it may also occur when there is no haematoma formation as a diffuse congestive type of swelling (see above). Cerebral oedema and large intracerebral haematomas may in turn give rise to **raised intracranial pressure.** The various types of pathology so far mentioned, for example contusions, haematomas, and the effects of raised intracranial pressure, may lead to **cerebral ischaemia.** This may be made worse by hypoxia and hypotension resulting from associated injury or infection of other parts of the body, for example the chest. Penetrating head injuries may allow the entry of organisms and therefore the development of infections such as meningitis and cerebral abscesses. Another recognized complicating factor is **fat embolism.**

Clinical features

The clinical features that result from the types of neuropathological change noted above have been described earlier in this chapter. So far as the mental manifestations of head injury are concerned, these can be conveniently classified into the acute and chronic sequelae.

The acute effects of head injury include: impairment of consciousness, or concussion; acute post-traumatic psychosis; and memory disorders evident following recovery of consciousness. Clinically, an important measure of memory impairment is the length of the **post-traumatic amnesia,** that is, the interval of time between the moment of head injury and the resumption of normal continuous memory. It has been found that the duration of the post-traumatic amnesia is a good indicator of the degree of psychiatric disablement (Lishman 1968), post-traumatic personality change (Steadman & Graham 1970), and neurological and cognitive impairment (Russell & Smith 1961, Smith 1961).

The chronic psychological sequelae of head injury

comprise: cognitive impairment, which may be focal (Newcombe 1983) or generalized (Jennett & Plum 1972); personality change, which is related to the site of damage, being particularly common following frontal lobe lesions; psychoses including schizophrenia and delusional disorders (Davison & Bagley 1969, Achté et al, 1967); and mood disorder (Symonds 1937, Parker 1957, Achté et al 1967); suicide (Vauhkonen 1959, Achté & Anttinen 1963); and neuroses (Ota 1969). Post-traumatic epilepsy commonly takes the form of complex partial seizures, which is in its own turn associated with psychological sequelae.

Missile head injury

Missile head injuries may cause a depressed skull fracture, they may penetrate the cranial cavity without leaving it, or they may cause a perforating injury. The types of neuropathological change are similar to those that occur with non-missile head injuries, with haemorrhage, laceration, and oedema occurring mainly in the sites damaged by the missile. In those who survive, the access for organisms means that infections are common, with cerebral abscesses occurring more often than in non-missile head injuries. Also occurring more often is post-traumatic epilepsy.

Punch-drunk syndrome

The punch-drunk syndrome is also known as post-traumatic dementia and boxing encephalopathy.

Aetiology

The punch-drunk syndrome develops in individuals such as boxers who have received repeated blows to the head.

Pathological features

A characteristic gross neuropathological feature of the punch-drunk syndrome is cerebral atrophy. The regions that are particularly affected are the cerebral cortex and the hippocampal-limbic region, and enlargement of the lateral ventricles is common. Other changes that commonly occur include perforation of the septum pellucidum and thinning of the corpus callosum.

Histological changes include cortical neuronal loss and neurofibrillary degeneration with the presence of a similar type of neurofibrillary tangle in the cortex and brain stem to that seen in Alzheimer's disease (see below); neuritic plaques, however, are not present.

Clinical features

The clinical features of the punch-drunk syndrome include: a progressive impairment of memory and intellect, without any confabulation; a deterioration of personality with irritability and reduced drive; features of cerebellar dysfunction; extrapyramidal signs; and pyramidal signs.

In a study of 17 ex-boxers, Johnson (1969) reported the following psychiatric features: a chronic amnesic state in 11 patients; progressive dementia in three cases; morbid jealousy (delusional disorder, jealous type) in five cases; and the occurrence of rage reactions in three patients.

DEMYELINATING DISEASES

Demyelinating diseases are disorders of the central nervous system in which the characteristic primary neuropathological change is demyelination without axonal degeneration; secondary axonal degeneration can occur later. In this section the commonest demyelinating disease, multiple sclerosis, is considered first, followed by a brief account of another disease which has important psychiatric aspects, the much rarer Schilder's disease. Demyelination can also result from disorders of sphingolipid metabolism, and these are discussed in the next section of this chapter. One of the sphingolipidoses, adrenoleucodystrophy, is very similar to Schilder's disease but in this chapter they are treated separately. This section ends with a very brief account of two further demyelinating diseases, central pontine myelinolyis and Marchiafava–Bignami disease, both of which are usually caused by chronic alcoholism.

For the sake of completion, it should be noted that demyelination can also occur secondary to carbon monoxide poisoning and toxic agents such as hexachlorophane and tin, and in post-viral and post-immunization encephalomyelitis.

Multiple sclerosis

Aetiology

The aetiology of multiple sclerosis (MS), also known as disseminated sclerosis, is unknown. One possibility is that it is caused by a viral infection giving rise to an abnormal immunological response.

Pathological features

Discrete patches or **plaques** of demyelination occur in the white matter of the brain and spinal cord. The commonest site for plaques are the optic nerves, brain stem, cerebellar peduncles, periventricular regions of the cerebral hemispheres, and the cervical part of the spinal cord.

Acute lesions are yellow-white and soft. Histologically there is evidence of **peri-axial demyelination** with degeneration of the myelin sheaths but preservation of the axons. Lipid phagocytes are present in the lesions. Perivascular cuffing by T-lymphocytes and immunoglobulin-containing plasma cells also occurs in the plaques. The number of oligodendrocytes is reduced in the plaques.

With time the lesions change. The degenerated myelin is completely phagocytozed and, following astrocytosis, **glial scarring** gives the plaques a more grey-white firmer appearance. It is at this stage that secondary axonal degeneration occurs in the chronic lesions. Prior to this, however, axonal conduction may improve owing to resolution of oedematous fluid. In time degeneration of the ascending and descending spinal cord tracts occurs in chronic multiple sclerosis.

Clinical features

Since plaques may occur in any place in the white matter of the central nervous system, the clinical presentation of multiple sclerosis can vary. The commonest presentations are as follows: cervical spinal cord involvement leading variously to motor, sensory, bladder, bowel and erectile dysfunction; retrobular neuritis secondary to involvement of the optic nerve; cerebellar signs; diplopia secondary to an ocular nerve palsy or an internuclear ophthalmoplegia; and vertigo secondary to vestibular nuclear complex involvement.

The course of multiple sclerosis may be one of progressive deterioration or, particularly in those with a younger onset age, there may be spontaneous partial remissions interwoven with periods of relapse. In most affected individuals there is eventually an accumulation of multiple neurological handicaps with the development of dementia, quadriplegia and blindness.

The psychiatric complications of multiple sclerosis include: intellectual impairment, usually leading eventually to dementia; abnormalities of mood and associated personality changes, with episodes occurring of depression and, particularly as the condition progresses, inappropriate euphoria (see, for example, Surridge 1969, Kahana et al 1971); and psychotic symptomatology including, rarely, a psychotic presentation of multiple sclerosis (Geocaris 1957, Matthews 1979). There are also reports of hysterical conversion reactions occurring in chronic multiple sclerosis (Brain 1930, Herman & Sandok 1967).

Schilder's disease

Schilder's disease is also known as encephalitis periaxalia diffusa and diffuse cerebral sclerosis.

Aetiology

As with multiple sclerosis, the cause of Schilder's disease is unknown.

Pathological features

A characteristic macroscopic neuropathological feature is a change in the white matter of the brain into softened greyish or brownish translucent regions. This change is usually most evident in the posterior parts of the cerebral hemispheres. The cerebral cortex is usually not directly affected. Histologically, Schilder's disease may resemble multiple sclerosis. Demyelination with axonal preservation occurs in acute cases, while in chronic cases there is secondary axonal degeneration.

Clinical features

Schilder's disease occurs more commonly in children than in adults. Because of occipital lobe involvement, cortical blindness is a common presentation. Other clinical features that may occur include: headache, vomiting and vertigo, thus making it difficult sometimes to differentiate Schilder's disease from a cerebral tumour; other causes of visual impairment such as optic atrophy and papilloedema; spastic paresis and sensory deficit; central deafness; and epilepsy.

So far as psychiatric manifestations are concerned, Schilder's disease may cause cognitive impairment which progresses to dementia, and hysterical reactions. Cases have been reported of Schilder's disease giving rise to a clinical picture indistinguishable from that of schizophrenia (for example, Ferraro 1934, 1943, Holt & Tedeschi 1943, Ramani 1981); the frontal lobes have been found to be more commonly affected by the above neuropathological changes in such cases.

Central pontine myelinolysis

This rare fatal demyelinating disease is usually caused by chronic alcoholism. The characteristic neuropathological change is progressive demyelination of central pontine structures, including pontocerebellar fibres, pyramidal tracts, and nuclei. Clinical features include vomiting, confusion, cranial nerve palsies (particularly the fourth, fifth and sixth nerves), pyramidal signs, dysfunction of vasomotor centres, and coma.

Marchiafava–Bignami disease

This is another rare fatal demyelinating disease that is usually caused by chronic alcoholism. The characteristic neuropathological change is widespread demyelination affecting the central corpus callosum, and often also the middle cerebellar peduncles, the white matter of the cerebral hemispheres, and the optic tracts. There is usually a clinical presentation of emotional disturbance and cognitive impairment followed by epilepsy, delirium, paralysis and coma.

METABOLIC DISORDERS

In this section the following metabolic disorders are considered: the sphingolipidoses; the mucopolysaccharidoses; the leukodystrophies (disorders of myelin metabolism); Wilson's disease (a disorder of copper metabolism); and the acute porphyrias (disorders of porphyrin metabolism).

Metabolic disorders without important psychiatric aspects or specific neuropathological changes are not discussed in this section. Disorders of amino acid metabolism are described in Chapter 3.

Sphingolipidoses

It will be recalled from Chapter 3 that there exist four main groups of sphingolipids: sphingomyelin, cerebrosides, gangliosides and sulphatides. Metabolic disorders involving the first three of these groups are described in this subsection.

Sphingomyelinosis

As mentioned in Chapter 3, the most important disorder of sphingomyelin metabolism is **Niemann–Pick disease.** Four main varieties are recognized (types A, B, C and D). They are all autosomal recessive disorders, and the classical type is the first type (type A) which is particularly likely to occur in Jewish children. The cause of this type of Niemann–Pick disease is deficiency of sphingomyelinase leading to abnormal reticuloendothelial system storage of sphingomyelin. A characteristic neurohistological feature is the presence of ballooned sphingomyelin-containing **foam cells** in the meninges, choroid plexus, and associated with blood vessels. Neuronal loss and reactive astrocytosis occur in due course. Foam cells are also found systemically and therefore seen in bone marrow or hepatic biopsies. Clinically, the following features may occur: the presence of a cherry red spot on the retinal macula; hepatosplenomegaly; intellectual impairment; and failure to thrive. Most cases die before the age of six years.

Cerebrosidosis

In Chapter 3 it is mentioned that galactocerebroside β-galactosidase deficiency is associated with Krabbe's disease, while Gaucher's disease is associated with β-glucocerebrosidase deficiency; Krabbe's disease is discussed in the subsection on the leukodystrophies below.

Gaucher's disease is an autosomal recessive systemic lipidosis which has the following clinical features in the infant (type II) and juvenile (type III) forms: hepatosplenomegaly; and progressive neurological and intellectual impairment. Neurological features are not seen in the adult (type I) form.

Pathologically, in the infant and juvenile forms of Gaucher's disease there is atrophy of the cerebral cortex and basal ganglia and demyelination of the white matter of the central nervous system. Histologically, neuronal loss and a reactive astrocytosis occur. In demyelinated regions large Gaucher cells may be seen together with lipid-laden macrophages.

Gangliosidosis

It will be recalled from Chapter 3 that the two most important types of ganglioside are known as G_{M1} and G_{M2}.

There are two main types of **generalized G_{M1} gangliosidosis**. Type I is an infantile form which has a similar clinical picture to that seen in Hurler's disease (see below) and is therefore also known as pseudoHurler's disease. In Type II there is a more gradual deterioration and the bony abnormalities and hepatosplenomegaly characteristic of both type I and Hurler's disease are usually absent. In both types, G_{M1} accumulates because of a deficiency of β-galactosidase. Histologically, in both types there is an accumulation of lipids in neurones which therefore become enlarged.

The most important examples of G_{M2} gangliosidosis are **Tay–Sachs' disease,** caused by deficiency of the enzyme N-acetyl-β-hexosaminidase A, and **Sandhoff's disease,** caused by deficiency of N-acetyl-β-hexosaminidases A and B. Both diseases are autosomal recessive disorders occurring in infancy; Tay–Sachs' disease occurs particularly in Ashkenazi Jewish populations. In both diseases the enzyme deficiency leads to an abnormal accumulation of G_{M2} in the grey matter of the brain and in retinal ganglion cells. Pathologically, atrophy of brain, cerebellum, and optic nerves may be seen, while secondary demyelination may lead to an abnormal appearance of the white matter of the brain. There may be secondary ventricular dilatation. Histologically, neuronal lipid accumulation leads to neuronal enlargement. Neuronal membranous cytoplasmic inclusion bodies may be seen. In later stages megalencephalopathy may occur macroscopically, while histologically there may be neuronal loss and a reactive astrocytosis. Clinically, the diseases lead to mental retardation, a cherry red macular spot, optic disc and retinal atrophy leading to blindness, convulsions, paralysis and death by the age of three years.

Mucopolysaccharidoses

The mucopolysaccharidoses are disorders of metabolism of glycosaminglycans, complex polysaccharides found in many parts of the body including the skin, bone, cartilage, tendons, cardiac valves and blood vessels. At least eight types of mucopolysaccharidosis are known, of which the commonest is **Hurler's syndrome,** also known as Hurler's disease, mucopolysaccharidosis type I, and gargoylism. It is an autosomal recessive disorder caused by a deficiency of α-L-iduronidase that is required for the metabolic breakdown of dermatan sulphate and heparan sulphate. As a consequence, both of these substances are abnormally stored in many regions of the body, including the central nervous system, and both can be detected in the urine of children with Hurler's syndrome. Pathologically, there is cerebral atrophy with loss of both the grey and white matter of the brain, and secondary ventricular dilatation. Histologically, there is neuronal ganglioside storage, neuronal loss, reactive astrocytosis, and demyelination. A characteristic ultrastructural change is the occurrence of neuronal lipid inclusion bodies known as **zebra bodies**, so-named because of the appearance of alternate dark and light lamellae under the electron microscope. Clinical features include: retarded physical and mental development; coarsened facial features; corneal clouding; an enlarged tongue; cardiac abnormalities; hepatosplenomegaly; an umbilical hernia; joint abnormalities; and death at an early age.

Another mucopolysaccharidosis, **Hunter's syndrome,** also known as mucopolysaccharidosis type II, is similar to Hurler's syndrome in its clinical and pathological features, being caused by accumulation of dermatan sulphate and heparan sulphate secondary to a deficiency of sulphoiduronate sulphatase. The most important differences between Hunter's syndrome and Hurler's syndrome are that Hunter's syndrome is an X-linked recessive disorder with milder clinical features and sparing of the cornea.

Leukodystrophies

The leukodystrophies are disorders of myelin metabolism that give rise to demyelination and which could therefore also be classed under the demyelinating diseases considered above. The more important leukodystrophies are now described briefly.

Metachromatic leukodystrophy

Metachromatic leukodystrophy or sulphatide lipidosis is the commonest leukodystrophy and is caused by a deficiency of the enzyme arylsulphatase A, which normally allows sulphate radicles to separate from sulphatide molecules. As a result of this deficiency there is an accumulation of metachromatic sulphatides in the glia and the myelin of the white matter of the central nervous system and peripheral nerves. This leads to the following neuropathological changes: demyelination of the white matter of the central and peripheral nervous system; loss of oligodendrocytes; the intracellular and extracellular presence in demyelinated regions of metachromatic granules in association with neurones, astrocytes, and macrophages. Four main types of this autosomal recessive disorder are recognized according to age of onset. The commonest type begins at the age of two or three years and its clinical features include: increasing dysfunction of the limbs progressing to paralysis; optic atrophy; mental retardation; and death within four years.

Krabbe's disease

Krabbe's disease is also known as globoid cell leukodystrophy or galactosylceramide lipidosis and is caused by a deficiency of the enzyme galactocerebroside β-galactosidase, leading to an accumulation of galactocerebroside in the central and peripheral nervous systems. Neurohistological features include: widespread demyelination of the white matter of the central nervous system with a reactive astrocytosis and relative axonal preservation; and the presence of epithelioid and multinucleated globoid cells around blood vessels in the regions of demyelination. The cerebral cortex is usually spared. Clinically, Krabbe's disease usually begins in the first year of life with features such as convulsions and irritability, followed by rigidity, tonic spasms precipitated by stimuli such as noise, and bulbar paralysis. Death usually occurs within a year or so.

Adrenoleukodystrophy

Adrenoleukodystrophy is also known as Addison–Schilder's disease and is an X-linked recessive disorder that has a neuropathological and clinical picture similar to that of Schilder's disease (see above) with the exception that in adrenoleukodystrophy there is also, as implied by its name, atrophy of the suprarenal or adrenal glands. The cause of adrenoleukodystrophy is not known but it is believed to be a disorder of the metabolism of long-chain fatty acids.

Wilson's disease

Aetiology

Wilson's disease, also known as hepatolenticular degeneration, is an autosomal recessive disorder of copper metabolism. The exact metabolic defect is unknown. However, there is a decrease in the plasma copper binding protein ceruloplasmin, an increase in the albumin-bound copper, and an increased urinary excretion of copper.

Pathological features

The pathological features of Wilson's disease are principally the result of abnormal deposition of copper in: hepatocytes, leading to cirrhosis; the limbus of the cornea, leading to a zone of golden-brown, yellow or green corneal pigmentation known as a Kayser–Fleischer ring; and the central nervous system.

Neuropathologically, copper deposition is widespread throughout the central nervous system but most markedly so in the basal ganglia. As a result, there is a brownish discolourment and atrophy of the basal ganglia. Histological changes in the basal ganglia and other regions affected, particularly the cerebral cortex and dendate nucleus, include: neuronal loss; changes in protoplasmic astrocytes (see Chapter 1) with the presence of large numbers of Alzheimer type II astrocytes, which have enlarged and deformed nuclei containing glycogen, and fewer Alzheimer type I astrocytes, which have hyperchromatic irregular nuclei and cytoplasmic granules; and the presence of Opalski cells, particularly in the globus pallidus, which are large globoid cells of uncertain origin with foamy cytoplasm and small dense eccentric nuclei.

Clinical features

The hepatic and ophthalmic features of Wilson's disease have been mentioned above. Renal copper deposition may give rise to aminoaciduria, glycosuria, phosphaturia, proteinuria, and a decreased glomerular filtration rate.

The neurological features are principally those of an extrapyramidal disorder, and include choreiform movements of the face and hands, athetoid movements of the limbs, rigidity and bradykinesia. The mask–like facies of parkinsonism may also occur. Occasionally, dysarthria is the presenting feature. Mental manifestations include loss of emotional control and dementia. In a prospective study of psychopathology in 31 cases of Wilson's disease, Dening and Berrios (1989) found personality changes in over half the cases, including psychopathic, schizoid and neurotic features. The alterations in personality and behaviour were found to be related to neurological disease. Depressive symptomatology was found to be associated with hepatic and biochemical abnormalities. No psychotic symptomatology was found in this study.

Acute porphyrias

Classification

The porphyrias can be classified into: those that are acute, including acute intermittent porphyria, variegate porphyria, and hereditary coproporphyria; and the non-acute porphyrias. The non-acute porphyrias are not associated with neuropsychiatric complications.

Porphobilinogen biosynthesis

In a reaction occurring mainly in the liver and bone marrow and catalyzed by the mitochondrial enzyme δ-ALA synthetase, δ-ALA (δ-amino-laevulinic acid) is formed from glycine and succinyl-CoA. Two molecules of δ-ALA then react to form a molecule of porphobilinogen. Molecules of porphobilinogen are the precursors of porphyrins pigments found in haemoglobin, myoglobin and cytochromes.

Aetiology

The porphyrias, acute and non-acute, are associated with increased activity of δ-ALA synthetase. The acute porphyrias are autosomal dominant disorders in which the increased δ-ALA synthetase activity leads to an abnormal accumulation of δ-ALA and porphobilinogen. Neuropsychiatric complications are believed to result partly from the binding of δ-ALA to central GABA receptors.

Pathological features

Neuropathological features include those of cerebrovascular accidents and retinal ischaemia, as a result of cerebral arterial spasm, and chromatolysis of spinal cord neurones (as a result of peripheral neuropathy) and motor nuclear neurones.

Clinical features

Acute episodes occur with: gastrointestinal features such as abdominal pain, vomiting, and constipation; cardiovascular features such as tachycardia, hypertension, and left ventricular failure; neurological and psychiatric features such as peripheral neuropathy, seizures, emotional disturbance, depression, delusional and schizophreniform psychoses, and delirium which may progress to coma. In the case of acute intermittent porphyria and variegate porphyria, acute episodes may be precipitated by drugs such as barbiturates, sulphonamides and the oral contraceptive pill, by acute infections, and by fasting or alcohol. Often the diagnosis is missed and in the past many such patients were diagnosed as suffering from hysteria.

WERNICKE-KORSAKOFF SYNDROME

Wernicke's encephalopathy and Korsakoff's psychosis were independently described in the last century and are closely related, with Wernicke's encephalopathy representing an acute neuropsychiatric reaction to thiamine deficiency and Korsakoff's psychosis representing a chronic residual reaction that can develop out of Wernicke's encephalopathy.

Aetiology

Wernicke's encephalopathy is caused by severe deficiency of thiamine (vitamin B_1), which may result from: chronic alcoholism; lesions of the stomach (for example gastric carcinoma), duodenum, or jejunum causing malabsorption; hyperemesis; and starvation.

In addition to the causes of Wernicke's encephalopathy, such as chronic alcoholism, from which it may emerge, Korsakoff's psychosis may also result from: carbon monoxide poisoning; head injury; anaesthetic accidents; heavy metal poisoning; cerebral neoplasia; and bilateral hippocampal damage, for example following neurosurgery.

As mentioned in Chapter 6, there is evidence that a transketolase defect inherited as an autosomal recessive disorder may give rise to a genetic component in the Wernicke–Korsakoff syndrome (Blass & Gibson 1979).

Pathological features

In Wernicke's encephalopathy there are symmetrical lesions in the following regions: the mammillary bodies; the mesencephalic periaqueductal grey matter; the walls of the third ventricle; the floor of the fourth ventricle; certain thalamic nuclei, particularly the medial dorsal nuclei, anteromedial nuclei, and the pulvinar; the terminal fornices; the brain stem; and the cerebellar superior vermis and anterior lobe. Macroscopically, the lesions show evidence of vascular engorgement and petechial haemorrhages. Histologically, there is parenchymal loss with demyelination but relative neuronal preservation in acute cases, astrocytosis, haemorrhage and capillary endothelial hyperplasia, and the presence of siderophages.

Lesions in the same sites occur in Korsakoff's psychosis, and histologically the glial and vascular reactions show signs of greater chronicity, with blood vessel proliferation being more common than in Wernicke's encephalopathy. In addition, atrophy of the cerebral cortex, particularly of the frontal lobes, and ventricular dilatation may occur. Atrophy of the mammillary bodies may be responsible at least in part for the memory impairment.

Clinical features

The most important clinical features of Wernicke's encephalopathy are: ophthalmoplegia and nystagmus; ataxia; clouding of consciousness; and peripheral neuropathy. In Korsakoff's psychosis there is a characteristic impairment of recent memory associated with disorientation in time. Patients make up for the memory gaps through confabulation. Other features include lack of insight, and apathy or euphoria. Hypothermia may result from involvement of the hypothalamus. Cases of sudden death have been reported.

PARKINSON'S DISEASE

Aetiology

The commonest cause of the Parkinsonian syndrome is idiopathic Parkinson's disease or paralysis agitans. Other causes that can give rise to parkinsonism include: drugs such as neuroleptics, methyldopa and reserpine; intoxication with poisons such as carbon monoxide, carbon disulphide, manganese and mercury; MPTP, a synthetic opiate analogue that was a contaminant of illicitly synthesized heroin in the United States in the 1980s; trauma; infection, with parkinsonism having commonly developed in survivors of the 1919 to 1924 pandemic of encephalitis lethargica — known as post-encephalitic parkinsonism; and the parkinsonism–dementia complex of Guam, which occurs in the population of Guam and is of uncertain origin. Cerebral arteriosclerosis is a controversial cause (Eadie & Sutherland 1964). Other diseases which may give rise to parkinsonism include cerebral palsy, Huntington's chorea, progressive supranuclear palsy and Wilson's disease.

Idiopathic Parkinson's disease is considered in this section, but many of the features outlined below also apply to parkinsonism resulting from other causes. The characteristic site of neuropathological change is the **substantia nigra** of the midbrain, leading to a reduction in the dopaminergic neurones of this part of the extrapyramidal motor system and therefore a reduction in the inhibitory dopaminergic action of the nigrostriatal pathway on striatal cholinergic neurones. Parkinsonism can also result from a functional deficiency of dopamine in this system caused, for example, by neuroleptic medication (as mentioned earlier in this book).

Pathological features

Macroscopically, there is depigmentation of the substantia nigra, particularly of the zona compacta. Histologically, there is neuronal loss, a reactive astrocytosis, the presence of Lewy intracytoplasmic hyaline inclusion bodies in surviving neurones, and the presence of melanin-containing macrophages. These histological changes are seen not only in the substantia nigra, but also in the pigmented cells of the locus coeruleus, reticular formation, and dorsal vagal nucleus. The core of Lewy bodies contain neurofilaments. Diffuse cortical atrophy may also occur.

Clinical features

The most important neurological features of Parkinson's disease are bradykinesia, a resting tremor, cogwheel rigidity and postural abnormalities. Other features include hypersalivation, seborrhoea, oculomotor abnormalities, fatigue, poor balance, micrographia, dysarthria and urinary disturbance. Psychiatric features include: cognitive impairment; personality changes; mood disorder, primarily depressive illness; and very rarely, schizophreniform psychosis (for example, Crow et al 1976).

DEMENTIA

Dementia is an acquired global impairment of intellect, memory and personality, without impairment of consciousness (Lishman 1987). There is a large number of causes of dementia, including degenerative disorders, vascular disorders, toxic causes, trauma,

drugs, neoplasia, hydrocephalus, infection, metabolic causes and vitamin deficiency.

Some causes have already been discussed in other sections of this chapter, for example, normal pressure hydrocephalus and Creutzfeldt–Jakob disease. In this section the following degenerative disorders that cause dementia are described: Alzheimer's disease; Pick's disease; Huntington's chorea. An important vascular cause, multi-infarct dementia, is also described.

Alzheimer's disease

Alzheimer's disease is the commonest cause of dementia in the population over the age of 65 years. The same pathological changes take place in both the senile and the presenile form of the condition. The precise causes of the neuropathological changes in Alzheimer's disease are uncertain; as mentioned in the last chapter, an inherited form exists with an Alzheimer's disease locus being located on chromosome 21.

Pathological features

Macroscopically, there is global atrophy of the brain which is shrunken with widened sulci and ventricular enlargement. The atrophy is usually most marked in the frontal and temporal lobes.

Histologically, there is neuronal loss, shrinkage of dendritic branching, and a reactive astrocytosis in the cerebral cortex. The layers of the cortex with the most marked cell loss are usually the molecular or plexiform layer (the most superficial layer), the external granular layer, and the external pyramidal layer. Other neuropathological features of Alzheimer's disease include: the abundant presence, particularly in the cerebral cortex, of neurofibrillary tangles; the presence mainly in the cortex of silver-staining neuritic plaques (formerly known as senile plaques); granulovacuolar degeneration, particularly in the middle pyramidal layer of the hippocampus; and Hirano bodies. Neurofibrillary tangles, granulovacuolar degeneration and Hirano bodies are described earlier in this chapter. The numbers of neurofibrillary tangles and neuritic plaques correlate with the degree of cognitive impairment.

Electronmicroscopy reveals that each neuritic plaque contains an amyloid core made of a protein known as A4 or beta amyloid. Abnormal neurites surround the amyloid. As mentioned in the previous chapter, the gene coding for the beta-amyloid precursor protein has been cloned and localized to the long arm of chromosome 21.

Biochemically, in the post mortem brain there is reduced activity of both acetylcholinesterase and choline acetyltransferase; these enzymes and the cholinergic hypothesis of memory dysfunction are described in Chapter 3. Other biochemical changes reported in Alzheimer's disease that are mentioned in Chapter 3 are reduced levels of brain GABA and cortical noradrenaline.

Clinical features

Alzheimer's disease is commoner in women and usually presents with memory loss. Other clinical features may include: apathy or lability of mood; progressive impairment of intellectual function; progressive deterioration of personality; features typical of parietal lobe dysfunction; paranoid features; parkinsonism; the mirror sign, in which individuals may fail to recognize their own reflection; disorders of speech such as logoclonia and echolalia; epilepsy; and aspects of the Klüver–Bucy syndrome which includes hyperorality, hypersexuality, hypermetamorphosis, hyperphagia and placidity.

Pick's disease

Pick's disease is a rare cause of dementia some cases of which appear be transmitted as an autosomal dominant disorder.

Pathological features

Macroscopically, there is selective asymmetrical atrophy of the frontal and temporal lobes which, because of its severity, causes the gyri to become very thin — this being known as knife-blade atrophy of the gyri.

As mentioned earlier in this chapter, a characteristic histological feature of Pick's disease is the presence of argyrophilic intracytoplasmic neuronal inclusion bodies known as **Pick's bodies** which consist of neurofilaments, paired helical filaments, and endoplasmic reticulum. Other histological changes include loss of neurones, most marked in the outer layers of the cerebral cortex, and a reactive astrocytosis. These changes may also occur in the basal ganglia, locus coeruleus, and substantia nigra.

Clinical features

Pick's disease is commoner in women, with a peak age of onset between 50 and 60 years. Clinical features include: personality deterioration with features of frontal lobe dysfunction; nominal aphasia; memory impairment; perseveration, and aspects of the Klüver–Bucy syndrome (see above).

Huntington's chorea

Huntington's chorea is also known as Huntington's disease and is an autosomal dominant disorder caused by a gene which has been localized to the most distal band of the short arm of chromosome 4 (see Chapter 6).

Pathological features

Macroscopically, the brain is usually small with reduced mass and there is marked atrophy of the corpus striatum of the basal ganglia, particularly the caudate nucleus, and of the cerebral cortex, particularly the gyri of the frontal lobes.

Histological changes include: neuronal loss in the cerebral cortex, particularly affecting the frontal lobes, and in the corpus striatum, particularly affecting GABA neurones with relative sparing of large neurones; and astrocytosis in the affected regions.

Biochemical changes include reduced levels of GABA and glutamic acid decarboxylase (see Ch. 3) and dopamine hypersensitivity.

Clinical features

Males and females are affected equally by Huntington's chorea and the average age of onset is in the thirties. It causes an insidious onset of choreiform movements and progressive global dementia. Other features include: personality change; ataxia; slurring of speech; extrapyramidal rigidity; and epilepsy. Psychiatric features include depression, increased risk of suicide, and schizophreniform and delusional disorders. Insight tends to be retained until a late stage. Death usually occurs within 15 years of onset of symptoms.

Multi-infarct dementia

Multi-infarct dementia is also known as arteriosclerotic dementia and is an ischaemic disorder caused by multiple cerebral infarcts, with the extent of cerebral infarction being related to the degree of cognitive impairment. Multi-infarct dementia is associated with chronic hypertension and arteriosclerosis.

Pathological features

Macroscopically, there are multiple cerebral infarcts, local or general atrophy of the brain with secondary ventricular dilatation, and evidence of arteriosclerotic changes in major arteries. In most cases in which cognitive impairment is detectable the volume of the infarcts is greater than 50 ml, while a volume of greater than 100 ml is particularly likely to be associated with dementia (Tomlinson et al 1970). The histological changes of infarction and ischaemia are seen.

Clinical features

Multi-infarct dementia is commoner in men and a history and clinical features of hypertension are usually present. The onset is usually acute, peaks in the sixties and seventies, and may be associated with a cerebrovascular accident. Other clinical features include: stepwise deterioration; focal neurological features; nocturnal confusion; fits; fluctuating cognitive impairment; and emotional incontinence and low mood.

CEREBRAL TUMOURS

In this section the most common types of primary cerebral tumour are considered with reference to their pathological features. The clinical features of cerebral neoplasia are also briefly outlined. General features such as the effects of raised intracranial pressure are described earlier in this chapter.

Types

In order of relative frequency, the main types of cerebral tumours are as follows: gliomas, which are tumours derived from glial cells and their precursors; metastases, particularly from primary neoplasia in the lung, breast, kidney, colon, ovary, prostate and thyroid; meningeal tumours; pituitary adenomas, derived from adenohypophyseal cells; neurilemmomas or Schwannomas, derived from Schwann cells; haemangioblastomas; and medulloblastomas.

Clinical manifestations

Neurological effects

The general neurological effects of cerebral tumours include the effects of intracranial expanding lesions described earlier in this chapter, including raised intracranial pressure and cerebral oedema.

Epilepsy occurs in approximately one-third of cases, and may be focal or generalized. Epilepsy is particularly common with tumours in the frontal and temporal lobes.

Depending on their location, cerebral tumours may cause varying focal neurological features. The clinical features of focal cerebral disorder have been described earlier in this chapter. Invasion of the meninges may give rise to signs of meningism.

Certain cerebral tumours give rise to particular manifestations which are not generally seen with other types of cerebral tumour. For example, pituitary adenomas may cause endocrinological effects, while neurilemmomas affecting the eighth cranial nerve (acoustic neuromas) cause progressive deafness and occasionally vertigo and tinnitus.

Psychiatric effects

The general psychological and psychiatric effects of cerebral tumours include: impairment of consciousness and other cognitive changes, both focal and generalized; mood changes, such as emotional dullness, apathy, and irritability; hallucinations in any modality (depending on the location of the tumours) and also in association with epilepsy; neurotic phenomena; and psychotic phenomena.

Gliomas

Gliomas include cerebral tumours derived from astrocytes, oligoendrocytes, and ependymal cells.

Astrocytomas

Astrocytomas are the commonest primary parenchymal cerebral tumours and occur mainly in the white matter of the cerebral hemispheres. They are usually solid pale coloured poorly demarcated infiltrative tumours that cause expansion and distortion of the adjacent cerebral tissue. Histologically, astrocytomas may contain fibrillary, protoplasmic, and gemistocytic astrocytic forms all in the same tumour. The most malignant form of astrocytoma is known as glioblastoma multiforme.

Oligodendrocytomas

Oligodendrocytomas usually occur in the cerebral hemispheres and tend to be better demarcated than astrocytomas. They grow relatively slowly and often contain cystic structures, areas of haemorrhage and areas of calcification.

Ependymomas

Ependymomas occur particularly during the first two decades of life and most often from the ependymal lining of the fourth ventricle. In the latter case hydrocephalus results. Ependymomas may spread via the cerebrospinal fluid.

Also occurring particularly during the first two decades of life are **choroid plexus papillomas** which are papillary tumours that may secrete cerebrospinal fluid. This secretion, or irritation and blockage of the cerebrospinal fluid pathways by cells shed by the tumour, may give rise to hydrocephalus.

Meningeal tumours

The commonest meningeal tumour is the **meningioma**, which is often derived from arachnoid granulations. Meningiomas grow very slowly and only rarely metastasize. Local invasion of the skull causes bone erosion and hyperostosis. Meningiomas are commonly calcified.

Much rarer meningeal tumours include meningeal sarcomas and primary malignant melanomas derived from pia-arachnoid melanocytes.

Pituitary adenomas

Adenomas can arise from the different cell types that normally elaborate the different hormones mentioned in Chapter 2. Therefore there is a wide variation in the histological appearance of pituitary adenomas. The main clinical effects of pituitary adenomas are twofold. An adenoma may secrete the hormone normally elaborated by the cell of origin, leading to increased levels of that hormone and corresponding clinical manifestations such as acromegaly in adults or gigantism in children (excessive somatotropin), Cushing's disease (excessive corticotropin), and amenorrhoea, galactorrhoea and increased body mass (excessive prolactin). The second main clinical effect is caused by local compression of the optic chiasma, thereby giving rise to visual field defects, and the hypothalamus.

Neurilemmomas

Neurilemmomas are also known as Schwannomas because of the cells of origin. They are the commonest nerve sheath tumours and can affect the cranial nerves, particularly the eighth cranial nerve, the posterior nerve roots of the spinal cord and peripheral nerves. They are slow-growing and histologically are often cystic.

Neurilemmomas affecting the eighth cranial nerve, known as **acoustic neuromas**, are the commonest tumours to occur in the cerebello-pontine angle. As mentioned above, they can give rise to progressive deafness, vertigo and tinnitus. As they increase in size, large tumours compress local structures and therefore also progressively give rise to: facial sensory loss and masticatory weakness (pressure on the ipsilateral fifth cranial nerve); facial palsy (pressure on the ipsilateral seventh cranial nerve); lateral rectus palsy (pressure on the ipsilateral sixth cranial nerves); ipsilateral cerebellar signs; and features of brain stem compression.

Haemangioblastomas

Haemangioblastomas are derived from blood vessels and usually occur in adulthood in the cerebellum. They may occur as part of the von Hippel–Lindau syndrome in which multiple cerebral haemangioblastomas are associated with retinal haemangioblastomas, pancreatic and renal cysts, and suprarenal tumours. Histologically, haemangioblastomas contain a dense vascular network, and cysts may also be present. Some haemangioblastomas secrete erythropoietin thereby giving rise to polycythaemia.

Medulloblastomas

Medulloblastomas are cerebellar tumours that are commonest in children; indeed they are the commonest childhood primary central nervous system tumours. Clinical effects include ataxia and brain stem signs. When the fourth ventricle is blocked, hydrocephalus results. Death usually occurs within three years.

BRAIN IMAGING

In this section the following types of brain imaging are outlined: radiography; X-ray computerized tomography; magnetic resonance imaging; magnetic resonance spectroscopy; cerebral blood flow studies; positron emission tomography; and single-photon emission computerized tomography. Other methods which are now rarely used, for example air encephalography, are not discussed.

Radiography

Skull radiography is mainly used in the assessment of trauma. However, it can occasionally be of help in indicating the presence of intracranial expanding lesions. With increasing age there is a greater likelihood of calcification of the pineal gland, and the presence of an intracranial expanding lesion may cause the pineal to be displaced from the midline and show up on radiography. Calcification may also occur in cerebral tumours. Raised intracranial pressure and suprasellar lesions may cause erosion of the posterior clinoid processes and flattening of the sella turcica.

The sella turcica may be expanded and decalcified in the presence of a pituitary tumour.

Computerized tomography

X-ray computerized tomography (CT), also known as computerized axial tomography (CAT) or computed tomography, is a development of radiography in which X-ray beams pass through a tissue plane in different directions and the emerging X-rays recorded by scintillation counters are reconstructed by computer into radiodensity maps. By repeating this procedure with successively adjacent planes, computerized tomographs that represent adjacent slices of, say, the brain, can be constructed. Computerized tomography is used extensively in clinical practice, and is able to demonstrate shifts of intracranial structures; intracranial expanding lesions, so long as they are not smaller than the power of resolution, such as cerebral tumours, haematomas, and abscesses; cerebral infarction; cerebral oedema; cerebral atrophy and ventricular dilatation; atrophy of other structures; and radiodensity changes caused by other neuropathological processes such as demyelination. Computerized tomography has also proved to be a useful psychiatric research tool, particularly in the investigation of schizophrenia (reviewed by Lewis 1990), alcoholism (Ron 1983), dementia (Jacoby 1981), mood disorder (Standish-Barry et al 1982, Targum et al 1983), and eating disorders (Swigar et al 1983).

Magnetic resonance imaging

Magnetic resonance imaging (MRI), known previously as nuclear magnetic resonance (NMR), also produces computerized images of adjacent slices of the body. The patient is placed in a strong static magnetic field which causes the proton spin axes of certain nuclei to align. The precession of the protons is augmented by the administration of radiofrequency pulses of the same frequency as that of the precession. At the end of each pulse, energy is emitted and the rate of its decay, the relaxation time, is measured. Two different types of relaxation times together with proton density are computed into a magnetic resonance image. Because relaxation times differ markedly between white matter, grey matter, and cerebrospinal fluid, MRI is particularly useful in differentiating these three types of tissue. Relaxation times are usually also changed by neuropathological processes thus making MRI useful in clinical studies, for example in the demonstration of cerebral and cerebellar atrophy, and subtle neuropathological changes such as the presence of small demyelinating lesions. The use of MRI as a research tool is reviewed by Krishnan (1990).

Magnetic resonance spectroscopy

Magnetic resonance spectroscopy (MRS) is a development of MRI that measures tissue spectra of isotopes such as phosphorus-31(^{31}P) and so offers an opportunity to explore living chemical processes. At the time of writing, MRS is being evaluated for its potential clinical applications, and it appears likely to become a useful imaging tool in the study of metabolic and phopholipid changes associated with neuropsychiatric disorders (Krishnan 1990, Lock et al 1990).

Cerebral blood flow studies

In cerebral blood flow (CBF) studies the γ radiation emitted from radioisotopes such as xenon-131 (^{131}Xe) that are introduced into the cerebral circulation, for example by intracarotid injection or through inhalation of radiolabelled gas, is measured in order to image regional cerebral blood flow (rCBF). CBF techniques have been carried out in neuropsychiatric research into schizophrenia, mood disorder and dementia (reviewed by Trimble 1985).

Positron emission tomography

Positron emission tomography (PET) is another method which, like MRS and CBF studies, primarily images neurofunctioning rather than, as in the case of CT and MRI, neurostructure. Radioactive substances that emit positrons, such as radioisotopes like fluorine-18 (^{18}F) and radiolabelled receptor-binding ligands, are introduced into the cerebral circulation, for example through intravenous injection or inhalation, and the positrons emitted from tissues in turn cause the emission of γ radiation through positron–electron interactions. Each positron causes the simultaneous release of γ radiation in two opposite directions. Computed measurement of simultaneous dual γ photons travelling in opposite directions is used to produce brain slice images which can demonstrate metabolic changes (as occurs in epileptiform foci, for example), regional cerebral blood flow, and ligand binding. The clinical and research applications of PET are reviewed by Krishnan (1990) and Bench et al (1990).

Single-photon emission computerized tomography

Single-photon emission computerized tomography (SPECT) is another neurofunctioning imaging technique and has the advantage of being relatively cheap compared with PET. Another difference between SPECT and PET is that the radioactive substances introduced into the cerebral circulation in SPECT, such as xenon-133 (^{133}Xe) and substances radio-labelled with iodine-123 (^{123}I) and technetium-99m (^{99m}Tc), emit single γ photons (not dual γ photons). Computed tomographic techniques allow the measurement of the γ photons to yield regional cerebral blood flow images (which can be combined to produce three dimensional rCBF images) and, through the use of radiolabelled receptor-binding ligands, ligand binding images. At the time of writing, metabolic changes cannot be imaged by SPECT. The application of SPECT in dementia studies is reviewed by Geaney and Abou-Saleh (1990).

REFERENCES

Adams J H, Corsellis J A N, Duchen L W 1984 Greenfield's neuropathology. 4th edn. Edward Arnold, London

Adams J H, Graham D I 1988 An introduction to neuropathology. Churchill Livingstone, Edinburgh

Govan A D T, Macfarlane P S, Callander R 1991 Pathology illustrated. 3rd edn. Churchill Livingstone, Edinburgh

Lishman W A 1987 Organic psychiatry: the psychological consequences of cerebral disorder. Blackwell Scientific, Oxford

Walton J 1985 Brain's diseases of the nervous system. 9th edn. Oxford University Press, Oxford.

8. Statistics

This chapter makes use of diagrams, tables, and worked examples in order to help foster an understanding of statistics and the use of statistical techniques. The subject is considered in the following order: basic concepts; data presentation; measures of location; measures of dispersion; probability; discrete probability distributions; continuous probability distributions; sampling, estimation, and hypothesis testing; the χ^2 test; Fisher's exact probability test; regression and correlation; multivariate analysis; and non-parametric tests.

BASIC CONCEPTS

In this section alternative meanings of the term statistics, and the concepts of variables, populations and samples are described. The section ends by outlining a conventional nomenclature system that can be used to differentiate between sample statistics and population parameters.

Statistics

Definition

The Shorter Oxford English Dictionary defines statistics as 'the department of study that has for its object the collection and arrangement of numerical facts or data, whether relating to human affairs or to natural phenomena.'

An alternative use of the term statistics is illustrated by the following example. Suppose one wished to know the average age of hospitalized psychiatric patients currently diagnosed as schizophrenic in a certain country. One method of finding this out would be to ascertain the age of all the hospitalized psychiatric patients currently diagnosed as schizophrenic in that country. One could then work out the required average age by dividing the sum of all these ages by the total number of such patients. However, this could prove a very long and tedious exercise. A more practical method could make use of statistics.

All the hospitalized psychiatric patients currently diagnosed as schizophrenic in the particular country can be said to form a **population**; this is the set of all the people (or objects etc.) about which information (in this case their average age) is required.

The average age of this population is known as a population **parameter**.

A random **sample** could be chosen from the population and used to obtain an estimate of the average age of the population; this estimate is known as a **statistic**.

In other words, a sample statistic can be used to estimate the value of the population parameter, as shown in Figure 8.1

Classification

Descriptive statistics are means of organizing data or information, and include tabular, diagrammatic, graphical, and numerical methods.

Inferential statistics, on the other hand, allow conclusions to be derived from the data.

Variables

Definition

A variable is a quantity or attribute which varies from one member of the population being studied to

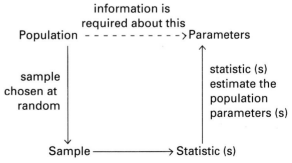

Fig. 8.1 The relationship of statistics and parameters.

another. For instance, in the above example the quantity of interest which varies from one member of the population being studied to another is the age. Therefore the age of the patients is the variable under study in that example.

Classification

Variables can be classified as being qualitative or quantitative. **Qualitative** variables refer to attributes such as gender or eye colour.

Quantitative variables refer to numerically represented data, which can be continuous or discrete. **Continuous** variables can take on any value within a defined range, for example body mass. Thus, an individual may have a body mass of 56.3 kg, to one decimal place, that is, a mass has been allocated to the range 56.25 kg to 56.35 kg (exclusive). **Discrete** variables, on the other hand, can only take on known fixed values. An example of a discrete variable is the number of admissions to a psychiatric in-patient unit each day. The value of this variable could be zero, or one, or two or three, and so on, but it could not take on a value in between these (such as 1.8).

Populations

Definition

As mentioned above, a population is the total collection of objects, people or data about which information is needed. In the above example, all the hospitalized psychiatric patients currently diagnosed as schizophrenic in the particular country under study can be said to form a population.

Classification

In the above example, because there is a limit to the number of hospitalized psychiatric patients diagnosed as schizophrenic, this population is said to be **finite**.

An example of an **infinite** population is the situation in which the results of repeated trials of an antidepressant are being measured, since in theory the trials could be continued for ever (assuming that at some stage in the future an event such as the extinction of the human race did not occur).

Parameter

A parameter is a summarizing value which describes a population. In the first example above, the actual average or mean age of the given population is a population parameter.

Samples

Definition

It is clearly not possible to find the value of a given variable for all the members of an infinite population. Similarly, it is usually not possible in practice to do so for a finite population. For instance, in the first example given above it would clearly be very difficult to actually determine the individual ages for the whole of the hospitalized psychiatric patient population currently diagnosed as schizophrenic in the particular country concerned, unless this population were small.

Instead, as indicated above, one can obtain values of the variables for members of a part or subset of the population; this subset of the population is known as a sample. A sample is usually chosen to be representative of the population with respect to the variables being studied.

Statistic

A statistic is a summarizing value which describes a sample. In the first example above, the individual ages in a representative randomly chosen sample could be measured and the mean of these values would be a statistic. Such a statistic can then be used to estimate the value of the corresponding parameter.

Nomenclature

A conventional way of distinguishing between statistics and their corresponding parameters is to denote sample statistics using English (Roman) letters, and to employ Greek letters for population parameters. This is the convention used in this chapter. Common examples of statistics and corresponding parameter are shown in Table 8.1; details of the mean, standard deviation and correlation coefficient are given later in this chapter.

Table 8.1 Nomenclature: examples of statistics and corresponding parameters

Statistics	Parameter
Sample mean: $\bar{x}$	Population mean: μ
Sample standard deviation: s	Population standard deviation: σ
Pearson's sample product moment correlation coefficient: r	Pearson's population product moment correlation coefficient: ρ

DATA PRESENTATION

There are various diagrammatic ways of presenting data and these can confer advantages over calculating just the numerical summarizing values discussed later in this chapter. One advantage is that an overall picture becomes available. Another is that a diagrammatic method such as graphical representation may enable one to gain an approximate value for a variable such as the correlation coefficient (see below); occasionally a mistake that has been made in calculating the value of the variable (for example by pressing the wrong key of a calculator or computer keyboard and entering an incorrect datum) can be recognized because of this approximate value.

Frequency distribution and frequency table

A frequency distribution is a systematic way of arranging data. When in the form of a frequency table the first column gives the possible values of a given variable, which may be qualitative or quantitative. The adjacent column gives the frequency with which each variable occurs.

Example

Table 8.2 is a frequency table based on a study of polydipsia and water intoxication in a mental handicap hospital (Bremner & Regan 1991). The first column gives the level of mental handicap while the adjacent columns give the frequency with which polydipsia was found for each level of handicap.

Relative frequency

The relative frequency of a variable is the proportion of the total frequency that corresponds to that variable, and it is calculated using the following ratio:

$$\frac{\text{frequency of the variable}}{\text{total frequency}}$$

For the data in Table 8.2 the relative frequency table of Table 8.3 can be constructed as shown. Relative frequencies can also be expressed as percentages by multiplying each relative frequency by 100. The sum of the relative frequencies is always one (or 100%); in tables, the sum may come to slightly more or slightly less than one owing to rounding errors.

Bar chart

The above frequency data can also be represented as a

Table 8.2 Frequency table: polydipsia and water intoxication in a mental handicap hospital

Level of mental handicap (ICD-9)	Tally	Frequency (no. of patients)
Borderline (IQ 70–85)	⋕⋕ II	7
Mild (IQ 50–69)	⋕⋕ IIII	9
Moderate (IQ 35–49)	⋕⋕ I	6
Severe (IQ 20–34)	⋕⋕ II	7
Profound (IQ < 20)	II	2
Unspecified		0
Total		31

Table 8.3 Relative frequency table based on the data of Table 8.2

Level of mental handicap (ICD-9)	Frequency (no. of patients)	Relative frequency
Borderline (IQ 70–85)	7	7/31 = 0.226
Mild (IQ 50–69)	9	9/31 = 0.290
Moderate (IQ 35–49)	6	6/31 = 0.194
Severe (IQ 20–34)	7	7/31 = 0.226
Profound (IQ < 20)	2	2/31 = 0.065
Unspecified	0	0/31 = 0
Total	31	31/31 = 1

bar chart as shown in Figure 8.2. The same bar chart can also be used to represent relative frequencies, in which case the vertical axis has a new scale from zero to a value which is never greater than one (since the maximum possible relative frequency is one). A bar chart representing the relative frequency data of Table 8.3 is shown in Figure 8.3. It can be seen from both Figures 8.2 and 8.3 that the length of each bar is directly proportional to the frequency (or relative frequency) of the variable.

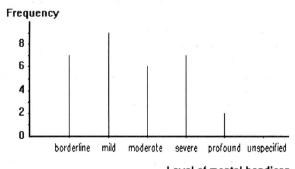

Fig. 8.2   A bar chart representing the frequency data of Table 8.2.

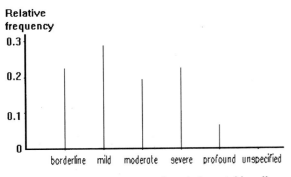

Fig. 8.3   A bar chart representing the relative frequency data of Table 8.3.

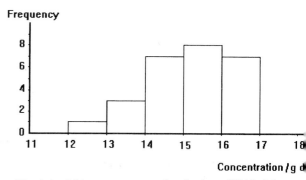

Fig. 8.4   A histogram representing the data of Table 8.4.

Histogram

Consider Table 8.4, which shows the haemoglobin levels found in a hypothetical randomly chosen group of 25 adult male psychiatric patients. These data could be represented by simply listing the frequency for each particular value. However, because of the large number of possible values it is more useful to divide the data into groups which will cover successive values. For example:

12.0 to 12.9 g dl^{-1}
13.0 to 13.9 g dl^{-1}
14.0 to 14.9 g dl^{-1}
15.0 to 15.9 g dl^{-1}
and so on

Table 8.4   Haemoglobin levels (g dl^{-1}) in a randomly chosen group of 25 male psychiatric patients

14.3	15.2	14.7	16.1	13.4
15.4	15.7	14.9	15.8	16.3
12.6	16.3	14.7	15.5	15.3
14.5	15.8	16.4	14.8	13.6
16.2	15.6	13.7	16.1	14.7

The range covered by each group is known as the **class interval**, and these results can be represented by the histogram shown in Figure 8.4

It can be seen from Figure 8.4 that the class intervals are represented on the horizontal axis of a histogram, and when the class intervals are all of equal size then the heights of the rectangles are equal to the class frequencies.

Frequency polygon

A frequency polygon is obtained by joining together the mid-points of the tops of the rectangles of a histo-gram with straight lines. Thus, for the histogram of Figure 8.4 the frequency polygon shown in Figure 8.5 can be constructed.

Frequency polygons can be useful when comparing more than one set of data. For example, a super-imposed frequency polygon for values of the haemo-globin levels from a random sample of 25 adult female psychiatric patients might appear as shown in Figure 8.6

Cumulative frequency

The cumulative frequency of a given value of a variable is the total frequency up to that particular value. For example, Table 8.5 shows the results, including the cumulative frequency, of a study of the age distribution of early-onset and late-onset depression in the elderly (Burvill et al 1989). Looking at the second line of data in this table it can be seen that, for the early-onset depressives, the cumulative frequency of 31 is the sum of 18 and 13; other cumulative frequencies are similarly derived.

These cumulative frequency data can also be repre-sented graphically in the form of **cumulative fre-quency curves**, as shown in Figure 8.7

Table 8.5   Age distribution of early-onset and late-onset depressives

Age (years)	Early-onset		Late-onset	
	Frequency	Cumulative frequency	Frequency	Cumulative frequency
60–64	18	18	5	5
65–69	13	31	5	10
70–74	13	44	11	21
75–79	6	50	12	33
80–84	1	51	11	44
85–89	1	52	6	50

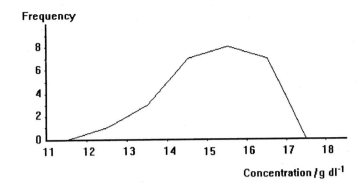

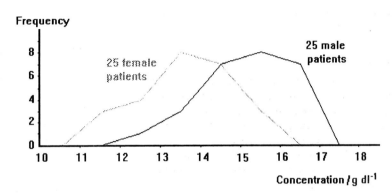

Fig. 8.5 A frequency polygon based on the histogram of Figure 8.4.

Fig. 8.6 The frequency polygon of Figure 8.5 with a superimposed frequency polygon representing haemoglobin levels of a random sample of 25 adult female psychiatric patients.

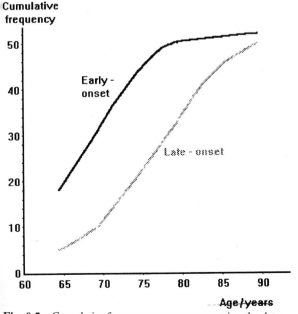

Fig. 8.7 Cumulative frequency curves representing the data of Table 8.5.

Pie diagram

A pie diagram, also known as a pie chart, circular diagram, or circular chart, is a circle which is divided into sectors the sizes of which correspond to the proportional frequencies of the classes represented. The angle made by a sector at the centre of the circle defines the size of the sector, and since the total angle at the centre of a circle is 360°, the angle required for a given sector is calculated using the following formula:

angle of sector =
(class frequency proportion of total) × 360°

This is demonstrated in the example shown in Table 8.6 which shows the categories of causes of spinal cord compression and the corresponding sector angles required for a pie diagram. The corresponding pie diagram is shown in Figure 8.8.

Pictogram

In a pictogram, a standard symbol — such as a picture — is used to represent the value of a variable; the number of times the symbol is repeated corresponds to the value of the variable. Consider, for example, the data given in Table 8.7: the number of people referred annually to a psychotherapy clinic over a 7-year peri-

Table 8.6   Causes of spinal cord compression

Cause	Proportion of total	Angle of sector required for pie diagram
Extradural (disorders of vertebrae)	45% = 0.45	0.45 × 360° = 162°
Intradural (meningeal disorders)	45% = 0.45	0.45 × 360° = 162°
Intramedullary (disorders of spinal cord)	10% = 0.1	0.1 × 360° = 36°

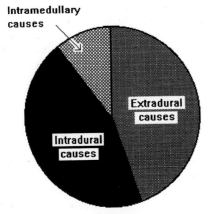

Fig. 8.8   A pie chart showing categories of causes of spinal cord compression.

od. By representing every 100 people with the given symbol the pictogram shown in Figure 8.9 can be constructed.

One-dimensional dot diagram

In a one-dimensional dot diagram a distribution is represented diagrammatically with each observation

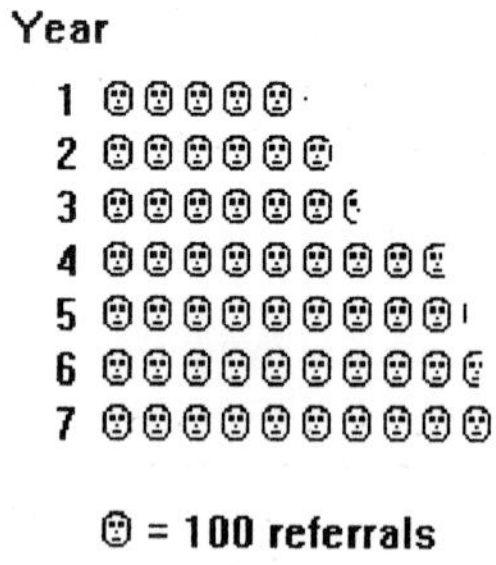

Table 8.7   The number of people referred annually to a psychotherapy clinic over a seven-year period

Year	Number of referrals
1	502
2	597
3	648
4	873
5	909
6	964
7	1010

shown as a dot on, or close to, a calibrated line. The line is usually horizontal but may, as in the example shown in Figure 8.10, be vertical. Dots corresponding to different sets of observations can be represented by different shading, different colours, or, as in Figure 8.10, by a spatial separation. When there is more than one observation with the same value the corresponding dots can be placed adjacent to each other, perpendicular to the appropriate point on the calibrated line, as shown in Figure 8.10 in which they are placed horizontally, perpendicular to the vertical left-hand calibrated line. The example shown in Figure 8.10 represents the Lithium Knowledge Test scores for three professional groups (Peet & Harvey 1991).

Fig. 8.9   A pictogram representing the data of Table 8.7.

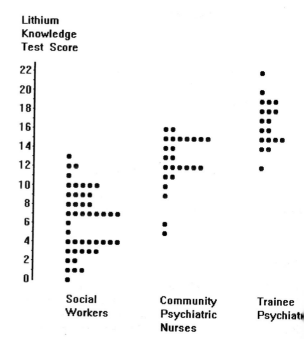

Fig. 8.10   The Lithium Knowledge Test scores for three professional groups (after Peet & Harvey 1991).

Scatter Diagram

In a scatter diagram or dot graph two perpendicular axes divide the space into a coordinate system in which observations can be placed. There can be more than one value on the vertical axis or **ordinate** for a given value on the horizontal axis or **abscissa**. Consider a study on untreated depressed patients in which low lumbar cerebrospinal fluid (CSF) samples are taken and analysed for the serotonin metabolite 5-hydroxyindoleacetic acid (5-HIAA) and for cortisol. Table 8.8 shows the hypothetical results for samples from eight patients. These results can be represented by the scatter diagram shown in Figure 8.11

Table 8.8 Cerebrospinal fluid (CSF) levels of 5-hydroxyindoleacetic acid (5-HIAA) and cortisol in eight untreated depressed patients

Patient	CSF 5-HIAA (nM)	CSF cortisol (nM)
1	100	38
2	52	25
3	50	27
4	48	18
5	124	29
6	167	39
7	147	26
8	112	35

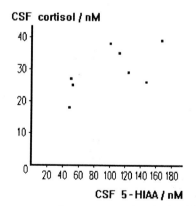

Fig. 8.11   A scatter diagram representing the data of Table 8.8.

Line graphs

A line graph is similar to a scatter diagram but with successive points, relative to the abscissa, being joined by lines. Normally a continuous variable, such as time, is represented on the abscissa. There are many types of line graph. In a time series, time is represented on the abscissa, while a line graph in which cumulative frequency is represented on the ordinate is known as an ogive or cumulative frequency polygon.

In a **log-linear graph** one axis, usually the ordinate, has a logarithmic scale, while the other axis is linear. On the linear scale (usually the abscissa), pairs of points are the same difference apart when their differences are equal. However, on the logarithmic scale, pairs of points are the same distance apart when their ratios are equal, as shown in Figure 8.12.

MEASURES OF LOCATION

Measures of central tendency attempt to give a value around which the distribution clusters; **quantiles** include values other than central ones. Measures of central tendency include the mean and the mode. Quantiles include quartiles, quintiles, deciles and percentiles. The median is both a measure of central tendency and a quantile.

Measures of central tendency

Arithmetic mean

The arithmetic mean is just one type of mean; others include the geometric mean and the harmonic mean — which are not described in this chapter.

The arithmetic mean of a set of numbers is the sum of the items divided by the number of items. It is also sometimes referred to as the **average** or simply the **mean**.

Table 8.9 shows the ages in years (to one decimal place) of 10 patients. The arithmetic mean of the ages of these 10 patients can be calculated as follows:

$$\text{arithmetic mean} = \frac{\text{sum of the items}}{\text{number of items}}$$

$$= \frac{248.1}{10} = 24.81 \text{ years}$$

So the arithmetic mean of the ages of the patients is 24.8 years (to one decimal place).

Consider the general case of a sample with a total of n items, and the following observations:

x_1, x_2, x_3, and so on to x_n.

Table 8.9 Ages (in years) of ten patients

13.4	45.8	23.1	25.6	15.6
16.7	23.0	12.9	43.1	28.9

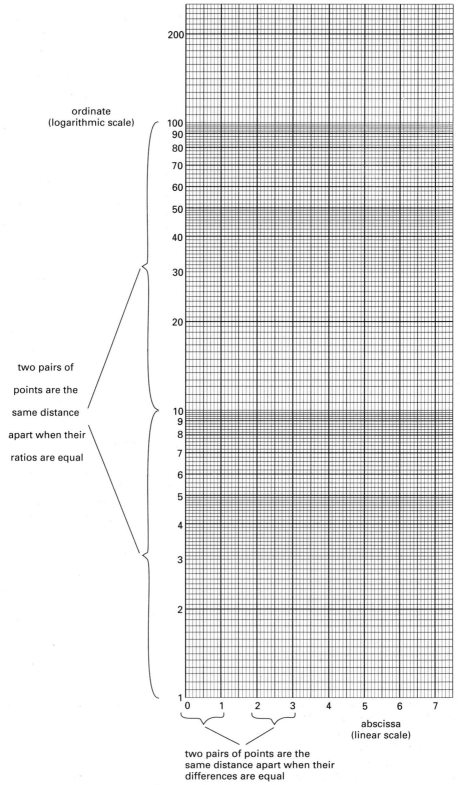

ordinate
(logarithmic scale)

two pairs of

points are the

same distance

apart when their

ratios are equal

abscissa
(linear scale)

two pairs of points are the
same distance apart when their
differences are equal

Fig. 8.12   Log-linear graph paper.

The arithmetic mean for a sample is conventionally given the symbol $\bar{x}$. (pronounced 'x bar') and is given by:

$$\text{arithmetic mean}, \bar{x} = \frac{\text{sum of the items}}{\text{number of items}}$$

$$= \frac{x_1 + x_2 + x_3 + \dots + x_n}{n}$$

$$= \frac{\text{sum of all the } xs}{n}$$

The upper case Greek letter sigma, Σ, can be used as a shorthand way of writing 'sum of'. So the sample arithmetic mean is given by:

$$\bar{x} = \frac{\Sigma x}{n}$$

Similarly, for a population of size N, the population arithmetic mean, μ, is given by:

$$\mu = \frac{\Sigma x}{N}$$

The arithmetic mean takes all the values into account, even the odd extreme value. Hence the arithmetic mean has the major disadvantage that it may yield a measure of central tendency around which few or no other actual values exist. In such cases the median is a measure of central tendency that is particularly useful.

Median

The median is the middle value of a set of observations ranked in order. After ranking the observations, if there is an odd number of observations then the median is the middle value. If there is an even number of observations, then the median is the arithmetic mean of the two middle values of the ranked observations.

For example, for the data in Table 8.9 the median is the arithmetic mean of 23.0 and 23.1, that is, 23.05 years.

Mode

The mode of a set of observations is the value of the observation occurring with the greatest frequency.

In the distribution shown in Figure 8.13, for example, the mode can be seen to be six, while for the frequency polygon shown in Figure 8.14, the mode is the value on the abscissa corresponding to the maximum frequency value on the ordinate.

If there are no repeated values then there is no mode. Like the median, the mode is a measure of central tendency which is not as susceptible to the effect of extreme values as is the arithmetic mean.

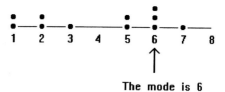

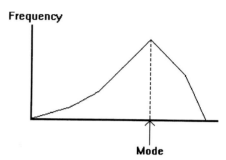

The mode is 6

Fig. 8.13 A one-dimensional dot diagram illustrating the mode.

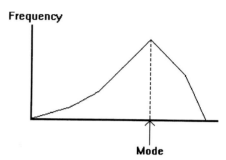

Fig. 8.14 A frequency polygon illustrating the mode.

A **unimodal distribution** has one maximum value, a bimodal distribution has two, a trimodal distribution has three, and so on.

Unimodal distribution curves

Symmetrical unimodal distribution

In a unimodal symmetrical distribution curve the mean, median and mode are equal, as shown in Figure 8.15. This is evident, for example, in the curve representing the Normal distribution (described later in this chapter). In a symmetrical distribution curve the arithmetic mean is a good measure of central tendency.

Skewed unimodal distribution

In a unimodal distribution curve that is **positively skewed** there is a longer **right tail**, as shown in

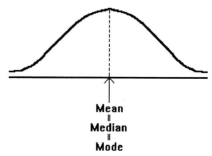

Mean
‖
Median
‖
Mode

Fig. 8.15 A unimodal symmetrical distribution curve.

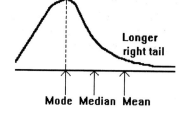

Fig. 8.16 A positively skewed unimodal distribution curve.

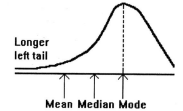

Fig. 8.17 A negatively unimodal distribution curve.

Figure 8.16, and the relationship between measures of central tendency is as follows:

mode < median < mean

In contrast, in a unimodal curve that is **negatively skewed** there is a longer **left tail**, as shown in Figure 8.17, and the relationship between measures of central tendency is as follows:

mean < median < mode

For both positively and negatively skewed unimodal distribution curves the median is a better measure of central tendency than the arithmetic mean, the latter being overly sensitive to outlying values.

Quantiles

As has been seen above, the median splits a distribution into two equal parts, and a distribution curve into two equal areas, so that 50% of values lie below the median and 50% lie above it.

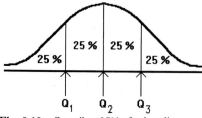

Fig. 8.18 Quartiles: 25% of values lie below Q_1 (the first quartile), between Q_1 and Q_2, between Q_2 and Q_3, and above Q_3 (the third quartile).

Similarly, other quantiles can be defined as values below or above which given percentages of the values lie. The three **quartiles** of a distribution are conventionally denoted as Q_1, Q_2, and Q_3, as shown in Figure 8.18. The 2nd quartile, Q_2, is equal to the median.

A distribution is similarly divided into equal portions of 20% by the four **quintiles**; into equal portions of 10% by the nine **deciles**, D_1 to D_9; and into equal portions of 1% by the 99 **percentiles**, P_1 to P_{99}. The 5th decile, D_5, and the 50th percentile, P_{50}, are both equal to the median.

MEASURES OF DISPERSION

Measures of location on their own, while helpful as statistical descriptions, do not describe the extent of dispersion of a distribution. Considered on their own, therefore, measures of location can be misleading. The measures of dispersion considered in this chapter are as follows: the range; measures relating to quantiles; and the standard deviation and variance.

Range

Definition

Of all the measures of dispersion the range is the easiest to calculate. It is simply the difference between the smallest and largest values in a distribution:

range = (largest value) – (smallest value)

Example

For the ages, in years, of ten patients given in Table 8.9:

range = 45.8 – 12.9 = 32.9 years

Disadvantages

The range does not give any information about the dispersion of values lying between the smallest and the largest. It is also clear that the range will, by definition, be unduly influenced by one extreme value. Furthermore, as the number of observations increases, the range does not decrease; it can only stay the same, or increase, if a new observation is added.

Measures relating to quantiles

Interquartile range

The interquartile range is $Q_3 - Q_1$. It contains the central 50% of ordered values.

Semi-interquartile range

The semi-interquartile range is $\dfrac{Q_3 - Q_1}{2}$

The semi-interquartile range is also known as the quartile deviation and contains 25% of values.

10 to 90 percentile range

The 10 to 90 percentile range is $P_{90} - P_{10}$. The 10 to 90 percentile range contains the central 80% of ordered values. It is also equal to $D_9 - D_1$, hence the alternative name of the **interdecile range**.

Standard deviation

The standard deviation is easier to calculate than measures relating to quantiles, and is the preferred measure of dispersion.

The standard deviation is based on deviations from the arithmetic mean. However, one cannot simply calculate the sum of all these deviations because, given the definition of the arithmetic mean, this sum would always be zero (the negative deviations cancel out the positive deviations). The method used in the calculation of the standard deviation to overcome this difficulty is to work with the *squares* of the deviations from the mean. This is because the square of a (real) number is never negative.

Standard deviation of a population

For a population of size N and mean μ, the mean of the squared deviations is given by:

$$\frac{\sum(x - \mu)^2}{N}$$

To compensate for the fact that squared deviations are being dealt with, the square root of the above expression is taken. The result is defined as the population standard deviation and is conventionally denoted by σ. The abbreviation s.d. is sometimes used for 'standard deviation':

population s.d., $\sigma = \sqrt{\dfrac{\sum(x - \mu)^2}{N}}$

Standard deviation of a sample

For a sample of size n, and mean $\bar{x}$, if the corresponding population mean, μ, is unknown, then it can be shown that in order to obtain a good estimate of the population standard deviation it is necessary to divide by $(n - 1)$ and not simply n, particularly when n is small (when n is large n and $(n - 1)$ are approximately equal). The conventional symbol for the sample standard deviation is s:

sample s.d., $s = \sqrt{\dfrac{\sum(x - \bar{x})^2}{n - 1}}$

When calculating the value of the sample standard deviation an easier formula to use, as shown in the example below, is the following:

sample s.d., $s = \sqrt{\dfrac{\sum(x^2) - \dfrac{(\sum x)^2}{n}}{n - 1}}$

It can be seen from the above formulae that the standard deviation has the same units as the original observations.

Example

Table 8.10 shows ten successive measurements of the two-hourly pulse of a patient over a period of 20 hours. Calculate the mean and standard deviations of these observations. The number of observations, n, = 10.

$$\sum x = 56 + 67 + 65 + 60 + 88 + 96 + 102 + 110 + 90 + 86 = 820$$

Therefore,

$$\text{mean, } \bar{x} = \frac{\sum x}{n}$$

$$= 820/10 = 82 \text{ min}^{-1}$$

$$\sum(x^2) = 56^2 + 67^2 + 65^2 + 60^2 + 88^2 + 96^2 + 102^2 + 110^2 + 90^2 + 86^2 = 70\,410$$

Using the second formula for calculating the standard deviation, we have:

$$\text{sample s.d., } s = \sqrt{\dfrac{\sum(x^2) - \dfrac{(\sum x)^2}{n}}{n - 1}}$$

$$= \sqrt{\dfrac{70\,410 - (820^2/10)}{9}}$$

$$= \sqrt{(3170/9)}$$

$$= 18.8 \text{ min}^{-1}$$

Table 8.10 Successive two-hourly pulses (pulse min-1) of a patient in 20 hours

56	67	65	60	88	96	102	110	90	86

Variance

The variance is simply the **square of the standard deviation**. Hence the formulae for the variance are as follows:

population variance, $\sigma^2 = \dfrac{\Sigma(x - \mu)^2}{N}$

sample variance, $s^2 = \dfrac{\Sigma(x - \bar{x})^2}{n - 1}$

As for the standard deviation, the alternative formula for easier calculation can be used:

sample variance, $s^2 = \dfrac{\Sigma(x^2) - \dfrac{(\Sigma x)^2}{n}}{n - 1}$

As in the case of the standard deviation, the variance has the advantage that all the observations are used in its calculation. Unlike the standard deviation, which has the same units as the observations, the variance has units which are the square of the units of the original observations. For example, the variance for the data in Table 8.10 is as follows:

sample variance, $s^2 = \dfrac{\Sigma(x^2) - \dfrac{(\Sigma x)^2}{n}}{n - 1}$

$$= 3170/9 = 352.2 \text{ min}^{-2}$$

Note that the result has the unit of min^{-2} which is the square of min^{-1}

PROBABILITY

The determination of the probability, or chance, of an event occurring is important in the area of inferential statistics, as is evident later in this chapter. In this section basic concepts of classical probability are described, followed by an outline of permutations and combinations.

Basic concepts

Basic formula

Consider one throw of a die. Clearly there is an equal chance that any of its six faces may land uppermost.

Since there are six such faces, the chance that any given face, e.g. a 'six', may be uppermost is one in six.

In classical probability the probability of an **event E** occurring is conventionally denoted by **Pr(E)**. If the event E may occur in n different ways out of a total of N equally likely possible ways, then

$$\text{Pr(E)} = \frac{n}{N}$$

Applying this formula to the above example, since there are six possible outcomes of a throw of a die, N is six. Therefore

$$\text{Pr (throwing a six)} = n/N = \frac{1}{6}$$

Probability of zero

Suppose the question were asked: 'What is the probability that a throw of a die results in a 'seven'?' Since the full set of possible outcomes of throwing a die does not include a 'seven', there is no possibility of throwing a 'seven', and n is zero, while N remains six. Therefore,

$$\text{Pr (throwing a seven)} = n/N = \frac{0}{6} = 0$$

In general, a probability of zero implies that an event **never** occurs.

Probability of one

Consider now the question 'What is the probability that a throw of a die results in a 'one', 'two', 'three', 'four', 'five' or 'six'?'. In this case, since the full set of possible outcomes includes each of 'one', 'two', 'three', 'four', 'five' and 'six', and no other possibility, n must have the value six. As before, N remains six. Therefore,

$$\text{Pr (throwing a one, two, three, four, five}$$
$$\text{and six)} = n/N = \frac{6}{6} = 1$$

In general, a probability of one implies that an event **always** occurs.

It is clear from the above that the probability of an event occurring can have a **minimum value of zero** (implying that it never occurs), and a **maximum value of one** (implying that it always occurs):

$$0 \leq \textbf{Pr (E)} \leq 1$$

If n represents the different ways in which an event E can occur out of a total of N equally likely possible ways, then it follows that $(N - n)$ represents the different ways in which E *cannot* occur. So the probability that E will *not* occur, conventionally denoted as Pr $(\bar{E})$, is given by:

$$\text{Pr}\,(\bar{E}) = \frac{N-n}{N}$$

$$= 1 - \frac{n}{N}$$

But,

$$\text{Pr}\,(E) = \frac{n}{N}$$

Therefore

$$\text{Pr}\,(\bar{E}) = 1 - \text{Pr}\,(E)$$

Thus

Pr (E) + Pr ($\bar{E}$) = 1

That is, the sum of the probabilities that an event will occur and that it will not occur is equal to one.

Mutually exclusive events

Events are said to be **mutually exclusive** when the occurrence of one of the events means that the other event(s) cannot occur. For example, throwing a 'one' and throwing a 'two' with one throw of a die are two mutually exclusive events because if a 'one' is thrown, then the die cannot simultaneously show a 'two' as the uppermost face, and vice versa.

It can be shown that when two events, E_1 and E_2, are mutually exclusive, the probability that one or the other will occur is equal to the **sum** of their probabilities:

Pr (E_1 or E_2) = Pr (E_1) + Pr (E_2)

where E_1 and E_2 are mutually exclusive. For instance, taking the example of the throw of a die, in calculating the probability of throwing a 'one' or a 'two', this event may occur in two different ways out of a total of six possibilities. That is, n is two, N is six, and the required probability is $2/6$ or $1/3$. Applying the above formula instead:

Pr (throwing a one or a two)
= Pr (throwing a one) + Pr (throwing a two)
= $1/6 + 1/6$
= $1/3$

This rule also applies to more than two mutually exclusive events:

Pr (E_1 or E_2 or ... E_k)
= Pr (E_1) + Pr (E_2) + ... + Pr (E_k)

where E_1, E_2, ... and E_k are mutually exclusive events.

Independent events

Events are said to be **independent** when the occurrence of one event does not in any way influence the probability with which the other(s) will occur. For example, if a die is thrown twice, then whatever the result of the first throw, the probability of throwing a 'six', say, on the second throw remains $1/6$. So these two throws are independent events. Similarly, when two dice are thrown simultaneously, the result of each does not in any way affect the result of the other.

It can be shown that when two events, E_1 and E_2, are independent, the probability that they will both occur is equal to the **product** of their probabilities:

Pr (E_1 and E_2) = Pr (E_1) x Pr (E_2)

where E_1 and E_2 are independent. For instance, taking the example of two throws of a die, or the throw of two dice simultaneously, consider the probability of throwing two 'sixes'. The total number of possible outcomes, N, is 6×6 or 36 (from 'one' 'one'; 'one' 'two'; and so on to 'six' 'six'). Only one of these outcomes ('six' 'six') is the required outcome of throwing two 'sixes', and so n is one. Thus the required probability is $1/36$. Applying the above formula instead:

Pr (throwing 2 sixes)
= Pr (throwing a six) $\times$ Pr (throwing a six)
= $1/6 \times 1/6$
= $1/36$

This rule also applies to more than two independent events:

Pr (E_1 and E_2 and...E_k)
= Pr (E_1) $\times$ Pr (E_2) $\times$ ... $\times$ Pr (E_k)
where E_1, E_2, ... and E_k are independent events.

Large numbers of trials

It has been shown that the probability of throwing a given number, a 'six' say, with a die is $1/6$. This does not mean that in a trial of 6 throws of the die one (and only one) of these throws will necessarily be a 'six'. However, after 120 throws the proportion of these throws that result in a 'six' will be approximately $1/6$. And after 1200 throws the proportion will be even nearer $1/6$. The graph in Figure 8.19 shows the typical results on throwing a die many times. Similarly, the greater the number of tosses of a coin, the closer does the proportion of 'heads' tend towards $1/2$.

In general, when a trial or experiment is repeated a large number of times, the proportion of times that a

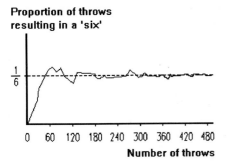

Proportion of throws resulting in a 'six'

Fig. 8.19 The relationship between the proportion of throws of a die that result in a 'six' and the number of throws.

given result occurs tends to approach the probability of that result occurring.

Permutations and combinations

Permutations and combinations are outlined briefly in this subsection because they can appear in the calculation of probabilities. For example, in the next section on the binomial distribution, a formula is given which incorporates a combination.

Permutations

In order to calculate a probability it may be necessary to consider permutations, that is, the number of ways of arranging objects.

Suppose, for example, that a psychiatric trainee preparing for a clinical examination in psychiatry has been given the advice that it would be useful to see four particular patients, A, B, C and D, on a psychiatric ward, in order to take a history and carry out a mental state examination. Assuming that all the patients give their permission to be interviewed, then one could ask the question 'How many different ways exist in which the psychiatric trainee could arrange the order of examining them?'.

Now the first patient can be any of the four patients, that is, any of A, B, C or D, and so can be chosen in four ways. Once chosen, this will leave three patients, so that the second patient can now be chosen in three ways. This then leaves two patients, so that the third patient can be chosen in two ways. Finally, there is just one patient left, so that the fourth patient can be chosen in only one way. Therefore the total number of permutations of four patients is $4 \times 3 \times 2 \times 1$, that is, 24. These 24 permutations are shown in Table 8.11.

Table 8.11 The 24 permutations of four patients (A, B, C and D)

ABCD	ACBD	BACD	BCAD	CABD	CBAD
ABDC	ADBC	BADC	BDAC	DABC	DBAC
ACDB	ADCB	CADB	CDAB	DACB	DCAB
BCDA	BDCA	CBDA	CDBA	DBCA	DCBA

In general, the number of permutations of n objects out of a total of n objects is

$$n \times (n-1) \times (n-2) \times (n-3) \times \ldots \times 2 \times 1$$

This is known as the **factorial** of n, and is given the symbol $n!$.

So, for example,

$$5! = 5 \times 4 \times 3 \times 2 \times 1 = 120$$
$$10! = 10 \times 9 \times 8 \times 7 \times 6 \times 5 \times 4 \times 3 \times 2 \times 1 = 3\,628\,800$$

Let us now suppose that the psychiatric trainee feels that there is time to see only two of the four patients. What now is the total number of ways of seeing two patients from four, paying attention to the order in which they are seen?

Once again, the first patient can be any of the four patients, and so can be chosen in four ways. Once chosen, this will leave three patients, so that the second patient can be chosen in three ways. So the total number of ways in which two patients can be chosen from four, paying attention to the order in which they are chosen, is 4×3, that is, 12. These 12 permutations are shown in Table 8.12. This is a permutation of two objects from four, and can be written as $^4P_2$.

Table 8.12 The 12 permutations of two patients from four (A, B, C and D)

| AB | AC | AD | BA | BC | BD | CA | CB | CD | DA | DB | DC |

In general, the number of permutations of r objects from n, ^nP_r, is given by:

$$^nP_r = n \times (n-1) \times (n-2) \times (n-3) \times \ldots \times (n-r+1)$$

This can be written more neatly as:

$$^nP_r = \frac{n!}{(n-r)!}$$

So, for example, the number of permutations of four objects from six is given by:

$$^6P_4 = \frac{6!}{(6-4)!}$$

$$= \frac{6!}{2!} = \frac{6 \times 5 \times 4 \times 3 \times 2 \times 1}{2 \times 1}$$

$$= 6 \times 5 \times 4 \times 3$$

$$= 360$$

Combinations

Combinations are similar to permutations except that the order in which objects are chosen is *not* important. What is important is just which objects are selected.

Returning to the example of the psychiatric trainee who wishes to pick two out of the four patients. If the order in which the patients are to be seen were unimportant to the psychiatric trainee, then, although AB and BA would count as two permutations, they would count as only one combination, and similarly with AC and CA, and so on. The number of combinations of two patients that can be made from four can be denoted by $^4C_2$. Because each of these combinations consists of two patients, it is clear that the number of ways in which they can be arranged is $2!$, or 2. So it follows that:

$2! \times {}^4C_2$ = number of permutations

$\therefore \; 2! \times {}^4C_2 = {}^4P_2$

$\therefore \; {}^4C_2 \quad = \dfrac{4 \times 3}{2 \times 1}$

$\qquad \qquad = 6$

These 6 combinations are AB, AC, AD, BC, BD and CD. Note that AB could equally be written as BA, and so on (that is, the order is not important).

In general, for a given number of selections or combinations of r objects from n, ^nC_r, each of the selections can be arranged in $r!$ ways. Therefore,

$r! \times {}^nC_r = {}^nP_r$

$\qquad = \dfrac{n!}{(n-r)!}$

$\therefore \; {}^nC_r = \dfrac{n!}{r!(n-r)!}$

Note that ^nC_r can also be written as:

$$\binom{n}{r}$$

Returning again to the example of the selection of patients, the clinical tutor who selected the four patients for the psychiatric trainee did so from a total of 15 patients with different conditions on the psychiatric ward. If we were now to ask in how many ways any four patients could be selected from the 15, the answer would be given by:

$$^{15}C_4 = \frac{15!}{4!\,(15-4)!}$$

$$= \frac{15!}{4! \times 11!}$$

$$= \frac{15 \times 14 \times 13 \times 12}{4 \times 3 \times 2 \times 1}$$

$$= 1365$$

Factorial of zero

Occasionally the use of the above formulae may involve $0!$. In order to give a value to $0!$ consider ^nC_n. Clearly, there is only one way of selecting n objects from n. Therefore,

$^nC_n = 1$

$\therefore \; \dfrac{n!}{n!\,(n-n)!} = 1$

$\therefore \; \dfrac{n!}{n! \times 0!} = 1$

$\therefore \; \mathbf{0! = 1}$

The value of 1 for $0!$ should be used when calculating permutations and combinations.

DISCRETE PROBABILITY DISTRIBUTIONS

In this section the concept of probability distributions is described in relation to two commonly used discrete probability distributions, the binomial and the Poisson distributions. Continuous probability distribution are described in the next section.

Binomial distribution

A binomial distribution is said to occur when there are just **two alternative outcomes**, A or B, for a trial, and a series of n trials is carried out.

Let $Pr(A) = p$

Then $Pr(B) = 1 - Pr(A)$
$\qquad \qquad = 1 - p$

Probability distribution

Consider the case of the number of daughters, D, born in a family with n children. It will be assumed that:

Pr (having a daughter) = Pr (having a son) = $\frac{1}{2}$

The case in which n = 0

If there were no children, that is, $n = 0$, then there would only be one possibility, namely:

number of daughters = 0

This possibility is certain, and so the probability is one that the number of daughters is zero when $n = 0$:

$$Pr (D = 0) = 1$$

The case in which n = 1

Continuing with the above example, if there were just one child in the family then this child could be either a daughter or a son. Thus there is a probability of $1/2$ that the child is a daughter, and there is a probability of $1/2$ that the child is a son. Therefore,

$$Pr(D = 0) = 1/2$$
$$Pr(D = 1) = 1/2$$

Sum of probabilities = 1

This is known as a probability distribution, and the sum of all the probabilities in it must be one. This probability distribution can be represented graphically as shown in Figure 8.20.

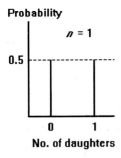

Fig. 8.20   Probability distribution for binomial distribution, $Pr(A) = Pr(B) = 1/2$, $n = 1$.

The case in which n = 2

With two children there are four possible events as shown in Table 8.13. Note that this table includes a **frequency distribution,** which is analagous to a probability distribution except that is refers to frequencies rather than the probabilities. Since each of the four events is equally likely it follows that:

$$Pr(D = 0) = 1/4$$
$$Pr(D = 1) = 2/4 = 1/2$$
$$Pr(D = 2) = 1/4$$

Once again, the sum of the probabilities has to be one.

Table 8.13   The four possible events for $n = 2$

Daughter Daughter	Daughter Son	Son Son	
	Son Daughter		total = 4
D = 2	D = 1	D = O	
1	2	1	frequency distribution

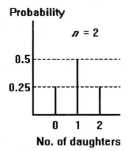

Fig. 8.21   Probability distribution for binomial distribution, $Pr(A) = Pr(B) = 1/2$, $n = 2$.

This probability distribution can be represented graphically as shown in Figure 8.21.

The case in which n = 3

In this case there are eight possible events, shown in Table 8.14. Since each of the eight events is equally likely it follows that:

$$Pr(D = 0) = 1/8$$
$$Pr(D = 1) = 3/8$$
$$Pr(D = 2) = 3/8$$
$$Pr(D = 3) = 3/8$$

Again, the sum of the probabilities is one. The prob-

Table 8.14   The eight possible events for $n = 3$

Daughter Daughter Daughter	Daughter Daughter Son	Daughter Son Son	Son Son Son	
	Daughter Son Daughter	Son Daughter Son		total = 8
	Son Daughter Daughter	Son Son Daughter		
D = 3	D = 2	D = 1	D = 0	frequency distribution
1	3	3	1	

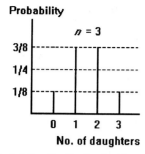

Fig. 8.22   Probability distribution for binomial distribution, Pr(A) = Pr(B) = 1/2, n = 3.

ability distribution is shown in graphical form in Figure 8.22.

Pascal's triangle

The numbers in the above frequency distributions can be readily derived from Pascal's triangle, shown in Figure 8.23. This is a triangle of numbers in which each line starts and ends with one, and in which the middle numbers are each formed by adding together the two numbers immediately adjacent to it on the line above. So in the last example the frequency distribution for, say, n = 3, can be simply found from the appropriate line in Pascal's triangle, as:

1 (D = 0)   **3** (D = 1)   **3** (D = 2)   **1** (D = 3)

The case in which Pr(A) ≠ Pr(B)

So far the use of Pascal's triangle has been considered in situations in which the probabilities of the two alternative outcomes are equal, that is,

$$Pr(A) = Pr(B) = 1/2$$

Binomial distributions can also apply when Pr(A) does not equal Pr(B). For example, when a die is thrown, if A represents the outcome of throwing a 'six', we have:

$$Pr(A) = p$$
$$= 1/6$$

$$Pr(B) = Pr(\text{not throwing a six})$$
$$= 1 - p$$
$$= 5/6$$

Pascal's triangle can be used in such a situation, so long as it is borne in mind that the numbers of the frequency distribution from the triangle now refer to the coefficients of the following probabilities:

$$p^n, p^{n-1}(1-p), p^{n-2}(1-p)^2, ..., p(1-p)^{n-1}, (1-p)^n$$

Consider the case in which there is a total of six throws of a die, that is, n = 6. From Pascal's triangle (Fig. 8.23) it can be seen that the frequency distribution corresponding to this value is given by:

1 6 15 20 15 6 1

Table 8.15   Binomial distribution coefficients for the number of 'sixes' thrown in six throws of a die; Pr(A) = 1/6; n = 6

1	(the coefficient for throwing 0 or 6 sixes)
6	(the coefficient for throwing 1 or 5 sixes)
15	(the coefficient for throwing 2 or 4 sixes)
20	(the coefficient for throwing 3 sixes)
15	(the coefficient for throwing 4 or 2 sixes)
6	(the coefficient for throwing 5 or 1 sixes)
1	(the coefficient for throwing 6 or 0 sixes)

This is shown in Table 8.15. It will have been noticed in the previous example that the numbers of the frequency distribution from Pascal's triangle could be applied the other way around (that is, starting with 1 for D = 3, and so on) without this making any difference. This is because Pascal's triangle is symmetrical. In the present example, with Pr(A) ≠ Pr(B), it is important to ensure that p^n corresponds with Pr(throwing 6 sixes (and *not* Pr(throwing 0 sixes)). The binomial distribution for the number of 'sixes' thrown in six throws of a die is shown in Table 8.16, and graphically in Figure 8.24

n	frequency distribution	total
0	1	1
1	1 1	2
2	1 2 1	4
3	1 3 3 1	8
4	1 4 6 4 1	16
5	1 5 10 10 5 1	32
6	1 6 15 20 15 6 1	64
7	1 7 21 35 35 21 7 1	128
n	1 7 21 35 35 21 7 1	2^n

Fig. 8.23   Pascal's triangle.

Table 8.16 Binomial distribution for the number of 'sixes' thrown in six throws of a die; Pr(A) = $^1/_6$; $n = 6$

Pr(6 sixes) = $1 \times p^6$	$= (^1/_6)^6$	= 0.000 (to 3 decimal places)
Pr(5 sixes) = $6 \times p^5(1-p)$	$= 6(^1/_6)^5(^5/_6)$	= 0.001
Pr(4 sixes) = $15 \times p^4(1-p)^2$	$= 15(^1/_6)^4(^5/_6)^2$	= 0.008
Pr(3 sixes) = $20 \times p^3(1-p)^3$	$= 20(^1/_6)^3(^5/_6)^3$	= 0.054
Pr(2 sixes) = $15 \times p^2(1-p)^4$	$= 15(^1/_6)^2(^5/_6)^4$	= 0.201
Pr(1 six) = $6 \times p(1-p)^5$	$= 6(^1/_6)(^5/_6)^5$	= 0.402
Pr(0 sixes) = $1 \times (1-p)^6$	$= (^5/_6)^6$	= 0.335
Sum of probabilities		= 1 (allowing for approximations to 3 decimal places in the above numbers)

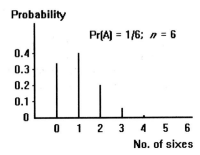

Fig. 8.24 Binomial distribution for the number of 'sixes' thrown in six throws of a die.

Formula

From Pascal's triangle in Figure 8.23 it can be seen that for n trials the number of combinations is 2^n. So, for example, for $n = 20$, the total number of combinations is 2^{20}, that is, 1 048 576. Clearly, Pascal's triangle provides a practical method for working out the binomial distributions for only relatively small values of n. A more practical formula is now provided.

Let the probability of success (that is, the probability of A) be p. Then the probability of failure (that is, the probability of B) is $(1 - p)$. If there are n independent trials with r successes, then the total number of failures must be $(n - r)$. So, applying the rule of the product of probabilities for independent events:

probability of a specific series of r successes occurring by chance = $p^r(1 - p)^{n-r}$

The probability of r successes is therefore the product of this term and the number of mutually exclusive ways in which r objects can be chosen from n. Hence:

$$\text{Pr(r successes)} = {}^nC_r \times p^r(1 - p)^{n-r}$$

$$\therefore \; \textbf{Pr(r successes)} = \frac{\textbf{\textit{n}}!}{\textbf{\textit{r}}! \, (\textbf{\textit{n}} - \textbf{\textit{r}})} \times \textbf{\textit{p}}^{\textbf{\textit{r}}}(1 - \textbf{\textit{p}})^{\textbf{\textit{n}} - \textbf{\textit{r}}}$$

Example

The above formula will now be applied, instead of Pascal's triangle, to the previous example of the die. Suppose the probability of throwing one 'six' in six trials (that is, six throws of the die) is required. We have:

n = number of trials
    = 6
r = number of 'sixes'
    = 1
p = Pr(throwing a six)
    = $^1/_6$

Therefore,

$$\text{Pr(throwing 1 six)} = \frac{n!}{r! \, (n - r)!} \times p^r(1 - p)^{n - r}$$

$$= \frac{6!}{1! \, (6 - 1)!} \times (^1/_6)^1 \, (^5/_6)^5$$

$$= \frac{6!}{5!} \times \frac{5^5}{6^6}$$

$$= \frac{6 \times 5^5}{6^6}$$

$$= (^5/_6)^5$$

$$= 0.402 \text{ (to 3 decimal places)}$$

This is the same result as was obtained using Pascals's triangle earlier in this section.

Mean and variance

For a large value of n, rather than state all the individual probabilities, it can be more useful simply to give summarizing data in terms of the mean (also known as the expected value) and variance (the square of the standard deviation) of a binomial distribution. Using the same notation as above, it can be demonstrated algebraically that:

mean, $\mu = np$
variance, $\sigma^2 = np(1 - p)$

For example, suppose one wishes to determine the mean or expected number of 'sixes' thrown in six throws of one die (or six dice thrown together). We have:

$n = 6, \, p = ^1/_6$

Therefore:

mean = $np = 6 \times (^1/6) = 1$

The variance can also be simply calculated:

variance = $np(1 - p) = 6 \times (^1/6) \times (^5/6)$
$= 5/6$

Poisson distribution

The binomial distribution is one of a number of discrete distributions, that is, distributions in which a set of discrete separate values is taken by a random variable. (The sum of the corresponding probabilities is one.) Although the binomial distribution is the discrete probability distribution most often used in medical statistics, others are sometimes used. One of these is the Poisson distribution.

Criteria

The Poisson distribution is used in situations in which events occur randomly in time or space. Other criteria that need to be satisfied before the Poisson distribution can be used are: the events are independent; two or more events cannot take place simultaneously; and the mean number of events per given unit of time or space is constant.

An example of a situation in which the Poisson distribution is applicable to events occurring in time is death from certain causes taking place independently and in a random manner in a given population. Similarly, an example in which the Poisson distribution is applicable to random independent events occurring in space is the number of cells of a given type per unit area on a microscope slide, when the cells have been placed on the slide in a dilute solution.

Formulae

The mathematics involved in deriving the formulae that follow have not been included. The interested reader is referred to the list of books for further reading at the end of this chapter. Instead, the formulae will simply be stated.

The probability of r events occurring in a given unit of time or space, $\Pr(r)$, is given by:

$$\mathbf{Pr}(r) = \frac{\lambda^r e^{-\lambda}}{r!}$$

where λ is the mean number of events in the same unit of time or space, and e is the exponential constant and base of natural logarithms (e = 2.718 to three decimal places).

The value of λ may be known or it may need to be estimated by using the arithmetic mean of a sample, $\bar{x}$.

The formulae for the mean and variance are as follows:

mean, $\mu = \lambda$
variance, $\sigma^2 = \lambda$

Since the mean of a Poisson distribution is the same as the variance, this provides a quick method of checking whether a given distribution is (or approximates to) a Poisson distribution.

Example

Suppose that the number of emergency admissions to a psychiatric unit were on average two per day and had been found to have a Poisson distribution. What is the probability that on a given day there will be (a) no emergency admissions; and (b) more than one emergency admission?

(a) No emergency admissions. Since λ, the mean number of events per day, has a value of two (the average number of emergency admissions per day), the required probability, $\Pr(r = 0)$, is given by:

$$\Pr(r = 0) = \frac{\lambda^r e^{-\lambda}}{r!}$$
$$= \frac{2^0 e^{-2}}{0!}$$

Now, 2^0 is equal to one, and it has been shown earlier in this chapter that 0! can also be taken as one. Therefore,

$\Pr(r = 0) = e^{-2}$
$= 0.135$ (to 3 decimal places)

(b) More than one emergency admission. Since the following sum of probabilities is equal to 1:

$\Pr(r = 0) + \Pr(r = 1) + \Pr(r = 2) + \Pr(r = 3) + ... = 1$

then it follows that

$\Pr(r = 0) + \Pr(r = 1) + \Pr(r > 1) = 1$
$\therefore \Pr(r > 1) = 1 - \Pr(r = 0) - \Pr(r = 1)$

$\Pr(r = 0)$ has been calculated, and so only $\Pr(r = 1)$ now needs to be calculated in order to obtain the required probability.

$$\Pr(r = 1) = \frac{\lambda^r e^{-\lambda}}{r!}$$
$$= \frac{2^1 e^{-2}}{1!}$$
$$= 2e^{-2}$$
$$= 0.2707 \text{ (working to 4 decimal places)}$$

Therefore the required probability is given by:

$$Pr(r > 1) = 1 - Pr(r = 0) - Pr(r = 1)$$
$$= 1 - 0.1353 - 0.2707 \text{ (working to 4 decimal places)}$$
$$= 0.594 \text{ (to 3 decimal places)}$$

Different values of λ

The Poisson distribution can be calculated for different values of λ; these are shown in graphical form in Figure 8.25 for λ having values of one, two and five. As would be expected, the sum of the lengths of all the lines in each graph is one (according to the scale of the ordinate), since the sum of all the probabilities of a discrete probability distribution must come to one.

It can be seen that the probability distribution becomes flatter and more symmetrical as λ increases.

CONTINUOUS PROBABILITY DISTRIBUTIONS

In this section continuous probability distributions are first considered in general, followed by a description of the Normal distribution and other continuous probability distributions such as the *t* and χ^2 distributions.

General considerations

Probability histogram

As has been mentioned previously, the probability distributions looked at in the previous section, the

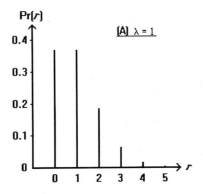

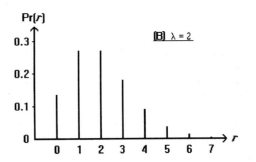

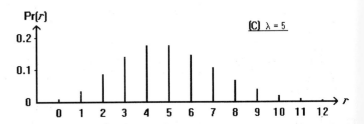

Fig. 8.25 Poisson distribution with different values of λ. (**A**) λ = 1; (**B**) λ = 2; (**C**) λ = 5.

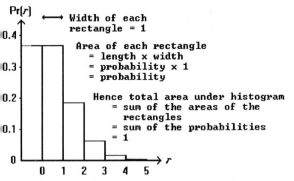

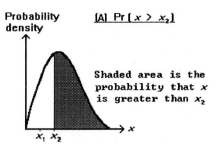

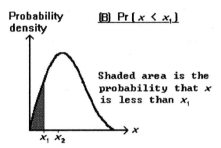

Fig. 8.26   Probability histogram representing the Poisson distribution with $\lambda = 1$.

Fig. 8.28   Areas of probability density function curve corresponding to (**A**) Pr $(x > x_2)$; and (**B**) Pr $(x < x_1)$.

binomial distribution and the Poisson distribution, are known as discrete probability distributions because the random variable, which we may call x, takes discrete values.

A discrete probability distribution can be represented as a line graph, as in Figures 8.24 and 8.25, in which case the sum of the lengths of all the lines is one (since the sum of all the probabilities is one). It can also be represented as a probability histogram, in which case the total area under the histogram is also one. The probability histogram for the Poisson distribution, with $\lambda = 1$, is shown in Figure 8.26, which also explains why the total area under the histogram is one.

Probability density function

In a continuous probability distribution the random variable, x, is continuous. This means that it can be represented graphically as a continuous curve (instead of as a probability histogram as for a discrete probability distribution). Such a curve is known as a probability density function. A generalized probability density function curve is shown in Figure 8.27, in which the area corresponding to the probability that x lies between x_1 and x_2 is shaded. Figure 8.28 shows the areas corresponding to (A) Pr$(x > x_2)$ and (B) Pr$(x < x_1)$.

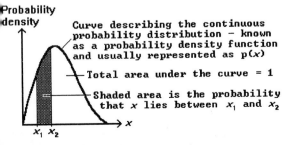

Fig. 8.27   Generalized probability density function curve.

Normal distribution

The Normal distribution has the form of the bell-shaped curve shown in Figure 8.29 and is a good approximation to many naturally occurring continuously variable distributions including height, body mass, white blood cell count, and many types of experimental error. Indeed, when such a distribution differs markedly from the Normal distribution this may have important implications. For example, the distribution of intelligence in a population follows the Normal distribution throughout most of the range. Indeed, intelligence tests which give an intelligence quotient or IQ are based on the assumption that intelligence is distributed in the population as a Normal distribution. However, in the lower range of IQ, there are more people than would be expected from the theoretical Normal distribution, as shown in Figure 8.30; these 'excess cases' are caused by deleterious environmental and social factors.

It should be noted that there is nothing intrinsically 'abnormal' about a continuous distribution which does not follow the Normal distribution. This is one reason why some prefer the alternative name **Gaussian distribution** instead of Normal distribution; Gauss lived from 1777 to 1855 and used this distribution in his theory of errors.

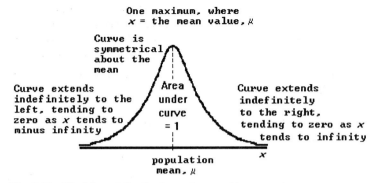

One maximum, where
x = the mean value, μ

Curve is
symmetrical
about the
mean

Area
under
curve
= 1

Curve extends
indefinitely to the
left, tending to
zero as x tends to
minus infinity

Curve extends
indefinitely
to the right,
tending to zero as x
tends to infinity

x

population
mean, μ

Fig. 8.29   The Normal distribution and some of its properties.

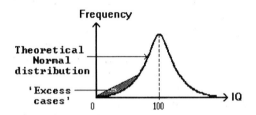

Frequency

Theoretical
Normal
distribution

'Excess
cases'

0   100

→ IQ

Fig. 8.30   Frequency distribution of intelligence quotients.

Properties

Some of the properties of the Normal distribution are shown in Figure 8.29 from which it can be seen that the Normal distribution curve is a unimodal curve symmetrical about the mean and extending indefinitely to the right and left; $p(x)$ tends to zero as x tends to infinity or minus infinity. Since this is a unimodal curve, the median and the mode each have the same value as the mean. As a probability distribution, the total area under the curve must be one.

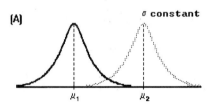

[A]   σ constant

μ_1 μ_2

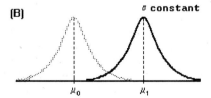

[B]   σ constant

μ_0 μ_1

Fig. 8.31   Effects of a Normal distribution of (**A**) increased mean, and (**B**) decreased mean.

From the above properties it follows that the Normal distribution may be an appropriate model for a distribution which is unimodal, continuous, symmetrical, and has frequencies which tend to zero as the variable moves in either direction from the mean.

Formula

The formula that describes the Normal distribution is:

$$f(x) = \frac{1}{\sigma\sqrt{2\pi}}\, e^{-\left(\frac{(x-\mu)^2}{2\sigma^2}\right)}$$

where μ is the mean and σ is the standard deviation.

The important thing to note from this formula is that, since π and e are constants, a given Normal distribution is fully described by two parameters: the mean, μ; and the standard deviation, σ (or variance σ^2).

Changes in parameters

When the mean of a Normal distribution increases from μ_1 to μ_2 (that is, μ_2 is greater than μ_1), with the standard deviation remaining constant, then the whole curve shifts along the horizontal x-axis (or abscissa) to the right, as shown in Figure 8.31(A). Figure 8.31(B) shows the opposite effect of a shift to the left that occurs when the mean decreases from μ_1 to μ_0 (that is μ_0 is less than μ_1) with the standard deviation remaining constant.

When the standard deviation of a Normal distribution decreases from σ_1 to σ_2 (that is, σ_2 is less than σ_1) with the mean remaining constant, then the whole curve becomes taller, more peaked, and thinner, as shown in Figure 8.32(A). Similarly, when the standard deviation increases from σ_1 to σ_0 (that is σ_0 is greater than σ_1) with the mean remaining constant, then the whole curve becomes flatter, less peaked, and fatter, as shown in Figure 8.32(B).

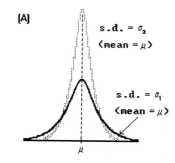

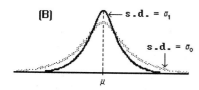

Fig. 8.32 Effects on a Normal distribution of (**A**) decreased standard deviation, and (**B**) increased standard deviation.

Area under the curve

For a given Normal distribution with mean μ and standard deviation σ, the area enclosed by $x = \mu - \sigma$, and $x = \mu + \sigma$, is 68.27% of the total area under the curve, as shown in Figure 8.33. Similarly, the interval two standard deviations either side of the mean encloses 95.45% of the total area, and the interval three standard deviations either side of the mean encloses 99.73% of the total area. Since the Normal distribution is an example of a continuous probability distribution, it follows that it can be stated that the probability (for a Normal probability density function) that x lies between $\mu - \sigma$ and $\mu + \sigma$ is approximately 0.68:

$$\Pr(\mu - \sigma < x < \mu + \sigma) \approx 0.68$$

Likewise:

$$\Pr(\mu - 2\sigma < x < \mu + 2\sigma) \approx 0.95$$

and so on.

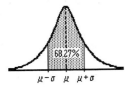

Fig. 8.33 Area under the Normal distribution curve lying between one standard deviation either side of the mean.

As is evident later in this chapter, there are occasions when a more accurate range corresponding to a probability of 0.95 is required for the Normal distribution. This is given by the range:

$$\mu - 1.96\sigma \text{ to } \mu + 1.96\sigma$$

That is

$$\Pr(\mu - 1.96\sigma < x < \mu + 1.96\sigma) = 0.95$$

Standardization

It has been seen that a given Normal distribution can be described completely by the values of the mean, μ, and the standard deviation, σ. It has also been shown that altering the value of μ or σ (or both) leads to changes in the shape of the Normal distribution curve. In order to make it easier to calculate areas under the curve for any Normal distribution curve, tables have been produced for one **standard Normal distribution**, having a **mean of zero** and a **standard deviation of one**. The areas for other Normal distribution curves can then be calculated from the table for the standard Normal distribution by changing to the scale of the latter via the process of standardization.

An abbreviated form of the table for the standard Normal distribution areas appears in the Appendix as Table I. For this special curve the random variable is known as z and not x. Table I gives the probability that a random variable with a standard Normal distribution has a value between 0 and z, as shown in Figure 8.34.

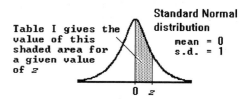

Fig. 8.34 Area under the standard Normal distribution curve given by Table I in the Appendix.

Conveniently, the value of z is simply the value of the number of standard deviations by which the variable on the horizontal axis differs from the mean value of zero. So a direct comparison can be made of the horizontal scale used for the standard Normal distribution (the z-scale) and that for any Normal distribution (the x-scale), as shown in Figure 8.35. It can be seen that the formula for changing the units of measurement on the x-scale into the **standard units** of the z-scale,

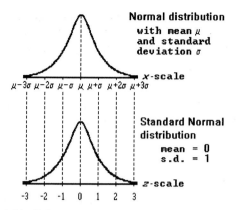

Normal distribution
with mean μ
and standard
deviation σ

x-scale

$\mu-3\sigma\ \mu-2\sigma\ \mu-\sigma\ \mu\ \mu+\sigma\ \mu+2\sigma\ \mu+3\sigma$

Standard Normal
distribution
mean = 0
s.d. = 1

z-scale

-3 -2 -1 0 1 2 3

Fig. 8.35 Relationship between a Normal distribution (mean μ, standard deviation σ) and the standard Normal distribution (mean zero, standard deviation one).

that is, the process of standardization, is given by:

$$z = \frac{x - \mu}{\sigma}$$

Example

A random variable is known to have a Normal distribution with a mean (μ) of 75 and a standard deviation (σ) of 4. What is the probability that the random variable has a value between 76.1 and 77.0?

The area required is that shaded in Figure 8.36(A). In order to use Table I (Appendix) the problem must first be converted into one for the standard Normal distribution by means of the process of standardization. We have:

$\mu = 75, \sigma = 4$

$$z = \frac{x - \mu}{\sigma}$$

Therefore

$$z = \frac{x - 75}{4}$$

Hence, for $x = 76.1$, $z = (76.1 - 75)/4 = 0.275$, and for $x = 77$, $z = (77 - 75)/4 = 0.5$. So, for the standard Normal distribution, the required area is that shaded in Figure 8.36(B). This area is in turn equal to the difference between the areas shown in Figure 8.36(C). So, the required area = 0.1915 – 0.10835 = 0.08315. Hence, the required probability is 0.083 to two significant figures.

Notation

It is convenient to use the notation N_z to represent the

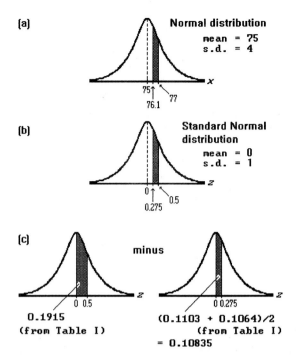

[a]

Normal distribution
mean = 75
s.d. = 4

x

75
76.1   77

[b]

Standard Normal
distribution
mean = 0
s.d. = 1

z

0
0.275   0.5

[c]

minus

z

0 0.5

0.1915
(from Table I)

z

0 0.275

(0.1103 + 0.1064)/2
(from Table I)
= 0.10835

Fig. 8.36 Steps in the calculation of the probability that a random variable having a Normal distribution (mean 75, standard deviation 4) has a value between 76.1 and 77.0 (see text).

area under the standard Normal distribution curve below *z*. So, N_0 represents the area under one-half of the curve, and has a value of 0.5. In the above example, (see Figure 8.36(C)) the required area could more conveniently have been written as:

$N_{0.5} - N_{0.275}$

which in turn is equal to:

$(0.1915 + N_0) - (0.10835 + N_0)$

which is equal to 0.1915 minus 0.10835, as before.

Percentage points

Later in this chapter it is seen how, for example in the calculation of confidence intervals, it is useful to be able to use the symmetric values of *z* (*z* and *−z*) which cut off two tails of equal areas (or probabilities), the sum of which is α, such that α has certain specified values; see Figure 8.37. Since the sum of the two shaded areas is α, each shaded area must have a value of α/2, and so the required values of *z* can be written as $z_{\alpha/2}$, for the lower value, and $z_{1-\alpha/2}$ for the higher value. (Since the curve is symmetrical, the lower value

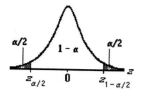

Fig. 8.37 Two-tailed 100α percentage points of the standard Normal distribution.

Table 8.17 Two-tailed percentage points of the standard Normal distribution

α	z
0.5	0.67
0.1	1.64
0.05	1.96
0.01	2.58
0.001	3.29

is also $-z_{1-\alpha/2}$). Table 8.17 shows the values of z corresponding to the most commonly required values of α. For example, to calculate the values of z which correspond to a total of 5% (that is $\alpha = 0.05$) of the area being cut off in two symmetrical tails (each of 2.5%), it can be seen from Table 8.17 that z has the values 1.96 and -1.96. Since two tails are cut off, these values of z are said to be two-tailed percentage points. For convenience, Table 8.17 is also reproduced in the Appendix as Table II.

In general, the two-tailed (or two-sided) 100α per cent point is the value of x such that there is a probability of α that a random variable is greater than or equal to x, or less than or equal to $-x$. To determine its value, standardization is first carried out, and then Table 8.17 or Table II is consulted. To illustrate the simple algebra used to convert back to a (non-standard) Normal distribution, let us use the example above with α being 0.05. Converting back to x for any Normal distribution we have:

$$z = \frac{x - \mu}{\sigma}$$

$$\therefore \quad 1.96 = \frac{x - \mu}{\sigma}$$

$$\therefore \quad x = \mu + 1.96\sigma$$

So, for any Normally distributed random variable, the two-tailed 5% point is given by $\mu + 1.96\sigma$.

For a standard Normal distribution the one-sided 100α percentage point is the value of z for which there

is a probability of α (or $100\alpha\%$) that the random variable is greater than or equal to z. Table III in the Appendix gives commonly used one-sided percentage points.

In fact, Tables II and III are not both required as one will suffice. This is because the value of z for a one-sided 100α percentage point is clearly also the value for the two-sided $2(100\alpha)$ percentage point.

Other distributions

t distribution

The t distribution is another important continuous probability distribution which, like the Normal distribution, is symmetrical about the mean, but has longer tails than the standard Normal distribution. The t distribution is used in a similar way to the Normal distribution, except that it is used in Normal (or approximately Normal) cases in which the sample size is small (less than 30, say) and in which the standard deviation is estimated, as is described later in this chapter. As the value of the sample size, n, increases, the shape of the t distribution approximates closer and closer to that of the standard Normal distribution. In order to determine which t distribution curve to use, tables give values for this distribution for different values of the number of **degrees of freedom**. For the purposes of this chapter the following formula can be used to calculate the number of degrees of freedom, d.f.:

number of degrees of freedom, d.f. $= n - 1$

Table IV in the Appendix gives the two-tailed 100α percentage points of the t distribution, that is, the values of t ($t_{\alpha/2}$ and $t_{1-\alpha/2}$) shown in Figure 8.38. Again, because this is a symmetrical distribution, the one-sided 100α percentage points are equal to the two-sided $2(100\alpha)$ percentage points.

From Table IV it can be seen that as the number of degrees of freedom tends to infinity, the values of the percentage points tend towards the same values as given by the standard Normal distribution. It can also be seen that for sample sizes greater than about 30,

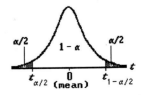

Fig. 8.38 Two-tailed 100α percentage points of the t distribution.

use of the standard Normal distribution will give figures that differ little from those obtained by using the t distribution.

χ^2 distribution

Unlike the standard Normal and t distributions, the χ^2 distribution is an asymmetrical distribution. Therefore, right-hand, one-tailed 100α percentage points only are given for this distribution in Table V in the Appendix, as shown in Figure 8.39. As with the t distribution, the shape of the curve varies with different values of the number of degrees of freedom. This use of the χ^2 distribution and the calculation of the number of degrees of freedom, d.f., for this distribution are described later in this chapter. Note that the Greek letter χ, chi, is pronounced with a hard ch, as in the first two letters of the word kite.

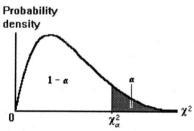

Fig. 8.39   The χ^2 distribution.

F distribution

Like the χ^2 distribution, to which it is related, the F distribution is asymmetrical, and tables are available giving the 100α percentage points (that is, the value of F_α) for different numbers of degrees of freedom. The F distribution is used in the analysis of variance when different samples are being compared. It is not considered further in this chapter.

SAMPLING, ESTIMATION AND HYPOTHESIS TESTING

Statistical aspects of random sampling, estimation of confidence intervals, and hypothesis testing, are considered in this section; some practical aspects also appear in the next chapter.

Random sampling

Methods of sampling

There are a number of methods of choosing random samples from populations. In **periodic sampling** every nth member of the population is chosen. For example, every third patient admitted to a ward might be chosen. A better method might be to use **random numbers** to determine which patients are chosen. For example, suppose we wish to choose every fifth patient out of a given population. Each patient could be assigned a random number, using a table of random numbers or a random number generator, and then all those with a random number ending in zero or five (that is, one in five of the patients) could be chosen. Sometimes such a procedure does not lead to a random choice because of some unforeseen underlying pattern. Another difficulty is that arising from non-response, with the responders differing from non-responders in ways which may be uncertain.

In **stratified random sampling** a given population is first divided into a number of strata. Random samples are then selected from each stratum, with the size of each sample chosen usually being proportional to the size of the stratum from which it is selected. This is a useful strategy to use when studying a disease, for example, since diseases tend to vary with respect to subdivisions such as age.

Other methods of sampling and practical details are beyond the scope of this chapter.

Sampling distributions

Suppose three random samples, A, B and C, are chosen from a population. Each of these samples yields sample statistics; for example, for sample A these include the sample mean, $\bar{x}_A$, the sample standard deviation, s_A, and so on. These sample statistics are themselves random variables; $\bar{x}_A$ is unlikely to be equal to $\bar{x}_B$, which in turn is unlikely to equal $\bar{x}_C$, and similarly for other statistics. The probability distributions of these sample statistics are known as **sampling distributions**.

So far as the sampling distribution of the sample mean, $\bar{x}$, is concerned, it can be shown that, as might be expected, $\bar{x}$ gives an unbiased estimated of the population mean, μ. That is, **the mean of the sampling distribution of $\bar{x}$ is the population mean μ.**

The standard deviation of the sampling distribution of $\bar{x}$ is known as the **standard error of the mean**. It can be shown that for a population of mean μ and standard deviation σ, the standard deviation of the sampling distribution of the sample mean ($\bar{x}$) is given by:

$$\textbf{standard error of the mean} = \frac{\sigma}{\sqrt{n}}$$

where n is the sample size. In the case of a finite population, of size N, from which samples are taken and not replaced, if n is relatively large (at least 50% of the value of N), then the standard error of the mean is given more accurately by multiplying the above expression by the square root of $((N - n)/(N - 1))$.

Other standard errors

The formulae for other standard errors are now stated.

For two independent samples, sizes n_1 and n_2, with respective sample means $\bar{x}_1$ and $\bar{x}_2$ and standard deviations s_1 and s_2(which should be similar), the standard error of the **difference between $\bar{x}_1$ and $\bar{x}_2$** is given by:

$$s\sqrt{\frac{1}{n_1} + \frac{1}{n_2}}$$

where the pooled standard deviation s is given by:

$$s = \sqrt{\frac{(n_1 - 1)\, s_1{}^2 + (n_2 - 1)s_2{}^2}{n_1 + n_2 - 2}}$$

Note that if n_1 and n_2 are equal, then the formula for the pooled standard deviation simplifies to $\sqrt{[(s_1{}^2 + s_2{}^2)/2]}$.

The standard error of a **proportion**, p, of subjects or items with a particular condition or feature in a random sample of size n is given by $\sqrt{(p(1 - p)/n)}$.

For two independent samples, sizes n_1 and n_2 with respective observed proportions p_1 and p_2, the standard error of the **difference between p_1 and p_2** is given by $\sqrt{[(p_1(1 - p_1)/n_1) + (p_2(1 - p_2)/n_2)]}$

Central limit theorem

When a sample of size n is taken from a population with a Normal distribution, then the sampling distribution of the sample mean, $\bar{x}$, is also Normal. According to the central limit theorem, even when the population is *not* Normal, as the sample size n increases, the sampling distribution of $\bar{x}$ tends towards a Normal distribution. So long as the population distribution does not have a peculiar shape it is usually assumed that values of n of at least 30 are large enough for it to be assumed that the sampling distribution of $\bar{x}$ is approximately Normal. The central limit theorem is of fundamental importance since it allows the properties of the Normal distribution to be applied to a large variety of statistical problems.

Estimation: Confidence intervals

From sample statistics the $100(1 - \alpha)\%$ confidence

intervals for the corresponding population parameters can be calculated as indicated in this subsection. Following a study, if a $100(1 - \alpha)\%$ confidence interval from a statistic or statistics is calculated, then this implies that if the study were repeated with other random samples drawn from the same population and $100(1 - \alpha)\%$ confidence intervals similarly individually calculated, the overall percentage of these confidence intervals which included the corresponding population parameter(s) would tend to $100(1 - \alpha)$. Less strictly, one can think of the $100(1 - \alpha)\%$ confidence interval from one study as giving a probability of α that the confidence interval does *not* include the estimated parameter(s).

The commonest confidence interval is the 95% one, for which α is 0.05 and, as shown earlier in this chapter, $z_{\alpha/2}$ is -1.96 and $z_{1-\alpha/2}$ is 1.96; if the sample size, n, is less than 30 then the t distribution should be used and the corresponding values of $t_{\alpha/2}$ and $t_{1-\alpha/2}$ vary with the number of degrees of freedom, d.f., and need to be looked up in Table IV in the Appendix. Tables II and IV provide the two-tailed 100α percentage points required for other confidence intervals; for example, for a 99% confidence interval α is 0.01.

Confidence interval for μ: large sample ($n \geqslant 30$)

When the sample size, n, is at least 30, the central limit theorem can be applied, assuming the population does not have a very unusual distribution, so that the sampling distribution of $\bar{x}$ can be taken as having an approximately Normal distribution. It has been seen above that in a standard Normal distribution curve the following area contains $100(1 - \alpha)\%$ of the total area: $z_{\alpha/2}$ to $z_{1-\alpha/2}$ (these are the two-tailed 100α percentage points). Now, for any Normal distribution we have:

$$z = \frac{x - \mu}{\sigma}$$

Substituting the above 100α two-tailed percentage points for z, $\bar{x}$ for x, and the standard error of the mean, $\sigma/\sqrt{n}$, for σ, yields the following $100(1 - \alpha)\%$ confidence interval inequality for μ:

$$\bar{x} + z_{\alpha/2} \cdot \frac{\sigma}{\sqrt{n}} < \mu < \bar{x} + z_{1-\alpha/2} \cdot \frac{\sigma}{\sqrt{n}}$$

For example, for the 95% confidence interval this becomes:

$$\bar{x} - 1.96 \cdot \frac{\sigma}{\sqrt{n}} < \mu < \bar{x} + 1.96 \cdot \frac{\sigma}{\sqrt{n}}$$

Similarly, from Table II in the Appendix it can be seen

that the 99% confidence interval for the mean is given by:

$$\bar{x} - 2.58 \cdot \frac{\sigma}{\sqrt{n}} < \mu < \bar{x} + 2.58 \cdot \frac{\sigma}{\sqrt{n}}$$

Example

A hospital laboratory is asked to carry out measurements of urinary lead levels for a research team looking at the association of lead intake and cognitive functioning. A series of 49 measurements has a a mean of 1.52 µmol (24 h)$^{-1}$ and a standard deviation of 1.3 µmol (24 h)$^{-1}$. What is the 95% confidence interval for the true mean?

Using conventional notation, and leaving out the units, we have: sample mean, $\bar{x}$ = 1.52; sample standard deviation, s = 1.3, and this will be used as an estimate of the population standard deviation, σ; and sample size, n = 49. Therefore, the 95% confidence interval for the true mean (that is, the population mean, μ) is given by:

$$\bar{x} - 1.96 \cdot \frac{s}{\sqrt{n}} < \mu < \bar{x} + 1.96 \cdot \frac{s}{\sqrt{n}}$$

$$\therefore \quad 1.52 - \frac{1.96 \times 1.3}{\sqrt{49}} < \mu < 1.52 + \frac{1.96 \times 1.3}{\sqrt{49}}$$

$$\therefore \quad 1.52 - \frac{1.96 \times 1.3}{7} < \mu < 1.52 + \frac{1.96 \times 1.3}{7}$$

$$\therefore \quad 1.52 - 0.364 < \mu < 1.52 + 0.364$$

$$\therefore \quad 1.16 < \mu < 1.88 \text{ (to two decimal places)}$$

Hence, the 95% confidence interval for the true mean is 1.16 to 1.88 µmol (24 h)$^{-1}$.

Confidence interval for µ: small sample (n < 30)

In this case, so long as it is valid to assume that the population is approximately Normal, the t distribution can be used. The population standard deviation is not usually known but can be estimated by using the sample standard deviation, and the required confidence interval is given by:

$$\bar{x} + t_{\alpha/2} \cdot \frac{s}{\sqrt{n}} < \mu < \bar{x} + t_{1-\alpha/2} \cdot \frac{s}{\sqrt{n}}$$

The number of degrees of freedom, d.f., is $(n-1)$

Example

Consider again the last example involving measurements of urinary lead levels. This time let us suppose

that a series of only 20 measurements has a mean of 1.54 µmol (24 h)$^{-1}$ and a standard deviation of 1.2 µmol (24 h)$^{-1}$. What is the 95% confidence interval for the true mean?

Using conventional notation, and leaving out the units, we have: $\bar{x}$ = 1.54; s = 1.2; and n = 20.

$\therefore$ degrees of freedom = $n - 1$ = 19

From Table IV in the Appendix it can be seen that when d.f. is 19 the value of t corresponding to a probability, α, of 0.05 (two-tailed) is 2.093. (Hence $t_{\alpha/2}$ is –2.093, and $t_{1-\alpha/2}$ is 2.093):

$$\bar{x} + t_{\alpha/2} \cdot \frac{s}{\sqrt{n}} < \mu < \bar{x} + t_{1-\alpha/2} \cdot \frac{s}{\sqrt{n}}$$

$$\therefore \quad 1.54 - 2.093 \cdot \frac{1.2}{\sqrt{20}} < \mu < 1.54 + 2.093 \cdot \frac{1.2}{\sqrt{20}}$$

$$\therefore \quad 1.54 - 0.562 < \mu < 1.54 + 0.562$$

$$\therefore \quad 0.978 < \mu < 2.102$$

Therefore, the 95% confidence interval for the true mean is 0.98 to 2.10 µmol (24 h)$^{-1}$ (to two decimal places).

Other confidence intervals

In general, the $100(1 - \alpha)$% confidence interval for a population parameter(s) estimated by a sample statistic(s) of the type described in this section (for example, the difference between two population means) is given by:

[(statistic + $t_{\alpha/2}$, (standard error)] to
[(statistic + $t_{1-\alpha/2}$, (standard error)]

or, when appropriate (for example, if the sample size is large enough):

[(statistic + $z_{\alpha/2}$, (standard error)] to
[(statistic + $z_{1-\alpha/2}$, (standard error)]

Since the standard Normal distribution and the t distribution are both symmetrical, the above two confidence intervals can also be written as:

[(statistic – $t_{1-\alpha/2}$ (standard error)]) to
[(statistic + $t_{1-\alpha/2}$ (standard error)]

and,

[(statistic – $z_{1-\alpha/2}$ (standard error)] to
[(statistic + $z_{1-\alpha/2}$ (standard error)])

Note that for proportions, since the sample size is not important, the standard Normal distribution should be used rather than the t distribution.

Example

The total time taken in each of two centres to complete a battery of psychological tests is compared. In centre 1, using a sample size of 12, the average time taken is 7.4 hours (standard deviation 0.8 hours), while at centre 2 with a sample size of 15 the average time taken is 8.3 hours (standard deviation 0.6 hours). Calculate the 95% confidence interval for the corresponding difference in the population means.

For centre 1 we have: sample size, $n_1 = 12$, sample mean, $\bar{x}_1 = 7.4$ hours, and sample standard deviation , $s_1 = 0.8$ hours. For centre 2: sample size, $n_2 = 15$, sample mean, $\bar{x}_2 = 8.3$ hours, and sample deviation, $s_2 = 0.6$ hours. The difference between the two sample means is 0.9 hours.

It has been mentioned above that for two independent samples, sizes n_1 and n_2 with respective sample means $\bar{x}_1$ and $\bar{x}_2$ and standard deviations s_1 and s_2 (which are similar in this example), the standard error of the difference between $\bar{x}_1$ and $\bar{x}_2$ is given by:

$$s \sqrt{\frac{1}{n_1} + \frac{1}{n_2}}$$

where the pooled standard deviation s is given by:

$$s = \sqrt{\frac{(n_1 - 1) s_1^2 + (n_2 - 1) s_2^2}{n_1 + n_2 - 2}}$$

$$= \sqrt{\frac{11(0.8)^2 + 14(0.6)^2}{25}}$$

$$= \sqrt{(12.08/25)}$$

$$= \sqrt{(0.4832)}$$

$$= 0.6951 \text{ hours}$$

Therefore, the standard error of the difference between the two sample means is given by:

$$0.6951 \times \sqrt{[(1/12) + (1/15)]}$$

$$= 0.6951 \times \sqrt{(0.15)}$$

$$= 0.2692 \text{ hours}$$

The number of degrees of freedom, d.f., is given by:

$$\text{d.f.} = n_1 + n_2 - 2$$

$$= 12 + 15 - 2$$

$$= 25$$

For the 95% confidence interval, α is 0.05, and the corresponding value of $t_{1 - \alpha/2}$ from Table IV is 2.060. Hence, the 95% confidence interval for the difference between the population means is given by:

[(statistic $- t_{1 - \alpha/2}$ (standard error)] to [(statistic $+ t_{1 - \alpha/2}$ (standard error)]

$$= 0.9 - (2.06)(0.2692) \text{ to } 0.9 + (2.06)(0.2692)$$

$$= 0.35 \text{ to } 1.45 \text{ hours (to two decimal places)}$$

Hypothesis testing

In the previous subsection one type of statistical inference, that of estimation, has been considered. In this subsection a different type of statistical inference, known as hypothesis testing (or, less precisely, carrying out a test of significance) is described. Hypothesis testing is used in decision making, as illustrated by the following example.

Decision making

Suppose a paper were published which stated that it had been found that the average concentration of a certain substance in the blood in a given population was 20 units, with a standard deviation of 5 units. On reading this paper a doctor working in a hosptial laboratory decides to check this claimed mean value by calculating the mean for a random sample of size 100. He decides, further, that he will accept the hypothesis that $\mu = 20$ units if his sample mean, $\bar{x}$, lies within the range [claimed mean $-$ 2(standard error of mean)] to [claimed mean $+$ 2(standard error of mean)]. Since the standard error of the mean is given by $\sigma/\sqrt{n}$, that is, $5/\sqrt{(100)}$ or $1/2$, it follows that the doctor will accept the claimed value if $\bar{x}$ lies within the range 19 to 21 units; if $\bar{x}$ is less than 19 units or greater than 21 units he will reject the claimed value.

Clearly the hypothesis is either true or false. Consider the following two instances:

(a) the hypothesis $\mu = 20$ units is true;
(b) the hypothesis $\mu = 20$ units is false, and in reality $\mu = 18.5$ units.

(a) Hypothesis true, $\mu = 20$ units. The Normal distribution can be used to calculate the probability that the hypothesis is accepted, and also the probability that the hypothesis is rejected, since:

Pr(hypothesis rejected) = 1 $-$ Pr(hypothesis accepted)

The application of the Normal distribution and the standard Normal distribution is shown in Figure 8.40. From Table I in the Appendix, it can be seen that for $z = 2$ the corresponding area or probability is 0.4772. Therefore, the unshaded area under the Normal distribution curve in Figure 8.40 has a value of 2×0.4772, that is, 0.9544.

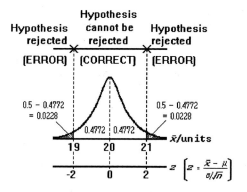

Fig. 8.40 Hypothesis true, $\mu = 20$ units (see text).

Therefore, the probability that the hypothesis is accepted is 95.4% (to one decimal place). Hence, the probability that the hypothesis is rejected is (1 − 0.9544), that is, 4.6% (to one decimal place), and since the hypothesis is in fact true, 4.6% also represents the probability of making an error.

(b) Hypothesis false, $\mu = 18.5$ units. This situation is illustrated in Figure 8.41. Since the area under the curve to the right of $\bar{x} = 21$ ($z = 5$) is negligible, the shaded area representing the probability that the hypothesis is accepted (which is an error since the hypothesis is false) is given by:

0.5 − (area between $z = 0$ and $z = 1$)
= 0.5 − 0.3413 (from Table I in the Appendix)
= 0.1587 or 15.9% (to one decimal place)

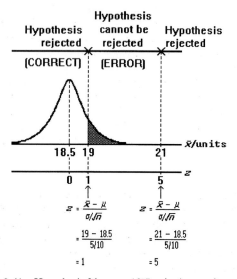

Fig. 8.41 Hypothesis false, $\mu = 18.5$ units (see text).

Table 8.18 Summary of the results for the examples in the text (a) hypothesis true, $\mu = 20$ units; (b) hypothesis false, $\mu = 18.5$ units

	Hypothesis accepted	Hypothesis rejected	Total probability
Hypothesis is true ($\mu = 20$ units)	95.4%	4.6% (error)	100%
Hypothesis is false ($\mu = 18.5$ units)	15.9% (error)	84.1%	100%

Hence, the probability that the hypothesis is (correctly) rejected is (1 − 0.1587) or 84.1% (to one decimal place).

The above probability results are summarized in Table 8.18.

Types of error

In general, for a hypothesis, H, Table 8.19, showing type I and type II errors, can be drawn up.

A **type I error** is the error of wrongly rejecting the hypothesis when it is true. The probability of making a type I error is conventionally given the symbol α. Thus for the previous example α is 0.046.

A **type II error** is the error of wrongly accepting the hypothesis when it is false. The probability of making a type II error is conventionally given the symbol β. Thus for the previous example β is 0.159.

Table 8.19 Types of error

	H accepted	H rejected	Total probability
H true	Decision correct	Type I error (probability = α)	1 (= 100%)
H false	Type II error (probability = β)	Decision correct	1 (= 100%)

Types of hypothesis

In our example it was relatively easy to calculate the value of α, the probability of making a type I error, because we were looking at the hypothesis that the value of μ was 20 units; that is, μ had just one value. This is an example of a **simple hypothesis**.

In our example, in order to calculate the value of β, the probability of making a type II error, in theory we should have looked at the following hypotheses: $\mu \neq$ 20 units; $\mu < 20$ units; and $\mu > 20$ units. These are known as **composite hypotheses**. Clearly in this case μ could have taken on a very large number of possible values, and we got around this only by choosing the alternative value of $\mu = 18.5$ units in an essentially arbitrary way.

It can be seen from the above that is far easier to deal with simple hypotheses than composite ones. Therefore, in practice, when we are working with statistical calculations involving a hypothesis, we usually try to state the hypothesis in the form of a simple hypothesis rather than a composite one. This is done even when the hypothesis may then make a statement diametrically opposed to that which we may be trying to prove.

For example, suppose we wished to discover whether a new treatment, X, for a given disease, were more effective than an existing treatment, Y. Rather than test the following composite hypothesis : X is more effective than Y, it would be easier to test the following simple hypothesis: there is *no difference* in the effectiveness of X and Y. Such a simple hypothesis is known as a **null hypothesis** because it is a hypothesis that postulates *no difference*. It can be seen that it may be formulated simply in order to be rejected. Conventionally a null hypothesis is denoted by H_0.

When a null hypothesis is formulated, an **alternative hypothesis** is also formulated which disagrees with the null hypothesis and is to be accepted if the null hypothesis is rejected. The alternative hypothesis is conventionally denoted by H_1.

Power

The power of a test is the probability that the null hypothesis is rejected when it is indeed false. From Tables 8.18 and 8.19 it is evident that

power = 1 – Pr (type II error)

$\therefore$ **power = 1 – β**

Significance level

The level of significance is defined as the probability of making a type I error.

$\therefore$ **level of significance = α**

Summary

Table 8.20 summarizes the types of error and their associated probabilities when a null hypothesis H_0, is tested.

A summary of a common method used to carry out

Table 8.20 Summarizing table

	Ho accepted	Ho rejected	Total probability
Ho true	Decision correct probability = $1-\alpha$ = 1–significance level	Type I error probability = α = significance level	1
Ho false	Type II error probability = β = 1–power	Decision correct probability = $1-\beta$ = power	1

hypothesis testing follows. H_0 and H_1 are formulated and the significance level, α, is specified. From the sampling distribution of the test statistic the test criterion for testing H_0 versus H_1 is created. (The sampling distribution of the test statistic may, for example, be N, t, χ^2 or F; the test statistic is the statistic based on sample values such as $\bar{x}$ on which the statistical test is being based.) The value of the test statistic is then calculated from the sample data. The difference, d, between the value of the test statistic from the sample and the value expected from the null hypothesis is next calculated, and then the probability of d occurring by chance, P, is calculated.

If P is less than the significance level, α, (that is, d is considered too great to be attributed to chance) the result is **statistically significant** at the level of α, and H_0 is rejected.

However, if $P \geqslant \alpha$, (that is, it is considered that d may be attributed to chance) then the result is not statistically significant at the level of α, and either H_0 is then accepted, or judgement is reserved. Reserving judgement is a way of avoiding a type II error.

The significance level, α, is usually specified at 0.05 (5%) or 0.01 (1%). Note that as α is made smaller, Pr(type I error) decreases and Pr(type II error) increases.

Although the example at the beginning of this subsection used a two-tailed significance test, there are occasions, for example with the use of the χ^2 distribution, when a one-tailed significance test is more appropriate; the χ^2 test is considered in the next section.

Example

Returning to the example of the comparison of the total time taken in each of two centres to complete a battery of psychological tests, test the null hypothesis, at the 5% level of significance, that the difference between the population means is zero.

It has already been shown that the 95% confidence interval for the difference between the population means does not include zero, but this result will be ignored for the purposes of this calculation. The t test used in this example is known as the **two sample t test**; non-parametric alternative tests are discussed later in this chapter.

Assuming that the populations have a Normal distribution and have the same standard deviation, we can use the t distribution to determine whether there is a difference in the corresponding population means, μ_1 and μ_2, between these two samples at the required 5% significance level:

$H_0: \mu_1 = \mu_2$
$H_1: \mu_1 \neq \mu_2$ (two-tailed)
$\alpha = 0.05$

As before, the number of degrees of freedom, d. f. = $n_1 + n_2 - 2 = 25$.
From Table IV it can be seen that when d. f. = 25 the two-tailed critical values for $\alpha = 0.05$ are 2.060 and −2.060.

The test statistic is given by:

$$t = \frac{\bar{x}_1 - \bar{x}_2}{\sqrt{\left(\frac{1}{n_1} + \frac{1}{n_2}\right) \cdot \frac{s_1^2 (n_1 - 1) + s_2^2 (n_2 - 1)}{n_1 + n_2 - 2}}}$$

$$\therefore t = \frac{7.4 - 8.3}{\sqrt{\left(\frac{1}{12} + \frac{1}{15}\right) \cdot \frac{(0.8)^2 (12 - 1) + (0.6)^2 (15 - 1)}{12 + 15 - 2}}}$$

$$= \frac{-0.9}{\sqrt{(0.0725)}} = -3.34$$

The null hypothesis, H_0, states that the means of the two populations from which the two samples come are equal. Since −3.34 is less than the critical value −2.06, we must reject the null hypothesis. That is, at the 5% level of significance the difference between the two sample means is *unlikely* to be caused by chance.

Advantages of estimation over hypothesis testing

Over the last decades the use of hypothesis testing has come to dominate the way in which statistical inference is used in medical and psychiatric research. However, it has been increasingly argued recently that testing the null hypothesis is often inappropriate for this sort of biological research, and that instead there should be a move towards the use of confidence intervals. Indeed, this move away from just calculating the values of P (which is then compared with α to test the null hypothesis) and instead, or additionally, giving confidence intervals, is now encouraged by a number of medical journals, including the Annals of Internal Medicine (Braitman 1988), the British Medical Journal (Gardner & Altman 1986), the Lancet (Lancet 1987) and the New England Journal of Medicine (Rothman 1978).

Disadvantages of using only hypothesis testing in reporting the results of psychiatric research have been summarized by Gardner and Altman (1990): ' ... tests of the null hypothesis begin by turning an investigator's idea (for example, that some new treatment will perform better than that currently being prescribed) upside down and substitute the notion of no effect or no difference. The test then evaluates the probability of the observed study result, or a more extreme result, occurring if this null hypothesis were in fact true... Also, proper understanding of a study result is obscured by transforming it ...onto a remote scale constrained from zero to unity. Nonetheless, obtaining a low P value, particularly < 0.05, is widely interpreted as implying merit, in some abstract sense, and leads to the findings being generally accepted as important and deemed publishable. In contrast, this status is often denied study results which have not achieved this magic, but arbitrary, characteristic... The P value, on its own, tells us nothing about the magnitude, or even the direction, of any difference between treatments. We do not suggest that only confidence intervals be given, nor that confidence intervals should be given for every comparison. We ... suggest ... that confidence intervals accompany the main results of a study.'

THE χ^2 TEST

When comparing the value of discrete variables belonging to one sample with those of another sample, the χ^2 test may be used. The populations from which these samples are derived do not have to have a Normal distribution or any other form of underlying distribution of parameter in order to use the χ^2 test. For this reason the χ^2 test is known as a **non-parametric test**.

The χ^2 test may be used in the analysis of contingency tables, as described below, and in trials that have more than two possible outcomes (multinomial rather than binomial trials).

The actual numbers have to be used when applying the χ^2 test described in this chapter. Proportions or percentages, for example, cannot be used.

More than one degree of freedom

Contingency table

The core of a hypothetical 4 × 2 (pronounced '4 by 2'

that is, 4 rows and 2 columns) contingency table is shown in Figure 8.42. The samples must be independent in order to use the χ^2 test. It has been seen earlier in this chapter that the shape of the χ^2 distribution varies with the number of degrees of freedom, d. f.; for a contingency table this is given by:

degrees of freedom = (number of rows – 1) × (number of columns – 1)

So for a 4 × 2 contingency table, d. f. is $(4 - 1) \times (2 - 1)$, that is, 3.

The contingency table shown in Figure 8.42 can be expanded by adding the totals for the rows and columns, as shown in Figure 8.43. Note that the addition of one extra column (of totals) and of one extra row (of totals) does *not* affect the number of degrees of freedom; this remains a 4 × 2 contingency table. The totals are not included when calculating the degrees of freedom, which can be thought of as being the minimum number of independent cells that needs to be filled in order to allow us to calculate the remaining cell values by looking at the totals. So, in this example, if we had the values of any three independent cells, we

	Sample$_1$	Sample$_2$	Total	
----	√	?	√	
----	√	?	√	degrees of
----	?	√	√	freedom
----	?	?	√	= 3
Total	√	√	√	

From the values ticked, including the values of any three independent cells, the values of the remaining cells can be calculated by subtraction from the marginal totals. Three is the minimum number of cells for which this is true, i.e. the number of degrees of freedom is three.

Fig. 8.44 4 × 2 contingency table showing the minimum number of independent cell values needed to calculate the remaining cell values from the marginal totals.

could work out the values of the remaining five from the marginal totals of the table, as shown in Figure 8.44

Null hypothesis

In order to work out whether the difference between the distribution of variables of two samples (sample$_1$ and sample$_2$) in a contingency table is statistically significant, we begin with the simple null hypothesis that there is no difference and test this hypothesis.

Since the totals for the columns for sample$_1$ and sample$_2$ are usually different from each other, the null hypothesis does not necessarily mean that the values of the cells of sample$_1$ are equal to the values of the cells of sample$_2$. Rather, the null hypothesis implies that the value of each cell of the sample$_1$ column is, as a *proportion* of the total for that column, equal to the corresponding proportionate values for the sample$_2$ column. Consider, for example, Table 8.21. According to the null hypothesis:

$$\frac{a}{a+b+c+d} = \frac{w}{w+x+y+z}$$

and $$\frac{b}{a+b+c+d} = \frac{x}{w+x+y+z}$$

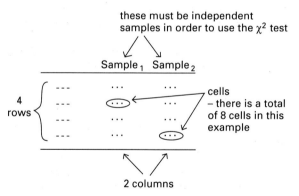

Fig. 8.42 The structure of a 4 × 2 contingency table.

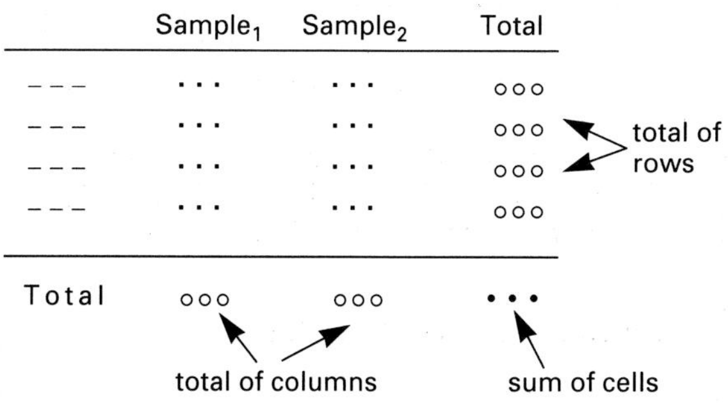

Fig. 8.43 4 × 2 contingency table with totals.

Table 8.21 Observed values

	Sample$_1$	Sample$_2$	Total
——	a	w	$a + w$
——	b	x	$b + x$
——	c	y	$c + y$
——	d	z	$d + z$
Total	$a + b + c + d$	$w + x + y + z$	$a + b + c + d + w$ $+ x + y + z$ (sum of cells)

and $\dfrac{c}{a+b+c+d} = \dfrac{y}{w+x+y+z}$

and $\dfrac{d}{a+b+c+d} = \dfrac{z}{w+x+y+z}$

Expected values

In order to test the null hypothesis a second table is drawn up the cells of which have the values they would have if the null hypothesis were indeed true. These are known as the expected values and they are given the symbols a', b', and so on, as shown in Table 8.22; note that as would be expected, the totals for the table of expected values remain the same as in the original table of observed values.

The expected values are calculated as follows :-

$a' =$ the proportion of $(a + b + c + d)$ that would be expected from the total of the first row

$= \dfrac{a+w}{a+b+c+d+w+x+y+z} \times (a+b+c+d)$

$= \dfrac{\text{(row total)} \times \text{(column total)}}{\text{(sum of cells)}}$

Similarly,

$b' = \dfrac{b+w}{\text{(sum of cells)}} \times \text{(total of first column)}$

$= \dfrac{\text{(row total)} \times \text{(column total)}}{\text{(sum of cells)}}$

and so on for the first column.

For the second column:

$w' = \dfrac{a+w}{\text{(sum of cells)}} \times \text{(total of second column)}$

and so on.

So in general:

$$\textbf{expected value} \atop \textbf{of a cell} = \dfrac{\textbf{(row total)} \times \textbf{(column total)}}{\textbf{(sum of cells)}}$$

Table 8.22   Values expected under the null hypothesis

	Expected value		
	Sample$_1$	Sample$_2$	Total
—	a'	w'	$a + w$ (row total)
—	b'	x'	$b + x$ (row total)
—	c'	y'	$c + y$ (row total)
—	d'	z'	$d + z$ (row total)
Total	$a+b+c+d$	$w+x+y+z$	$a+b+c+d+$ $w+x+y+z$
	(column total)	(column total)	(sum of cells)

Calculation of χ^2

Continuing with the above example, the next stage is to compare the observed values (Table 8.21) with the values expected under the null hypothesis (Table 8.22). This is carried out by calculating the following sum:

$$\chi^2 = \Sigma \, \dfrac{(\text{observed value} - \text{expected value})^2}{\text{expected value}}$$

This can be abbreviated to:

$$\chi^2 = \Sigma \, \dfrac{(O - E)^2}{E}$$

where O = observed value and E = expected value. This is shown in Figure 8.45.

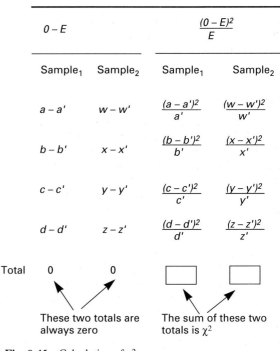

Fig. 8.45   Calculation of χ^2.

Critical value

The critical value for χ^2 with a given number of degrees of freedom is obtained from Table V in the Appendix; this is $\chi^2{}_\alpha$. If χ^2 is greater than $\chi^2{}_\alpha$ (and therefore P is less than α as indicated in Figure 8.46) then the null hypothesis is rejected. If $\chi^2{}_\alpha$ is less than $\chi^2{}_\alpha$ (and P is greater than α) the null hypothesis cannot be rejected. This is illustrated in Figure 8.46.

For example, for $\alpha = 0.05$ and with six degrees of

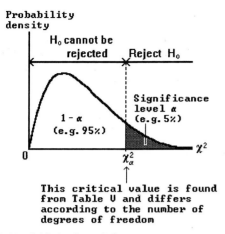

Fig. 8.46 Critical values of χ^2.

freedom the critical value of χ^2 is 12.592. Thus if we were to obtain a value for χ^2 of 14.57 (with six degrees of freedom) we should have to reject the null hypothesis.

Small expected values

The χ^2 test is not valid if the expected values are too small. Criteria based on those put forward by Cochran (1954) are commonly applied: when d.f. is greater than one (2×2 contingency tables are considered later in this section), in order for the χ^2 test to be valid, all the expected values should be greater than or equal to one, and at least 80% of the expected cell values must be greater than five.

If these criteria are not met, the overall sample size is usually not large enough to allow the χ^2 probability distribution to be a suitable approximation to the sampling distribution of χ^2. It may be possible to satisfy the criteria by combining one or more classes, or by omitting one or more classes; however, this leads to a loss of information being analyzed. It may be possible to fulfil the criteria by increasing the sample size; although this is a better alternative, it is clearly not always a practical one. Another possibility is the use of a computer program based on a test similar to Fisher's exact probability test described for 2×2 tables in the next section (Agresti & Wackerly 1977).

Example

In a retrospective study of suicide during pregnancy the data in Table 8.23 were obtained (Appleby 1991). This table shows the observed numbers of postnatal suicide during pregnancy in England and Wales (1973

Table 8.23 Observed and expected numbers of postnatal suicides in England and Wales, 1973 to 1984 (Appleby 1991)

Age group (years)	Observed number of postnatal suicides	Expected number of postnatal suicides	Total
15 to 19	7	19.6	26.6
20 to 24	24	127.6	151.6
25 to 29	25	165.9	190.9
30 to 34	13	92.3	105.3
35 to 39	6	35.8	41.8
40 to 44	1	8.4	9.4
Total	76	449.6	525.6

Table 8.24 Expected cell values under the null hypothesis

Age group (years)	Expected cell values Observed number of postnatal suicides	Expected number of postnatal suicides	Total
15 to 19	3.846	22.754	26.6
20 to 24	21.921	129.679	151.6
25 to 29	27.604	163.297	190.901
30 to 34	15.226	90.074	105.3
35 to 39	6.044	35.756	41.8
40 to 44	1.359	8.041	9.4
Total	76.0	449.601	525.6

to 1984) and the expected numbers of such suicides. Is the difference between these two sets of numbers statistically significant?

The expected values if the null hypothesis, that there is no difference between the two sets of numbers, were true, are shown in Table 8.24, having been calculated using the formula:

$$\text{expected value of a cell} = (\text{row total}) \times (\text{column total}) / 525.6$$

For example, the top left-hand expected cell has a value of (26.6×76) divided by the sum of the cells, 525.6; this comes to 3.846, working to three decimal places. As a check, it should be noted that the totals in this table should be the same as the corresponding totals in Table 8.23 (although sometimes slight discrepancies may arise as a result of rounding errors). It can be seen that the following criteria are fulfilled:

all expected values $\geqslant 1$
at least 80% of the expected values > 5

Therefore, in this case the χ^2 test for a 6×2 table can be used.

Next, Table 8.25, showing the values of $(O - E)$, can be constructed from Tables 8.23 and 8.24. Two further useful checks are provided at this stage; the sum for each sample (in this case, for each column in Table 8.25) should be zero, and the value of a given cell for the first sample is the negative of the value of the corresponding adjacent cell of the other sample in the second column (or the second row if, as in the original paper by Appleby (1991) the tables were laid out in the form 2×6).

Finally, Table 8.26, showing $(O - E)^2/E$, can be constructed from Tables 8.25 and 8.24; rather than drawing up four separate tables, in practice some of Tables 8.23 to 8.25 may conveniently be combined. From Table 8.26 it is clear that χ^2 has a value of 3.450 + 0.582, that is, 4.032.

The number of degrees of freedom
= (no. of rows – 1) × (no. of columns – 1)
= (6 – 1) × (2 – 1)
= 5

Note that in calculating the degrees of freedom the row and column of totals are not included.

If a level of significance, α, is chosen as 0.05, then it can be seen from Table V in the Appendix that, for d.f. = 5, χ^2_α = 11.070. Since the value of χ^2 obtained,

4.032, is less than $\chi^2_{0.05}$, it follows that the null hypothesis cannot be rejected; at the 5% level of significance there is no statistically significant difference between the two samples of the original table.

2×2 contingency tables

Simplified formula

In the special case in which a contingency table has just two rows and two columns, with cell values denoted as in Table 8.27, then the following simplified formula can be used instead of the previous calculations to work out the value of χ^2:

$$\chi^2 = \frac{(az - by)^2 (a + b + y + z)}{(a + b) (y + z) (a + y) (b + z)}$$

Degrees of freedom

For a 2×2 table (also known as a four-fold table) there is always one degree of freedom, since:

d. f. = (no. of rows – 1) × (no. of columns – 1)
= (2 – 1) × (2 – 1) = 1

Table 8.25 Cell values of $(O–E)$, constructed from Tables 8.23 and 8.24

Age group (years)	$O - E$ Observed number of postnatal suicides	Expected number of postnatal suicides	Total
15 to 19	3.154	–3.154	0
20 to 24	2.079	–2.079	0
25 to 29	–2.604	2.603	0.001
30 to 34	–2.226	2.226	0
35 to 39	–0.044	0.044	0
40 to 44	–0.359	0.359	0
Total	0.000	0.001	0.001

Table 8.26 $(O–E)^2/E$, constructed from Table 8.25 and 8.24

Age group (years)	$(O–E)^2/E$ Observed number of postnatal suicides	Expected number of postnatal suicides
15 to 19	2.587	0.437
20 to 24	0.197	0.033
25 to 29	0.246	0.041
30 to 34	0.325	0.055
35 to 39	0.000	0.000
40 to 44	0.095	0.016
Total	3.450	0.582

Table 8.27 2×2 Contingency table

			Total
	a	y	$a + y$
	b	z	$b + z$
Total	$a + b$	$y + z$	$a + b + y + z$

Small values

For a 2×2 table all the expected values should have a value of at least five (and therefore the overall total should be at least 20) in order for the use of the χ^2 test with the above formula to be valid. If this criterion is not fulfilled, then Fishers' exact probability test, described in the next section, may be used. Even when the criterion is fulfilled, if the overall total is less than 100 then in order to provide a better fit with the χ^2 probability distribution for small samples **Yate's continuity correction** for 2×2 tables, described next, should be used. Indeed, some statisticians argue that Yate's correction should be used for all 2×2 tables, but this is an opinion open to debate; nevertheless, it is probably wise to use Yate's correction if there is an observed value of less than ten.

Yate's continuity correction

Using the notation of Table 8.27:

χ^2 **with Yates' correction**

$$= \frac{[\,|az - by| - \frac{1}{2}\,(a + b + y + z)]^2\,(a + b + y + z)}{(a + b)\,(y + z)\,(a + y)\,(b + z)}$$

$|az - by|$ is the *modulus* of $(az - by)$, and means that the positive numerical value of $(az - by)$ is used. For example, $|2| = 2$, $|-3| = 3$, and so on.

It can be seen from the above formula that Yates' correction has the effect of *decreasing* the value of χ^2. Therefore, if a value of χ^2 (without the correction) is obtained which is less than the critical value of χ^2, we would *not* reject the null hypothesis, and there would be no need to calculate the value of χ^2 with Yates' correction since this would simply diminish the value of χ^2 further.

Example

Table 8.28 shows the results after four weeks of treatment of a trial comparing the efficacy of phenelzine and a new type of antidepressant, X, in depressed patients. Use the χ^2 test to determine whether the difference between the two samples is statistically significant at the 1% level, and calculate the 95% confidence interval for the percentage difference in improvement between the two treatments.

Since the value of one of the cells is less then ten, and the overall total (99) is less than 100, Yates' correction is applied:

χ^2 **with Yates' correction**

$$= \frac{[\,|az - by| - \frac{1}{2}(a + b + y + z)]^2\,(a + b + y + z)}{(a + b)\,(y + z)\,(a + y)\,(b + z)}$$

$$= \frac{[\,|(10)(4) - (25)(60)| - \frac{1}{2}(99)]^2\,(99)}{(35)(64)(70)(29)}$$

$$= \frac{[\,|-1460| - 49.5]^2\,(99)}{(35)(64)(70)(29)}$$

Table 8.28 Hypothetical results after four weeks of a trial comparing the efficacy, in depressed patients, of phenelzine with a new antidepressant, X

Response after 4 weeks	Treatment		
	Phenelzine	X	Total
Improvement	10	60	70
No improvement	25	4	29
Total	35	64	99

$$= \frac{[1460 - 49.5]^2\,(99)}{(35)(64)(70)(29)}$$

$$= 43.31$$

For a 1% level of significance (that is, $\alpha = 0.01$), with one degree of freedom, the critical value χ^2_α is found from Table V (in the Appendix) to be 6.635. So our value of χ^2 is greater than this critical value, implying that the null hypothesis should be rejected. In fact, from Table V we can see that our value for χ^2 is also greater than the critical value corresponding to $\alpha = 0.001$, that is, $P < 0.001$. Hence the difference between the two samples can be said to be highly statistically significant.

The proportion of the first sample (treated with phenelzine) showing improvement, p_1, is 10/35 or 0.286, while the proportion of the second sample (treated with X) showing improvement, p_2, is 60/64 or 0.938. Hence the increased percentage improvement with treatment X compared with phenelzine is:

$$93.8 - 28.6 = 65.2$$

that is, 65% (to two significant figures)

Earlier in this chapter it is stated that for two independent samples, sizes n_1 and n_2, with respective observed proportions p_1 and p_2, the standard error of the **difference between p_1 and p_2** is given by $\sqrt{[(p_1(1 - p_1)/n_1) + (p_2(1 - p_2)/n_2)]}$. Therefore the standard error of this difference is:

$$\sqrt{[(p_1(1 - p_1)/n_1) + (p_2(1 - p_2)/n_2)]}$$
$$= \sqrt{[(0.286 \times 0.714/35) + (0.938 \times 0.062/64)]}$$
$$= 0.082$$

As mentioned earlier in this chapter, because we are dealing with proportions, the required 95% (that is, $\alpha = 0.05$) confidence interval is given by:

[statistic $- z_{1 - \alpha/2}$ (standard error)] to
[statistic $+ z_{1 - \alpha/2}$ (standard error)]
$$= [(0.652 - (1.96)(0.082)] \text{ to } [0.652 + (1.96)(0.082)]$$
$$= [0.652 - 0.161] \text{ to } [0.652 + 0.161]$$
$$= 0.491 \text{ to } 0.813$$

So, to two significant figures, the required confidence interval for the 65% difference in improvement is 49% to 81%. This confidence interval could be narrowed further by increasing the sample sizes.

FISHER'S EXACT PROBABILITY TEST

Fisher's exact probability test has already been mentioned in the previous section as a test that can be used for 2 × 2 tables in which the numbers are too small for the χ^2 test to be valid. The exact probability test allows the determination of exact probability values. The formula used involves factorials, which are mentioned earlier in this chapter. In particular, it will be recalled that 0! is unity.

Formula

Using the notation of Table 8.27:

Exact probability of table

$$= \frac{(a+y)! \, (b+z)! \, (a+b)! \, (y+z)!}{(a+b+y+z)! \, a! \, b! \, y! \, z!}$$

To test the null hypothesis that there is no difference between two samples, calculating the probability of a given table of observations is not sufficient. We also have to calculate the probabilities of more extreme tables occurring by chance. This can be carried out in two ways, as shown in the worked example which follows.

Example

In a study of amnesia for criminal offences in a group of 35 offenders the relationship is studied between depressive features, independent of diagnosis, and repeated amnesia for the offence. The data shown in Table 8.29 are obtained. Is the difference between the two groups of offenders statistically significant at the 5% level?

Exact probability of table

$$= \frac{(a+y)! \, (b+z)! \, (a+b)! \, (y+z)!}{(a+b+y+z)! \, a! \, b! \, y! \, z!}$$

$$= \frac{4! \, 31! \, 17! \, 18!}{35! \, 3! \, 14! \, 1! \, 17!}$$

$$= \frac{4! \, 31! \, 18!}{35! \, 3! \, 14!}$$

$$= \frac{4 \times 31! \times 18 \times 17 \times 16 \times 15}{35!}$$

$$= \frac{4 \times 18 \times 17 \times 16 \times 15}{35 \times 34 \times 33 \times 32}$$

$$= 0.2338$$

Now, the smallest row or column total in the Table 8.29 is four. This can be made up in the following ways:

(1) 4 + 0 = 4
(2) 3 + 1 = 4
(3) 2 + 2 = 4
(4) 1 + 3 = 4
(5) 0 + 4 = 4

Table 8.29 Hypothetical results of a study of the relationship between depressive features, independent of diagnosis, and repeated amnesia for the offence in a group of 35 offenders

Memory	Depressed mood	Mood not depressed	Total
Amnesic	3	1	4
Non-amnesic	14	17	31
Total	17	18	35

Table 8.30   Tables (1) to (5)

(1)	4	0	4	(2)	3	1	4	(3)	2	2	4
	13	18	31		14	17	31		15	16	31
	17	18	35		17	18	35		17	18	35

(4)	1	3	4	(5)	0	4	4
	16	15	31		17	14	31
	17	18	35		17	18	35

It follows that there are only five possible ways in which the observed values in the table can be rearranged so as to still have the same row and column totals. These are shown as (1) to (5) in Table 8.30, with table (2) being the one actually observed.

The exact probability for each table can be calculated:

Table (1)
Exact probability

$$= \frac{4! \, 31! \, 17! \, 18!}{35! \, 4! \, 13! \, 0! \, 18!} = \frac{31! \, 17!}{35! \, 13!}$$

$$= \frac{17 \times 16 \times 15 \times 14}{35 \times 34 \times 33 \times 32} = 0.0455$$

Table (2)
This is the observed table, and its probability has already been calculated as being 0.2338

Table (3)
Exact probability

$$= \frac{4! \, 31! \, 17! \, 18!}{35! \, 2! \, 15! \, 2! \, 16!}$$

$$= 0.3974$$

Table (4)
Exact probability

$$= \frac{4! \, 31! \, 17! \, 18!}{35! \, 1! \, 16! \, 3! \, 15!}$$

$$= 0.2649$$

Table (5)
Exact probability

$$= \frac{4! \ 31! \ 17! \ 18!}{35! \ 0! \ 17! \ 4! \ 14!}$$

$= 0.0584$

As a check, it would be expected that the sum of the probabilities for the five possible tables should be one, and this is indeed the case.

Method 1
Assuming that the more extreme tables are less probable than the observed table (2), it can be seen that the probabilities attached to tables (1) and (5) are less than that of (2). Adding these probabilities gives the total probability for the level of significance:

Pr(table (2)) + Pr(table (1)) + Pr(table (5))
$= 0.2338 + 0.0455 + 0.0584 = 0.3377$

Since $0.3377 > 0.05$, it follows that the null hypothesis cannot be rejected at the 5% level of significance. That is, the difference between the two groups of offenders is not statistically significant at the five per cent level. In this case, since the probability attached to the observed table, 0.2338, was itself greater than 0.05, in practice we need not have carried out the additional calculations.

Method 2
An alternative method of calculating the significance level is to add to the probability of the observed table only probabilities of extreme tables which differ in the same direction as the original observed table ('one-tailed'). Then to make the result 'two-tailed', it can be doubled.

In this example only table (1) is more extreme in the same direction as the observed table (2). So, the significance level is:

$2 \times$ [Pr(table (2)) + Pr(table (1))]
$2 \times$ [0.2338 + 0.0455]
0.5586

Again this is greater than 0.05 and so the null hypothesis cannot be rejected at the 5% level of significance.

Comparison of the two methods

Of the two methods just demonstrated, the second is clearly easier to carry out, and is the one recommended by some statisticians (for example, Armitage 1971). The reader will have noticed that the two methods give different results. In practice this does not usually affect whether or not the null hypothesis is rejected.

REGRESSION AND CORRELATION

Correlation

Line of best fit

Scatter diagrams were mentioned earlier in this chapter. They represent a graphical method of showing an association, if any, between two variables for a population, conventionally symbolized by x for the independent variable and y for the dependent variable (see below). Figure 8.47(A) shows a typical line of best fit, known as the **linear regression line** (see next subsection); Figure 8.47(B) shows a scatter diagram in which the line of best fit is not a straight line but is curved.

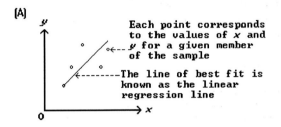

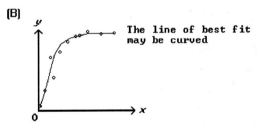

Fig. 8.47 Scatter diagrams and lines of best fit. (**A**) Linear regression line, and (**B**) a curved line of best fit.

Correlation coefficient

In order to determine the degree of correlation, a correlation coefficient is defined, and in the case of **Pearson's product moment correlation coefficient,** is conventionally given the symbol r. Pearson's product moment correlation coefficient is usually referred to simply as the correlation coefficient, and is the ratio of the sum of the products of the differences of the two variables from the mean to the square root of the product of the sum of the squares of these differences; this is easier to understand from the formula for r which is given below. The numerical value of the correlation coefficient corresponds to the degree of correlation between the two variables (x and y, say)

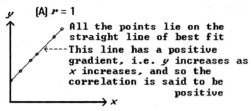

[A] $r = 1$

All the points lie on the straight line of best fit

This line has a positive gradient, i.e. y increases as x increases, and so the correlation is said to be positive

$r = 1$ implies perfect positive correlation

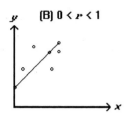

[B] $0 < r < 1$

$0 < r < 1$ implies fair (but not perfect) positive correlation

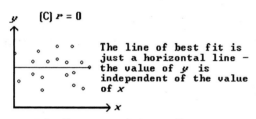

[C] $r = 0$

The line of best fit is just a horizontal line — the value of y is independent of the value of x

$r = 0$ implies no correlation at all

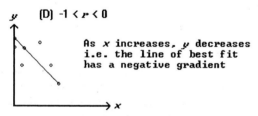

[D] $-1 < r < 0$

As x increases, y decreases i.e. the line of best fit has a negative gradient

$-1 < r < 0$ implies fair (but not perfect) negative correlation

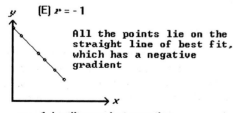

[E] $r = -1$

All the points lie on the straight line of best fit, which has a negative gradient

$r = -1$ implies perfect negative correlation

Fig. 8.48 Typical scatter diagrams and linear regression line gradients for all the possible values of r. (**A**) $r = 1$; (**B**) $0 < r < 1$; (**C**) $r = 0$; (**D**) $-1 < r < 0$; and (**E**) $r = -1$.

and varies between −1 and 1 (inclusive):

$$-1 \leqslant \mathbf{r} \leqslant 1$$

Figure 8.48 illustrates the typical scatter diagrams and linear regression line gradients for all the possible values of r: (a) $r = 1$; (b) $0 < r < 1$; (c) $r = 0$; (d) $-1 < r < 0$; and (e) $r = -1$. It can be seen that the degree of correlation varies from perfect positive correlation when r is 1, to perfect negative correlation when r is −1.

Formula

The formula for calculating the value of Pearson's product moment correlation coefficient, r, is given by;

$$r = \frac{\sum (x - \bar{x})(y - \bar{y})}{\sqrt{\sum (x - \bar{x})^2 \sum (y - \bar{y})^2}}$$

In making this calculation manually, it is useful to note that:

$$\sum (x - \bar{x})^2 \quad = \quad \sum x^2 - \frac{(\sum x)^2}{n}$$

$$\sum (y - \bar{y})^2 \quad = \quad \sum y^2 - \frac{(\sum y)^2}{n}$$

$$\sum (x - \bar{x})(y - \bar{y}) = \quad \sum xy - \frac{(\sum x)(\sum y)}{n}$$

The value of r can easily be calculated by entering the values of x and y into many types of calculator with statistical functions. However, when using a calculator or computer, it is still a useful check to plot the values on a scatter diagram to see if the value of r looks right. The following example demonstrates the manual calculation of r.

Example

Table 8.31 shows the elimination half-life of the drug zopiclone and the corresponding serum albumin level in nine cirrhotic patients (after Parker & Roberts 1983). Calculate the correlation coefficient to two significant figures.

As a check, these data are plotted as a scatter diagram, shown in Figure 8.49, from which it can be seen that $-1 < r < 0$.

Let x be the serum albumin level, and y be the elimination half-life. In Table 8.32 the values of x^2, y^2, and xy are calculated. Ignoring units, we have:

$\sum x = 312.1$
$\sum x^2 = 11\ 243.77$
$\sum y = 77.7$
$\sum y^2 = 718.01$
$\sum xy = 2572.54$
sample size, $n = 9$

Table 8.31 The elimination half-life of zopiclone and the corresponding serum albumin level in nine cirrhotic patients (after Parker & Roberts 1983)

Patient	Serum albumin level (gl^{-1})	Elimination half-life of zopiclone (h)
1	29.2	10.0
2	25.1	11.6
3	27.5	9.2
4	27.5	12.6
5	40.0	6.2
6	42.4	5.3
7	41.7	7.8
8	36.6	7.2
9	42.1	7.8

Table 8.32 Calculation based on the data in Table 8.31; x is the serum albumin level and y is the elimination half-life

Patient	x	x^2	y	y^2	xy
1	29.2	852.64	10.0	100.00	292
2	25.1	630.01	11.6	134.56	291.16
3	27.5	756.25	9.2	84.64	253
4	27.5	756.25	12.6	158.76	346.5
5	40.0	1600.00	6.2	38.44	248
6	42.4	1797.76	5.3	28.09	224.72
7	41.7	1738.89	7.8	60.84	325.26
8	36.6	1339.56	7.2	51.84	263.52
9	42.1	1772.41	7.8	60.84	328.38
Total	312.1	11243.77	77.7	718.01	2572.54

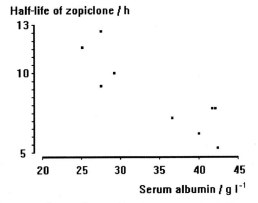

Half-life of zopiclone / h

Serum albumin / g l^{-1}

Fig. 8.49 Scatter diagram based on the data in Table 8.31.

$$= \Sigma xy - \frac{(\Sigma x)(\Sigma y)}{n}$$

$$= 2572.54 - \frac{(312.1)(77.7)}{9}$$

$$= 2572.54 - 2694.46 = -121.92$$

Substituting into the formula for the correlation coefficient:

$$r = \frac{\Sigma(x - \bar{x})(y - \bar{y})}{\sqrt{\Sigma(x - \bar{x})^2 \Sigma(y - \bar{y})^2}}$$

$$= \frac{-121.92}{\sqrt{(420.84)(47.20)}}$$

$$= -121.92/\sqrt{(19\,863.65)} = -121.92/140.94$$

$$= -0.86505$$

Hence the correlation coefficient is –0.87 to two significant figures. This is within the range we would expect from the scatter diagram, and implies a strong, though not perfect, negative correlation between the two variables in this example. Note that the correlation coefficient does not have any units.

Care in the interpretation of the correlation coefficient

In interpreting the value of the correlation coefficient, r, it is important to bear the following points in mind: correlation does not imply causation; x and y are interchangeable; and extrapolation is not valid outside the range used to calculate r. These points are now explained further.

Correlation does not imply causation, even if $r = 1$, say. For example, over a three-year period it may be found that the annual amount spent on food by a group of psychiatric trainees is highly correlated with

Therefore,

$$\Sigma(x - \bar{x})^2$$

$$= \Sigma x^2 - \frac{(\Sigma x)^2}{n}$$

$$= 11\,243.77 - \frac{(312.1)^2}{9}$$

$$= 11\,243.77 - 10\,822.93 = 420.84$$

$$\Sigma(y - \bar{y})^2$$

$$= \Sigma y^2 - \frac{(\Sigma y)^2}{n}$$

$$= 718.01 - \frac{(77.7)^2}{9}$$

$$= 718.01 - 670.81 = 47.20$$

$$\Sigma(x - \bar{x})(y - \bar{y})$$

the annual amount spent on psychiatry textbooks and examination revision books. Clearly, however, neither of these two types of expenditure is the cause of the other. There may be indirect connections between two correlated factors; the rate of inflation may be one such factor in this example.

Conventionally for paired data when one variable depends on the other, the **dependent variable**, or outcome variable, is represented graphically by the vertical or y-axis, and the variable it depends on, the **independent variable**, or predictor variable, by the horizontal or x-axis. In cases where there may be a causative relationship between two variables, the correlation coefficient does not give any indication of which is the dependent and which the independent variable. In calculating the value of r, it does not matter which variable is assigned as x, and which as y. From the formula for r given above it can be seen that the value for r remains unaltered if x and y are swapped with each other.

When a linear correlation is found, it cannot be assumed that the line of best fit can be extrapolated to encompass values of x and y outside the range used in establishing the correlation. The reason for this is illustrated in Figure 8.50

Estimation of ρ

Fisher's z transformation allows the sampling distribution of r, which is not Normal even if both x and y have a Normal distribution, to be transformed into a variable with an approximately Normal distribution, by using $z_r = \textbf{arctanh } r$, that is:

$$z_r = \frac{1}{2} \ln \frac{1 + r}{1 - r}$$

If n is the number of matched (x,y) pairs, then the estimated standard error of z_r can be shown to be $1/\sqrt{(n - 3)}$. Hence, the $100(1 - \alpha)\%$ confidence interval for the population correlation coefficient, ρ, in terms of z_r, is given by:

$$\textbf{z}_\textbf{r} - (\textbf{z}_{1 - \alpha/2})/\sqrt{(\textbf{n} - 3)} \text{ to } \textbf{z}_\textbf{r} + (\textbf{z}_{1 - \alpha/2})/\sqrt{(\textbf{n} - 3)}$$

To convert from z_r back to r, tables often used to be employed in the past. However, given the widespread availability of electronic calculators, it is a simple matter of using either $r = \tanh z_r$, or the algebraic formula of r in terms of z_r:
Since

$$z_r = \frac{1}{2} \ln \frac{1 + r}{1 - r}$$

$$\therefore \ e^{2z_r} = \frac{1 + r}{1 - r}$$

$$\therefore \ e^{2z_r} - re^{2z_r} = 1 + r$$

Therefore,

$$r = \frac{e^{2z_r} - 1}{e^{2z_r} + 1}$$

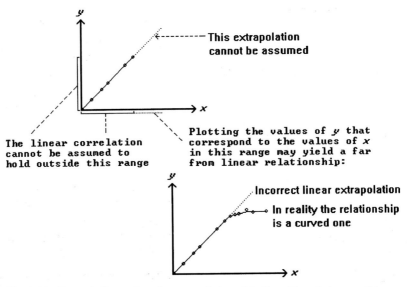

Fig. 8.50   Example illustrating why extrapolation of the line of best fit is not valid outside the range used to calculate the correlation coefficient.

Hypothesis testing

Pearson's product correlation coefficient described in this section is a parametric measure of the degree of correlation between two variables; a non-parametric equivalent, Spearman's rank correlation, is described later in this chapter. In order for hypothesis testing about the population correlation coefficient, ρ, to be valid using the formula given in this subsection, at least one of the variables from which r is derived must be from a Normal distribution. So long as this criterion is fulfilled, the t test can be used to test the null hypothesis that there is no correlation.

H_0: $\rho = 0$

The formula used is:

$$t = r\sqrt{\frac{n-2}{1-r^2}}$$

with $(n-2)$ degrees of freedom.

For a given value of the level of significance, α, $P < \alpha$ implies the correlation is statistically significant, and the null hypothesis is rejected, while $P > \alpha$ implies the correlation could be caused by chance, and the null hypothesis cannot be rejected.

Example

For the previous example calculate whether the correlation coefficient is statistically significant at the one per cent level, and calculate the 95% confidence interval for the population correlation coefficient.

(a) Hypothesis test. From the previous example:

$r = -0.86505$
$n = 9$

The null hypothesis is that the population correlation coefficient is zero:

$H_0 : \rho = 0$

The direction of the difference is assumed to be unimportant and therefore a two-tailed t test is used. Therefore,

$$t = r\sqrt{\frac{n-2}{1-r^2}}$$

$$= -0.86505\sqrt{\frac{9-2}{1-(-0.86505)^2}}$$

$$= -0.86505\sqrt{\frac{7}{0.25169}}$$

$$= -0.86505 \times \sqrt{(27.812)}$$

$$= -0.86505 \times 5.2737$$

$$= -4.56 \text{ (to 3 significant figures)}$$

Number of degrees of freedom $= n - 2$
$$= 9 - 2$$
$$= 7$$

From Table IV in the Appendix (t distribution) this corresponds to a two-tailed probability value of $0.01 > P > 0.001$. Therefore, with $P < 0.01$, the null hypothesis (H_0: $\rho = 0$) is rejected and the sample correlation is statistically significant at the 1% level.

(b) Estimation of ρ.

$$z_r = \frac{1}{2}\ln\frac{1+r}{1-r}$$

$$= \frac{1}{2}\ln[(1 + (-0.86505))/(1 - (-0.86505))]$$

$$= \frac{1}{2}\ln[(1 - 0.86505)/(1 + 0.86505)]$$

$$= \frac{1}{2}\ln(0.072357)$$

$$= -1.31307$$

Alternatively, the value of arctanh r (also written tanh^{-1} r) can be keyed directly into a calculator with scientific functions:

$z_r = $ arctanh (-0.86505)
$\quad = -1.31307$

The standard error of z_r is given by:

$1/\sqrt{(n-3)}$
$= 1/\sqrt{(9-3)}$
$= 0.40825$

In terms of z_r the 95% confidence interval is given by:

$z_r - (z_{1-\alpha/2})/\sqrt{(n-3)}$ to $z_r + (z_{1-\alpha/2})/\sqrt{(n-3)}$
$= -1.31307 - 1.96/\sqrt{6}$ to $-1.31307 + 1.96/\sqrt{6}$
$= -2.11323$ to -0.51290

For the first value, $e^{2z_r} = e^{2(-2.11323)} = 0.014603$
For the second value, $e^{2z_r} = e^{2(-0.51290)} = 0.35851$
Since

$$r = \frac{e^{2z_r} - 1}{e^{2z_r} + 1}$$

then, converting back to the scale used for r, the required 95% confidence interval is:

$[(0.014603 - 1)/(0.014603 + 1)]$
to $[(0.35851 - 1)/(0.35851 + 1)]$
$= -0.97121$ to -0.47220

Alternatively, the conversion can be made simply by using $r = $ tahn z_r, so that the confidence interval in the scale used for r is given by:

tanh (−2.11323) to tanh (−0.51290)
= −0.97121 to −0.47220

Hence the required confidence interval is −0.971 to −0.472, to three significant figures.

Other types of correlation are discussed in the section on non-parametric methods later in this chapter, and in the next chapter.

Linear regression

Straight line equation

From Figure 8.51, which represents a typical straight line, it can be seen that any straight line can be represented by the equation:

$$y = a + bx$$

where a = intercept on the y-axis, and b = gradient of the line.

In Figure 8.51 the gradient of the line is positive, and so b is positive. However, if the line were to slope in the other direction, as shown in Figure 8.52, the

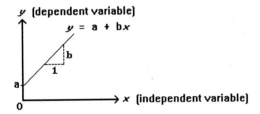

The gradient of the line
$$= \frac{\text{change in } y}{\text{corresponding change in } x}$$
$$= b/1$$
$$= b$$

The intercept on the y axis $= a$.
At this point $x = 0$

Fig. 8.51   Graph of $y = a + bx$.

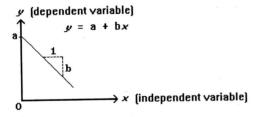

For an increase in y of 1, there is a decrease (or negative change) in x. Therefore the gradient is negative and so the value of b is negative.

Fig. 8.52   Straight line ($y = a + bx$) with a negative gradient.

gradient of the line, and therefore the value of b, would be negative.

Transformation

When the scatter diagram gives a non-linear relationship it may be possible to transform this into a linear relationship by using mathematical functions such as logarithms, reciprocals and square roots. A detailed discussion of the choice of transformation is outside the scope of this chapter.

Linear regression line

It has been mentioned above that the linear regression line is the straight line of best fit in a scatter diagram. To calculate the values of a and b that give the straight line, $y = a + bx$, that best fits the points on a scatter diagram, the method of **least squares fit** is used, as shown in Figure 8.53.

It can be shown that the required values of b and a are given by:

$$b = \frac{\sum(x - \bar{x})(y - \bar{y})}{\sum(x - \bar{x})^2}$$

$$a = \bar{y} - b\bar{x}$$

The gradient, b, is also known as the sample **regression coefficient** (for the population regression coefficient, the Greek equivalent letter, β, is used). When r is positive, so too is b; when r is zero, so too is b, and the regression line in this case is a horizontal line intercepting the y-axis at $y = a$; when r is negative, so too is b.

From the formula for the intercept on the y-axis, a, it can be seen that the linear regression line passes through the mean point $(\bar{x}, \bar{y})$.

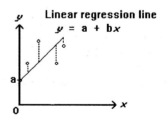

The values of a and b chosen are such that the sum of the squares of the vertical distances shown by dashed lines is a minimum

Fig. 8.53   Method of least squares.

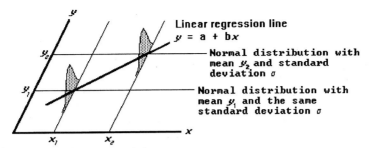

Fig. 8.54   Diagram illustrating some of the assumptions made in using linear regression (see text).

Units

From the equation $y = a + bx$ it can be seen that the units of a are the same as those of y, while the units of b are the units of y divided by those of x.

Assumptions

A number of assumptions need to be made in order to use linear regression. These include: there is no error in the observed values of x; for any value of x, there is a Normal distribution of values of y, as shown in Figure 8.54; the magnitude of scatter of values of y for each value of x is the same throughout the range of the linear regression line, that is, the Normal distribution curves in Figure 8.54 have the same standard deviation; and the true mean value of y from the Normal distribution of values of y corresponding to each x lies on the linear regression line (for example, in Figure 8.54, y_1 and y_2 lie on the regression line).

Example

Consider again the example above comparing the elimination half-life of a drug with the corresponding serum albumin level in nine cirrhotic patients (see Tables 8.31 and 8.32). The linear regression line for the data given in that example is now calculated.

The serum albumin level is the independent variable, x, and the elimination half-life is the dependent variable, y, since it depends on x. In the following calculations the units have been omitted until the final answers.

Calculation of b

Omitting units, $\sum(x - \bar{x})(y - \bar{y})$ has already been calculated above, as –121.92. Similarly, it has been shown that $\sum(x - \bar{x})^2$ is 420.84. Therefore:

$$b = \frac{\sum(x - \bar{x})(y - \bar{y})}{\sum(x - \bar{x})^2}$$

$$= \frac{-121.92}{420.84}$$

$$= -0.2897 hg^{-1}l$$

Calculation of a

Omitting units and using the values of $\sum x$ and $\sum y$ found above, we have:

$$\bar{x} = \frac{\sum x}{n}$$

$$= \frac{312.1}{9}$$

$$= 34.68$$

$$\bar{y} = \frac{\sum y}{n}$$

$$= \frac{77.7}{9}$$

$$= 8.63$$

Therefore

$$a = \bar{y} - b\bar{x}$$

$$= \frac{77.7}{9} - \frac{(-0.2897)(312.1)}{9}$$

$$= \frac{77.7 + 90.42}{9}$$

$$= \frac{168.12}{9}$$

$$= 18.68\,h$$

Linear regression equation

Substituting the values for b and a into the equation $y = a + bx$ gives the following linear regression equation:

$$y = 18.68 - 0.2897x$$

Calculators with statistical functions automatically

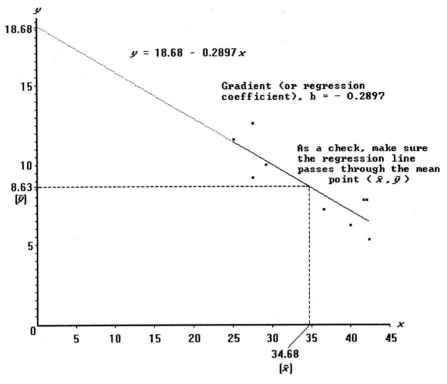

Fig. 8.55   Linear regression line (see text).

compute the values of b, $\bar{x}$, $\bar{y}$, and a when the original data are entered. But it is still useful to plot the scatter diagram and linear regression line as a check. In this example the linear regression line is as shown in Figure 8.55

Prediction

The linear regression line can be used to predict values of y for values of x, so long as the latter lie within the range of x used to determine the line in the first place (see discussion above.)

Example

Calculate (a) the predicted value of the elimination half-life in the last example when the serum albumin level is 30.0 g 1^{-1}; and (b) the serum albumin level corresponding to an elimination half-life of 8.0 h.

(a) We have $x = 30$,

$$\therefore \ y = 18.68 - 0.2897x$$
$$= 18.68 - (0.2897 \times 30)$$
$$= 9.989$$

So the required elimination half-life is 10.0 h (to three significant figures).

(b) We have $y = 8$, $y = a + bx$

$$\therefore \ x = \frac{y - a}{b}$$

$$= \frac{y - 18.68}{-0.2897}$$

$$= \frac{8 - 18.68}{-0.2897}$$

$$= \frac{-10.68}{-0.2897}$$

$$= 36.87$$

So the required serum albumin level is 36.9 g 1^{-1} (to three significant figures).

Testing β for significance

If the symbol y_p is used to represent a value of y predicted by the linear regression line, then because the least squares fit method is used to calculate the equation of the regression line, it follows that $\Sigma(y - y_p)^2$ has the minimum value possible.

The variance of y about the linear regression line, $\sigma^2_{y/x}$, is estimated from $s^2_{y/x}$, which can be shown to be given by:

$$s^2_{y/x} = \frac{\Sigma(y - y_p)^2}{n - 2}$$

$$= \frac{\Sigma y^2 - (\Sigma y)^2/n - b(\Sigma xy - \Sigma x \Sigma y/n)}{n - 2}$$

$$= \frac{(n - 1)(s^2_y - b^2 s^2_x)}{n - 2}$$

where s^2_y is the standard deviation of y, and s^2_x is the standard deviation of x.

The value of the corresponding standard deviation, $s_{y/x}$, is the square root of the above expression.

The calculation of the sample regression coefficient, b, has been described above. The value of b is usually not zero, and in order to test whether the population regression coefficient, β, is actually zero, and b has a value other than zero by chance, the null hypothesis tested is:

$$H_0: \beta = 0$$

This is analogous to testing the null hypothesis $H_0: \rho = 0$ for the correlation coefficient (see above). Again, the t distribution is used:

$$t = \frac{b\sqrt{(\Sigma(x - \bar{x})^2)}}{s_{y/x}}$$

d.f. $= n - 2$

Estimation

In order to calculate confidence intervals for β, α (the population parameter corresponding to a), and the predicted true value of y, the t distribution with $(n - 2)$ degrees of freedom is used so that, as before, the confidence interval is given by:

[statistic $- t_{1-\alpha/2}$ (standard error)] to
[statistic $+ t_{1-\alpha/2}$ (standard error)]

The relevant standard errors are now stated:

for β: $s_{y/x}/[s_x\sqrt{(n - 1)}]$

for α: $s_{y/x}\sqrt{[(1/n) + (\bar{x}^2/(s^2_x(n - 1)))]}$

for the predicted mean true value of y, for a given value of x, x_0:

$s_{y/x}\sqrt{[(1/n) + ((x_0 - \bar{x})^2/(s^2_x(n - 1)))]}$

Note that the units of α and β are the same of those of a and b, respectively.

Example

For the last example (Tables 8.31 and 8.32) test the null hypothesis that the elimination half-life of zopiclone is independent of its serum albumin level, and calculate the 95% confidence interval for the true regression coefficient.

Hypothesis test

The null hypothesis to be tested is $H_0: \beta = 0$. Omitting units, from Tables 8.31 and 8.32 we can calculate the standard deviations for x and y, and obtain:

$s_x = 7.2529$
$s_y = 2.4290$

Therefore,

$$s^2_{y/x} = \frac{(n - 1)(s^2_y - b^2 s^2_x)}{n - 2}$$

$$= 8[(2.4290)^2 - (-0.2897)^2(7.2529)^2]/7$$

$$= 1.6973$$

$\therefore s_{y/x} = 1.3028$

From a previous calculation we have:

$\Sigma(x - \bar{x})^2 = 420.84$

Therefore,

$$t = \frac{b\sqrt{(\Sigma(x - \bar{x})^2)}}{s_{y/x}}$$

$$= [(-0.2897)\sqrt{(420.84)}]/1.3028$$

$$= -4.5617$$

d.f. $= n - 2 = 7$

Since the t distribution is symmetrical, we can look up $t = -4.5617$, d.f. $= 7$, in Table IV (in the Appendix) from which we obtain:

$0.01 > P > 0.001$

Hence the null hypothesis that the elimination half-life of zopiclone is independent of its serum albumin level, $H_0: \beta = 0$, is rejected, and it can be stated that the population regression coefficient is statistically significantly different from zero at the 1% level of significance.

Confidence interval for β

The appropriate standard error is

$s_{y/x}/[s_x\sqrt{(n - 1)}]$
$= 1.3028/[(7.2529)\sqrt{8}]$
$= 0.063507$

For d.f. $= 7$, and $\alpha = 0.05$ (two-tailed), $t_{1-\alpha/2} = 2.365$

(from Table IV in the Appendix). Hence the 95% confidence interval for the population regression coefficient, β, is given by:

$[b - t_{1-\alpha/2}$ (standard error)] to
$[b + t_{1-\alpha/2}$ (standard error)]

$= [-0.2897 - (2.365)(0.063507)]$ to
$\quad [-0.2897 + (2.365)(0.063507)]$

$= -0.440$ to -0.140 hg^{-1}1 to three significant figures.

MULTIVARIATE ANALYSIS

In the last section linear relationships between two variates, or random variables, has been discussed; this is an example of bivariate analysis. However, in statistical psychiatric modelling it is usually much more likely that for each element of a population at least three variates need to be measured, yielding data that is known as multivariate. Such multivariate data is conveniently represented in the form of matrices. However, a description of the principles of matrix algebra is outside the scope of this chapter. Unlike the presentation in the preceding part of this chapter, the statistical techniques used in different forms of multivariate analysis are not described in detail in this section. Instead, some of the different forms of multivariate analysis used in psychiatric research are outlined briefly.

Multivariate regression analysis

Rather than using a bivariate model, $y = a + bx$, a better fit to reality may be provided by a linear multivariate regression equation of the form:

$y = a + bx_1 + cx_2 + dx_3 + ...$

in which b, c, d,...are regression coefficients. The multiple correlation coefficient is the maximum correlation between the dependent variable and multiple non-random independent variables, $x_1, x_2, x_3, ... x_n$, using a least squares method. Assumptions in the determination of the multiple correlation coefficient include: the dependent variable has a Normal distribution; the independent variables have fixed non-random values; and random errors in the model have an overall arithmetic mean of zero and have a Normal distribution.

Path analysis

This statistical technique uses a series of multiple regression analyses to allow hypotheses of causality between variables to be modelled and tested. These variables and their hypothetical relationships can be expressed graphically in the form of a path diagram which shows the variables and arrows between them, with regression coefficients known as path coefficients being associated with the arrows. The model is tested using χ^2 tests. Among the assumptions made in using path analysis are that: there is a linear relationship between variables; observations used are error free; and dependent variables are functions of only the variables incorporated in the model and of no others.

Factor analysis

The statistical technique of factor analysis can be used to try to reduce the interrelationships within a large set of variables to a small number of underlying statistically independent factors or dimensions which are related to the former variables in a linear manner. Factor analysis was formulated initially by psychologists and includes a number of methods, such as maximum likelihood analysis and principal factor analysis. As well as assuming that the relationship between the variables and the factors or dimensions is linear, it is assumed that the variables are liable to random errors. Another assumption is that unlike the original variables, the factors or dimensions cannot be directly observed.

Principal component analysis

This method is similar to factor analysis, particularly the method of principal factor analysis, in that it can be used to try to produce a reduced number of principal components or dimensions from a large number of original variables. A description of the main differences between the two techniques, which relate mainly to the factor model assumptions in factor analysis and the direction of transformation (from variables to principal components in principal component analysis: from factors to variables in factor analysis), is beyond the scope of this chapter.

Canonical correlation analysis

Canonical correlation analysis may be regarded as an extended form of multivariate regression, except that in addition to an x-set of independent variables, there is also a y-set of dependent variables such that the number of dependent variables is no longer confined to being one, but is now *at least* one. The maximum correlation between the two sets of variables, known as the canonical correlation, is determined and provides information about interrelationships among the variables.

Discriminant analysis

Discriminant analysis allows individual items or people to be allocated to different groups. It is particularly useful in studies involving psychiatric diagnostic classifications.

Cluster analysis

As its name implies, cluster analysis allows individual items or people to be grouped into clusters which are said to be homogeneous in the sense that individual members of a given cluster are similar to each other but different to members of another cluster generated by the analysis. Hierarchical methods can be applied to the clusters.

NON-PARAMETRIC TESTS

Non-parametric tests are statistical tests that do not necessitate any assumptions concerning the parameters or distribution of the population from which the samples are drawn. For this reason, non-parametric tests are also sometimes referred to as distribution-free tests. It follows that non-parametric tests can be applied in more general conditions than can parametric tests. Another advantage is the fact that so long as the sample size, n, is less than about 50, non-parametric tests tend to be very easy to use.

A major disadvantage of non-parametric tests, however, is that when the assumptions required for a parametric test are fulfilled, then by using a non-parametric test one is discarding useful information which can be derived from the data. The use of a non-parametric test in a situation in which a parametric test is available may lead to a loss of power, that is, there is a greater probability of making a type II error. Furthermore, although it is possible to make estimations (and derive confidence intervals) with non-parametric tests, the methods needed can be laborious; non-parametric tests are essentially significance tests. The calculation of confidence intervals is not considered in this section.

It has already been mentioned earlier in this chapter that the χ^2 test is a non-parametric method. In the comparison of two samples, the parametric method employs the t distribution (see the worked example using the two-sample t test earlier in this chapter), while non-parametric equivalent tests include: the Wilcoxon rank sum test, the Mann–Whitney U test, the Wilcoxon signed rank test, the Wald–Wolfowitz test, and the Kolmogorov–Smirnov test. A non-parametric equivalent of the parametric analysis of variance (not described in this chapter) is the Kruskal–Wallis test;

both of these tests allow the comparison of more than two groups. Non-parametric equivalents of the parametric Pearson product moment correlation coefficient, r, include Spearman's rank correlation and Kendall's rank correlation. In the remainder of this section worked examples are provided for the following non-parametric tests: the Wilcoxon rank sum test, the Wilcoxon signed rank test, and Spearman's rank correlation; details of how to calculate confidence intervals for non-parametric tests are not given in this chapter and the interested reader is referred to a text such as that by Conover (1980) for further details.

Wilcoxon rank sum test

This test uses two independent samples to compare two populations with respect mainly to location. The null hypothesis is that there is no difference, with respect to location, between the two distributions of the samples of the two populations. The sizes of the two samples do not have to be equal, but the samples must be independent and random. A worked example follows.

Consider Table 8.33 which shows the fear ratings of two groups of patients treated in different ways. The fear rating schedule used gives a score from zero to 100.

Table 8.33 Fear ratings of two groups of patients treated in different ways

Sample 1	Sample 2
40	10
30	35
15	15
15	80
25	70
60	

The observations are *ranked together* in ascending order, as shown in Table 8.34. If more than one observation has the same value they are each given the appropriate mean rank; since there are three observations with the value 15 in Table 8.34, they are ranked the mean of second, third and fourth, that is, third.

The sum of the ranks in the smaller group (sample 2 in our example) is called T. When both samples are of the same size it does not matter which total is chosen as T. Therefore in this example T is 32. By convention, n_1 is the size of the smaller sample, and n_2 is the size of the other sample. Therefore, in this sample n_1 is

Table 8.34 Ranking of the data in Table 8.33

	Sample 1	Rank		Sample 2	Rank
	40	8		10	1
	30	6		35	7
	15	3		15	3
	15	3		80	11
	25	5		70	10
	60	9			
Total	$n_2 = 6$	34		$n_1 = 5$	32

5 and n_2 is 6. From an appropriate table (which is not reproduced in this book, but is readily available in texts dealing with non-parametric methods) it is found that for these values of n_1 and n_2 the corresponding critical values at the 5% level of significance ($\alpha = 0.05$; two-tailed) are 18 and 42. If the value of T found lies within the critical range formed by the critical values (in this example the critical range is T = 18 to T = 42 ($\alpha = 0.05$; two-tailed)), the null hypothesis cannot be rejected; if T lies outside of the critical range the null hypothesis is rejected.

Since T = 32, this means that the difference between samples 1 and 2 is not statistically significant at the 5% level, and the null hypothesis cannot be rejected.

Wilcoxon signed rank test

This test can be used to compare non-qualitative matched-pair data. If the data are quantitative, then the actual values are not used; this is an example of the wastage of information, mentioned earlier, which can occur with non-parametric tests. A worked example is now given.

A series of 10 epileptic patients is assessed for the number of grand mal seizures each suffers over a period of four weeks while on each of treatments A and B. The results are shown in Table 8.35

The difference between the values for treatments A and B are calculated and ranked in ascending order (ignoring the sign), leaving aside any case(s) where there is no difference. This is shown in Table 8.36 in which it can be seen that the mean rank of third has been allocated to those ranks that are tied first to fifth (inclusive).

The sum of the ranks of the positive differences, called T_+, is given by 3 + 3 + 3 + 3 + 6, that is, 18. Likewise, the sum of the ranks of the negative difference, T_-, is 7 + 3, that is, 10.

T is defined as being the smaller of T_+ and T_-. Therefore in this example T is 10. N is defined as the number of non-zero differences, and in this case is 7.

Table 8.35 Number of grand mal seizures over a four-week period in each of ten patients while on different treatments

Patient	Grand mal seizures	
	Treatment A	Treatment B
1	3	2
2	0	0
3	2	1
4	1	4
5	2	2
6	4	3
7	2	1
8	3	3
9	3	1
10	2	3

Table 8.36 Ranking of the differences for the data in Table 8.35

Treatment A	Treatment B	Difference	Rank
3	2	1	3
0	0	0	
2	1	1	3
1	4	-3	7
2	2	0	
4	3	1	3
2	1	1	3
3	3	0	
3	1	2	6
2	3	-1	3

The null hypothesis is that there is no difference between the two groups, and in the case of the Wilcoxon signed rank test a significant result is one which is *smaller* than the critical value of T. From an appropriate table it is found that for N = 7, at the 5% level of significance ($\alpha = 0.05$), the critical value of T is 2.

Since T is 10, it follows that in this example the null hypothesis cannot be rejected, that is, the difference between the two treatment responses is not statistically significant at the 5% level of significance.

Spearman's rank correlation

The differences between pairs of ranks are used in this test, rather than quantitative data as in the parametric equivalent of the correlation coefficient. A worked example based on Table 8.31 follows.

The first step is to rank each set of values. In addition to giving the elimination half-life of the drug zopiclone and the corresponding serum albumin level in nine cirrhotic patients, Table 8.37 also shows the rank

Table 8.37 Ranking of the serum albumin levels, and of the drug elimination half-lives, for the data in Table 8.31

Patient	Serum albumin level (gl⁻¹)	Rank	Elimination half-life of zopiclone (h)	Rank
1	29.2	4	10.0	7
2	25.1	3	11.6	8
3	27.5	1.5	9.2	6
4	27.5	1.5	12.6	9
5	40.0	6	6.2	2
6	42.4	9	5.3	1
7	41.7	7	7.8	4.5
8	36.6	5	7.2	3
9	42.1	8	7.8	4.5

Table 8.38 Values of d, the difference between the ranks, and d^2, for Table 8.37

Rank for serum albumin level	Rank for elimination half-life	d	d^2
4	7	−3	9
3	8	−5	25
1.5	6	−4.5	20.25
1.5	9	−7.5	56.25
6	2	4	16
9	1	8	64
7	4.5	2.5	6.25
5	3	2	4
8	4.5	3.5	12.25
Total			213

of each set of values. Once again, for tied ranks the mean rank is designated.

The next step is to determine the difference between the ranks, conventionally denoted by d, and then to square this difference. This is shown in Table 8.38. From the last column the sum of the values of d^2 is calculated:

$$\Sigma d^2 = 213$$

Spearman's rank correlation is conventionally given the symbol r_s to distinguish it from the correlation coefficient, r, and is calculated using the following formula:

Spearman's rank correlation, $r_s = 1 - \dfrac{6\Sigma d^2}{n(n^2 - 1)}$

As with r, r_s has the range

$$-1 \le r_s \le 1$$

Again

$r_s = 1$ implies perfect positive correlation

$0 < r_s < 1$ implies fair (but not perfect) positive correlation

$r_s = 0$ implies no correlation at all

$-1 < r_s < 0$ implies fair (but not perfect) negative correlation

$r_s = -1$ implies perfect negative correlation

In our example we have:

$$\Sigma d^2 = 213$$

number of patients, $n = 9$.

Substituting into the formula for r_s:

$$r_s = 1 - \frac{6\Sigma d^2}{n(n^2 - 1)}$$

$$= 1 - \frac{6(213)}{9(9^2 - 1)}$$

$$= -0.775$$

Tables of critical values are available which allow the significance of this result to be tested. From such a table it is found that at the five per cent level of significant ($\alpha = 0.05$) the two-tailed critical value is 0.683. Hence the null hypothesis that there is no association between the two variables cannot be rejected if r_s lies within the critical range $r_s = -0.683$ to 0.683 ($\alpha = 0.05$)

Hence for our example, since r_s is −0.775, this result is statistically significant at the 5% level of significance. At the 1% level of significance, however, the critical range is $r_s = -0.833$ to 0.833, and so the result would not be significant at this level (in the case of the correlation coefficient, r, it has been shown earlier that the result is significant at the 1% level).

An alternative way of assessing the significance of association is to use the t test as described previously for the correlation coefficient:

$$t = r_s \sqrt{\frac{n - 2}{1 - r_s^2}}$$

with $(n - 2)$ degrees of freedom. However, this t test should be used only if $n \ge 10$.

Kendall's rank correlation

Kendall's rank correlation is a similar non-parametric method which gives a coefficient, τ (tau), which has the same range as r and r_s:

$$-1 \le \tau \le 1$$

Once again, as with r and r_s, $\tau = 1$ implies perfect positive correlation, and so on. Because the method used to calculate τ is different to that used for r_s, the values of τ and r_s usually differ somewhat.

FURTHER READING

Chou, Y-L, 1975 Statistical analysis, 2nd edn. Holt, Rinehart & Winston, New York

Daniel W 1987 Biostatistics: a foundation of analysis in the health sciences, 4th edn. John Wiley, Chichester

Gardner, M J, Altman D G 1989 Statistics with confidence. British Medical Journal, London

Hill A B 1966 Principles of medical statistics, 8th edn. Lancet, London

Schefler W C 1979 Statistics for the biological sciences. 2nd edn. Addison-Wesley, Reading, Massachusetts

9. Research methodology

In this chapter the basic principles of research methodology are outlined. However, it needs to be emphasized at the start that very few research projects are fundamentally the same, and principles should only be a guide to choice, not absolute requirements. In each section the special difficulties that often arise in psychiatric research will be identified and ways of overcoming them outlined.

GENERAL APPROACH

Psychiatry is a discipline that is multi-factorial and one of its attractions to practitioners is its unpredictability and many facets. Research, however, prefers, or is forced by necessity, to focus on issues that are broken down into simpler questions which are then tested by appropriate research designs. Ideally, a research study confines itself to answering one question at a time and constructs a base of knowledge in small but consistent steps. This is much more difficult in psychiatry than almost any other subject because psychiatry does not have the same substrate of basic science supporting its assertions and has to rely largely on empiricism (i.e., is based on the results of observation and experiment only). In this context it is worth reminding ourselves that empiricism is also a term of criticism that used to be regarded in medicine as indicative of bad practice and quackery. In fact empiricism is perfectly respectable as a scientific theory, provided that its methods of observation and experiment are sound. For this reason, research methodology often assumes more importance in psychiatry than in many other disciplines.

In planning research it is important to understand that the methodology chosen should be appropriate for the state of knowledge of that subject. The first level of investigation needs to be extremely broad and covering a large area (and is well exemplified by the phrase 'flying a kite') whereas the last type of investigation into a subject about which much is already known consists of a tight and careful design that answers a highly specific question focusing on one small area of interest. It is worth examining this progression from broad interest to specific question in stages to establish the main features of research methodology.

Observation

Looking and describing are among the most valuable of research attributes. Observation is best approached from an unbiased viewpoint. Bias, an unfair inclination to one view rather than another, lies at the heart of much bad research in psychiatry. It is an unfortunate fact that most people are not sufficiently dispassionate to disregard or compensate for bias through their own efforts, with the result that research which is largely dependent on subjective judgment is usually self-fulfilling and of relatively little value.

Good observation is exemplified in descriptions of either a single patient or of only a few patients. The eponymous syndromes of Cotard, de Clérambault, Ganser, Down and Capgras relied on careful and accurate description without the need for statistics or any other research tools. Detailed and systematic description of one member of a population can often confirm the characteristics of the group as a whole.

Observation is still important when larger numbers are being studied. For example, Kraepelin's famous studies suggesting the separation of manic-depressive psychosis from schizophrenia (1883) relied on the meticulous observation of large numbers of psychiatric in-patients; those with a symptom pattern of manic-depressive psychosis had episodes of illness characterized by manic or depressive symptoms and with normal function between, whereas those with schizophrenia (or dementia praecox) tended to be chronically impaired with no periods of remission and in whom normal function was seldom, if ever, maintained.

The major problem in relying on observation with larger numbers is that once the observer has decided on a hypothesis (in Kraepelin's case that manic-depressive and schizophrenic psychoses showed fundamentally different natural histories) further observations are subjected to bias which tends to reinforce the original hypothesis. This is natural but unfortunate. It tends to lead to clean divisions that are not present in fact. This is well illustrated by further studies on manic-depressive psychosis and schizophrenia in which the original separation described by Kraepelin has proved not to be nearly so clear as he described (Post 1971, Brockington & Leff 1979). Even the best of investigators can be swayed by the magnetism of their own hypotheses. To take another well-known example, Gregor Mendel, after carrying out his famous experiments on peas which illustrated his two laws of heredity, described his findings in published form in such exact proportions that the results could not have arisen entirely by chance. As he demonstrated that inherited characteristics arise by random combination and recombination of genes some disparity in the findings must have been present but was ironed out before publication. Mendel's original findings have never been replicated! This is one of the many forms of '**data massage**', the adjustment of data to demonstrate the findings that you want to prove. For this and many other reasons it is necessary to rely on more than observation when moving on to the next stage of methodology.

Observation, either in the clinic or in other settings (e.g., of mortality data in an office) is usually the first step in research work. Occasionally in psychiatry it is possible for good research ideas to develop de novo but the empirical tradition of the subject makes this less likely than in other disciplines. All these sources help the investigator to construct an experiment to develop this idea further. The first task is to measure the phenomenon accurately.

SPECIFIC APPROACH

Measurement

Measurement is concerned with the recording of observations in numerical form. Once this has been carried out appropriately it is much easier to relate one set of findings to others because the rules of mathematics can be brought into operation. This allows different sets of data to be compared and contrasted and can allow unlimited possibilities of merging and separating data that will allow the essential elements to be identified. This can take place only if the measurement is an appropriate one. Whether or not it is appropriate is determined by the twin standards of reliability and validity.

Reliability

Reliability describes the level of agreement between sets of observations. Under most circumstances agreement (usually measured by a correlation coefficient (see Ch. 8)) needs to be high if the measurement is regarded as satisfactory, but there are exceptions. There are several kinds of reliability with different implications.

Inter-rater reliability

Inter-rater reliability describes the level of agreement between assessments made by two or more assessors at roughly the same time. It is commonly used in assessments of aspects of mental state. In interpreting the results it is important to realize that good to excellent levels of agreement are required before the measurement is regarded as satisfactory, as the assessments are made from observations of the same material. A minor variation, **intra-rater reliability,** describes the assessments by different raters of the same material presented at two or more different times. Thus, interviews that are videoed or tape recorded can be used for this purpose. This is particularly helpful when many different raters are being assessed, including those from different cultures and countries.

Test–retest reliability

This describes the level of agreement of observers assessing the same material under similar conditions but at two different times. Most test–retest reliability refers to assessments made relatively close together (within a few days) but this can be extended to many months or years. Under the latter circumstances the comparison of data is often termed **temporal reliability** and may be particularly important in assessing conditions that are alleged to be consistent over long periods (e.g., personality disorder).

There are two forms of reliability that could be applied to a measuring instrument that has several parts. **Alternative form reliability** describes two supposed similar forms of the measurement used to assess the same material either at the same time or immediately consecutively. Again the agreement between them should be very high if they are measuring the same feature. **Split-half reliability** characteristically involves dividing a measurement into two halves and using each half to assess the same material under similar circumstances. Because some instruments (e.g. questionnaires) have different sections describing different aspects it is sometimes more appropriate to use alternative questions in assessing split-half reliability so that the balance of each half is the same.

In assessing the extent of agreement different statistics can be used. The simplest (but most unsatisfactory) method is to measure the extent of agreement in percentage terms. The second method, which is most frequently used, is the **product–moment correlation coefficient** (r) but this often gives spuriously high results especially if many of the values agree by chance. For example, if 100 assessments made by two different raters yield a score of nil in half of them and the remaining 50 have scores between 1 and 10, spuriously high levels of agreement may be found if the raters consistently agree over the nil scores even though they may disagree considerably on the others. To correct for this it is more appropriate to use the **kappa statistic** (k), preferably weighted (k_w) in which allowance is made for chance agreement. Nor surprisingly, this yields lower levels of agreement than simpler correlation measures but a value of 0.7 or greater is normally necessary before satisfactory reliability can be said to be demonstrated (Landis & Koch 1977). The kappa statistic is most appropriate when different categories of measurement are being recorded. If agreement is measured for several items that can be regarded as part of a continuum or dimension the **intra-class correlation coefficient** (R_I) is probably more appropriate (Bartko 1966). This also makes allowances for chance agreement.

Validity

Validity is the term used to describe whether an instrument measures what it purports to measure. This is an essential question that is often difficult to answer in psychiatry because there is no independent yardstick which can be regarded as a true measurement. For example, in assessing whether or not an instrument actually measures depressed mood there is no acceptable agreed measurement with which a rating scale for depression can be compared. Most validity testing of instruments in psychiatry therefore compares one (new) measurement with another established one, the latter being generally accepted to be a true measure of the item in question but which has not been definitely shown to be such.

Much depends on which type of validity is being considered. It is possible to define validity in such a way that more accurate measurements are made. **Face validity** is not really validity at all. It refers to the subjective judgment as to whether the measurement in question appears on the surface to measure the feature in question. It is of relevance only when an instrument appears to have absolutely no connection to the subject being measured (e.g., a questionnaire on mania which contains 20 questions on dietary habits has no inherent face validity) or on other occasions when face validity is deliberately obscured because it is felt that social desirability may affect response patterns. **Content validity** also refers to the same issue but is less superficial. It examines whether the specific measurements aimed for by the instrument are assessing the content of the measurement in question. It may be particularly relevant in psychiatry because there is considerable overlap between different types of measurement. For example, the well known Hamilton Rating Scale for anxiety contains an item concerned with depression. It could be argued that this item should be excluded for purposes of content validity of 'anxiousness' but it could be equally well argued that the item should stay in the schedule because it is extremely common for anxious patients to feel depressed. Thus the rating scale could be used for assessing 'anxiety states' rather than anxiety as a pathological symptom. There are no actual measurements of content validity but it is one of the more important aspects of the subject (Morley & Snaith 1989).

Predictive validity is a relatively uncommon measurement. It could equally well be named predictive reliability as it determines the extent of agreement between a present measurement and one in the future. Thus, for example, a measurement predicting successful suicide would be validated if all individuals satisfying the criteria for positive scores subsequently committed suicide. Although predictive validity could have very important implications in practice there are no measures in psychiatry that demonstrate this form of validity satisfactorily.

Concurrent validity describes the comparison of the measure being tested with an external valid yardstick at the same time. Again, this is rarely established satisfactorily since there are so few independent psychiatric measurements that can be regarded as truly valid. However, when, for example, a simple measure is being used to evaluate a more complex one the measure of concurrent validity may well be appropriate. Thus, the measurement of pulse rate as an index of anxiety could be tested by the extent of agreement between pulse rate and a rating on an established rating scale for anxiety. A high level of agreement indicates good concurrent validity. Concurrent and predictive validity together are sometimes described as **criterion validity**.

The term **incremental validity** is used to indicate whether the measurement is superior to other measurements in approaching true validity. Thus, for example, if a structured interview schedule taking four hours is no better able to distinguish cases of a particular psychiatric diagnosis than a self-rating questionnaire taking three minutes the interview schedule has no incremental validity and cannot be recommended.

The generalizability of measurements has to be regarded as suspect until they have been tested under many different conditions. It is sometimes possible to obtain good reliability and apparent validity under tightly controlled testing conditions, but until similar results are obtained using different assessors in different settings, and at different times with comparable groups of subjects, the level of generalizability is low.

Cross-validity describes validation of a measurement which has its criterion validity established for one sample and then is retested on another sample. The use of these measures in practice is illustrated later in this chapter in the section describing the construction of a rating scale.

DESIGN

Once armed with an idea and a satisfactory measurement the investigator is usually able to create an appropriate framework to test the idea in more formal terms. All research methodology should include a description of this framework, commonly described as the study design. The form of the design, which is often helped by having a statistical adviser, enables the hypothesis under investigation to be tested in the most appropriate way. The design covers all parts of a research investigation and has three main components; definition of aims, construction of a hypothesis, and creating an experiment which tests this satisfactorily.

Definition of aims

Research investigations always start with an idea or concept and this has to be re-worded and modified in such a way that a specific hypothesis can be tested by the research. This is the correct form of research methodology. Unfortunately there is sometimes a second one, which is a variety of Parkinson's second law, that 'work expands to fill the time available'. In this less satisfactory research the methodology (and often the technology) is available to do certain investigations, and hypotheses are developed which fit the methodology in question. In deciding on the aims of the investigation it is important to ask whether the study proposed is worth doing for its intrinsic value rather than other motives (e.g., it will pay my salary for

one year; it will look good on my curriculum vitae). The aims are formulated in general terms that are easily understood (e.g., I suspect that depressive illness is more common than might be expected in thyrotoxicosis. If it is, there may be some relationship between thyroid function and depressive illness. I would like to investigate this further).

Construction of a hypothesis

A hypothesis is any conjecture about the relationship between two or more concepts or structures. Most hypotheses are too general to lend themselves to scientific examination and so most **testable hypotheses** are limited in their definition so that they can be so investigated. Scientific method is based primarily on disproving hypotheses rather than proving them. This philosophy has been largely fostered by the work of Karl Popper (1963) and argues that there is no such thing as scientific truth and that all we can attempt to do is to provide the best working hypotheses in the light of information available. We can do this by disproving hypotheses that are unsound until we reach the point that we have the best explanation currently available. However, this too must remain suspect and, when new theories and information become available, it is likely that this too will fail the test and be replaced by a new working hypothesis.

For this reason null hypotheses are normally tested in research design. The **null hypothesis** states that there is in reality no difference (or association) between two or more sets of data. It is therefore commonly expressed in the opposite way from the aims of a study. Thus the relationship between depression and thyrotoxicosis suspected in the earlier section could be tested by at least two null hypotheses: (i) depressive illness is no more common at the time of diagnosis of thyrotoxicosis than at the time of diagnosis of disease X (to be chosen as similar in physical morbidity to thyrotoxicosis but not involving thyroid hormones), or (ii) raised levels of thyroid hormones are not associated with raised levels of depressed mood. The hypothesis is a more careful negative formulation of one of the aims of the investigation. If it is disproved it does not mean that the positive form of the hypothesis is proved; it merely becomes more acceptable than the null hypothesis until further information is available.

Creating an experiment

The construction of an appropriate experiment, either using retrospective data (i.e., data already collected) or creating data prospectively, is the main part of research methodology. In general, prospective studies are superior to retrospective ones because they allow suitable experimental conditions to be created which are designed to test the hypothesis in question. It is extremely unlikely that retrospective data will satisfy such requirements. However, past data have the advantage of being much greater in number than most prospective studies are able to obtain, and techniques are available to combine the results of these (e.g., **meta-analysis**, which creates a 'statistical summary' of differences between the same variables compared in a range of experiments and calculates an overall **effect size**). Such results can give support to prospective studies but will rarely be a replacement for them.

Before constructing a definitive experiment it may be appropriate to carry out a **pilot study** to test that the study is feasible. For large and expensive investigations this is usually desirable.

The experimental protocol

The protocol of an investigation is designed to falsify the null hypothesis. Just as the aim of the study has to be formulated carefully and specifically with clear limits the same applies to the construction of the protocol. This includes:

1. selection of **inclusion and exclusion criteria** for patients or other variables
2. carefully defined **conditions** of the study (e.g, treatment schedules, settings, times)
3. **measurements** and forms of evaluation
4. statistical **design**
5. allocation of groups
6. ethical issues (including consent from patients)
7. size of investigation (including **power** calculations)
8. external control of investigation (including monitoring and administration)
9. recording and storing of data
10. statistical analyses
11. communication of findings.

Selection criteria

Inclusion criteria define the population that should be considered for testing the hypothesis in question. This allows the recruitment of a relatively homogeneous group, but this usually needs to be reduced further by the introduction of exclusion criteria which reduce the total numbers. Because it is extremely rare for any research investigation to look at the total population with the condition being investigated it is common to investigate a sample (i.e., a representative group from the total population). In psychiatry the selection of an appropriate sample is often very important because there is much more variation in psychiatric conditions than in other medical disorders. Thus, for example, an investigation of a sample of patients with Wilson's disease, an inherited metabolic disorder in which there is excess absorption of copper, can be easily diagnosed and in no way can be confused with other types of disorder. Thus the study of clinical phenomena and their associations in this condition, or response to treatment, can be examined by choosing any population which satisfies the diagnostic criteria for the condition. The results are likely to be generalizable to all other patients with Wilson's disease because the main selection criterion is the diagnosis. In this particular example, it would be unwise to have too many selection criteria in any case because the condition is rare and it is difficult to recruit sufficient subjects. If, however, a condition such as depression is being examined the investigator has to consider selection by diagnosis (since there are many ways of classifying depressive disorders), age (as there are important differences between the very young and the old when compared to others), the setting where the patients are seen (since depression in the community is very different from depression in psychiatric out-patient clinics or in-patient wards), and possibly many other factors such as duration of symptoms, current and previous treatment, and co-occurrence of other diagnoses. Fortunately the use of common diagnostic systems has improved in recent years and many of the difficulties that prevented generalizability in the past no longer apply today.

In deciding on the list of **exclusion criteria** there is often conflict between the needs of expediency and science. In order to ensure the homogeneous population many exclusion criteria can be introduced but, if these are too many, there is a danger that many patients seen with the condition will not qualify for the study because the exclusion criteria are too comprehensive. This means that there would be difficulty in getting sufficient numbers to test the hypothesis in question and there is also the danger that the results would have little generalizability if they only apply to a highly selected sample of the total population with the condition being tested.

Defining conditions

Most testable hypotheses are simple ones that address one or two questions but seldom more. In order to ensure that these questions only are being asked it is necessary to control other variables as much as possible. The most common example of this is when the conditions of the study are also part of the hypothesis being tested. The commonest example of this is treatment. In order to test the null hypothesis that treatment x and treatment y show no difference in efficacy in a named disorder it is important to ensure that x and y are given under constant and optimum conditions for efficacy. Thus, in the case of a drug treatment it is necessary to define the form in which the drug is to be given, how it is to be administered, the frequency and total daily dosage, the policy about increasing or reducing dosage in response to both beneficial and unwanted effects, the duration of treatment and the measurement of **compliance** (the extent to which the patient takes the treatment as prescribed). All these need to be specified in the design. The same applies to other forms of treatment, including psychological ones, although in practice adherence to this approach is less rigorous than in comparing biological therapies. Thus, for example, in the case of cognitive therapy it is possible to specify the form in which the treatment is given, by which personnel and with what degree of training, the duration and frequency of each treatment as well as its total duration, and allowed modifications to treatment in response to benefits and apparent side-effects. In fact, when comparing drug and psychological treatments it is desirable to set similar rules for both types of treatment if they are to be compared adequately.

It is equally important to set rules for other factors directly or indirectly relating to treatment during such studies. Ideally no other treatment should be given when two or more specific ones are being compared but in practice combined treatment is common and it may be considered more appropriate to adopt a pragmatic trial rather than an explanatory one (see below). Certain treatments may therefore be allowed but these have to be specified, together with their limits, and the setting in which patients receive care needs to be considered. Thus subjects requiring treatment as inpatients during the study may sometimes have to be regarded as drop-outs from the point they reach inpatient status.

Although treatments are most commonly considered under this heading the issue is equally relevant to all forms of research inquiry. The problems are made easier in investigations that take place at a single point in time. Such studies are described as cross-sectional ones and are easier to perform. Thus, for example, to test the hypothesis that bulimia nervosa is an autoimmune disease it would be appropriate in the first instance to measure an appropriate biological variable (e.g. anti-nuclear factor) in patients with bulimia

nervosa and an appropriately matched control group. To the best of our knowledge anti-nuclear factor is not affected by the setting in which the test is undertaken so it would not matter where the blood sample was taken. However, it would be necessary to ensure that the estimations of anti-nuclear factor were made in the same laboratory under conditions where bias could not affect the results.

In studies in which comparison is made between groups over a period of time it is possible to control most additional variables for a short time (e.g. several weeks) but for many reasons, ethical, clinical and pragmatic, it is impossible to do this over a much longer time scale. In such **longitudinal studies** the degree of external control of the studies is lessened. In some studies it may be considered appropriate to abandon to any degree control of the conditions in which the study is undertaken so as to make it as representative of the condition or disorder in actual practice. Thus **cohort studies** are designed specifically to record the baseline characteristics of a population and then to follow up that population at regular intervals in order to record predetermined characteristics. In the case of a disorder for which no satisfactory treatment is known, such studies enable the **natural history** of the disorder to be determined.

Measurements and evaluation

Measurement is much more of a problem in psychiatric research than in other disciplines and this is emphasized in the initial part of this chapter. Some measurements in psychiatry are unequivocal and brook no argument. Demographic variables such as age, sex, marital status and social class can in most cases (but not invariably) be identified clearly, and associated ones such as height, weight and eye colour can similarly be identified. Biological variables used in other forms of medicine can also be measured accurately in the presence of suitable technological equipment. These include the measurement of chemicals in the body fluids (including hormones), measurement of enzyme activity, and physiological measurements. Some of the latter change markedly in response to psychological state and are commonly called **psychophysiological** measures. They include cardiovascular measures such as pulse rate and blood flow, respiratory rate and depth, measures of sweating (skin conductance and galvanic skin response), salivation, eye movements, electromyography (EMG) and electrocephalography (EEG). All these measurements are unequivocally valid if carried out correctly with appropriate equipment and in theory can be described as 'objective'. However, if the assessment of these variables is not carried out by objective criteria then subjective bias can enter into their interpretation. For example, interpretation of the electrocephalogram (EEG) can be extremely difficult and varying results can be obtained from trained and untrained assessors by simple examination of the traces (often called the '**eyeball approach**' to measurement). If the examination is confined to the proportion of different EEG frequencies in the total record this can be carried out using a computer programme and will give more 'objective' findings.

Recent advances in technology have allowed changes in the brain to be studied in man much more easily than in the past particularly through the development of imaging methods including computerized tomography (CT) and magnetic resonance imaging (MRI), and positron emission tomography (PET) scanning. Again, however, it is important to emphasize that the interpretation of some of these findings can be subjective and it is dangerous to assume that because an assessment is biological it is therefore truthful.

Although biological measurements have an important place in psychiatry most of the changes that are important in clinical psychiatric practice have no obvious measurable biological substrate. Investigators therefore have to rely on other psychological means of measurement. By far the most common of these is the **rating scale.** The rating scale records the characteristic (usually a descriptive or qualitative one) in numerical form so that it can be assessed quantitatively. These are divided into **self-rating scales**, also called **questionnaires**, **observer rating scales** carried out by assessors who have normally been trained for the task, and more complex instruments such as **interview schedules**, both **structured**, with careful adherence to well-defined instructions, and **semistructured**, in which the interviewer has a framework round which questions are asked but there is some discretion allowed to the interviewer so that, for example, supplementary questions not in the schedule can be asked to elicit more information. The form of a rating scale varies between the **categorical** scale, which, at its simplest, contains only two categories (e.g. 0 = no diagnosis, 1 = diagnosis), to interval rating scales and **visual analogue** scales (VAS). Interval rating scales have anchor points connecting the scores on the scale. Sometimes for statistical purposes it is important to know whether the intervals between scores are equal; if this is the case the scale is sometimes described as a **Likert** scale. A visual analogue scale is a continuous one which is represented by a single line of exact length (usually 10 cm)

in which only the extremes of the scale are specified (e.g. 'very severe symptoms' at one extreme and 'no symptoms at all' at the other). In these scales the subject is asked to make a mark across the scale at the approximate point appropriate to the question.

The choice of scale type depends on many factors. Clearly if the characteristic in question can be conceived of in clear categories then the categorical scale is most appropriate. However, much of psychiatric work involves the assessment of dimensions in which the use of categories is somewhat artificial. In general, assuming equal value of anchor points, the reliability of the scale improves steadily to seven points but increasing the number beyond this leads to no further improvement in reliability (Cicchetti et al 1985). Much depends on whether the investigation being carried out is one between groups of individuals (the **nomothetic** approach) or within individuals over time (**idiographic** approach).

In comparing groups it is important to have common measurements that can be applied equally to all subjects, where as in idiographic studies a set of ratings can be chosen which could be unique to that individual; these are then recorded, often repeatedly, before analysis. The nomothetic approach is generalizable whereas the idiographic one does not intend to be. If in studies being carried out into symptoms in a sample of depressed patients the results are intended to have implications for these depressed patients as a whole then it is better to use an established rating scale with good reliability in a variety of settings and with cross-validity as well as criterion validity. If, however, the investigator is interested in measuring a subject's feelings over a short period (e.g. five hours) or changes in the quality of relationships with other individuals then it is much less important to have a scale which is generalizable. Thus, a set of simple analogue scales may be constructed which cover the territory of measurement, or similar analogue scales used from previous studies (e.g. Bond & Lader 1974, Tyrer 1976). For those idiographic 'person-centred' enquiries that make repeated measurements of unique variables (e.g. the relationship between two subjects) semantic differential and similar scales can be used, although they are less popular now than in the past. Much of the stimulus for this approach has come from the work of Kelly (1955).

The choice of rating scale is not easy as there are many hundreds available in psychiatry. In addition to the important issue as to whether the scale is appropriate for the aims of the study, other issues such as ease of administration and scoring, suitability for repeated use and sensitivity in detecting change are all important. Although fuller discussion of these issues and descriptions of individual instruments are found elsewhere (Ferguson & Tyrer 1989, Murphy & Tyrer 1989, Thompson 1989, Peck & Shapiro 1990), it might be useful to illustrate the issues involved by describing the stages of formation of a rating scale.

Construction of a rating scale

If no rating scale or questionnaire is available for a variable being measured then it may be necessary to create one. Let us, for example, take a possible study in which it is felt necessary to measure the symptom or experience of **depersonalization** accurately.

This experience is not easy to define and is often misunderstood by patients. The content validity of the proposed instrument first needs to be examined. Comparison of the latest official classifications (the 10th revision of the International Classification of Diseases (ICD-10) and the revised third revision of the Diagnostic and Statistical Manual of Mental Disorders (DSM-III-R) shows there are three common features of the depersonalization syndrome: (a) a feeling of detachment from the self, (b) a qualitative change in the perception of one's experiences, and (c) altered perception of one's surroundings, which are often described 'as if in a dream'. The last of these is described as derealization and in ICD-10 is normally regarded as part of the depersonalization syndrome but in DSM-III-R is excluded (Tyrer 1989, p. 108). It is important to stress that these experiences should be spontaneous and not created by external influences such as drugs or delusional experiences (e.g., passivity experiences in schizophrenia).

These are difficult symptoms to describe, and in the first instance a set of questions, suitable for administration in either questionnaire or interview form, will need to be constructed. These can be shown to colleagues to determine their face validity before testing formally. It may be necessary to decide at an early stage whether a categorical or dimensional instrument is necessary but this could be left to the results of later testing. If depersonalization as an experience is being compared between groups of patients with other psychiatric disorders a categorical diagnosis may be preferred. If, on the other hand, it is wished to study depersonalization phenomena in depth it is better to allow a greater range of scoring, probably with an interval scale separating seven or eight points. If changes in depersonalization are being recorded frequently over time (e.g., during the treatment of severe depression) it may be better to choose an analogue scale. Examples of possible scales are shown in Figure 9.1.

Once the rating scales have been selected it is necessary to test them formally before using it in a study. Content and face validity may be accepted but it could be helpful to show the scales to one or more experienced clinicians or research workers to confirm that all components of the symptom complex are addressed.

Reliability is best tested next, as although a highly reliable test may have poor validity, a highly unreliable test is certainly not valid. As depersonalization is a difficult symptom to assess it may be appropriate to link studies of reliability with concurrent validity. This can be done by comparing the instruments in a population which is considered by experienced clinicians to demonstrate depersonalization, and one which is similar but shows no features of depersonalization. The assessments would be carried out with the interviewer unaware of the two population groups. This could yield the results shown in Table 9.1.

Analysis by t-tests for independent samples show that all three of these differences are significant but that for the interview schedule shows marginally better findings (i.e. lower probability values of $P < 0.001$), mainly because the standard error is lower for this sample. Test–retest reliability of the questionnaire shows $k_w = 0.78$, of the analogue scale 0.46, and inter-rater reliability of the interview schedule shows $k_w = 0.67$. As the distribution of scores on an analogue scale tends to show an unusual distribution it is better to convert the scores using an arcsin transformation (Aitken 1969) and after this the agreement improves to $k_w = 0.52$.

The investigator now has to decide which form is best for the purposes of the proposed investigation. It looks from these findings that the interview schedule or the questionnaire is better than the analogue scale. Problems in scoring these will have been identified

Table 9.1 Assessments of depersonalization

	Population A (depression plus depersonalization) ($n = 25$)	Population B (depression minus depersonalization) ($n = 36$)
Questionnaire (mean score)	3.1	2.3
Interview (mean score)	4.5	2.2
Analogue scale	68.4	32.5

Questionnaire format

In the past two weeks have you had any of the following feelings. Please tick one box only for each question. Do not include feelings you have had when under the influence of drugs or alcohol.

	Score: 2 Yes, definitely	1 Yes, possibly	0 No
1. The feeling as though you were detached from your body.			
2. The feeling that your body, or part of it, seemed different from normal.			
3. The feeling that things around you were unreal or changed from normal.			

Total score range : 0 — 6

Interview format (semi-structured)

Similar questions asked as above but with the option of adding further questions about feelings of remoteness or loss of control, in response to the patient's answers. Scored as (a) definite depersonalisation, (b) possible depersonalisation, and (c) no depersonalisation, or as an interval scale as below.

```
    0        1        2        3        4        5        6        7
    |_____|_____|_____|_____|_____|_____|_____|
  No                Mild                     Moderate            Severe
  depers            depers                    depers              depers
```

The anchor points could each be specified in more detail if preferred, or the individual questions could be scored separately.

Range of scores : 0 – 7 (or 0 – 21 if three scales used)

Analogue scale

Please put a vertical mark across the line below at the point which represents your feeling at present.

No _____ Really
unreality   severe/
(depersonalisation)*   unreality
                                                      (depersonalisation)

Range of scores : 0 – 100 (assuming the line is 10 cm long)

* defined separately in written instructions to the subject.

Fig. 9.1 Examples of different scales constructed for the assessment of depersonalization.

during testing, and improvements to the chosen scale can be made before testing further or going ahead with the investigation. If the latter is decided then it is advisable to carry out a separate test of reliability of the scale during the study (e.g., inter-rater reliability testing on every third patient).

STATISTICAL DESIGN

In choosing the overall design of the protocol it is essential to incorporate an appropriate statistical design that allows the hypothesis to be tested adequately once data have been collected. For this reason it is appropriate to have independent statistical advice before the protocol is agreed. In most investigations comparisons are being made between groups or within individuals and the design chosen allows these differences to be tested. Even in cohort studies in which the progress of a group of subjects is followed up over time without any other population being studied some comparisons will necessarily be made.

The most important of these is the difference between the initial assessments and subsequent ones (commonly known as **pre-post testing**). These are usually unsatisfactory measures because no allowance is made for the effect of time. Sometimes it is possible to use measurements not recorded in the study from past work for comparative purposes. Past data for comparison purposes are commonly described as historical controls and can sometimes be useful. For example, if a cohort of patients with bulimia nervosa was followed up, and 10% found to develop an autoimmune disease, then it might be reasonable to use the known incidence of autoimmune disease in the general population in this age group for comparative purposes. In most cases, however, it is unwise to use historical controls, particularly in the evaluation of new treatments (Pocock 1983, pp. 55–60). In general, studies with historical controls tend to exaggerate the value of a new treatment and so can be extremely misleading. In particular, it is often not appreciated how much impact experimenters can have on the responses of subjects in research studies. Positive social interaction between experimenters and subjects leading to improvement that is independent of the variables being tested is known as the **Hawthorne effect** and offers another reason why comparisons with historical controls are suspect.

The most common design (and certainly the simplest study) is the **parallel group** design. This is most commonly used in clinical trials, which are planned experimental studies involving comparison of treatments in human subjects. In parallel group studies there are two or more groups, usually including a control group for comparison purposes. In parallel group studies subjects remain in the same group and their progress is compared, often over a long time-scale, with other groups 'in parallel'. In a **cross-over design** subjects change groups on one or more occasions during the study. They are usually used in comparisons of treatments and reduce the total numbers necessary to be recruited. Whereas in parallel designs differences are compared between the patients in each group, in the cross-over design the treatment differences within the same subjects are compared. Cross-over designs are suspect when there

is continuation of the effects of one condition into the second. If the extent of this 'carry-over' is significant the results may be completely misleading. For example, comparison of two treatments for depression, one a standard anti-depressant and the other placebo, in a six-week cross-over study in which anti-depressant and placebo were each taken for three weeks, might well lead to results in which the full benefit of the anti-depressant treatment might not be shown until the assessments made during placebo treatment. This might give the spurious impression that placebo was superior to anti-depressants. In order to compensate for carry-over effects it is common for the order of treatments to be randomized so that, for the example described above, half the patients would start on placebo and the other half on the anti-depressant. In cases where three or more treatments are being compared order effects require more complicated designs. The most common is the **Latin square design** in which each treatment is followed by each other treatment an equal number of times (Fig. 9.2).

If a Latin square or similar design is being used it is necessary to recruit subjects in blocks so the conditions for Latin square design are met. Thus for three treatments blocks of six are required (Fig. 9.2) but this increases markedly with more treatments, so that blocks of sixteen for four treatments and 100 for ten treatments are needed. In some cases it is not always possible for all patients to receive the treatments in question, and other designs, termed **balanced incomplete block designs**, are used in which each subject receives a proportion of the treatments only.

In **fixed sample size** trials the number of patients to be included is decided initially, often using power

Fig. 9.2   A Latin square design to compare the effects of three treatments, A, B and C, in a cross-over design.

calculations (see section on size of investigation). In these studies recruitment continues until the predetermined number in the fixed sample is reached. In sequential designs a study continues until a predetermined point. The analysis of a fixed sample size trial is delayed until the end of the study, whereas in sequential trials it is repeated on many occasions throughout the study. In fully sequential designs (Armitage 1975) analyses are carried out after every patient has completed treatment. In a typical closed sequential trial in which two treatments are being compared in a crossover design each patient receives both treatments in random order and a choice is made between them after completion. The nature of the treatment is recorded by an independent investigator and the study ends when one of the boundaries of the design is crossed, indicating that one of the treatments is superior to the other or alternatively that there is no difference between the treatments. A good example of this approach is the first study of the positive value of beta-blocking drugs in anxiety states (Granville-Grossman & Turner 1966). The advantage of sequential designs is that the smallest possible number of patients is included. This can have important ethical considerations. For example, if two treatments are being compared for a potentially fatal disease and one improves the mortality rate more than the other it is important to know this as soon as possible so that the new treatment can be incorporated into clinical practice. It is ethically wrong to continue trials until the fixed sample size has been reached if the new treatment turns out to be so much better than originally expected. One way of resolving this difficulty is to use **group sequential designs** in which analyses are repeated at regular intervals throughout a fixed sample size trial. This can have statistical implications, particularly with regard to choosing appropriate levels of probability, but are quite feasible (Pocock 1977).

Allocation of groups

The most satisfactory allocation of subjects is the randomized one. In **random** allocation people are distributed to two or more groups entirely by chance. By this technique the groups being tested are comparable on all factors at the time of allocation apart from the independent variables being studied. Complete randomization may lead to markedly unequal numbers in different groups (because it is a chance procedure). Techniques such as **stratification** and **minimization** can be used to reduce this. Stratification describes the categorization of patients into groups defined by one or more important variables affecting outcome. This can

be decided before randomization (**pre-stratification**) or at the time of analysis when the actual baseline variables affecting prognosis will be known (**post-stratification**). Minimization describes a number of ways of allocating patients by taking account of prognostic variables through prestratification so that randomization of similar patients is achieved (Johnson 1989). **Constrained randomization** is similar in that it ensures after that certain numbers of patients (e.g. 20, 40) have been entered there will be similar proportions in each group.

Because there is still the possibility of bias entering into assessments of groups who have been allocated at random it is useful to make the assessment of response as objective as possible. In the case of some treatments, particularly drugs, it is possible to conceal the nature of treatment from both assessor and patient. Under these circumstances all drugs (and placebos) used in a study are delivered in tablets or capsules of identical appearance and the nature of the drug can be obtained only by breaking the code. This technique is described as **double-blind** procedure and is recommended in trials of drug treatment. In many studies double-blind procedure is not always used. For example, it is common to use a **placebo washout** period in the evaluation of new drugs before allocation of treatments. Under these circumstances the investigator knows that the patient will receive placebo but the patient may often be ignorant. This is described as **single-blind** assessment. When psychological treatments are being compared it is impossible to prevent the patient from knowing which treatment he or she is receiving but the assessor may make an evaluation without this information being disclosed. This is another example of single-blind procedure. Because it is not always possible to completely eliminate information that could give a clue to the nature of treatment being given it is frequent for the word **masking** to be used to describe some of these procedures; this implies that the subject and/or investigator is not completely unaware of the nature of treatment being given. Under circumstances where bias may enter into evaluation of the results and complete double-blind procedure is not possible it is sometimes useful for either patient or investigator to guess which treatment has been allocated at the end of the study and to record the accuracy of this information in subsequent communications about the project.

Triple-blind procedure is not an exact term and refers to extensions of double-blind procedure. Thus, for example, a study may be carried out with neither investigator nor subject being aware that data are being recorded and compared (this helps to reduce the

Hawthorne effect mentioned earlier), concealing the nature of treatments to all parties to the investigation (including those involved in analysis) until all these have been completed, and more complicated forms of deception such as experimental data being recorded at times when the investigators are not expecting it and ignoring data collected during the investigators' experimental times. These usually have little positive value as well as being very difficult to arrange, and for most purposes the double-blind procedure is regarded as the 'gold standard' for investigators.

Other designs include simple description of the effects of a treatment over time, commonly described as **open studies**. Open studies are invariably more efficacious than double-blind ones because of the absence of satisfactory control populations and the introduction of bias that tends to operate in favour of the new treatment. Other designs can be used when the number of patients is severely limited, sometimes reduced to one. If the disorder is a chronic one then it is possible to introduce a new condition (such as treatment) at regular intervals and rate the response. Ideally changes should be made under double-blind conditions so that neither investigator nor patient knows when the change has occurred. These are sometimes called **ABAB designs** where **A** describes no treatment or placebo, and **B** describes a period of active treatment. In multiple-base line designs several problems are assessed simultaneously at base line and treatments are given for one problem but with simultaneous measurement of others. These are often appropriate for psychiatry in which many problems coexist but are to some extent correlated (Peck 1989).

Ethical issues

Almost all research investigations involving patients need to be approved by a local ethical committee and, in addition to a full protocol, a separate application would need to this committee for approval. In constructing this it is important to use simple language and to try to anticipate ethical problems in advance. In particular the ethical committee will wish to know what risks subjects will run if they take part in a study compared to the risks of alternative approaches. The committee will also like to be reassured that undue pressure is not applied to the subjects to take part in the study. One of the important ethical aspects is patient consent. Although in the past verbal consent was often considered acceptable written consent now is considered mandatory in many instances. This includes a short account of the study and confirmation that it has been explained to the subject, together with the additional information that patients are free to withdraw at any time if they so wish.

The phrase **informed consent** is often used in this context but it is extremely difficult to be certain exactly how informed such consent should be. For example, if in the evaluation of a commonly used drug such as aspirin it was necessary to list all the possible side-effects to patients very few subjects would wish to take part in the study. Despite this, patients do consume aspirin in large quantities because, in practice, most of the side-effects are uncommon and have no significant consequences. However, if the drug can lead to an unusual side-effect relatively frequently it is wrong to hide this from the patient. For example, the antipsychotic drug, clozapine, is thought to be a good drug for resistant schizophrenia, but in approximately 2% of cases it leads to depression of the bone marrow and a fall in the white cell count. For this reason it is necessary to monitor the white cell count regularly during treatment with the drug. Any study that involves the use of clozapine therefore needs a patient information form that confirms that this risk has been explained fully to the patient and accepted. There are major ethical problems if the patient is not considered able to give informed consent. This can apply with the mentally handicapped, the very young, or patients under compulsory orders of the Mental Health Act. There are no general rules that can be applied in such cases, although in psychiatry the American Psychiatric Association (1985) has faced the issue fairly and squarely.

Size of investigation

In most instances it is necessary to have some idea of the size of the sample chosen for investigation from the beginning of the study. The main exception is the fully sequential design discussed earlier in which the outcome of individual patients determines the termination point. In other studies it is desirable in advance to calculate the number of patients required before a designated difference in response between groups becomes significant. This is achieved using the power calculation. This is important because of two common errors that are often made in inferring conclusions from research investigations (Fig. 9.3).

The power of a test measures the likelihood of the test to reject the null hypothesis when the null hypothesis is incorrect. This can be calculated but depends partly on clinical judgment as this determines the criteria for the power calculation. Figure 9.3 represents the four possible results of comparisons between two conditions (e.g., two treatments for a psychiatric disorder).

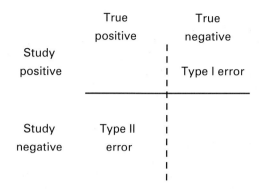

	True positive	True negative
Study positive		Type I error
Study negative	Type II error	

Fig. 9.3 Type I and type II errors.

Two of them are satisfactory; they represent agreement on a true difference between the conditions that is also found in the study and a true absence of difference that is also found in the study. The two other (discordant) categories are commonly called the type I and type II error. The **type I error** is the probability of detecting a 'significant difference' when the treatments are really equally effective and is therefore a false-positive result, whereas the **type II error** is the probability of not detecting a significant difference when there really is an important difference between the two treatments (i.e., a false-negative result). By deciding in advance what is a clinically acceptable difference between effectiveness between two treatments (e.g., a difference of four on the Hamilton Scale for depression) it is possible to calculate the numbers of patients required to minimize the likelihood of these errors.

It is not always possible to obtain sufficient patients to satisfy power requirements, and in such instances it may be necessary to redesign the study. In practice most studies in psychiatry employ insufficient numbers to avoid type II errors in particular. As a consequence, many treatments currently being used for psychiatric disorders on the basis that they are no different from standard treatments are being prescribed under false pretences. They may indeed be equivalent in efficacy, but the studies necessary to show this have not been carried out.

External control of investigation

This is an aspect of research methodology that is often underestimated at the beginning of the study. A lot of hard work goes into establishing the design of the study but it is insufficiently appreciated that many patients (and investigators) will not follow the study protocol satisfactorily and decisions have to be made about the data derived from patients in such instances. A decision has to be made at the beginning about inclusion of incomplete data. In the past it was common to have rigid protocols in which all patients who failed to satisfy the requirements at or during the study would be regarded as drop-outs and not included in analysis. This approach, commonly described as the **explanatory trial** adopts the strict scientific approach that controls (i.e., keeps stable or excludes) all factors liable to affect response apart from the specific one under question. Although such explanatory trials are important and necessary they are not always appropriate in clinical practice, particularly in psychiatry. The reason is that it is extremely difficult to generalize from a group of patients which is already highly selected and becomes even more reduced during such a study because of deviations from the protocol. Thus for example, out of 1 000 patients screened it is not uncommon to have only 50 or 60 completing a study. It is therefore not justified to generalize from the results of the 50 patients to the 950 who have been excluded for various reasons. The 950 probably represent the population as a whole much more accurately than the 50 completed ones. In **pragmatic trials** a different approach is adopted. Clinical practice involves treatment policies more than individual treatment types. Therefore in a pragmatic trial the consequences of employing two or more treatment policies is compared and data analysed on all patients randomized to the treatment groups irrespective of non-compliance or additional treatments. Thus, the only basis for including patients is 'the intention to treat' because this is representative of practice. In a pragmatic trial protocol deviations (departures from agreed procedures) are anticipated and to some extent ignored. In explanatory trials errors can often creep into evaluation. Thus it is possible to show that a high dose of a drug is more effective than a low dose but if three times as many patients drop out of the dose group because of side-effects compared with the low-dose one this is clinically very important. If the withdrawn patients are not included in the analysis, the high dose would then be recommended as the best treatment whereas in reality it would be inadvisable to make this recommendation because the high incidence of side-effects is likely to lead to non-compliance in other subjects.

Recording and storing of data

It is advisable in all research studies to record data, preferably on a computer, immediately after collection. This first involves transferring on to coding sheets to allow mistakes and omissions to be checked before entering data on to the computer. The coding key allows the data to be stored in a standard way and in an appropriate form for subsequent statistical analysis. Suitable scores need to be included for missing data and all information checked before analyses are undertaken. There is a tendency in many psychiatric studies to record too many variables on too few patients. In personal research supervision we almost invariably find that investigators record between three and ten times as many variables as subjects in studies. In order to avoid the type I error of chance positive findings from multiple statistical analyses it is much better to have a ratio of subjects to variables of at least five to one and some authorities argue that the ratio of ten to one is desirable. It is therefore useful to examine the variables being measured and tested carefully before final inclusion to avoid unnecessary recording.

Statistical analyses

These are described in detail in Chapter 8, and additional references are given in the reading list at the end of this chapter.

Communication of findings

This is an important part of research methodology and should be considered at the beginning when writing a protocol. Although it is wrong to anticipate the outcome of a research study it is nonetheless extremely common for investigators to find the results they expect to find. This is such an important issue that it is wise to record the expectations of the investigators before a study ever begins. If a research study is worth doing it is worth communicating to others and so anticipating writing up the project is appropriate. In designing the protocol, past work in the field will need to be considered and a library search undertaken to determine what other work has been published. Clearly if the new study is worth publishing it is to do something more than just repeat the procedure of a previous study (although there are some exceptions when this may sometimes be justified, such as for example repeating a pioneer experiment). It is often useful to consider all possible outcomes of a study in advance of its execution. If a study is worth doing, the findings are worth communicating no matter what the outcome, so examination of each of these should help the investigators to decide whether a study is worth starting.

When writing up a study it is important to include all the relevant findings, including those that are apparently contradictory to others. Because most journals have a bias towards (a) reports of positive findings and (b) short articles, there is a tendency to omit embarrassing inconsistencies in the findings that detract from the main message of the paper. In retrospect, these inconsistencies may constitute the most important findings and it is important to place them in the record of the study.

FURTHER READING

American Psychiatric Association 1985 The principles of medical ethics. APA, Washington DC

Johnson T 1989 Methodology of clinical trials in psychiatry. In: Freeman C, Tyrer P (eds) Research methods in psychiatry: a beginner's guide. Gaskell, London, pp 12–45

Kelly G A 1955 The psychology of personal constructs. Norton, New York

Peck D F, Shapiro C M (eds) 1990 Measuring human problems: a practical guide. John Wiley, Chichester

Pocock S J 1983 Clinical trials: a practical approach. Chichester, John Wiley

Thompson C (ed) 1989 The instruments of psychiatric research. John Wiley, Chichester

10. Epidemiology

Psychiatric epidemiology is concerned with the study of the incidence, prevalence, distribution and determinants of psychiatric disorders with respect to the populations within which they occur, and the ways in which these change over time and with respect to population subgroups. Psychiatric epidemiology is also used in classifying psychiatric disorders. In this chapter the order in which topics are considered is as follows: epidemiological studies; measurements used in descriptive studies; measurements of risk; morbidity and mortality statistics; and identification of cases. Statistical and research methods are described in the previous two chapters and are therefore only briefly mentioned in this chapter.

EPIDEMIOLOGICAL STUDIES

Research methods are described in the previous chapter and only a brief summary is therefore given in this section. **Descriptive epidemiological studies** are based on observations of naturally occurring phenomena and can be followed by either **analytical epidemiological studies**, in which measurements of risk are calculated to test hypotheses concerning risk factors, or **intervention studies** in which an intervention is performed on an experimental group which is then compared with a control group. Intervention studies are also known as experimental studies. Epidemiological studies may be **prospective**, in which a group is followed-up, or **retrospective**, in which historical data (obtained, for example, by interview or from case notes) are used. Thus, for example, cohort studies (see previous chapter) are both analytical and prospective.

MEASUREMENTS USED IN DESCRIPTIVE STUDIES

This section begins by describing two important measurements of disease frequency, prevalence and incidence, followed by a consideration of the need for adjustment owing to confounding variables.

Disease Frequency

The concepts of frequency distribution, frequency table, relative frequency, frequency polygon, cumulative frequency and cumulative frequency curve, are described in the section on data presentation in Chapter 8.

Prevalence

The prevalence of a disease (or disorder) is the proportion of a defined population that has the disease at a given point in time. The population used may be defined variously, for example in terms of demography, geography, in relation to index cases of the same of another disease, and in relation to social, psychological, genetic and environmental factors. For example, one may wish to study the prevalence of a given psychiatric disorder in the sons of alcoholic mothers. Different types of prevalence are now defined.

The **point prevalence** is the proportion of a defined population that has a given disease at a given point in time, and is obtained by dividing the number of people having the disease at the given point in time by the total population at that same point in time.

Similarly, the **period prevalence** is the proportion of a defined population that has a given disease during a given time interval (for example annual prevalence sets the period of one year). In calculating the period prevalence the numerator is the sum of the number of existing cases and new cases diagnosed during the given time interval.

The **lifetime prevalence** is the proportion of a defined population that has or has had a given disease (at any time during each person's lifetime so far) at a given point in time.

The **birth defect rate** is the proportion of live births that has a given disease. The birth defect rate is a type of prevalence, and not a measure of incidence; incidence is defined below. Likewise, the **disease rate at post mortem**, which is the proportion of bodies on which post mortems are carried out that has a given disease, is also a measure of prevalence.

Because it is a ratio of two numbers, prevalence does not have any units, and may be expressed as a percentage by multiplication by 100. For example, Jablensky and Sartorius (1975) estimated the annual prevalence (a period prevalence) of schizophrenia to be two to four per 1 000 (or 0.2–0.4%).

Incidence

The incidence of a disease is the rate of occurrence of new cases in a defined population over a given period of time. It is calculated by dividing the number of new cases over the given time period by the total **population at risk** during the same time period. The population at risk does not include people who have already had the disease by the time of the commencement of the specified time period; instead, it includes only individuals free of the disease who are at risk of becoming new cases. In calculating the incidence, individuals who have already had the disease (and therefore cannot become new cases) must also be excluded from the numerator.

Morbidity and mortality rates are special types of incidence defined below in the section on morbidity and mortality statistics.

The unit of incidence is (specified time period)$^{-1}$. For example, the incidence of schizophrenia found by Dunham (1965) was approximately 0.22 per 1 000 per year (or 22 per 100 000 per year).

Relationship between prevalence and incidence

In the steady state, in which the incidence of a disease remains constant during a given period of time, and the time between the onset of caseness (the time of diagnosis) and its ending (at recovery or death) also remains constant, the point prevalence, P, of the disease is directly proportional to the incidence, I:

P = ID

where D is the **chronicity** or average duration of the disease, and has the inverse units of I (for example, if I is the rate per year, then D is measured in years).

Confounding

Age standardization is used in the comparison of measurements of prevalence and incidence between populations having different age distribution profiles. This is necessary because the prevalence and incidence of most disorders, such as schizophrenia and mood disorders, vary with age. Because age is a variable which affects other variables in calculating measurements used in descriptive epidemiological studies, it is known as a **confounder**. Age is one of the most important confounders, particularly in the calculation of morbidity and mortality rates (see below); other confounders that may need to be compensated for in some studies include gender and race. The statistical details of how data adjustment may be carried out to compensate for confounding are not described in this book.

MEASUREMENTS OF RISK

Analytical epidemiological studies

If the determination of descriptive measurements indicates the possibility of risk factors, prospective (such as cohort studies) or retrospective (including case-control studies) analytical epidemiological studies may be carried out in order to investigate the association of the disease with such risk factors. In this section measurements of risk used in such studies are considered.

Relative risk

The relative risk of a disease with respect to a given risk factor is the ratio of the incidence of the disease in people exposed to that risk factor to the incidence of the disease in people not exposed to the same risk factor.

The relative risk may be calculated in prospective epidemiological studies as follows. Using the notation of Table 10.1, which is a 2 × 2 contingency table used in analytical studies, the incidence of the disease in people exposed to the risk factor is $a/(a + b)$, and the incidence of the disease in people not exposed to the risk factor is $c/(c + d)$. Therefore the relative risk is the former divided by the latter:

relative risk = $a(c + d)/[c(a + b)]$

A value of one for the relative risk implies no causation, and this is the null hypothesis in hypothesis testing. When the relative risk has a value that is

Table 10.1 2 × 2 Contingency table used in analytical epidemiological studies

Exposed to risk factor	Disease	No disease	Total
Yes	a	b	$a + b$
No	c	d	$c + d$
Total	$a + c$	$b + d$	$a + b + c + d$

statistically significantly different to one this is taken as evidence of an association (positive if the relative risk is greater than one, and negative if it is less than one) between the disease and the relevant risk factor. Such an association does not, however, necessarily imply a causative or preventative role for the risk factor.

Odds ratio

The relative risk as calculated above cannot usually be used in retrospective epidemiological studies because a sample of cases and a sample of controls (people without the disease) are selected first and it is not usually possible to estimate the rates of development of the disease with and without exposure to the risk factor under study. If the disease is relatively rare, however, then:

$a \ll b \therefore a + b \approx b$
$c \ll d \therefore c + d \approx d$

Hence the formula given above for relative risk can be approximated by the following formula, known as the odds ratio:

Odds ratio = $ad/(bc)$

Attributable risk

The attributable risk, also known as the risk difference or absolute excess risk, is the incidence of the disease in the group exposed to the risk factor being studied, minus the incidence of the disease in the group not exposed to this risk factor. Using the notation of Table 10.1:

attributable risk = $[a/(a + b)] - [c/(c + d)]$

A value of zero for the attributable risk implies no association, while positive association is indicated by a positive value for the attributable risk.

MORBIDITY AND MORTALITY STATISTICS

Morbidity rate

The morbidity rate of a disease is the rate of occurrence of new non-fatal cases of the disease in a defined population at risk over a given period of time. It is a measurement of incidence and therefore has the same units as incidence.

The **standardized morbidity rate** is the morbidity rate adjusted to compensate for a confounder; if the adjustment is carried out to compensate for the confounding effect of age, the morbidity rate is more precisely known as the **age-standardized morbidity rate**.

In comparing the standardized morbidity rate in a defined population under study with the rate in a comparable standard population, the **standardized morbidity ratio** is used. It is the ratio of the observed standardized morbidity rate (from the population under study) to the expected standardized morbidity rate for the same disease (from the standard population), and is usually expressed as a percentage by multiplying this ratio by 100. This ratio is particularly useful in retrospective studies.

Sources of information for morbidity rates include hospitals, family doctors, and questionnaires.

Mortality rate

The mortality rate is the number of deaths in a defined population during a given period of time divided by the number in the population during the same time period. For the mortality rate from a given disease, the numerator of the ratio is replaced by the number of deaths from that disease. As with the morbidity rate, the mortality rate is a measurement of incidence and therefore has the same units as incidence.

The **standardized mortality rate** is the mortality rate adjusted to compensate for a confounder; if the adjustment is carried out to compensate for the confounding effect of age, the mortality rate is more precisely known as the **age-standardized mortality rate**.

In comparing the standardized mortality rate in a defined population under study with the rate in a comparable standard population, the **standardized mortality ratio**, conventionally abbreviated to SMR, is used. It is the ratio of the observed standardized mortality rate (from the population under study) to the expected standardized mortality rate (from the standard population), and is usually expressed as a percentage by multiplying this ratio by 100. As with the standardized morbidity ratio, the SMR is particularly useful in retrospective studies.

The major source of information for mortality rates in developed countries is from records of death registration. In England and Wales, for example, mortality statistics are available from the Office of Population Censuses and Surveys.

Life expectancy

The life expectancy is a measure of the mean length of time an individual can be expected to live, if mortality rates used in its calculation were to remain constant, and is calculated from the ratio of the total length of time a hypothetical group of people is expected to live to the number of people in that group. The total expected life-span of the group is calculated by applying mortality rates for each age (age-specific rates) to each individual in the group, and extrapolating forwards in time, applying the appropriate age-specific rates with each change in the hypothetical age of the individuals, until the model predicts that all the individuals in the group would have died.

IDENTIFICATION OF CASES

In applying many of the above epidemiological concepts in psychiatric research it is clearly important to be able to identify cases of a given psychiatric disorder correctly. Case definition and screening are two important aspects of this subject and are now considered.

Case Definition

Caseness

As is the case with many biological variables, psychiatric disorders tend to lie on a continuum rather than existing as discrete entities. For example, in clinical practice disorders such as mood disorders exhibit different degrees of severity between different affected patients. Furthermore, two patients both said to be suffering from depression, say, may not manifest the same psychopathological phenomena, let alone the same severity of such phenomena. In psychiatric research, both epidemiological and otherwise, it is necessary to have a standardized method of defining caseness. An overall threshold ideally needs to be defined to differentiate cases from non-cases. Such a threshold may contain thresholds for individual psychopathological phenomena, each of which may in itself be somewhat arbitrary, and overall since a given

condition may manifest different phenomena, in order to count as a case the standardized case criteria may not require all the individual component thresholds to be passed. Standardized classifications used in psychiatry are now considered, while screening is considered later in this section.

Classification

In 1987 the American Psychiatric Association published its revised third edition of the Diagnostic and Statistical Manual of Mental Disorders, commonly abbreviated to DSM-III-R (American Psychiatric Association 1987). This is a multi-axial classification system providing specific operational diagnostic criteria as guides for making each psychiatric diagnosis. Although originally intended primarily for use in the United States, DSM-III-R, and its predecessor DSM-III (American Psychiatric Association 1980), have come to be widely used in psychiatric research internationally. This is partly because the equivalent international classification of psychiatric disorders contained in the ninth edition of the World Health Organization's International Classification of Diseases, ICD-9, published in 1978, does not provide similar diagnostic criteria, although ICD-10 (World Health Organization 1990) does, in the form of diagnostic guidelines. In discussing the merits of these classification systems one has to be aware of the need to obtain good international agreement over diagnostic practice, a measure of reliability (discussed in the previous chapter), and diagnostic accuracy (a measure of validity). Operational criteria for diagnosis may improve reliability at the expense of validity (for example by introducing rigid time scales for the duration of a symptom before it can be regarded as a diagnosis). As psychiatrists prefer a categorical system of classification to a dimensional one, the dividing line between a diagnosis and no diagnosis (that is, implied health) is somewhat arbitrary and inevitably open to dispute. Nevertheless, there is no doubt that the introduction of better case identification has improved diagnostic practice overall.

Case registers

In Chapter 6 the usefulness of the Swedish and Danish twin registers in carrying out twin studies is mentioned. Similarly, psychiatric case registers, containing records of individuals who have been treated for psychiatric disorders in, for example, certain hospitals or hospital catchment areas, can also be useful in carrying out epidemiological and other psychiatric research. Such records may be particularly useful in longitudinal studies since all contacts with the individuals may be recorded. Furthermore, using a psychiatric case register individuals with a given disorder may be readily identified, and if necessary contacted, if research on that disorder is to be carried out. Limitations of psychiatric case registers arise from the geographical mobility of the registered individuals, particularly in some geographical areas such as some inner-city areas, and from registers not being kept up to date for other reasons, although this is now less likely with the increasing use of computerized record keeping.

Screening

Screening is used to identify cases of psychiatric disorder in general or a specific disorder in a population.

Instruments

The development of psychiatric assessment instruments, such as standardized psychiatric interview schedules (which are not strictly screening instruments but are discussed in this subsection because of their use in case identification) and screening questionnaires, have greatly aided the identification of cases for research purposes. Standardized psychiatric interview schedules may be structured, in which case all interviewees are asked the same questions, or unstructured, in which a more open interview is carried out with the direction each interview takes being dependent on the clinical assessment by the interviewer of the answers given. Screening questionnaires may be observer-rated or self-rated. (Reliability and validity have been discussed in the previous chapter). Commonly used psychiatric assessment instruments are now described.

A commonly used standardized psychiatric interview schedule developed in Britain is the Present State Examination (PSE) (Wing et al 1974) which consists of a 140 item semi-structured interview with probes being suggested for individual symptoms. The reliability of this instrument is increased by training interviewers in its use. The results of the PSE can be entered into a computer program called CATEGO (Wing & Sturt 1978) which then provides a CATEGO and ICD-9 classification. The PSE is particularly useful in the diagnosis of schizophrenia but has a poor reliability so far as the diagnosis of personality disorder, organicity, alcoholism and mental retardation are concerned. While the PSE is primarily a diagnostic instrument, a development of it known as the Index of Definition (ID) (Wing et al 1978) can be used to identify cases of psychiatric disorder. The PSE data are used to assign an ID between one (no symptoms) and eight, with an ID of five being the threshold for caseness. If an individual scores five or more, then the PSE data can be used as before to give a diagnosis. The Schedule for Affective Disorders and Schizophrenia (SADS) (Endicott 1978) was developed in the United States and is a structured interview which, in addition to being useful in diagnosing cases of mood disorder and schizophrenia, generally has a better reliability than the PSE in identifying cases of personality disorder and alcoholism. A questionnaire commonly used for screening for psychiatric disorder is the General Health Questionaire (GHQ) (Goldberg 1972) which is self-rated and consists of 60 items each with four possible answers. It is particularly useful in the assessment of anxiety and depression in outpatient and primary care medical patients and is not meant to be used to identify psychotic cases.

Sensitivity, specificity and predictive values

The sensitivity and specificity of an assessment instrument or any other test give a measure of its effectiveness. They are functions of the test and not of the tested individuals.

Consider the generalized diagnostic and test results shown in Table 10.2. The number of individuals for whom the test result is **true positive**, that is, the test result is positive and the true diagnosis is also positive, is a. Similarly, the number of individuals for whom the test is **true negative**, that is, the result is negative and the true diagnosis is also negative, is d. A **false-positive** result occurs when the test result is positive but the true diagnosis is negative, whereas a **false-negative** result occurs when the test result is negative but the true diagnosis is positive. In Table 10.2 the numbers of individuals for whom the tests are false positive and false negative are b and c, respectively. Sensitivity and specificity are measures of the true positive and true negative rates, respectively.

$$\text{sensitivity} = \frac{\text{true positive}}{\text{true positive} + \text{false negative}}$$

$$= \frac{a}{a+c}$$

$$\text{specificity} = \frac{\text{true negative}}{\text{true negative} + \text{false positive}}$$

$$= \frac{d}{d+b}$$

The sensitivity and specificity of a test are often expressed as percentages by multiplying the above

Table 10.2 Generalized diagnostic and test results

Test result	True diagnosis	
	Positive	Negative
Test positive	a	b
Test negative	c	d

expressions by 100. It can be seen from the above expressions that the sensitivity of a test is an index of how well it detects what is being looked for (for example, caseness), whereas specificity is an index of how well what is not being looked for (true negatives) is excluded. Ideally a test should have a high sensitivity and a high specificity. In a test for caseness, if the caseness threshold is raised, this will result in a reduced number of false positives, and therefore increased specificity, but there will also be a decrease in the number of true positives and therefore a decreased sensitivity. Conversely, a reduction in the caseness threshold leads to increased sensitivity but decreased specificity. Since sensitivity and specificity are proportions, the formulae for the standard error of a proportion and the calculation of confidence intervals for proportions described in Chapter 8 can be used.

The **predictive value** of a positive test result is the proportion of the positive results that is true positive, that is, $a/(a + b)$. Similarly, the predictive value of a negative test result is the proportion of the negative results that is true negative, that is, $d/(d + c)$. The **efficiency** of a test is the proportion of all the results that is true (true positive or true negative), that is, $(a + d)/(a + b + c + d)$. Predictive values and efficiency are also often expressed as percentages by multiplying by 100, and again ideally a test should have high predictive values and a high efficiency.

FURTHER READING

Hennekens C H, Buring J E 1987 Epidemiology in medicine. Little, Brown, Boston

MacMahon B, Pugh T F 1970 Epidemiology: principles and methods. Little, Brown, Boston

Susser M 1973 Causal thinking in the health sciences. Oxford University Press, London

Wald N J (1990) The epidemiological approach. In: Souhami R L, Moxham J (eds) Textbook of medicine. Churchill Livingstone Edinburgh

11. Psychology

In this chapter the following topics are covered: psychological development; basic psychological processes in the adult; social psychology; and psychological assessment.

PSYCHOLOGICAL DEVELOPMENT

In this section some aspects of human psychological development are considered with particular emphasis on the periods of infancy and childhood, adolescence, and old age.

Infancy and Childhood

In infancy and childhood the sequence in which developmental stages occurs in different individuals is the same, although the rate of progress through these stages may differ markedly. Each stage differs from its adjacent stage(s) in a qualitative way.

Attachment and bonding

Attachment refers to the tendency of infants (and other young mammals) to try to be, and remain, close to certain people (or certain other mammals of the same species) with whom they share strong emotional ties of a positive nature, particularly in the presence of stimuli that induce fear. When an infant becomes attached to one individual, usually the mother, the attachment is called **monotropic**. Attachment to more than one individual, called **polytropic**, is less common. Whereas attachment essentially takes place from infant to mother (usually), **bonding** takes place in the opposite direction and both processes can begin immediately following birth. In addition to humans, attachment can be readily observed in other mammals, for example in the way kittens stay close to their mothers, particularly in the presence of fear-inducing stimuli, and the clinging behaviour towards their mothers of infant monkeys.

An early theory of **behaviourism**, a school of psychology concerned with the study of observable behaviour, considered attachment to be the result of **conditioning** (described later in this chapter in the next section under learning), with the mother acting as a conditioned reinforcer since she sees to the infants needs, for example by providing food.

The behaviourist theory that attachment was the result of learned behaviour was challenged by the work of Harlow (1959) who offered infant rhesus monkeys a choice of spending time with one of two artificial surrogate mothers. One, made of exposed wire mesh, provided food from a component feeding bottle, whereas the other did not provide food but was cuddlier and softer to touch. It was found that the infant monkeys spent most of the time clinging to the cuddlier surrogate mother, occasionally venturing to the adjacent wire mother for feeding, often while still holding on to the cuddlier mother with a foot, for example. Similarly, the monkeys appeared to be less afraid to explore other parts of their environment so long as they were still in contact with the cuddlier mother. On the basis of this work Harlow concluded that attachment, at least in these monkeys, is a function of requirement to be in contact with a soft object, that is, **contact comfort,** and provides **security** for the infant. Similar studies also showed that infant monkeys preferred warm artificial surrogate mothers to colder ones, and rocking to still surrogates.

An ethological theory based on the work of Lorenz (1962) considered attachment to result from the process of **imprinting**. This is the name given to a process demonstrated by Lorenz in, for example, geese, whereby the first nearby moving object encountered during a critical period soon after hatching is then persistently followed around. Normally goslings will follow their mothers, but when Lorenz caused the first moving object to be encountered to be an object other than the real mother, the goslings were found to follow this object. Among objects different experimental groups of newly-hatched goslings became attached to were Lorenz himself, the wellington boots he was wearing, and even a watering can (which was pulled along by a string). Whenever the goslings were frightened, for example by loud noises, they would hurry back to this object. Imprinting has been demonstrated in other birds, for example in ducklings, and in puppies and lambs. There is no evidence it occurs in primates.

A psychoanalytic understanding of attachment in humans has been provided by Bowlby (1973, 1980) who stated that it takes place in the context of a warm, intimate and continuous relationship with the mother (or other care-giver) in which both infant and mother find satisfaction and enjoyment. It has been suggested that the attachment process, which is said to take an average of six months to become fully established, and more particularly the reciprocal process of bonding, may be stronger if there is tactile contact as soon as possible following birth (Stevens & Mathews 1978). Historically this was a major factor in influencing the maternity hospitals of many countries to adopt a policy of encouraging the mother to hold her baby very soon after birth, and of allowing or encouraging the father to attend the birth. However, there is no evidence that infants who do not experience this early emotional interaction with their parents cannot later form a healthy attachment to their parents or indeed with adoptive parents (Grossman et al 1981). Attachment behaviour on the part of the mother is reinforced by infant activities that promote a maternal behavioural response; such activities include smiling, movement, and signs of distress such as crying (Bowlby 1958). Once attachment has been established an infant will be visibly distressed, usually crying, when the mother leaves his or her presence. With increasing maturity the intensity of this reaction decreases and the growing child can tolerate increasing physical and temporal maternal separation. Bowlby considered that attachment behaviour was never actually outgrown, even in adulthood. Attachment behaviour, including that to non-parental attachment figures such as teachers, has been observed in adolescents and adults (Ainsworth 1985).

The fear that an infant shows of being separated from his or her mother or other care-giver is known as **separation anxiety** and is a manifestation of attachment. Separation anxiety has been described in various disparate societies and has been found to begin at around the age of six months and decrease visibly by three years of age (Kagan et al 1978). Bowlby described a sequence of childhood behaviours that is often seen, particularly in very young children, following prolonged maternal separation, for example because of maternal or childhood hospitalization. These effects are collectively termed **maternal deprivation,** and were at one stage believed to be an important factor, in the case of long-term maternal deprivation, in the formation later in life of an antisocial personality (Bowlby 1958, 1969). The stages are: firstly, **protest,** including crying and searching behaviour; a second stage of **despair,** in which the child is apathetic and miserable and seems to believe that his or her mother may not return; and finally **detachment,** in which the child appears to have distanced himself or herself from his or her mother emotionally, and when the mother returns there is indifference on the part of the child. The link between long-term maternal deprivation and antisocial personality is no longer held to be as important as previously (Rutter 1972), although poor growth, sometimes referred to as **deprivation dwarfism**, may be associated with such chronic deprivation.

Stranger anxiety

Stranger anxiety refers to the fear of strangers exhibited by infants usually between the ages of eight months and one year, although it may first be detected several weeks earlier (Emde et al 1976). Stranger anxiety may occur independently of the associated separation anxiety described above. For example, even though an infant is in physical contact with his or her mother, stranger anxiety may still occur.

Gender role development

Acceptable gender characteristics and behaviours, for example with respect to clothing and occupation, often vary between societies and within given societies with respect to time. The process by which individuals acquire a sense of gender and gender-related cultural traits appropriate to the society and age into which they are born is termed **gender typing** or sex typing. It usually begins at an early age with male and female infants being treated differently, with, for example, the choice of their clothing varying according to gender. The degree to which an individual believes that he or she conforms to the view of gender role of his or her society and time, that is, to prevailing **gender–role stereotypes,** is known as gender-role identity, or more simply, **gender identity**. A sense of gender identity may be present as early as two years of age, and usually by the age of three years. This in turn influences the type of play, for example with respect to the choice of toys, that the child engages in. Factors that influence childhood gender role development include observation of adults, including parental (or other care-giver) roles, the influence of children of a similar age, the media (particularly, in developed countries, the influence of television), and children's stories.

Cognitive development

The most influential theory of cognitive development is that of Piaget (1963, 1971), a Swiss psychologist who identified the four stages of cognitive development now outlined. Piaget believed that infantile and childhood intellectual development entails interaction with the outside world, for example through play. Such interactions, in turn, were held to lead to the construction of new cognitive structures incorporating new information and known as **schemes**, or, in the presence of suitable existing schemes, either the incorporation of the new information into appropriate existing schemes, a process called **assimilation**, or the modification of one or more existing schemes, a process called **accommodation.**

The first stage of Piaget occurs from birth to two years of age and is known as the **sensorimotor stage**. Repeated voluntary motor activities, such as shaking a toy, occur from the age of around two months and are called **circular reactions**. Whereas circular reactions have no apparent purpose until the age of around five months, and are therefore termed primary, from that age until approximately nine months of age secondary circular reactions take place with experimentation and purposeful behaviour gradually being manifested. Tertiary circular reactions occur between the ages of around one year and 18 months and include the creation of original behaviour patterns and the purposeful quest for novel experiences. During the sensorimotor stage the infant also comes to distinguish himself or herself from the environment and, after the age of around 18 months, fully develops the concept of **object permanence,** whereas until about the age of six months an infant believes that an object hidden from view no longer exists.

In the second stage, known as the **pre-operational stage** and occurring from the age of two years to seven years, the child learns to use the symbols of language. Thought processes exhibited during this stage include: **egocentrism**, which is also a feature of the sensorimotor stage, the child believing that everything happens in relation to him or her; **animism**, whereby life, thoughts and feelings are attributed to all objects, including inanimate ones; **precausal reasoning** which is based on internal schemes rather than the results of observation so that, for example, the same volume of a liquid poured from one container to another with a different height and a different diameter

may be thought to have changed volume in spite of this being contradicted by observation; **artificialism**, in which natural events such as the sun shining are attributed to the action of people; and **authoritarian morality,** whereby the child believes that wrongdoing, including breaking the rules of a game, should be punished according to the degree of the damage caused, whether accidental or not, rather than according to motive, while negative events are perceived as being punishments.

The third stage is known as the **concrete operational stage** and occurs between the ages of seven and 12 years. The child demonstrates logical thought processes and more subjective moral judgements. An understanding of the **laws of conservation** of, initially, number and volume, and then weight, is normally achieved during this stage.

The final stage is known as the **formal operational stage** and occurs from the age of around 12 years onwards. It is characterized by the achievement of being able to think in the **abstract,** including the ability to systematically test hypotheses.

Erikson's stages of psychosocial development

The first four of the eight stages of psychosocial development proposed by Erikson (1963) are held normally to take place from birth up to the age of 12 years. According to Erikson's theory, in each of these stages, and the last four stages to old age (see below), the individual has to cope with a **developmental crisis** related to social and environmental interactions. An individual may or may not resolve each crisis in a satisfactory manner.

The first stage, of **basic trust versus mistrust** takes place in the first year of life. Infants may develop a basic trust in others to help them with respect to their needs, leading to basic trust, or if they are often let down in this regard mistrust of others and the world in general develops.

The second stage is **autonomy versus shame and doubt** and takes place between one and three years of age. The expression of parental disapproval of the healthy development of autonomy in the child can lead instead to feelings of shame and doubt.

Similarly, in the third stage, **initiative versus guilt,** which usually takes place between the ages of three and six years, the expression of parental disapproval of the healthy development of initiative in the child, including an interest on the part of the child in his or her sexuality, can lead instead to feelings of guilt.

The final childhood psychosocial developmental stage is that of **industry versus inferiority**, and usually takes place between the ages of six and 12 years. During these school years, if the child is unable to start to gain adult-like skills and achieve an appropriate level of competency in one or more areas then feelings of inferiority may develop.

Adolescence

Following the onset of puberty marked hormonal changes in male and female adolescents lead to a reawakening of sexual curiosity (following that in childhood) and a sex drive of varying intensity. This is usually accompanied by a wish for greater independence from parents or other care-givers and others in positions of authority. These developing emotional drives may conflict with the standards of behaviour that are acceptable to the society and age in which the growing adolescent lives and the successful resolution of these tensions leads to the development of an adult **ego identity.** The latter has been seen by Erikson as a sense of identity that allows the adolescent to have an understanding of who he or she is and of his or her future role in society. It also enables present and future difficulties, which may not be of the individual's own making but may be unfairly caused by others, to be coped with in a mature, appropriate and successful way. In Erikson's fifth psychosocial developmental stage of **identity versus role confusion,** occurring between the ages of 12 and 18 years, an unsuccessful resolution of the tensions of adolescence and the questions of ego identity lead to role confusion, with confusion permeating the adolescent's thoughts about who he or she is, and his or her sexuality and future role in society.

Old age

Prior to the onset of old age parents usually witness their children leaving home, particularly in Western cultures. The **empty nest syndrome** has been used to describe the difficulties some parents encounter on being left on their own. According to Erikson the stages of psychosocial development that occur in young and middle adulthood are, respectively, **intimacy versus isolation,** in which either intimate, including sexual, relationships with others develop, or isolation occurs, and **generativity versus stagnation,** in which the alternatives are a concern with future generations or a feeling of personal stagnation.

Erikson's eighth and final stage of psychosocial development is that occurring in old age and is known as **ego integrity versus despair.** Successful resolution of the psychosocial crisis of this age leads to an integrated view on the part of the individual of his or her life, its meaning, its achievements (both for self and others including future generations), and the ways in which difficulties were coped with. There is an acceptance of one's mortality, a feeling that one's life has been lived in a satisfactory way, and a readiness to face death. The alternative is despair, both on reflection of how life has been lived and the way in which others have been treated, and also on looking to the future and the sense of transience that is felt on facing the end of life. Rather than having a sense of contentment and completion, there is despair at the prospect of death.

If it is believed that one's death is near, an individual may pass through stages that are similar to those recognized as occurring in the terminally ill: denial, anger, bargaining, depression, and finally, acceptance (Kübler-Ross 1969).

BASIC PSYCHOLOGICAL PROCESSES IN THE ADULT

In this section the following basic psychological processes occurring in adult humans are considered: perception, learning, remembering, and motivation and emotion.

Perception

Perception is an active process involving the awareness and interpretation by the brain of sensations received through sensory organs such as the eye. Modalities of sensation that can be perceived include vision, auditory sense, olfaction, taste, touch, pressure, temperature, pain, kinesthetic sense and vestibular sense. Relevant neuroanatomical pathways have been described in Chapter 1. In this subsection a few aspects of the principles of perception of particular relevance to some types of research in psychiatry and psychology are outlined briefly.

Absolute threshold

The minimum energy required to activate a sensory organ is the absolute threshold of that sensory modality. It is the minimum stimulation needed to allow the sensory modality to be experienced by the appropriate sensory organ. For vision, for example, it is the faintest amount of light that can be seen, while for auditory sense it is the faintest sound that can be heard. In practice, the value of the absolute threshold may vary widely between different individuals, and in the same individual tested at different times. In psychometric testing the absolute threshold is taken as the minimum value of the energy required to activate a sensory organ in 50% of trials. The value of the absolute threshold will also vary with other factors concerned with the stimulus itself which therefore need to be controlled for when comparing tests. For vision, such factors include the wavelength of the light and its brightness.

Difference threshold

The **difference threshold** of two sources of a sensory modality is the minimum difference that has to exist between the intensities of the two sources to allow them to be perceived separately. Again, since the value of the difference threshold varies it is taken, in psychometric tests, as being the minimum change in stimulus intensity between two sources, for example in the wavelengths of two sources of light in the case of visual perception, that allows a difference to be noted in 50% of trials. This is also known as the **just noticeable difference,** usually abbreviated to *jnd.*

Weber's law

According to Weber's law, the increase in stimulus intensity, ΔI, needed to allow two sources of intensity I and $(I + \Delta I)$ to be perceived as being different, is directly proportional to the value of I. The constant of proportionality, $\Delta I/I$, is called the **Weber's constant** and for the brightness of light, for example, has a value of approximately 1/60.

Fechner's law

Weber's law is only an approximation which fails to hold over a large range of stimulus intensity. A better, though again not perfect, approximation is provided by **Fechner's law** which holds that sensory perception is a logarithmic function of stimulus intensity so that as the latter increases there is a corresponding increase in the value of the *jnd.*

Signal detection theory

Signal detection theory holds that perception does not depend solely on stimulus intensity but is also a function of biophysical factors and psychological factors such as motivation, previous experiences, and expectations (Green & Birdsall 1978). For example, a mother is able to detect the cry of her baby against a background noise of much greater intensity. Signal detection theory allows that part of perception dependent on stimulus intensity to be separated from other factors influencing perception; the former is conventionally given the symbol d'.

Learning

Learning is a relatively permanent change in behaviour brought about as a result of prior experience. It does not include types of behaviour change caused by maturation or temporary conditions of the person such as the effect of drugs or fatigue.

Learning may occur through associations being made between two or more phenomena. Two forms of such **associative learning** are recognized: classical conditioning and operant conditioning. **Cognitive learning** is a more complex process in which current perceptions are interpreted in the context of previous information in order to solve unfamiliar problems. Evidence that learning can also take place through the observation and imitation of others has led to the development of the **social learning theory.**

Classical conditioning

Classical conditioning or respondent learning was first described by the Russian physiologist Pavlov (1927). In the laboratory setting it was arranged that the rate of salivary flow of a dog could be measured. If a light were switched on the dog would not salivate initially. A few seconds later food was delivered to the dog, and would, as expected, initiate copious salivation. Following several repetitions of this pairing of light followed by food it was found that just switching on the light without following this with the presentation of food would lead to salivation. In other words, the dog has been conditioned to associate the light with food. Food was acting as the **unconditioned stimulus,** eliciting the reflex response of salivation without new learning being involved. The response to the unconditioned stimulus is known as the **unconditioned response**. The light would normally not have elicited the response of salivation, but was now a **conditioned stimulus** that had elicited the response through its association with an unconditioned stimulus. The **conditioned response** is the learned or acquired response to a conditioned stimulus. Thus, in Pavlov's experiments, salivation was both an unconditioned response prior to conditioning, and a conditioned response following conditioning.

The **acquisition** stage of conditioning is the period during which the association is being acquired between the conditioned stimulus and the unconditioned stimulus with which it is being paired. In **delayed conditioning** the onset of the conditioned stimulus precedes that of the unconditioned stimulus and the conditioned stimulus continues until the response occurs. Delayed conditioning has been found to be optimal when the delay between the onset of the two stimuli is around 0.5 seconds. The conditioning becomes less effective as the delay increases above this figure. Less successful than delayed conditioning is **simultaneous conditioning** in which the onset of both stimuli is simultaneous and again the conditioned stimulus continues until the response occurs. In **trace conditioning** the conditioned stimulus terminates before the onset of the unconditioned stimulus and the conditioning becomes less effective as the delay between the two stimuli increases.

Extinction is the gradual disappearance of a conditioned response and occurs when the conditioned stimulus is repeatedly presented without the unconditioned stimulus. For example, in the case of the Pavlovian experiment above, extinction would occur if the light were to be switched on repeatedly without the presentation of food being paired to this. Experiments have demonstrated that extinction does not entail the complete loss of the conditioned stimulus. Following extinction, if an experimental animal is allowed to rest, a weaker conditioned response re-emerges; this is known as **partial recovery.**

Generalization is the process whereby once a conditioned response has been established to given stimulus, that response can also be evoked by other stimuli that are similar to the original conditioned stimulus. For example, in the above experiment, the dog would be found to respond to light of a slightly different wavelength to that of the original conditioned stimulus. In higher learning, generalization is a theory used to help explain how a child learns similarities.

Discrimination is the differential recognition of and response to two or more similar stimuli. With a large enough difference between two similar stimuli, an animal can be conditioned to respond to one stimulus and not to the other. For example a dog can learn to respond differentially to the sound of different bells. Similarly, in human development, children learn to discriminate between the faces (common stimulus) of different people.

Watson and Rayner (1920) described the experimental induction of a phobia, using classical conditioning, in an 11-month-old boy known as Little Albert. Before the experiment, Little Albert did not fear white rats. However, following several episodes of pairing in which the presentation of a white rat was accompanied by a loud noise, the boy developed a fear of the rat in the absence of the frightening noise. This was then repeated with a rabbit, and then generalized to any furry mammal. In the development of childhood phobias in general, it has been suggested that similar learned responses are of aetiological significance.

Operant conditioning

Operant conditioning or instrumental learning is particularly associated with Skinner (1938) although much of the groundwork for the development of the underlying theory was carried out earlier by Thorndike (1911). It is a form of learning in which a voluntary behaviour is engaged in because its occurrence is **reinforced** by being rewarded. Such behaviour is independent of stimuli and was termed **operant behaviour** by Skinner. An alternative type of behaviour termed **respondent behaviour** by Skinner refers to behaviour that is dependent on known stimuli, for example pupillary meiosis in response to bright light and the unconditioned responses described above under classical conditioning.

Thorndike described experiments in which hungry cats were placed in puzzle boxes. To escape from such a box and reach some food visible outside the box a cat would typically need to step on a pedal or pull a string which led to the door to the box being unlocked. By chance, in time a cat would effect such an escape and reach the food and in subsequent trials the time taken by the same cat to perform the same behaviour would diminish; this is called **trial-and-error learning** or behaviour. After several trials a cat might almost immediately voluntarily engage in the behaviour required to allow it to escape and gain its reward, the latter acting as a reinforcer. This observation was incorporated in Thorndike's **law of effect** which held that voluntary behaviour that is paired with subsequent reward is strengthened.

Operant conditioning can be demonstrated in mammals and birds placed in a special cage devised by Skinner and called a Skinner box. A typical Skinner box might incorporate a lever with a dish for food underneath it and a small light bulb above it. It can be arranged that every time the lever is pressed a pellet of food is released into the food dish, with a cumulative record of lever pressing by the experimental animal being recorded. If hungry rats are placed in such a Skinner box, random trial-and-error learning will lead to the rat pressing the lever, the **conditioned response,** to obtain the reinforcing stimulus of the reward of food pellets. If, after many repetitions of the pairing of this conditioned response with its subsequent reinforcement, it is then arranged that pressing the lever is no longer followed by the release of a food pellet, no matter how many times the animal presses the lever, that is, the conditioned response is no longer reinforced, then the conditioned response abates. In other words, as in classical conditioning, the phenomenon of **extinction** can occur in operant conditioning. Again as in classical conditioning, following extinction, if an experimental animal is allowed to rest, and then finds itself in circumstances the same as or similar to those in which the operant behaviour had previously been reinforced, a weaker conditioned response re-emerges; this is known as **partial recovery.** Similarly, the phenomenon of **discrimination** can also occur in operant conditioning. For example, if it is arranged that food pellets are released into the food dish only when both the lever is pressed and the overhead light is on, but not when the light is off, then the experimental animal becomes conditioned to pressing the lever only when the light is on; that is, the light is acting as a discriminative stimulus.

Skinner distinguished between a **positive reinforcer,** a reinforcing reward stimulus which increases the probability of occurrence of the operant behaviour, and a **negative reinforcer,** an aversive stimulus the removal of which increases the probability of occurrence of the operant behaviour. In animal experiments it has been seen that positive reinforcers include stimuli such as food and water that allow basic drives to be fulfilled, whereas an example of a negative reinforcer is an electric shock. For example, a Skinner box may be arranged such that, in order to avoid an aversive stimulus such as an electric shock, the animal must press a lever. Learning this response is called **avoidance conditioning.** In humans, money is an example of a positive reinforcer and fear is an example of a negative reinforcer. Negative reinforcement should not be mistaken for **punishment,** which is the situation that occurs if an aversive stimulus is presented whenever a given behaviour occurs; as would be expected, the probability of occurrence of this response will be diminished by the punishment. For children, a smack is an example of a punisher. A punisher is related to a negative reinforcer in so far as the removal of the same aversive stimulus then allows the latter to act as a negative reinforcer rather than a punisher.

The concepts of primary and secondary reinforcement are also used in operant conditioning. A **primary reinforcer** is a reinforcer that has not been learnt, such as those that act by satisfying basic drives, for example the need for food and water. On the other hand, a **secondary reinforcer** is a stimulus that is dependent on previous learning associating it with existing reinforcers. Examples in humans of secondary reinforcers or conditioned reinforcers as they are also known include money and career status.

Different **schedules of reinforcement** can be used in operant conditioning. In contrast to **continuous reinforcement,** in which reinforcement takes place following every conditioned response and which leads to the maximum response rate, in **partial reinforcement** only some of the conditioned responses are reinforced. The rate of acquisition of the conditioned response in partial reinforcement varies with different schedules of reinforcement. In a **fixed interval** schedule, reinforcement occurs after a fixed interval of time, for example every five minutes. Animal experiments demonstrate that this schedule is poor at maintaining the conditioned response; the animal will typically show the maximum response rate only when it expects the reinforcement to occur. In a **variable interval** schedule, reinforcement occurs after variable intervals of time. This schedule is very good at maintaining the conditioned response. In a **fixed ratio** schedule, reinforcement occurs after a fixed number of responses, for example after every five responses. This schedule is good at maintaining a high response rate with experimental animals responding as if aware of the need to respond rapidly the required number of times in order for each reinforcement to occur. In a **variable ratio** schedule, reinforcement occurs after a variable number of responses. This schedule is very good at maintaining a high response rate and is the type of partial reinforcement that occurs in many forms of gambling, for example the use of fruit machines, in which a win after a variable number of attempts is sufficient to maintain the response of the gambler.

Animals can be taught to perform different behaviours, including complicated routines as sometimes seen in circuses, through a process of reinforcing only those variations in behaviour that begin to approximate to the desired goal. By reinforcing successively closer approximations the desired behaviour can be achieved and this process is called **shaping.**

Applications of associative learning

Wolpe (1950) held that relaxation inhibits anxiety, so that the two are mutually exclusive. This neurological concept of reciprocal inhibition does not in fact hold true. Nevertheless, it can be used analogously in the psychological treatment of patients suffering from conditions, such as phobias, associated with anticipatory anxiety (Wolpe 1958). Patients identify increasingly greater anxiety evoking stimuli, to form an **anxiety hierarchy,** and in the procedure known as **systematic desensitization** the patient is successfully exposed, either in reality or in imagination, to these stimuli in the hierarchy, beginning with the least anxiety evoking stimulus, with each exposure being paired with relaxation. Different methods of relaxation may be employed, with deep muscular relaxation usually being the aim. Note that a strict hierarchical presentation is not essential; the most important element of systematic desensitization is exposure to the feared stimulus itself.

In experiments involving animals repeatedly subjected to unavoidable aversive stimuli, the animal learns that there is no behaviour that will allow it to avoid such stimuli and it may then move very little and look and act in a helpless manner, for example by curling up in a corner of its cage. Such **learned helplessness** may become generalized to other later circumstances and it has been suggested that some of its features, such as reduced voluntary movement and a belief that one has no control over the environment, occur in cases of depression (Seligman 1975).

In **aversion therapy** aversion conditioning is applied to the treatment of certain disorders, including alcoholism and psychosexual disorders, using association with an aversive stimulus such as drug induced nausea and electric shocks. This form of behaviour therapy was used more extensively in the past than it is at present.

Biofeedback is an application of operant conditioning in which human subjects are given feedback on their autonomic nervous system functioning (for example sweating) or motor function (for example muscle tension) and can be trained to exert some control over it, for example to lower systemic blood pressure.

Shaping finds application clinically in the management of behavioural disturbances in mentally retarded patients.

Cognitive learning

Cognition involves the reception, organization, and utilization of information so that cognitive learning is an active form of learning in which mental cognitive structures known as **cognitive maps** are formed by the individual. Cognitive maps allow mental images, not necessarily visual, to be formed which allow meaning and structure to be given to the internal and external environment. They include the schemes described earlier in this chapter. Cognitive learning can occur in a number of ways. It may occur apparently out of the blue, because of an understanding of the relationship between various elements pertinent to a given problem, and is then known as **insight learning.** In **latent learning,** cognitive learning takes place but is not manifested

except in certain circumstances such as the need to satisfy a basic drive. As its name implies, **observational learning** refers to the learning of behaviours and skills that can occur by observation. For example, children may acquire skills essential to being able to cook by watching their parents in the kitchen or by watching television cookery programmes.

Cognitive therapy attempts to treat disorders held to be caused by or associated with cognitive dysfunction (Meichenbaum 1977, Beck 1976). Psychiatric disorders that can be treated with cognitive therapy include depression, anxiety neurosis, and eating disorders (Gelder 1985). In depression, for example, manifestations of cognitive dysfunction include: irrational automatic intrusive thoughts; inflexible primary assumptions concerning the individual oneself and/or his or her relationship with others; and cognitive distortions (arbitrary inference, selective abstraction, overgeneralization, dichotomous thinking, and magnification/ minimization). These components have been summarized in a cognitive triad that occurs in depression, consisting of a negative view of oneself, a negative interpretation of one's present experience, and a negative expectation of the future (Beck 1963, 1964).

Social learning theory

Social learning theory holds that the process of acquiring characteristics, attitudes and behaviours that are in harmony with the society in which one lives, a process known as **socialization,** is the result of individuals interacting with others. Other people whom one likes are used as role models and aspects of classical conditioning, operant conditioning, and cognitive learning (particularly observational learning) are used in this process (Bandura 1977).

Remembering

Memory can be considered to be made up of the following types: sensory memory, short-term memory and long-term memory.

Sensory memory

The sensory memory is a very short-lived memory trace of the sensory input and has been demonstrated to occur with visual information by Sperling (1960). He showed that it is probable that on first encountering new visual information this is briefly entirely retained as a mental image called an **icon.** The sensory memory also exists for other sensory modalities. That for auditory information, for example, is called an **echoic** memory, while that for information from touch is called a **haptic** memory.

Short-term memory

Whereas most of the information or memory trace held briefly in the sensory memory rapidly **decays** and is forgotten, those aspects of this sensory information that are the object of active attention by the individual are transferred into a temporary working memory called the short-term memory. This is the memory used to temporarily hold a name and address or a telephone number, for example, while they are being transferred from a letter to one's address book, or while they are being used (for example in dialling the telephone number). Short-term memory consists of a small finite number of registers (Atkinson & Shiffrin 1968) which can be filled only by data entering one at a time. It has been known since the last century that the number of such registers in the adult human is on average seven with a range usually of between five and nine as can be demonstrated by studying how many meaningless or unconnected syllables, digits or words can be recalled by subjects following one presentation (Ebbinghaus 1964 (originally published in 1885), Miller 1956). Such experiments also demonstrate the **displacement principle** whereby when the registers are full the addition of a new datum leads to the displacement and loss of an existing one. In order to increase the capacity of each register a device called **chunking** can be used, whereby an otherwise internally meaningless string of digits or letters, for example, can be broken down into chunks of digits or letters, with each chunk having a special meaning to the individual and therefore being capable of being stored in one register. For example, British trainee and qualified psychiatrists would probably not need eight registers in order to remember the letters MRCPSYCH; similarly with chunks of letters such as DSM, ICD, EEG, and chunks of figures such as one's date of birth.

In order to maintain information in the short-term memory for longer than a few seconds it is necessary to **rehearse** the information. For example, if one wished to memorize a telephone number containing seven unfamiliar and unconnected digits for one minute, it would be necessary to keep repeating these digits to oneself during that time interval. In experiments in which rehearsal is prevented, for example by giving the subject an additional arithmetic task to perform mentally, it has been demonstrated that over 75% of the information stored in the short-term memory has usually been forgotten by the time nine seconds have elapsed

following the initial data presentation (Murdock 1961). Furthermore, the probability of correctly recalling an item of information is greater if it is one of the first items to be encountered (even if more than seven items have been presented), the **primacy effect,** or one of the most recent items to be encountered, the **recency effect.** Those items having an intermediate serial position are least likely to be recalled accurately, and this overall phenomenon is referred to as the **serial position effect** (Murdock 1962). Whereas the recency effect can be accounted for in terms of the comparatively short interval of time elapsing before recall, the primary effect is more difficult to explain, and may be caused by greater rehearsal of these first items (Rundus 1971).

Rehearsal is not as necessary in approximately 5% of children possessing a photographic memory, known in psychology as **eidetic imagery**, in whom a detailed visual image can be retained for over half a minute (Haber 1969).

Long-term memory

Long-term memory stores information more or less permanently, and it seems probable that its capacity cannot be exceeded over the course of a human lifespan. Indeed, there is evidence that all perceptions and thoughts may be stored in the long-term memory; for example, in Chapter 2 the recall of memories following direct brain electrical stimulation is described.

So far as verbal information is concerned, there is evidence that the long-term memory stores not the exact words, but their meaning. Evidence of such **semantic encoding** comes from experiments which demonstrate that it is easier to remember words paired with meanings (Bower 1972) and the recall of synonymous words to those in a given list (Sachs 1967). It follows that semantic encoding is a more efficient way than simple rehearsal of transferring information from the short-term memory to the long-term memory.

Retrieval of information from the long-term memory is improved by using a **hierarchical network** to organize the storage of information (Bower et al 1969, Collins & Quillian 1969). For example, a new psychiatric trainee who wished to learn a list containing the major psychotropic drugs should find it easier to use a hierarchical classification, including divisions such as anti-depressants, antipsychotics, and so on, and subdivisions such as tricyclic antidepressants, tetracyclic antidepressants, 5-HT reuptake inhibitors, and so on, rather than simply trying to learn a list of drugs by heart. Retrieval of events and emotions is more likely to be successful if it occurs in the same context as that in which the original events and emotions occurred; this is known as **state-dependent learning** (Estes 1972).

Motivation and emotion

Theories of motivation

To account for the motivation of the wide variety of human behaviours, including social behaviours, theories prevalent at the turn of this century based on instincts (James 1890, McDougall 1908) were replaced in the years between the two world wars by a **drive-reduction theory** in which the motivation of behaviour is to reduce the level of arousal associated with a physiological or **basic drive** or biological requirement, for example the need for food, in order to maintain homeostatic control of the internal somatic environment (see Hull 1952). Once such a need has been satisfied the arousal level associated with the particular basic drive is diminished and usually only slowly begins to rise again as the need re-emerges. For example, after eating, the hunger stimulus is satiated and may not re-emerge for several hours. Other basic drives include thirst, pain avoidance, and, in the adult, the sex drive. According to Hull, motivation could be defined as being the product of drive and learning (or habit).

In the sixth decade of this century another dimension, that of **incentives**, was added to the drive-reduction theory to account more comprehensively for motivation (Hull 1952). For example, the presentation of the positive incentive of seeing another person eating a delicious meal may induce hunger in someone who has only recently eaten a meal. An example of a negative incentive is an experience or object associated with a memory of pain. Negative incentives are avoided. In his updated model of motivation, the latter was defined by Hull as being the product of drive, learning, and incentive.

In addition to physiological or basic drives, psychological learned needs, called **social motives** can also be considered as important motivation factors (Murray 1938). Social motives include the need to achieve, to have friends, to attain power over others and control organizations, and to have understanding.

Social learning theory is considered earlier in this chapter and can lead to motivation. For example, new behaviour patterns may result from observational learning.

According to **psychoanalytic theory** as first formulated by Freud, motivation results from instinctual drives that comprise the life instinct, with its associated libido, and the death instinct or thanatos (Freud 1986).

Types of emotion

An emotion is a mental feeling or affection having cognitive, physiological and social concomitants. Plutchik (1980) has classified emotions into eight **primary emotions** which are found not only in humans but also many infrahuman species, and which can each be placed between adjacent similar primary emotions on a wheel or circle. Beginning arbitrarily with one of them, disgust, the other seven emotions are: anger, anticipation, joy, acceptance, fear, surprise and sadness. The wheel or circle is then complete, and disgust again follows sadness in the same direction. Opposing emotions are found opposite each other on this wheel: disgust–acceptance; anger–fear; anticipation–surprise; joy–sadness. Any two adjacent emotions can give rise to a **secondary emotion.** For example, the secondary emotion of love is derived from the primary emotions of joy and acceptance. Similarly, submission results from acceptance and fear, disappointment from surprise and sadness, contempt from disgust and anger, and so on. The wheel of primary emotions can also be extended into three dimensions using the dimension of intensity. Any one of the eight primary emotion can exist in a form of reduced or greater intensity. For example, a less intense form of anger is annoyance, while a more intense form is the emotion of rage. Similarly: distraction (less intense)–surprise (the primary emotion)–amazement (more intense); apprehension– fear–terror; and so on for the other primary emotions.

Theories of emotion

According to the **James–Lange theory** the experience of emotion is secondary to the somatic responses (such as sweating, increased cardiac rate, increased arousal, and action such as fight or flight) to the perception of given emotionally important events (James 1890, Lange 1967 (translation of an 1885 monograph)) . For example, according to this theory if an arachnophobic individual becomes aroused, experiences increased activity of the sympathetic nervous system (see Ch. 2) and runs away after seeing a spider, the feelings of anxiety and fear are the result of the increased sympathetic activity and running away, and not primarily because of the emotion-evoking stimulus.

Cannon (1927) criticized the James–Lange theory on a number of grounds. It was argued that similar physiological changes can accompany emotions A and B, say, when A and B are different. Moreover pharmacologically-induced simulation of such physiological changes is usually not accompanied by either emotion A or emotion B. Furthermore, the experience of emotions can be shown to be independent of somatic responses, occurring in certain cases before the somatic responses.

An alternative theory of emotions put forward by Cannon (1927) and Bard (1934), and known as the **Cannon–Bard theory,** holds that following the perception of an emotionally important event both the somatic responses and the experience of emotion occur together. In neurophysiological terms, the perceived stimulus undergoes thalamic processing, and signals are then relayed to both the cerebral cortex, leading to the experience of emotion, and other parts of the body, such as the autonomic nervous system, leading to somatic responses.

The Cannon–Bard theory can also be criticized on the basis of the observation that there are stimuli, such as sudden danger, which can lead to increased sympathetic activity before the emotion is experienced. Conversely, as mentioned above, there exist occasions when the experience of emotions occurs before the somatic response. These observations are accommodated within the **cognitive–physiological theory** of emotion, particularly associated with Schachter (1971), according to which the conscious experience of an emotion is a function of the stimulus, of somatic or physiological responses, and of cognitive factors such as the cognitive appraisal of the situation and input from long-term memory. The influence of cognitive factors on the conscious experience of emotion was demonstrated in an experiment by Schachter and Singer (1962) in which two groups of subjects were injected with adrenaline (which leads to increased sympathetic activity), but told that the injection was of a vitamin. Neither group of subjects was informed of the true expected actions (of increased arousal) of the injection; the first group was told nothing about its actions while the second group were wrongly informed that the injection would cause effects such as itching. The subjects were asked to wait in pairs, each with someone who had apparently also received the same injection. In reality the second member of each pair was associated with the experimenters and deliberately acted either as if euphoric or as if very angry. It was found that subjects from both groups exposed to the euphoric speech and actions were themselves also more likely to act and feel euphoric, whereas those subjects from both groups exposed to anger were more likely to have feelings of anger. It was therefore argued that in these subjects cognitive appraisal of the current situation, based on observation of others, influenced the jconscious experience of emotion. As a check, it was found that two further groups of subjects, one receiving the adrenaline but given accurate information about its actions of increased arousal (and not expecting to experience any particular emotion), and the other receiving injections of saline (as a control) and given no information, were not as susceptible to the induction of euphoria or anger.

Optimal arousal

Consider the case of a psychiatric trainee taking a psychiatry examination. If the trainee is extremely worked up and anxious during the examination, with a high level of arousal, the performance in the examination may well suffer, with the brain being bombarded with too many stimuli so that the candidate is unable to focus his or her attention on the task at hand. Again, the performance of the candidate will also suffer if the candidate has too low a level of arousal and is not alert enough (indeed, if the level of arousal is too low the candidate would be asleep). It is clearly best for the candidate to have an optimum moderate level of arousal to achieve the optimum performance. In general, as shown in Figure 11.1, a moderate level of arousal leads to an optimum degree of alertness and interest, and therefore to a comparatively high efficiency of performance. Figure 11.1 is sometimes referred to as the **Yerkes–Dodson curve.**

SOCIAL PSYCHOLOGY

Social psychology is concerned with the study of factors that influence the thoughts, feelings, and actions of humans in social situations. The general principles of social psychology are considered in this section from the viewpoints of both the individual and social situations.

Individual social behaviour

Attitudes

The study of attitudes is fundamental to social psychology. The definition of an attitude has been given variously as: 'a mental and neural state of readiness, organized through experience, exerting a directive or dynamic influence upon the individual's response to all objects and situations with which it is related' (Allport 1935); and 'an enduring organization of motivational, emotional, perceptual, and cognitive processes with

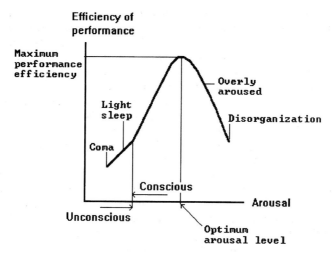

Fig. 11.1   The relationship between level of arousal and efficiency of performance.

respect to some aspect of the individual's world' (Krech & Crutchfield 1948).

The origin of attitudes can be by means of the processes of learning considered in the previous section: classical conditioning, operant conditioning, and observational learning. Superimposed on these types of learning are cognitive processes such as appraisal and modification in the light of new information.

Once formed, attitudes can be modified by either central pathways, entailing the consideration by the individual of new information, or by peripheral pathways involving the presentation of cues such as rewards (Petty & Cacioppo 1981). Advertising is a social influence that uses both pathways.

The **balance theory** of Heider (1958) holds that each individual attempts to organize his or her attitudes, perceptions, and beliefs so that they are in harmony or balance with each other. This balance is challenged, for example, if the individual finds he or she disagrees with someone, B, say, whom he or she likes, with respect to some situation, person, or concept, X, say. If B is also disliked, however, then the cognitive balance is not challenged. Again, cognitive balance is challenged if B is disliked and they both agree about X. If B is disliked and there is disagreement concerning X, cognitive balance again exists. Cognitive imbalance can be resolved in a number of ways, such as changing whether B is liked or disliked.

A related theory is that of **cognitive dissonance**, first formulated by Festinger (1957). According to this

theory cognitive dissonance or discomfort occurs when an individual holds two or more cognitions, including beliefs, attitudes, and self-perceptions, that are inconsistent, or when there is inconsistency between cognitions and behaviour. When cognitive dissonance occurs the individual feels uncomfortable, may experience increased arousal, and is motivated to achieve cognitive consistency (Croyle & Cooper 1983). This may occur by changing one or more of the cognitions involved in the dissonant relationship, changing the behaviour which is inconsistent with the cognition(s), or adding new cognitions which are consonant with pre-existing ones. Another important way in which cognitive consistency can be achieved when attitude and behaviour are inconsistent, called **attitude-discrepant behaviour,** is alteration of attitude. This was demonstrated in an experiment carried out by Festinger and Carlsmith (1959) in which one group of subjects was paid $1 each, and a second group $20 each (a relatively large sum in the late 1950s), to tell waiting subjects that a dull task that had just been completed was in fact interesting. Hence both groups were being financially induced to carry out attitude-discrepant behaviour. A third control group did not have to deceive waiting individuals. After carrying out the required behaviour it was found that the group that found the task most enjoyable and that was most willing to participate in another similar experiment was the group paid just $1 each. This otherwise rather surprising result is consistent with that predicted by the cognitive dissonance theory. Those paid $20 could use this relatively large financial inducement as sufficient motivation to carry out the experiment. However, those paid just $1 had insufficient justification to carry out the experiment and therefore can be considered to have experienced cognitive dissonance which they reduced by altering their attitudes and actually believing that the task was enjoyable after all.

An alternative explanation of the experimental findings just considered is provided by the **self-perception theory,** according to which an individual infers what his or her attitude must be by observation of his or her own behaviour, in a similar way to how other people infer his or her behaviour (Bem 1965). Thus, in the above experiment, whereas the subjects paid $20 each could consider that their behaviour was secondary to the financial reward, and therefore the issue of their attitude did not arise, the subjects paid $1 each would have inferred, from their behaviour, that the task involved must be enjoyable; there was no cognitive dissonance involved according to the self-perception theory. A review by Fazio et al (1977) of cognitive dissonance theory versus self-perception theory explanations for other similar experiments concluded that both theories are of use in explaining behaviour, with cognitive dissonance theory being more appropriate if the behaviour is markedly at variance with the initial attitude of the individual, and self-perception theory being more appropriate if the behaviour lies within the general range of behaviours acceptable to the individual.

Interpersonal attraction

Humans tend to demonstrate a wish to seek the company of others, to whom they are attracted, particularly when in difficult situations. This may be because this allows one to assess situations in the light of social comparisons by taking note of the opinions of the other people; this has been termed the **social comparison theory** (Festinger 1954). An alternative theory is that such behaviour leads to **arousal reduction** (Epley 1974).

There are a number of theories of interpersonal attraction. According to the **reinforcement theory** people are attracted to those who reinforce the attraction with rewards, and this process is a reciprocal one with rewards also passing in the opposite direction and further reinforcing the interpersonal attraction (Newcomb 1956). Conversely, punishments diminish the probability of interpersonal attraction. The **social exchange theory** holds that people have a preference for relationships that appear to offer an optimum cost–benefit ratio; that is, minimum costs such as time spent with the other, and maximum benefits such as love (Homans 1961). A modification is offered by the **equity theory** according to which the preferred relationships, particularly those of an intimate nature, between any two given people are those in which each feels that the cost–benefit ratio of the relationship for each person is approximately equal (Hatfield & Traupmann 1981).

Factors that predispose to interpersonal attraction include proximity, familiarity, similarity, and physical attractiveness. According to the **matching hypothesis,** heterosexual pairing tends to occur in such a way that individuals seek others who have a similar level of physical attractiveness rather than those who are the most attractive. This has been borne out by a number of studies (for example, Berscheid & Walster 1974) and appears to occur because although ideally a given individual would prefer to pair with the most attractive people (Huston 1973), in reality he or she feels that by seeking a partner who has a similar level of attractiveness there is a greater probability of acceptance by the other person, and therefore less likelihood of rejection.

Social influence

Social facilitation

Social facilitation refers to the phenomenon, observed in both humans and infrahuman species, whereby tasks and responses are facilitated when carried out in the presence of others (Allport 1920; Harlow 1932). In order for social facilitation to occur the other members of the species do not necessarily have to be engaging in the same task. Facilitation also occurs if the others are simply observing; this has been called the **audience effect** (Dashiell 1930).

Conformity

Two important types of conformity of individuals to the actions and opinions of others have been identified (Duetsch & Gerard 1955). In the first type, known as **informational social influence,** an individual conforms to the consensual opinion and behaviour of the group both publicly and in his or her own thoughts. This is particularly likely to be the case when individuals are presented with ambiguous stimuli. For example, experiments involving ambiguous perceptual tasks such as those involving optical illusions show that whereas when alone individuals may give widely varying estimates of a parameter such as the distance moved by a spot of light, the same individuals give closely converging estimates when they are together and know each other's estimates (Sherif 1936).

The second type of conformity is known as **normative social influence** and refers to situations in which an individual publicly conforms to the consensual opinion and behaviour of the group but has a different, often firmly-held, view in his or her own mind. The individual conforms to the group under social pressure, to avoid **social rejection.** An illustration was provided by a series of experiments by Asch (1955) in which a subject joined around six to nine others, who, unknown to the subject, had previously been briefed to give certain answers, and the group was asked to carry out a series of unambiguous tasks involving choosing one line out of three which had the same length as a given standard line. In those pre-arranged trials, in which almost all of the others gave the same incorrect answer before the subject was to answer, it was found that the social pressure to conform led to approximately three-quarters of subjects so tested giving an obviously incorrect answer at least once. By varying the size of the group it was found that having just three or four others in the group was sufficient to cause social conformity (Asch 1958).

In spite of the fact that individuals usually conform to majority opinions, it should be borne in mind that there have been notable occasions when what is initially a minority opinion, sometimes that of just one person, has caused the whole course of history for a nation to change. This has occurred in the case of many of the dictatorships which have blighted the twentieth century history of certain nations. One method used by such people is message repetition which, as in advertising, can be a persuasive influence leading to attitude change (Cacioppo & Petty 1979).

Obedience to authority

A defence that was used by many Nazis tried for horrific acts of barbarism, so-called crimes against humanity, was that they were only obeying orders. This subject of obedience to authority figures has been studied by social psychologists. One of the most important, and disturbing, set of experiments was carried out by Milgram (1963, 1974). Subjects were instructed to administer increasingly larger electrical shocks to people who had to learn words read out to them. The subjects were informed that this was part of a study into the effects of punishment on learning and, unknown to them, the electrical shock generating machines did not really administer shocks to the learners. The learners were instructed beforehand to simulate increasing pain when certain levels of

supposed shocks were administered, and the machine used by the subject gave warnings about danger for the final levels of voltage indications. It was found that most subjects would obey the experimenter's orders to administer what they believed to be increasingly powerful shocks to the learners, right up to the maximum voltage available on the machine. Factors that increased the rate of obedience by the subjects were found to include the actual presence of the experimenter, the implication that the subject was bound to continue to keep to the agreement to continue the experiment, and increasing distance from the apparently suffering learner.

Persuasion

Persuasion is used both in peace-time, as in advertising, and during wars, for example propaganda efforts. As mentioned above, message repetition can be a persuasive influence leading to attitude change. Other factors that have been found to increase the power of persuasion include: the credibility of the communicator, which is increased by for example apparently putting forward arguments that are not beneficial to oneself (Walster et al 1966); the attractiveness of the communicator, including physical attractiveness and likeability, which can lead to audience identification with the communicator (Kelman 1961); and views of reference groups, that is groups which one may not necessarily belong to but which play a role in allowing one to identify social norms and which regulate or interpret events, for example one's colleagues (Kelley & Woodruff 1956).

PSYCHOLOGICAL ASSESSMENT

This section begins by considering methods for describing and measuring behaviour. Psychometric methods of assessing intelligence and personality are then described; psychometrics refers to the measurement of psychological characteristics. Reliability and validity are discussed in Chapter 9.

Description and measurement of behaviour

Methods that can be used to describe and measure behaviour include interviews, self-predictions, psychophysiological techniques, and naturalistic observations.

Interviews

Aspects of interviewing and psychometric instruments are discussed in the previous two chapters. In this subection **sources of error,** arising from interview responses, are briefly outlined. An important source of error that is common to all types of interviews is the **response set,** whereby some subjects tend always to agree or to disagree with the questions asked. Another source of error also common to all interviews occurs because of the tendency of certain subjects always to avoid extreme responses. As a result, there is an excess choice of middle responses, and this type of interview response is known as **bias towards the centre** or bias towards the middle. The opposite tendency of selecting extreme responses is known as **extreme responding.**

Social desirability, also known as social acceptability, is characterized by the choice of responses that the subject believes are socially acceptable or the ones that the subject believes the interviewer desires. For instance, in an interview rating the action of a drug under test, the subject may only give answers that he or she believes the interviewer wishes, even when these are not the true answers. **Defensiveness** is another source of error in which the subject avoids giving entirely true answers because of a defensive wish not to give too much self-related information to the interviewer. Both social desirability and defensiveness may occur consciously or unconsciously, and are more likely to occur with questionnaires that are self-administered. Social desirability may to some extent be reduced through the inclusion of **lie scales,** which ask questions such that the socially desirable responses are highly unlikely to be true in practice. Another method used to reduce social desirability is the **forced-choice technique** whereby for each question the subject must choose from at least two items, each having an equivalent degree of social desirability.

Two further important sources of error are associated with the observer. The **halo effect** refers to a source of error in which the observer allows his or her preconception, based on a feature of general attitude, to influence the responses so that they tend to fit in with this preconception. For example, the subject may be judged to be suffering from a particular psychiatric diagnosis and responses that fit this diagnosis may then be more likely to be chosen than would be the case if a more objective assessment were being carried out by a different observer. The **Hawthorne effect** refers to the influence of the social interactions of the subject with the observer on the responses.

Self-predictions

A self-prediction is a direct method of measuring behaviour in which the subject is asked to give his or her own prediction concerning the behaviour under question. For example, former smokers can be asked to give their own predictions as to the likelihood that they will remain non-smokers. Self-predictions can be used in combination with **self-recording,** in which subjects record their own behaviour, for example the number of cigarettes smoked per day in the case of smokers. In general, both self-prediction and self-recording appear to be valuable techniques of measuring behaviour, with self-predictions being as good as more objective methods at predicting some types of behaviour such as the likelihood of resuming smoking after having stopped (Tiffany et al 1986). A notable exception is with respect to situations in which the subject has a strong motive for not being honest, as might occur in the case of prisoners asked to predict the likelihood of re-offending.

Psychophysiological techniques

Psychophysiological techniques involve the direct use of physiological measurements in assessing behaviour. For instance, some studies on sexual arousal in rapists and non-sex-offending male adults exposed to various stimuli have used direct measurements of penile tumescence (for example, Quinsey et al 1984).

Naturalistic observations

Naturalistic observations entail the assessment of behaviour as it occurs with minimum interference by the observer. Various methods of observation can be used, such as observation through one-way mirrors, direct observation of patients by the staff present at the time, and the use of video cameras. A difficulty with this method is that if, as is usually the case, the subject knows he or she is being observed, this can lead to the Hawthorne effect mentioned above. Moreover, observer bias may also occur, although this can be reduced through the use of independent observers whenever possible. **Time-sampling** techniques, in which the subject is observed during given time intervals at given times of the day or night, are used when, as is usually the case, it is not possible to carry out observations continuously.

Naturalistic observations of problem behaviours can be used to carry out a **functional analysis** of such behaviour in order to determine which variables the problem behaviour may be considered to be a function of. In order to carry out a functional analysis the following observations are analyzed: antecedents occurring prior to the episodes of problem behaviour; the actual episodes of problem behaviour; and the consequences of such behaviour. This method is sometimes referred by the acronym ABC (for Antecedents, Behaviours, Consequences). Functional analysis is used by clinical psychologists in the assessment of patients with mental retardation and disorders such as anorexia nervosa and bulimia nervosa (for example, Slade 1982), and more recently in other neurotic disorders.

Psychometric methods of assessing intelligence

Mental age scale

The French psychologist Binet devised the concept of mental age (MA) as being the average intellectual ability, as measured by the level of problem-solving and reasoning, and he published a mental age scale for use with children early in this century. The scale was devised such that the average range of scores corresponds to the chronological age (CA) of the subjects. Children with a higher than average level of intelligence have a MA which is greater than their CA, whereas the MA is less than the CA in children with lower than average intelligence.

A number of derivations of Binet's original and revised mental age scale have been devised, including the **Stanford–Binet test** created in 1916. In the 1972 revision of the Stanford–Binet test six items are allocated to each year, up to the age of 15, so that each item, when passed correctly, corresponds to two months on the mental age score. For example, for the age of two there is an item which asks the child to give the names of different parts of the body of a paper doll, while at the age of eight an item tests how well the child can remember a story. The reliability and validity of this test are both acceptable.

Intelligence quotient

The intelligence quotient (IQ) is the ratio of the mental age to the chronological age, expressed as a percentage:

$$IQ = \frac{MA}{CA} \times 100$$

It can be seen from this formula that the average IQ, when MA and CA are equal, is 100. An IQ of less than 100 (MA < CA) implies less than average intelligence, whereas an IQ of greater than 100 (MA > CA) implies greater than average intelligence. The IQ distribution is Normal, except in the lower range of IQ, as shown in Figure 8.30, with a mean of 100 and a standard deviation of 15. It will be recalled from

Chapter 8 that in the lower range of IQ there are more people than would be expected from the theoretical Normal distribution and that these 'excess cases' result from deleterious environmental and social factors.

The highest value of CA in the above formula is 15 since it is not possible to measure increases in intellectual ability beyond the age of 15 years on the Stanford-Binet test. To enable an IQ to be assigned to people older than 15 years the IQ of adults corresponds to their percentile positions (see Ch. 8) with respect to the general population. For example, an average IQ of 100 corresponds to the 50th percentile.

The IQ of an otherwise well adult tends to remain relatively fixed with age. Mental retardation corresponds to an IQ two standard deviations or more below the mean (that is, IQ ≤ 70); DSM-III-R allows a measurement error of five points so that an IQ of 70 is considered to represent a band of 65 to 75.

Wechsler Adult Intelligence Scale

Rather than giving one mental age score, as in the case of the Stanford–Binet test described above, the well standardized Wechsler Adult Intelligence Scale (WAIS) gives scores on a verbal scale, called the verbal IQ, and a performance scale, called the performance IQ. The WAIS consists of a total of eleven subtests, as shown in Table 11.1. The WAIS has come to be used widely since it was originally published in 1939 and has a relatively high reliability and validity.

Other related intelligence tests include a version of the WAIS suitably modified for children between the ages of five and 15 years known as the Weschler Intelligence Scale for Children (WISC), and the Weschler Preschool and Primary Scale of Intelligence (WPPSI) for children between the ages of four years

Table 11.1 Subtests of the WAIS

Scale	Subtest
Verbal	Information
	Comprehension
	Arithmetic
	Similarities
	Digit span
	Vocabulary
Performance	Picture completion
	Block design
	Picture arrangement
	Object assembly
	Digit symbol

and six years and six months. Note that the suffix '-R' is sometimes added to a test acronym in order to denote a revised version; for example, WISC-R is a more recent revision of WISC.

Group ability test

Unlike the Stanford–Binet, WAIS, and WISC-R, which are used by examiners with individual subjects, group ability tests can be used to allow one examiner to assess the intellectual ability and aptitude of a group of people. They can be used for screening purposes by employers. An example is the Armed Services Vocational Aptitude Battery (ASVAB) used by the armed forces of the United States in recruitment and placement.

Psychometric methods of assessing personality

Objective tests

The items presented to subjects in objective tests have limited responses. Because of their objective nature the reliability and validity of such tests can be assessed. While they have a place in psychiatric and psychological research, personality inventories are used little in British psychiatric practice.

The most commonly used standardized self-report personality inventory is the **Minnesota Multiphasic Personality Inventory** (MMPI). It consists of around 550 statements concerning attitudes, emotional reactions, physical symptoms, psychological symptoms and previous experiences which are presented in a 'true/false/cannot say' format, and it has been widely tested since its invention in 1937. The MMPI has the following scales to increase the test validity: a question scale, which is the number of items not answered; a lie scale derived from 15 items; a frequency scale derived from 64 items which measures random answering or exaggeration; and a correction scale derived from the answers to 30 items which gives a measure of how defensive the subject is in revealing his or her problems. A high score on the frequency scale may invalidate the overall personality profile that the MMPI provides. The MMPI provides scores for the 10 clinical scales shown in Table 11.2. These clinical labels do not imply that a subject scoring highly on any of them is necessarily suffering from the corresponding psychiatric disorder; rather, they should be viewed as being functions of personality attributes. In addition to analysis by a psychologist, the answers to the items can be analyzed by computer to provide both a graphical printout showing the scores on the validity scales and

Table 11.2 The clinical scales of the MMPI

Clinical Scale	MMPI abbreviation
1. Hypochondriasis	Hs
2. Depression	D
3. Hysteria	Hy
4. Psychopathic deviate	Pd
5. Masculinity–femininity	Mf
6. Paranoia	Pa
7. Psychasthenia	Pt
8. Schizophrenia	Sc
9. Hypomania	Ma
10. Social introversion–extraversion	Si

clinical scales, and a written printout of the MMPI interpretations for professional use.

The **California Psychological Inventory** (CPI) uses some of the same items as the MMPI. The CPI is used particularly in psychological research and allows the measurement of 18 traits that are part of normal personality, such as achievement, dominance, self-acceptance and sociability.

The **Eysenck Personality Questionnaire** (EPQ) contains 90 items to be answered in a 'true/false' format, and incorporates a lie scale. Subjects are rated on the following dimensions: extraversion; introversion; and neuroticism.

The **Hostility and Direction of Hostility Questionnaire** (HDHQ) (Caine et al 1967) is a well-known instrument used to measure relationships that could be affected by personality status.

In general, because of evidence that mental state markedly affects scoring of questionnaires, they have been replaced by interview schedules and other observer ratings (Ferguson & Tyrer 1988).

Projective tests

In contrast to objective psychological tests of personality, the items presented to subjects in projective tests have no one correct answer. Instead, the items take the form of ambiguous stimuli such as inkblots. The theory underlying such projective tests is that the subject will project his or her personality onto the ambiguous stimuli. The reliability and validity of these tests have not been established, and indeed the nature of these tests makes the assessment of validity very difficult, and this is one of the most important reasons why they are rarely used in British psychiatric practice.

The **Rorschach Inkblot Test** uses as its ambiguous

stimuli a number of inkblots none of which has any definite meaning. The standard test published by the Swiss psychiatrist Rorschach in 1921 consists of 10 such inkblots, each on a separate card, with five being monochromatic and the other five in colour. The cards are presented in a specific order and the subject asked to relate what the stimuli resemble or to give other responses based on them. The verbal responses of the subject to the presentation of the cards are recorded and the examiner then goes over these responses in turn, asking questions about details of the responses, including a clarification of which parts of each inkblot gave rise to specific aspects of the initial responses. A common method of scoring the responses is according to: location, that is, how much of each inkblot was used; determinants, that is, the features of the blot such as its form, colour, and shading, that the subject was responding to; the content of each response; and how popular, in tests on many subjects, a similar response to a given inkblot is.

The **Thematic Apperception Test** (TAT) was devised by the psychologist Murray in the 1930s in the United States. A number of drawings, each containing ambiguous elements, are presented in turn to the subject who is asked to relate any story associated with each card.

In the **Sentence Completion Test** (SCT) the subject is presented with a number of stems which he or she is asked to complete in turn. Typically there may be as many as 100 such stems, such as 'My greatest fear is ....'. The subject may be asked to give the first completion phrase that comes to mind, and one way of enabling this to occur is for the examiner to present the stems verbally to the subject.

FURTHER READING

Atkinson R L, Atkinson R C, Smith E E 1990 Introduction to psychology. 10th edn. Harcourt Brace Jovanovich, San Diego

Bandura A, Walters R H 1963 Social learning and personality development. Holt, Rinehart & Winston, New York

Beaumont J G 1983 Introduction to neuropsychology. Blackwell Scientific, Oxford

Cronbach L 1960 Essentials of psychological testing. Harper & Row, New York

Rathus S A 1987 Psychology. 3rd edn. Holt, Rinehart & Winston, New York

Spear P D, Penrod S D, Baker T B 1988 Psychology: perspectives on behaviour. John Wiley, New York

12. Social Studies

In this final chapter a brief introduction is given to some basic concepts in social studies, including social groups. For a more comprehensive and detailed discussion of these and other topics in sociology the reader is referred to the list of books given at the end of the chapter.

BASIC CONCEPTS

Social Class

Human societies tend to have hierarchical subdivisions that are somewhat reminiscent of the hierarchies that have been observed by ethologists to exist in many infrahuman primates, other mammals, and birds. One measure of the position of an individual in the hierarchy of a Western society is their social class. The determinants of social class include education, financial status, occupation, type and geographical area of residence, and leisure activities. While these determinants can be weighted and used to give an assessment of social class, the most widely used definition in British psychiatry is that given by the Office of Population Censuses and Surveys, previously the Office of the Registrar-General, which is based on occupational groups as shown in Table 12.1. This classification was first used in the 1911 census.

Those of the same social class are more likely to

spend leisure time with each other and enjoy similar interests. Until the 1979 General Election in Britain, the way in which individuals voted in British General Elections showed a good association with social class. The traditional allegiance of many manual workers to the socialist Labour Party in Britain changed during the 11 years of Margaret Thatcher's prime ministership from 1979, with many such individuals choosing instead to vote for the Conservative Party. A similar social change also took place during the same years in the United States, with the Democrats losing many traditional manual worker voters to the Republican Party under Ronald Reagan and then George Bush. So far as marriage is concerned, members of one social class are probably still more likely to marry someone in the same or an adjacent class.

The incidence and prevalence of many diseases, including psychiatric disorders, have been found to vary with social class, being greater for some classes than others. However, such an association does not necessarily imply causation in the direction of social class to the disease. For example, schizophrenia has a greater representation in the lower social classes. However, this may be caused partly by the disorder itself; schizophrenic children were found by Goldberg and Morrison (1963) to have a lower average social class than their fathers.

Status, role and culture

The status of an individual is his or her position in the above-mentioned social hierarchy. While a person may be considered to be born with a given status, his or her status may increase, for example through educational achievement, or it may decrease for example through mental illness.

The role of an individual in society is the pattern of behaviour in given social situations expected of him or her in relation to his or her social status. According to Parsons (1937), societies have needs that can be

Table 12.1 Social classes

Social class	Occupation	Example
I	Professional, higher managerial, landowners	Hospital consultant
II	Intermediate	Nurse
III	Skilled, manual and clerical	Secretary
IV	Semi-skilled	Postman
V	Unskilled	Cleaner

satisfied only by individuals, and individuals are persuaded to adopt a pattern of behaviour that allows such needs to be satisfied through the culture of a society, a culture being the society's set of beliefs and values.

Individual illness and society

Parsons (1951) defined the **sick role** as being the role given by society to a sick individual. According to Parsons the sick role carries the following two rights for the sick individual: exemption from blame for the illness; and exemption from obligations and responsibilities such as the need to go to work while sick. Similarly, he described two obligations that the sick individual has: the obligation to wish to recover; and the obligation to accept the appropriate help, for example from doctors. In this model the doctor is able to define and legitimize illness, and is able to offer appropriate help. Such legitimization is particularly important in the case of many psychiatric disorders. Whereas with physical injuries, for example, it is clear that an individual is sick and therefore unable to carry out his or her responsibilities such as going to work, this can be more difficult for the general public, and even relatives, to understand in the case of a psychiatric disorder such as depression in which the individual may for instance be lethargic and appear not to wish to work.

The four components outlined above do not always hold true in all cases of illness. For example, the individual may not be exempt from blame for the illness in the case of a drug-induced psychosis or drug overdose.

The behaviour adopted by sick individuals can be described in terms of a set of stages known as **illness behaviour** (Mechanic 1978). The initially well individual begins to experience symptoms of the illness and then asks the opinion of his or her immediate social contacts, such as relatives and close friends. Contact with a doctor or doctors is made, the illness is legitimized by the doctor(s), and the individual adopts the sick role outlined above. The dependent stage of the sick role is ultimately given up upon recovery, by entering a rehabilitation stage, or in the event of death.

The sick role and the particular form illness behaviour takes are culturally determined. In some societies it is not a doctor but rather a spirit healer, witchdoctor, or shaman, say, who both defines and legitimizes illness. Even in Western societies there exist groups, such as members of certain religious cults, for which a non-medically trained person is the one whose help is asked for by sick individuals.

SOCIAL GROUPS

The Family

Basic Concepts

The family can be considered to be a fundamental unit of society since all individuals, including those born to single mothers, begin life as part of a family. However, as mentioned below, thereafter the individual may be brought up in an alternative system to the family.

Kinship is said to exist between two individuals if they belong to the same family. In addition to biologically related individuals kinship also exists between an individual and his or her adopted relative, and between the individual and his or her in-laws.

Types of family

In Western societies the commonest form of the family is the **nuclear family**, consisting of a man and woman who are married to each other, and their child or children. The **extended family** is an extension of the nuclear family, for example the nuclear family together with the parents of one or both married partners of the nuclear family. Extended families are common in certain countries, for example in South East Asia.

If the family is dominated by the father it is termed **patriarchal**. A family dominated by the mother is termed **matriarchal**.

Alternatives to the nuclear and extended family

In many Western societies the occurrence of **one-parent families**, in which one parent, usually the mother, brings up the child or children, is on the increase, and is particularly common in Scandinavian countries.

In the **kibbutzim** of Israel the children are brought up collectively. They have shown a tendency not to marry each other, with non-biologically related male and female children brought up together appearing to treat each other more like biologically related siblings. Similarly, religious and other groups have set up **communes** in which parents and children live together and have common ownership of property and facilities.

In the early years of the **Soviet Union**, between the 1917 Revolution and the time of Stalin, the nuclear family was systematically undermined by the Soviet state through measures such as making divorce extremely easy, teaching young children to put the state and its needs before family loyalty, and encouraging children to spy on their parents (a phenomenon that also occurred in Nazi Europe). The reason for wishing to end the traditional family system was partly that, according to Marxist theory, the family is considered to be a fundamental unit of capitalism, with parents being viewed as agents passing on capitalist philosophies to their children. This policy had disastrous social effects in those parts of the Soviet Union most affected, with widespread disaffection among the youth, alcoholism, vandalism and the like. The policy was reversed during the time of Stalin.

Types of marriage

A taboo against incest, for example brother–sister marriage, appears to exist in all known societies. Historically, there have been exceptions, as in the case of incestuous marriages by the ancient Pharaohs, but these appear to have been sanctioned by the societies concerned, for example because the people allowed to engage in incest were considered to be 'gods' who should mate with others also carrying 'divine' blood.

In many societies, including those influenced by the Judaeo–Christian tradition, **monogamy**, entailing the marriage of one man to one woman, is the only form of marriage accepted by society. **Polygamy** includes both **polygyny**, in which a man has more than one wife at the same time (as sanctioned in Islam and preexilic Judaism), and the rarer **polyandry**, in which one woman has more than one husband at the same time.

Endogamy is the practice of allowing marriage only within the bounds of one's clan or tribe. The opposite custom, of allowing marriage only to someone outside one's group, is known as **exogamy**.

Communities and nations

Groups of individuals and families living according to common customs and sharing common traditions comprise communities. The common goals and group consciousness of a community may only be openly manifested at times of stress, such as during a war or following a natural disaster. Although communities usually live in a common geographical area, this does not necessarily have to be so, as in the case of the widespread members of the Jewish community during the Diaspora.

The social grouping known as the nation refers, from a sociological point of view, to a group of people or communities who consciously feel they belong to a larger community with agreement on a common government and set of laws. Again, a common geographical area does not have to exist. For example, during the Diaspora many Jews considered themselves to be part of the nation of Israel. Similarly, at the time of writing, many Kurds in countries such as Iraq, Iran, Syria and Turkey, feel that they belong to the nation of Kurdistan which currently does not exist in international law.

According to Durkheim (1964, originally published in 1893) a healthy society has a common set of values, its collective conscience, which if disrupted or undermined, for example because of a political revolution, leads to a condition known as **anomie** in which individuals no longer feel constrained in their behaviour by societal regulation. Durkheim (1970, originally published in 1897) further proposed that anomie can lead to **anomic suicide**.

Institutions

A social institution is an established and sanctioned form of relationship between social beings. Examples of institutions include the family, and political and religious groups. A **total institution** is an organization in which 'a large number of like-situated individuals, cut off from the wider society for an appreciable period of time, together lead an enclosed formally administered round of life' (Goffman 1961). Examples of total institutions include the older large mental hospitals, prisons, monasteries and large ships.

From his study of the large St Elizabeth's Hospital, in Washington D.C., Goffman (1961) found that most of the daily life of the patients was highly regulated by the staff who appeared to live in a different world to the patients; Goffman termed this phenomenon binary management. Items needed to sustain life were provided without the requirement to work for them. Goffman also noted the existence of a hierarchy of social status within the total institution. Furthermore, whereas, normally, life consists of a balance between work, home-life and leisure-time, in the total institution these three distinct entities were found by Goffman not to exist in the total institution with patients instead experiencing what Goffman termed batch living.

The process whereby an individual becomes an inhabitant of a total institution was termed the **mortification process** by Goffman. For psychiatric patients Goffman described the process as beginning with a betrayal funnel through which relatives, via doctors, ultimately send the individual to a psychiatric hospital. In the latter the patient was described as undergoing role-stripping and was metaphorically baptized into the patient role by being processed through the admissions procedure, which would usually include being physically stripped naked for the purposes of a physical examination and bathing before being given institutional clothing. Patients were said to show various possible reactions to this mortification process, including: rebellion; withdrawal; colonization, whereby the patient pretends to show acceptance; and actual acceptance both outwardly and inwardly, that is, institutionalization.

Barton (1959) used the term **institutional neurosis** to describe the apathetic, submissive, withdrawn state of patients in total institutions. Wing (1967, 1978) used the term **secondary handicap** to include both institutional neurosis, and similar features

occurring in individuals living outside total institutions. The primary handicap in such individuals may be psychiatric illness, somatic illness, or social difficulties. The secondary handicap results from the unfortunate way in which other people react to the primary handicap, both inside and outside total institutions.

The above criticisms have influenced changes that have occurred and are planned in psychiatric institutions; the discussion of these is beyond the scope of this book.

FURTHER READING

Clare A 1980 Psychiatry in dissent: controversial issues in thought and practice. 2nd edn. Tavistock, London
Goffman E 1961 Asylums: essays on the social situation of mental patients and other inmates. Doubleday, New York

Hollingshead A B, Redlich F C 1958 Social class and mental illness. John Wiley, New York
Mechanic D 1978 Medical sociology. 2nd edn. Free Press, Glencoe
Parsons T 1951 The social system. Free Press, Glencoe

Appendix

Table I Areas under the standard Normal distribution curve

z	.00	.01	.02	.03	.04	.05	.06	.07	.08	.09
0.0	.0000	.0040	.0080	.0120	.0160	.0199	.0239	.0279	.0319	.0359
0.1	.0398	.0438	.0478	.0517	.0557	.0596	.0636	.0675	.0714	.0753
0.2	.0793	.0832	.0871	.0910	.0948	.0987	.1026	.1064	.1103	.1141
0.3	.1179	.1217	.1255	.1293	.1331	.1368	.1406	.1443	.1480	.1517
0.4	.1554	.1591	.1628	.1664	.1700	.1736	.1772	.1808	.1844	.1879
0.5	.1915	.1950	.1985	.2019	.2054	.2088	.2123	.2157	.2190	.2224
0.6	.2257	.2291	.2324	.2357	.2389	.2422	.2454	.2486	.2517	.2549
0.7	.2580	.2611	.2642	.2673	.2704	.2734	.2764	.2794	.2823	.2852
0.8	.2881	.2910	.2939	.2967	.2995	.3023	.3051	.3078	.3106	.3133
0.9	.3159	.3186	.3212	.3238	.3264	.3289	.3315	.3340	.3365	.3389
1.0	.3413	.3438	.3461	.3485	.3508	.3531	.3554	.3577	.3599	.3621
1.1	.3643	.3665	.3686	.3708	.3729	.3749	.3770	.3790	.3810	.3830
1.2	.3849	.3869	.3888	.3907	.3925	.3944	.3962	.3980	.3997	.4015
1.3	.4032	.4049	.4066	.4082	.4099	.4115	.4131	.4147	.4162	.4177
1.4	.4192	.4207	.4222	.4236	.4251	.4265	.4279	.4292	.4306	.4319
1.5	.4332	.4345	.4357	.4370	.4382	.4394	.4406	.4418	.4429	.4441
1.6	.4452	.4463	.4474	.4484	.4495	.4505	.4515	.4525	.4535	.4545
1.7	.4554	.4564	.4573	.4582	.4591	.4599	.4608	.4616	.4625	.4633
1.8	.4641	.4649	.4656	.4664	.4671	.4678	.4686	.4693	.4699	.4706
1.9	.4713	.4719	.4726	.4732	.4738	.4744	.4750	.4756	.4761	.4767
2.0	.4772	.4778	.4783	.4788	.4793	.4798	.4803	.4808	.4812	.4817
2.1	.4821	.4826	.4830	.4834	.4838	.4842	.4846	.4850	.4854	.4857
2.2	.4861	.4864	.4868	.4871	.4875	.4878	.4881	.4884	.4887	.4890
2.3	.4893	.4896	.4898	.4901	.4904	.4906	.4909	.4911	.4913	.4916
2.4	.4918	.4920	.4922	.4925	.4927	.4929	.4931	.4932	.4934	.4936
2.5	.4938	.4940	.4941	.4943	.4945	.4946	.4948	.4949	.4951	.4952
2.6	.4953	.4955	.4956	.4957	.4959	.4960	.4961	.4962	.4963	.4964
2.7	.4965	.4966	.4967	.4968	.4969	.4970	.4971	.4972	.4973	.4974
2.8	.4974	.4975	.4976	.4977	.4977	.4978	.4979	.4979	.4980	.4981
2.9	.4981	.4982	.4982	.4983	.4984	.4984	.4985	.4985	.4986	.4986
3.0	.4987	.4987	.4987	.4988	.4988	.4989	.4989	.4989	.4990	.4990

Table II Two-tailed percentage points of the standard Normal distribution

α	*z*
0.5	0.67
0.1	1.64
0.05	1.96
0.01	2.58
0.001	3.29

Table III One-tailed percentage points of the standard Normal distribution

α	*z*
0.5	0
0.25	0.67
0.1	1.28
0.05	1.64
0.025	1.96
0.01	2.33
0.005	2.58
0.001	3.09
0.0005	3.29

Table IV Two-tailed *t* distribution [1]

	Probability												
d.f.	.9	.8	.7	.6	.5	.4	.3	.2	.1	.05	.02	.01	.001
1	.158	.325	.510	.727	1.000	1.376	1.963	3.078	6.314	12.706	31.821	63.657	636.619
2	.142	.289	.445	.617	.816	1.061	1.386	1.886	2.920	4.303	6.965	9.925	31.598
3	.137	.277	.424	.584	.765	.978	1.250	1.638	2.353	3.182	4.541	5.841	12.924
4	.134	.271	.414	.569	.741	.941	1.190	1.533	2.132	2.776	3.747	4.604	8.610
5	.132	.267	.408	.559	.727	.920	1.156	1.476	2.015	2.571	3.365	4.032	6.869
6	.131	.265	.404	.553	.718	.906	1.134	1.440	1.943	2.447	3.143	3.707	5.959
7	.130	.263	.402	.549	.711	.896	1.119	1.415	1.895	2.365	2.998	3.499	5.408
8	.130	.262	.399	.546	.706	.889	1.108	1.397	1.860	2.306	2.896	3.355	5.041
9	.129	.261	.398	.543	.703	.883	1.100	1.383	1.833	2.262	2.821	3.250	4.781
10	.129	.260	.397	.542	.700	.879	1.093	1.372	1.812	2.228	2.764	3.169	4.587
11	.129	.260	.396	.540	.697	.876	1.088	1.363	1.796	2.201	2.718	3.106	4.437
12	.128	.259	.395	.539	.695	.873	1.083	1.356	1.782	2.179	2.681	3.055	4.318
13	.128	.259	.394	.538	.694	.870	1.079	1.350	1.771	2.160	2.650	3.012	4.221
14	.128	.258	.393	.537	.692	.868	1.076	1.345	1.761	2.145	2.624	2.977	4.140
15	.128	.258	.393	.536	.691	.866	1.074	1.341	1.753	2.131	2.602	2.947	4.073
16	.128	.258	.392	.535	.690	.865	1.071	1.337	1.746	2.120	2.583	2.921	4.015
17	.128	.257	.392	.534	.689	.863	1.069	1.333	1.740	2.110	2.567	2.898	3.965
18	.127	.257	.392	.534	.688	.862	1.067	1.330	1.734	2.101	2.552	2.878	3.922
19	.127	.257	.391	.533	.688	.861	1.066	1.328	1.729	2.093	2.539	2.861	3.883
20	.127	.257	.391	.533	.687	.860	1.064	1.325	1.725	2.086	2.528	2.845	3.850
21	.127	.257	.391	.532	.686	.859	1.063	1.323	1.721	2.080	2.518	2.831	3.819
22	.127	.256	.390	.532	.686	.858	1.061	1.321	1.717	2.074	2.508	2.819	3.792
23	.127	.256	.390	.532	.685	.858	1.060	1.319	1.714	2.069	2.500	2.807	3.767
24	.127	.256	.390	.531	.685	.857	1.059	1.318	1.711	2.064	2.492	2.797	3.745
25	.127	.256	.390	.531	.684	.856	1.058	1.316	1.708	2.060	2.485	2.787	3.725
26	.127	.256	.390	.531	.684	.856	1.058	1.315	1.706	2.056	2.479	2.779	3.707
27	.127	.256	.389	.531	.684	.855	1.057	1.314	1.703	2.052	2.473	2.771	3.690
28	.127	.256	.389	.530	.683	.855	1.056	1.313	1.701	2.048	2.467	2.763	3.674
29	.127	.256	.389	.530	.683	.854	1.055	1.311	1.699	2.045	2.462	2.756	3.659
30	.127	.256	.389	.530	.683	.854	1.055	1.310	1.697	2.042	2.457	2.750	3.646
40	.126	.255	.388	.529	.681	.851	1.050	1.303	1.684	2.021	2.423	2.704	3.551
60	.126	.254	.387	.527	.679	.848	1.046	1.296	1.671	2.000	2.390	2.660	3.460
120	.126	.254	.386	.526	.677	.845	1.041	1.289	1.658	1.980	2.358	2.617	3.373
∞	.126	.253	.385	.524	.674	.842	1.036	1.282	1.645	1.960	2.326	2.576	3.291

[1] (Reproduced from Fisher & Yates (1963) Statistical tables for biological, agricultural and medical research, Longman.)

Table V The χ^2 distribution [1]

d.f.	.99	.98	.95	.90	.80	.70	.50	.30	.20	.10	.05	.02	.01	.001
							Probability							
1	$.0^3157$	$0.^3628$	.00393	.0158	.0642	.148	.455	1.074	1.642	2.706	3.841	5.412	6.635	10.827
2	.0201	.0404	.103	.211	.446	.713	1.386	2.408	3.219	4.605	5.991	7.824	9.210	13.815
3	.115	.185	.352	.584	1.005	1.424	2.366	3.665	4.642	6.251	7.815	9.837	11.345	16.266
4	.297	.429	.711	1.064	1.649	2.195	3.357	4.878	5.989	7.779	9.488	11.668	13.277	18.467
5	.554	.752	1.145	1.610	2.343	3.000	4.351	6.064	7.289	9.236	11.070	13.388	15.086	20.515
6	.872	1.134	1.635	2.204	3.070	3.828	5.348	7.231	8.558	10.645	12.592	15.033	16.812	22.457
7	1.239	1.564	2.167	2.833	3.822	4.671	6.346	8.383	9.803	12.017	14.067	16.622	18.475	24.322
8	1.646	2.032	2.733	3.490	4.594	5.527	7.344	9.524	11.030	13.362	15.507	18.168	20.090	26.125
9	2.088	2.532	3.325	4.168	5.380	6.393	8.343	10.656	12.242	14.684	16.919	19.679	21.666	27.877
10	2.558	3.059	3.940	4.865	6.179	7.267	9.342	11.781	13.442	15.987	18.307	21.161	23.209	29.588
11	3.053	3.609	4.575	5.578	6.989	8.148	10.341	12.899	14.631	17.275	19.675	22.618	24.725	31.264
12	3.571	4.178	5.226	6.304	7.807	9.034	11.340	14.011	15.812	18.549	21.026	24.054	26.217	32.909
13	4.107	4.765	5.892	7.042	8.634	9.926	12.340	15.119	16.985	19.812	22.362	25.472	27.688	34.528
14	4.660	5.368	6.571	7.790	9.467	10.821	13.339	16.222	18.151	21.064	23.685	26.873	29.141	36.123
15	5.229	5.985	7.261	8.547	10.307	11.721	14.339	17.322	19.311	22.307	24.996	28.259	30.578	37.697
16	5.812	6.614	7.962	9.312	11.152	12.624	15.338	18.418	20.465	23.542	26.296	29.633	32.000	39.252
17	6.408	7.255	8.672	10.085	12.002	13.531	16.338	19.511	21.615	24.769	27.587	30.995	33.409	40.790
18	7.015	7.906	9.390	10.865	12.857	14.440	17.338	20.601	22.760	25.989	28.869	32.346	34.805	42.312
19	7.633	8.567	10.117	11.651	13.716	15.352	18.338	21.689	23.900	27.204	30.144	33.687	36.191	43.820
20	8.260	9.237	10.851	12.443	14.578	16.266	19.337	22.775	25.038	28.412	31.410	35.020	37.566	45.315
21	8.897	9.915	11.591	13.240	15.445	17.182	20.337	23.858	26.171	29.615	32.671	36.343	38.932	46.797
22	9.542	10.600	12.338	14.041	16.314	18.101	21.337	24.939	27.301	30.813	33.924	37.659	40.289	48.268
23	10.196	11.293	13.091	14.848	17.187	19.021	22.337	26.018	28.429	32.007	35.172	38.968	41.638	49.728
24	10.856	11.992	13.848	15.659	18.062	19.943	23.337	27.096	29.553	33.196	36.415	40.270	42.980	51.179
25	11.524	12.697	14.611	16.473	18.940	20.867	24.337	28.172	30.675	34.382	37.652	41.566	44.314	52.620
26	12.198	13.409	15.379	17.292	19.820	21.792	25.336	29.246	31.795	35.563	38.885	42.856	45.642	54.052
27	12.879	14.125	16.151	18.114	20.703	22.719	26.336	30.319	32.912	36.741	40.113	44.140	46.963	55.476
28	13.565	14.847	16.928	18.939	21.588	23.647	27.336	31.391	34.027	37.916	41.337	45.419	48.278	56.893
29	14.256	15.574	17.708	19.768	22.475	24.577	28.336	32.461	35.139	39.087	42.557	46.693	49.588	58.302
30	14.953	16.306	18.493	20.599	23.364	25.508	29.336	33.530	36.250	40.256	43.773	47.962	50.892	59.703
32	16.362	17.783	20.072	22.271	25.148	27.373	31.336	35.665	38.466	42.585	46.194	50.487	53.486	62.487
34	17.789	19.275	21.664	23.952	26.938	29.242	33.336	37.795	40.676	44.903	48.602	52.995	56.061	65.247
36	19.233	20.783	23.269	25.643	28.735	31.115	35.336	39.922	42.879	47.212	50.999	55.489	58.619	67.985
38	20.691	22.304	24.884	27.343	30.537	32.992	37.335	42.045	45.076	49.513	53.384	57.969	61.162	70.703
40	22.164	23.838	26.509	29.051	32.345	34.872	39.335	44.165	47.269	51.805	55.759	60.436	63.691	73.402
42	23.650	25.383	28.144	30.765	34.157	36.755	41.335	46.282	49.456	54.090	58.124	62.892	66.206	76.084
44	25.148	26.939	29.787	32.487	35.974	38.641	43.335	48.396	51.639	56.369	60.481	65.337	68.710	78.750
46	26.657	28.504	31.439	34.215	37.795	40.529	45.335	50.507	53.818	58.641	62.830	67.771	71.201	81.400
48	28.177	30.080	33.098	35.949	39.621	42.420	47.335	52.616	55.993	60.907	65.171	70.197	73.683	84.037
50	29.707	31.664	34.764	37.689	41.449	44.313	49.335	54.723	58.164	63.167	67.505	72.613	76.154	86.661
52	31.246	33.256	36.437	39.433	43.281	46.209	51.335	56.827	60.332	65.422	69.832	75.021	78.616	89.272
54	32.793	34.856	38.116	41.183	45.117	48.106	53.335	58.930	62.496	67.673	72.153	77.422	81.069	91.872
56	34.350	36.464	39.801	42.937	46.955	50.005	55.335	61.031	64.658	69.919	74.468	79.815	83.513	94.461
58	35.913	38.078	41.492	44.696	48.797	51.906	57.335	63.129	66.816	72.160	76.778	82.201	85.950	97.039
60	37.485	39.699	43.188	46.459	50.641	53.809	59.335	65.227	68.972	74.397	79.082	84.580	88.379	99.607
62	39.063	41.327	44.889	48.226	52.487	55.714	61.335	67.322	71.125	76.630	81.381	86.953	90.802	102.166
64	40.649	42.960	46.595	49.996	54.336	57.620	63.335	69.416	73.276	78.860	83.675	89.320	93.217	104.716
66	42.240	44.599	48.305	51.770	56.188	59.527	65.335	71.508	75.424	81.085	85.965	91.681	95.626	107.258
68	43.838	46.244	50.020	53.548	58.042	61.436	67.335	73.600	77.571	83.308	88.250	94.037	98.028	109.791
70	45.442	47.893	51.739	55.329	59.898	63.346	69.334	75.689	79.715	85.527	90.531	96.388	100.425	112.317

For odd values of d.f. between 30 and 70 the mean of the tabular values for d.f.–I and d.f.+I may be taken. For larger values of d.f., the expression $\sqrt{2\chi^2}-\sqrt{2\text{d.f.}-I}$ may be used as a normal deviate with unit variance, remembering that the probability for χ^2 corresponds with that of a single tail of the normal curve.

[1] (Reproduced from Fisher & Yates (1963) Statistical tables for biological, agricultural and medical research, Longman.)

References

Abrams R, Schwartz C M 1985 Electroconvulsive therapy and prolactin release: relation to treatment response in melancholia. Convulsive Therapy 1: 38

Achté K A, Anttinen E E 1963 Suizide bei Hirngeschädigten des Krieges in Finland. Fortschritte der Neurologie, Psychiatrie 31: 645

Achté K A, Hillbom E, Aalberg V 1967 Psychoses following war brain injuries. Acta Psychiatrica Scandinavica 45: 1

Adachi D K, Kalivas P W, Schenk J O 1990 Neurotensin binding to dopamine. Journal of Neurochemistry 54: 1321

Adler N T (ed) 1981 Neuroendocrinology of reproduction. Plenum Press, New York

Adolfsson R, Gottfries C G, Roos B E, Winblad B 1979 Changes in the brain catecholamines in patients with dementia of Alzheimer type. British Journal of Psychiatry 135: 216

Adrian E D, Yamagiwa K 1935 The origin of the Berger rhythm. Brain 58: 323

Ågren H 1980 Symptom patterns in unipolar and bipolar depression correlating with monoamine metabolies in the cerebrospinal fluid: I. General patterns. Psychiatry Research 3: 211

Agresti A, Wackerly D 1977 Some exact conditional tests of independence for r × c cross-classification tables. Psychometrika 42: 111

Aguilar L, Lisker R, Hernandez-Peniche J, Martinez-Villar C 1978 A new syndrome characterized by mental retardation, epilepsy, palpebral conjunctival telangiectasias and IgA deficiency. Clinical Genetics 13: 154

Ainger L E, Kelley V C 1955 Familial athyreotic cretinism: report of 3 cases. Journal of Clinical Endocrinology 15: 469

Ainsworth M D S 1985 Attachments across the life span. Bulletin of the New York Academy of Medicine 61: 792

Aitken R C B 1969 Measurement of feelings using visual analogue scales. Proceedings of the Royal Society of Medicine 62: 989

Akesson H O 1965 Cutis verticis gyrata, thyroaplasia, and mental deficiency. Acta Geneticae Medicae et Gemellologiae 14: 200

Alexander W S 1949 Progressive fibrinoid degeneration of fibrillary astrocytes associated with mental retardation in a hydrocephalic infant. Brain 72: 373

Allport F H 1920 The influence of the group upon association and thought. Journal of Experimental Psychology 3: 159

Allport G W 1935 Attitudes. In: Murchison C (ed) A handbook of social psychology. Clark University Press, Worcester, Massachusetts

American Psychiatric Association 1980 Diagnostic and Statistical Manual of Mental Disorders. 3rd edn (DSM-III). American Psychiatric Association, Washington D.C.

American Psychiatric Association 1985 The principles of medical ethics. APA, Washington D.C.

American Psychiatric Association 1987 Diagnostic and Statistical Manual of Mental Disorders. 3rd ed — revised (DSM-III-R). American Psychiatric Association, Washington D.C.

Ampola M G, Elfron M L, Bixby E M, and Meshorer E 1969 Mental deficiency and a new aminoaciduria. American Journal of Diseases of Children 117: 66

Anch A, Salamy J G, McCoy G F, Somerset J S 1982 Behaviorally signalled awakenings in relationship to duration of alpha activity. Psychophysiology 19: 528

Anch A M, Browman C P, Mitler M M, Walsh J K 1988 Sleep: a scientific perspective. Prentice Hall, New Jersey

Anden N E 1975 Animal models of brain dopamine function. In: Birkmayer W, Hornykiewicz O (eds) Advances in Parkinsonism. Roche, Basle

Aperia B, Thoren M, Zettergren M, Wetterberg L 1984 Plasma pattern of adrenocorticotropin and cortisol during electroconvulsive therapy in patients with major depressive illness. Acta Psychiatrica Scandinavica 70: 361

Appleby L 1991 Suicide during pregnancy and in the first postnatal year. British Medical Journal 302: 137

Applezweig M H, Baudry F D 1955 The pituitary-adrenocortical system in avoidance learning. Psychological Reports 1: 417

Arakawa T, Tamura T, Ohara K, Narisawa K, Tanno K, Honda Y, Higashi O 1968 Familial occurrence of formiminotransferase deficiency syndrome. Tohoku Journal of Experimental Medicine 96: 211

Arana G W, Baldessarini R J 1987 Clinical use of the dexamethasone suppression test in psychiatry. In: Meltzer H Y (ed) Psychopharmacology: the third generation of progress. Raven Press, New York

Arendt J 1988 Melatonin. Clinical Endocrinology 29: 205

Arendt J 1989 Melatonin: a new probe in psychiatric investigation? British Journal of Psychiatry 155: 585

Arendt J, Aldhous M, English J et al 1987 Some effects of jet lag and their alleviation by melatonin. Ergonomics 30: 1379

Armitage P 1971 Statistical methods in medical research. Blackwell, Oxford

Armitage P 1975 Sequential medical trials. 2nd edn. Blackwell, Oxford

Åsberg M, Cronholm B, Sjöqvist F, Tuck D 1971 Relationship between plasma level and therapeutic effect of nortriptyline. British Medical Journal iii: 331

Åsberg M, Bertilsson L, Måtensson B, Scalla-Tomba G-P, Thorén P, Träskman-Bendz L 1984 CSF monoamine metabolites in melancholia. Acta Psychiatrica Scandinavica 69: 201

Asch S E 1955 Opinions and social pressure. Scientific American 103: 31

Asch S E 1958 Opinions and social pressure. Scientific American 103: 31

Asch S E 1958 Effects of group pressure upon modification and distortion of judgments. In: Maccoby E E, Newcomb T M, Hartley E L (eds) Readings in social psychology. 3rd edn. Holt, Rinehart & Winston, New York

Aschauer-Treiber G, Aschauer H N, Isenberg K E et al 1990 No evidence for linkage between chromosome 5 markers and schizophrenia. Human Heredity 40: 109

Aschoff J 1969 Desynchronization and resynchronization of human circadian rhythms. Aerospace Medicine 40: 847

Aserinsky E, Kleitman N 1953 Regularly occurring periods of eye motility and concomitant phenomena during sleep. Science 118: 273

Asher R 1949 Myxoedematous madness. British Medical Journal ii: 555

Ashton C H, Rawlins M D, Tyrer S P 1990 A double-blind placebo-controlled study of buspirone in diazepam withdrawal in chronic benzodiazepine users. British Journal of Psychiatry 157: 232

Aston-Jones G, Bloom F E 1981 Norepinephrine-containing locus coeruleus neurons in behaving rats exhibit pronounced responses to non-noxious environmental stimuli. Journal of Neuroscience 1: 887

Atkinson R, Shiffrin R M 1968 Human memory: a proposed system and its control processes. In: Spence K W, Spence J T (eds) The psychology of learning and motivation vol 2. Academic Press, New York

Aula P, Autio S 1980 Salla disease. In: Eriksson A W, Forsius H R, Nevanlinna H R, Workman P L, Norio R K (eds) Population structure and genetic disorders. Academic Press, New York

Ayuso-Gutierres J L, Cabranes J A, Garcia-Camba E, Almoguera I 1987 Pituitary-adrenal disinhibition and suicide attempts in depressed patients. Biological Psychiatry 22: 1409

Baar H S, Hickmans E M 1956 Cephalin-lipidosis: a new disorder of lipid metabolism. Acta Medica Scandinavica 155: 49

Baile C A, McLaughlin C L, Della-Ferra M A 1986 Role of cholecystokinin and opioid peptides in control of food intake. Physiological Reviews 66: 172

Bandura A 1977 Social learning theory. Prentice-Hall, Englewood Cliffs, New Jersey

Banki C M, Bissette G, Arato M, Nemeroff C B 1988 Elevation of immunoreactive CSF TRH in depressed patients. American Journal of Psychiatry 145: 1526

Bard P 1934 The neurohumoral basis of emotional reactions. In: Murchison C (ed) A handbook of general experimental psychology. Clark University Press, Worcester, Massachusetts

Barker W A, Goth J, Eccleston D 1987 The Newcastle chronic depression study — results of a treatment regime. International Clinical Psychopharmacology 2: 261

Barnes T R E, Braude W M 1985 Akathisia variants and tardive dyskinesia. Archives of General Psychiatry 42: 874

Baron M, Risch N, Hamburger R et al 1987 Genetic linkage between X chromosome markers and manic depression. Nature 326: 806

Bartko J J 1966 The intraclass correlation coefficient as a measure of reliability. Psychological Reports 34: 1

Barton W R 1959 Institutional neurosis. Wright, Bristol

Basner R, von Figura K, Glossl J, Klein U, Kresse H, Mlekusch W 1979 Multiple deficiency of mucopolysaccharide sulfatases in mucosulfatidosis. Pediatric Research 13: 1316

Bassett A S, McGillivray B C, Jones B D, Pantzar J T 1988 Partial trisomy chromosome 5 cosegregating with schizophrenia. Lancet i: 799

Beal M F, Growdon J H, Mazurek M F, Martin J B 1986 CSF somatostatin-like immunoreactivity in dementia. Neurology 36: 294

Bear D M 1986 Hemispheric asymmetries in emotional function: a reflection of lateral specialization in cortical-limbic connections. In: Doane B K, Livingston K E (eds) The limbic system: functional organization and clinical disorders. Raven Press, New York

Bear D M, Fedio P 1977 Quantitative analysis of interictal behaviour in temporal lobe epilepsy. Archives of Neurology 34: 454

Beard J D, Sargent W Q 1979 Water and electrolyte metabolism following ethanol intake and during acute withdrawal from ethanol. In: Majchrowicz E, Noble E P (eds) Biochemistry and pharmacology of ethanol, vol 2. Plenum Press, New York

Beck A T 1963 Thinking and depression: I. Idiosyncratic content and cognitive distortions. Archives of General Psychiatry 9: 324

Beck A T 1964 Thinking and depression: II. Theory and therapy. Archives of General Psychiatry 10: 561

Beck A T 1976 Cognitive therapy and the emotional disorders. International Universities Press, New York

Bem D J 1965 An experimental analysis of self-persuasion. Journal of Experimental Social Psychology 1: 199

Ben-Ami E, Burstein I, Cohen B E, Szeinberg A 1973 Cysteine peptiduria in a mentally retarded patient. Clinica Chimica Acta 45: 335

Bench C J, Dolan R J, Friston K J, Frackowiak R S J 1990 Positron emission tomography in the study of brain metabolism in psychiatric and neuropsychiatric disorders. British Journal of Psychiatry 157 (suppl. 9): 82

Berger H 1929 Über das electroenkephalogram des menschen. Archiv für Psychiatrie und Nervenkrankheiten 87: 527

Berger M, Pirke K M, Doerr P, Krieg J C, von Zerssen D 1984 The limited utility of the dexamethasone suppression test for the diagnostic process in psychiatry. British Journal of Psychiatry 145: 372

Berger P A, Faull K F, Kilkowski J et al 1980 CSF monoamine metabolites in depression and schizophrenia. American Journal of Psychiatry 137: 174

Bergner P-E E 1977 Bulletin of Mathematical Biology 39: 167

Berlin C I 1961 Congenital generalized melanoleucoderma associated with hypodontia, hypotrichosis, stunted growth and mental retardation occurring in two brothers and two sisters. Dermatologica 123: 227

Berman W, Haslam R H A, Konigsmark B W, Capute A J, Migeon C J 1973 A new familial syndrome with ataxia, hearing loss, and mental retardation. Archives of Neurology 29: 258

Bernoulli C, Seigfried J, Baumgartner G et al 1977 Danger of accidental person-to-person transmission of Creutzfeldt–Jakob disease by surgery. Lancet i: 478

Berrios G E, Katona C 1983 EEG and ECT, the current position. Journal of Neurology and Psychiatry (Lima). Presented at the Panamerican Congress of Psychiatry, 1983

Berscheid E, Walster E 1974 Physical attractiveness. In: Berkowitz L (ed) Advances in experimental social psychology. Academic Press, New York

Bertelsen A, Harvald B, Hauge M 1977 A Danish twin study of manic depressive disorders. British Journal of Psychiatry 130: 330

Bertoni J M, von Loh S, Allen R J 1979 The Aicardi syndrome: report of 4 cases and review of the literature. Annals of Neurology 5: 475

Betts T A, Smith J, Pidd S, Macintosh J, Harvey P, Finucane J 1976 The effects of thyrotropin releasing hormone on measures of mood in normal women. British Journal of Clinical Pharmacology 3: 469

Birau N, Alexander D, Bertholdt S, Meyer C 1984 Low nocturnal melatonin serum concentration in anorexia nervosa — further evidence for body weight influence. IRCS Medical Science 12: 477

Birkeland A J 1982 Plasma melatonin levels and nocturnal transitions between sleep and wakefulness. Neuroendocrinology 34: 126

Bissette G, Widerlöv E, Walléus H et al 1986 Alterations in cerebrospinal fluid concentrations of somatostatinlike immunoreactivity in neuropsychiatric disorders. Archives of General Psychiatry 43: 1148

Blackwell B 1963 Hypertensive crisis due to monoamine oxidase inhibitors. Lancet ii: 849

Blass J P, Gibson G E 1979 Genetic factors in Wernicke–Korsakoff syndrome. Alcoholism — Clinical and Experimental Research 3: 126

Blass J P, Schulman J D, Young D S, Hom E 1972 An inherited defect affecting the tricarboxylic acid cycle in a patient with congenital lactic acidosis. Journal of Clinical Investigation 51: 1845

Bloom F, Segal D, Ling N, Guillemin R 1976 Endorphins: profound behavioral effects in rats suggest new etiological factors in mental illness. Science 194: 630

Bohman M 1978 Some genetic aspects of alcoholism and criminality: a population of adoptees. Archives of General Psychiatry 35: 269

Bonafede R P, Beighton P 1978 The Dyggve–Melchior–Clausen syndrome in adult siblings. Clinical Genetics 14: 24

Bond A J, Lader M H 1974 The use of analog scales in rating subjective feelings. British Journal of Medical Psychology 47: 211

Bondy B, Ackenheil M, Birzle W, Elbers R, Frohler M 1984 Catecholamines and their receptors in blood. Biological Psychiatry 19: 1377

Bonn J A, Turner P, Hicks D C 1972 Beta-adrenergic-receptor blockade with practolol in treatment of anxiety. Lancet i: 814

Borjeson M, Forssman H, Lehmann O 1962 An X-linked, recessively inherited syndrome characterized by grave mental deficiency, epilepsy, and endocrine disorder. Acta Medica Scandinavica 171: 13

Bowen D, Smith C, White P, Dawson N 1976 Neurotransmitter-related enzymes and indices of hypoxia in senile dementia and other abiotrophids. Brain 33: 459

Bowen P, Armstrong H B 1976 Ectodermal dysplasia, mental retardation, cleft lip-palate and other anomalies in three sibs. Clinical Genetics 9: 35

Bower G H 1972 Mental imagery and associative learning. In: Gregg L W (ed) Cognition in learning and memory, Wiley, New York

Bower G H, Clark M, Winzenz D, Lesgold A 1969 Hierarchical retrieval schemes in recall of categorized word lists. Journal of Verbal Learning and Verbal Behavior 8: 323

Bowers M B 1973 5-HIAA and HVA following probenecid in acute psychotic patients treated with phenothiazines. Psychopharmacology 28: 309

Bowlby J 1958 The nature of the child's tie to his mother. International Journal of Psychoanalysis 39: 350

Bowlby J 1969 Attachment and loss vol 1: attachment. Basic Books, New York

Bowlby J 1973 Attachment and loss, vol 2: separation, anxiety and anger. Basic Books, New York

Bowlby J 1980 Attachment and loss, vol 3: loss, sadness and depression. Basic Books, New York

Braestrup C, Nielsen M 1982 Anxiety. Lancet ii: 1030

Brain W R 1930 Critical review: disseminated sclerosis. Quarterly Journal of Medicine 23: 343

Brain W R, Strauss E B 1945 Recent advances in neurology and neuropsychiatry. Churchill, London

Braitenberg V 1977 Cortical architectonics, general and areal. In: Brazier M A B, Petsche H (eds) Architectonics of the Cerebral Cortex. Raven Press, New York

Braitman L E 1988 Confidence intervals extract clinically useful information from data. Annals of Internal Medicine 108: 296

Brandon S, McClelland M A, Protheroe C 1971 A study of facial dyskinesia in a mental hospital population. British Journal of Psychiatry 118: 171

Bremner A J, Regan A 1991 Intoxicated by water. Polydipsia and water intoxication in a mental handicap hospital. British Journal of Psychiatry 158: 244

Broca P 1878 Anatomie comparée des circonvolutions cerebrales. Le grand lobe limbique et la scissure limbique dans la série des mamifères. Revue d'Anthropologie 1: 385

Brockington I F, Leff J P 1979 Schizoaffective psychosis: definition and incidence. Psychological Medicine 9: 91

Brodal A 1981 Neurological Anatomy. 3rd edn. Oxford University Press, Oxford

Brown F W 1942 Heredity in the psychoneuroses. Proceedings of the Royal Society of Medicine 35: 785

Brown J H, Fabre L F, Farrell G L, Adams E D 1972 Hyperlysinuria with hyperammonaemia. American Journal of Diseases of Children 124: 127

Brown G M, Woolever C A, Tsui H W, Grota L J 1981 24-Hour profile of melatonin in patients with amenorrhoea (abstract). Neuroendocrinology Letters 3: 103

Budd M A, Tanaka K R, Holmes L B, Efron M L, Crawford J D, Isselbacher K J 1967 Isovaleric acidaemia: clinical features of a new genetic defect of leucine metabolism. New England Journal of Medicine 277: 321

Bundey S, Griffiths M I 1977 Recurrence risks in families of children with symmetrical spasticity. Developmental Medicine and Child Neurology 19: 179

Bunney W E, Davis J M 1965 Norepinephrine in depressive reactions. A review. Archives of General Psychiatry 13: 483

Burke R E, Kang U J 1988 Tardive dystonia: clinical aspects and treatment. Advances in Neurology 49: 199

Burton B K 1979 Recurrence risk for congenital hydrocephalus. Clinical Genetics 16: 47

Burton B K, Ben-Yoseph Y, Nadler H L 1977 Lactosylceramidosis: a deficiency of neutral beta-galactosidase. American Journal of Human Genetics 29: 26A

Burvill P W, Hall W D, Stampfer H G, Emmerson J P 1989 A comparison of early-onset and late-onset depressive illness in the elderly. British Journal of Psychiatry 155: 673

Butcher L L, Engel J 1969 Behavioral and biochemical effects of l-dopa after peripheral decarboxylase inhibition. Brain Research 15: 233

Byrne E J, Arie T 1990 Controversies in therapeutics: Are drugs targeted at Alzheimer's disease useful? 2 Insufficient evidence of worthwhile benefit. British Medical Journal 300: 1132

Cacioppo J T, Petty R E 1979 Effects of message repetition and position on cognitive response, recall and persuasion. Journal of Personality and Social Psychology 37: 97

Cadoret R J 1978 Evidence of genetic inheritance of primary affective disorder in adoptees. American Journal of Psychiatry 135: 463

Cadoret R J, Gath A 1978 Inheritance of alcoholism in adoptees. British Journal of Psychiatry 132: 252

Caine T M, Foulds G A, Hope K 1967 Manual of the hostility and direction of hostility questionnaire (HDHQ). University of London Press, London

Camerino G, Mattei M G, Mattei J F, Jaye M, Mandel J L 1983 Close linkage of fragile X-mental retardation syndrome to haemophilia B and transmission through a normal male. Nature 306: 701

Cannon W B 1927 The James–Lange theory of emotions: a critical examination and an alternative theory. American Journal of Psychology 39: 106

Cannon W B, Britton S W 1925 Studies on the conditions of activity in endocrine glands. XV. Pseudaffective medulliadrenal secretion. American Journal of Physiology 72: 283

Cantu J M, Hernandez A, Larracilla J, Terejo A, Macotela-Ruiz E 1974 A new X-linked recessive disorder with dwarfism, cerebral atrophy, and generalized keratosis follicularis. Journal of Pediatrics 84: 564

Carey G, Gottesman I I 1981 Twin and family studies of anxiety, phobic, and obsessive disorders. In: Klein D F, Rabkin J (eds) Anxiety: new research and changing concepts. Raven Press, New York

Carlsson A 1988 The current status of the dopamine hypothesis of schizophrenia. Neuropsychopharmacology 1: 179

Carroll B J 1982 The dexamethasone suppression test for melancholia. British Journal of Psychiatry 140: 292

Carroll B J, Steiner M 1978 The psychobiology of premenstrual dysphoria: the role of prolactin. Psychoneuroendocrinology 3: 171

Carroll B J, Martin F I R, Davies B 1968 Resistance to suppression by dexamethasone of plasma 11-O.H.C.S. levels in severe depressive illness. British Medical Journal iii: 285

Carroll B J, Feinberg M, Greden J F et al 1981 A specific laboratory test for the diagnosis of melancholia. Archives of General Psychiatry 38: 15

Carson N A J, Scally B G, Neill D W, Carre I J 1968 Saccharopinuria: a new inborn error of lysine metabolism. Nature 218: 679

Cassidy S L, Henry J A 1987 Fatal toxicity of antidepressant drugs in overdose. British Medical Journal 245: 1021

Catalan J 1988 Psychosocial and neuropsychiatric aspects of HIV infection: review of their extent and implications for psychiatry. Journal of Psychosomatic Research 32: 237

Chang K J, Hazum E, Cuatrecasas P 1980 Multiple opiate receptors. Trends in Neurosciences 3: 160

Charney D S, Heninger G R (1986) Abnormal regulation of noradrenergic function in panic disorders. Effects of clonidine in healthy subjects and patients with agoraphobia and panic disorder. Archives of General Psychiatry 43: 1042

Checkley S A, Slade A P, Shur E 1981a Growth hormone and other responses to clonidine in patients with endogenous depression. British Journal of Psychiatry 138: 51

Checkley S A, Slade A P, Shur E, Dawling S 1981b A pilot study of the mechanism of action of desipramine. British Journal of Psychiatry 138: 248

Checkley S A, Winton F, Corn T H et al 1989 Neuroendocrine studies of the mechanism of action of phototherapy. In: Thompson C, Silverstone T (eds) Seasonal affective disorders. CHS Neuroscience, London

Cicchetti D V, Showalter D, Tyrer P 1985 The effects of number of rating scale categories on levels of inter-rater reliability: a Monte Carlo investigation. Applied Psychological Measurement 9: 31

Clark W E LeG, Meyer M 1950 Anatomical relationships between the cerebral cortex and the hypothalamus. British Medical Bulletin 6: 341

Clement-Jones V, McLoughlin L, Tomlin S, Besser G M, Rees L H, Wen H L 1980 Increased β-endorphin but not metenkephalin levels in human cerebrospinal fluid after acupuncture pain. Lancet ii: 946

Clow A, Jenner P, Marsden C D 1978 An experimental model of tardive dyskinesias. Life Sciences 23: 421

Coccaro E F 1989 Central serotonin and impulsive aggression. British Journal of Psychiatry 155 (suppl. 8): 52

Cochran W G 1954 Some methods for strengthening the common χ^2-test. Biometrics 10: 417

Collins A M, Quillian M R 1969 Retrieval time from semantic memory. Journal of Verbal Learning and Verbal Behavior 8: 240

Conover W J 1980 Practical non-parametric statistics. Wiley, New York

Cook L L, Bissette G, Dole K, Nemeroff C B 1989 A critical evaluation of cysteamine as a tool to deplete somatostatin in the rat central nervous system. Endocrinology 124: 855

Coppen A, Wood K 1978 Tryptophan and depressive illness. Psychological Medicine 8: 49

Coppen A, Ghose K, Montgomery S et al 1978a Amitriptyline plasma concentration and clinical effect. A World Health Organisation collaborative study. Lancet i: 63

Coppen A, Ghose K, Montgomery S, Rama Rao V A, Bailey J, Jorgensen A 1978b Continuation therapy with amitriptyline in depression. British Journal of Psychiatry 133: 28

Coppen A, Swade C, Wood K 1978c Platelet 5-hydroxytryptamine accumulation in depressive illness. Clinica Chimica Acta 87: 165

Costa E, Brodie B B 1964 Concept of the neurochemical transducer as an organized molecular unit at sympathetic nerve endings. Progress in Brain Research 8: 168

Cotes P M, Crow T J, Johnstone E C, Bartlett W, Bourne R C 1978 Neuroendocrine changes in acute schizophrenia as a function of clinical state and neuroleptic medication. Psychological Medicine 8: 657

Cotman C W, Monaghan D T, Ganong A H 1988 Excitatory amino acid neurotransmission: NMDA receptors and Hebb-type synaptic plasticity. In:Cowan W M, Shooter E M, Stevens G F, Thompson R F (eds) Annual review of neuroscience vol 11. Annual Reviews, Palo Alto

Crawshaw J A, Mullen P E 1984 A study of benzhexol abuse. British Journal of Psychiatry 145: 300

Creese I, Burt D R, Snyder S H 1976 Dopamine receptor binding predicts clinical and pharmacological potencies of antischizophrenic drugs. Science 192: 481

Crick F 1982 DNA Today. Perspectives in Biology and Medicine 25: 512

Crome L, Duckett S, Franklin A W 1963 Congenital cataracts, renal tubular necrosis and encephalopathy in two sisters. Archives of Disease in Childhood 38: 505

Cross H E, McKusick V A 1967 The Mast syndrome: a recessively inherited form of presenile dementia with motor disturbances. Archives of Neurology 16: 1

Cross H E, McKusick V A, Breen W 1967 A new oculocerebral syndrome with hypopigmentation. Journal of Pediatrics 70: 398

Cross J A, Cheetham S C, Crompton M R, Katona C L E, Horton R W 1988 Brain GABA$_B$ binding sites in depressed suicide victims. Psychiatry Research 26: 119

Crow T J 1978 Viral causes of psychiatric disease. Post-graduate Medical Journal 54: 763

Crow T J, Johnstone E C, McClelland H A 1976 The coincidence of schizophrenia and parkinsonism: some neurochemical implications. Psychological Medicine 6: 227

Crowe R R, Noyes R, Pauls D L, Slymen D 1983 A family study of panic disorders. Archives of General Psychiatry 40: 1065

Croyle R T, Cooper J 1983 Dissonance arousal: physiological evidence. Journal of Personality and Social Psychology 45: 782

Daly D, Siekert R G, Burke E C 1959 A variety of familial light sensitive epilepsy. Electroencephalography and Clinical Neurophysiology 11: 141

Dashiell J F 1930 An experimental analysis of some group effects. Journal of Abnormal and Social Psychology 25: 190

Davies D L 1949 Psychiatric changes associated with Friedrich's ataxia. Journal of Neurology, Neurosurgery and Psychiatry 12: 246

Davies P, Maloney A 1976 Selective loss of central cholinergic neurones in Alzheimer's disease. Lancet ii: 1403

Davis H, Davis P A, Loomis A L, Harvey E N, Hobart G 1937 Changes in human brain potentials during the onset of sleep. Science 86: 448

Davis K L, Davidson M, Yang R K et al 1988 CSF somatostatin in Alzheimer's disease, depressed patients, and control subjects. Biological Psychiatry 24: 701

Davison K, Bagley C R 1969 In Herrington R N (ed) Current problems in neuropsychiatry. British Journal of Psychiatry special publication no. 4. Headley Brothers, Ashford, Kent

de Myer W E 1972 Megalencephaly in children: clinical syndromes, genetic patterns, and differential diagnosis from other causes of megalocephaly. Neurology 22: 634

de Souza E B, Whitehouse P J, Kuhar M J, Price D L, Vale W W 1986 Reciprocal changes in corticotropin-releasing factor, CRF-like immunoreactivity and CRF receptors in cerebral cortex of Alzheimer's disease. Nature 319: 593

de Wied D 1984 Neurohypophyseal hormone influences on learning and memory processes. In:Lynch G, McGaugh J L Weinberger N M (eds) Neurobiology of Learning and Memory. Guildford Press, New York

de Wied D, Jolles J 1982 Neuropeptides derived from pro-opiocortin: behavioral, physiological, and neurochemical effects. Physiological Reviews 62: 976

Debicka A, Adamczak P 1979 Przypadek dziedziczenia zespolu Sturge'a-Webera. Klinika Oczna 81: 541

DeJong R A, Gold P, Rubinow D et al 1990 Cerebrospinal fluid levels of somatostatin, corticotropin-releasing hormone and corticotropin in alcoholism. Acta Psychiatrica Scandinavica 82: 44

Dekaban A S, Klein D 1968 Familial mental retardation. Acta Genetica et Statistica Medica 18: 206

Delabar J-M, Goldgaber D, Lamour Y et al 1987 Beta-amyloid gene duplication in Alzheimer's disease and karyotypically normal Down's syndrome. Science 235: 1390

DeMyer M K, Shea P A, Hendrie H C et al 1981 Plasma tryptophan and five other amino acids in depressed and normal subjects. Archives of General Psychiatry 38: 642

Dening T R, Berrios G E 1989 Wilson's disease: a prospective study of psychopathology in 31 cases. British Journal of Psychiatry 155: 206

Detera-Wadleigh S D, Berretin W H, Goldin L R, et al. 1987 Close linkage of C-Harvey ras-1 nad and the insulin gene to affective disorder is ruled out in three North American pedigrees. Nature 325: 806

Deutsch M, Gerard H G 1955 A study of normative and informational social influence on individual judgement. Journal of Abnormal and Social Psychology 51: 629

Dewhurst K 1969 The neurosyphilitic psychoses today: a survey of 91 cases. British Journal of Psychiatry 115: 31

Dias R D, Perry M L S, Carrasco M A, and Izquierdo I 1981 Rapid communication: effect of electroconvulsive shock on beta-endorphin immunoreactivity of rat brain, pituitary gland, and plasma. Behavioral and Neural Biology 32: 265

Diehl S, Su Y, Aman M, et al. 1989 Linkage studies of schizophrenia in Irish pedigrees. Cytogenetics and Cell Genetics 51: 989

Dooling E C, Schoene W C, and Richardson E P 1974 Hallervorden-Spatz syndrome. Archives of Neurology 30: 70

Drucker-Colin R, Bowersox S, and McGinty D 1982 Sleep and medial reticular unit responses to protein synthesis inhibitors: effects of chloramphenicol and thiamphenicol. Brain Research 252: 117

Drugs and Therapeutics Bulletin 1990 L-tryptophan and the eosinophilia-myalgia syndrome. Drugs and Therapeutics Bulletin 28: 37

Duffy P, Wolf J, Collins G, de Voe A G Streeten B, Cowen D 1974 Possible person-to-person transmission of Creutzfeldt–Jakob disease. New England Journal of Medicine 290: 692

Duffy F H, Burchfiel J L, Lombroso C T 1979 Brain electrical activity mapping (BEAM): a method for extending the clinical utility of EEG and evoked potential data. Annals of Neurology 5: 309

Duffy F H, Denckla M B, Bartels P H, Sandini G 1980a Dyslexia: regional dyslexia in brain electrical activity by topography mapping. Annals of Neurology 7: 412

Duffy F H, Denckla M B, Bartels P H, Sandini G, Kiessling L S 1980b Dyslexia: automated diagnosis by computerized classification of brain electrical activity. Annals of Neurology 7: 421

Dunham H W 1965 Community and schizophrenia: an epidemiological analysis. Wayne State University Press, Detroit

Dupont E, Christensen S E, Hansen A P, de Fine Olivarius B, Orskov H 1982 Low cerebrospinal fluid somatostatin in Parkinson disease: an irreversible abnormality. Neurology 32: 312

Durkheim E 1964 The division of labour in society. (Originally published in 1893.) Free Press, New York

Durkheim E 1970 Suicide: a study in sociology. Translated by Spaulding J A, Simpson G, and originally published in 1897. Routledge & Kegan Paul, London

Eadie M J, Sutherland J M 1964 Arteriosclerosis in parkinsonism. Journal of Neurology, Neurosurgery and Psychiatry 27: 237

Early T S, Posner M I, Reiman E M, Raichle M E 1989 Hyperactivity of the left-striato-palliodal projection. Psychiatric Developments 2: 85

Ebbinghaus H 1964 Memory: a contribution to experimental psychology. Translated by Ruger H A, Bussenius C E, and originally published in 1885. Dover, New York

Eccles J C 1964 The physiology of synapses. Springer-Verlag, Berlin

Egeland J A, Gerhard D S, Pauls D L et al 1987 Bipolar affective disorder linked to DNA markers on chromosome 11. Nature 325: 783

Ehrensing R H, Kastin A J 1974 Melanocyte-stimulating hormone release inhibiting hormone as an antidepressant: a pilot study. Archives of General Psychiatry 35: 63

Ehrensing R H, Kastin A J 1978 Dose-related biphasic effects of prolyl-leucyl-glycinamide (MIF) in depression. American Journal of Psychiatry 135: 562

Eison M S 1990 Serotonin: a common neurobiologic substrate in anxiety and depression. Journal of Clinical Psychopharmacology 10 (suppl. 3): 26S

Eldjarn L, Jellum E, Stokke O, Pande H, Waaler P E 1970 Beta-hydroxyisovaleric aciduria and beta-methylcrotonylglycinuria: a new inborn error of metabolism. Lancet i: 521

Emde R, Gaensbauer T, Harmon R 1976 Emotional expression in infancy: a biobehavioral study. Psychological Issues (Monograph 37) 10: 3

Endicott J 1978 A diagnostic interview: the schedule for affective disorders and schizophrenia. Archives of General Psychiatry 35: 837

Epley S W 1974 Reduction of the behavioral effects of aversive stimulation by the presence of companions. Psychological Bulletin 81: 271

Erikson E H 1963 Childhood and society 2nd edn. Norton, New York

Estes W K 1972 An associative basis for coding and organization in memory. In: Melton A W, Martin E (eds) Coding processes in human memory. Winston, Washington D C.

Ettigi P, Nair N P V, Lal S, Cervantes P, Guyda H 1976 Effects of apomorphine on growth hormone and prolactin secretion in schizophrenic patients, with or without oral dyskinesia, withdrawn from chronic neuroleptic therapy. Journal of Neurology, Neurosurgery and Psychiatry 39: 870

Evans J I, MacLean A W, Ismail A A, Love D 1971 Concentration of plasma testosterone in normal men during sleep. Nature 229: 261

Evans B M, Bridges P K, Bartlett J R 1982 Electroencephalographic changes as prognostic changes

after psychosurgery. Journal of Neurology, Neurosurgery and Psychiatry 44: 444

Fairburn C G, Hope R A 1988 Disorders of eating and weight. In: Kendall R E, Zealley A K (eds) Companion to psychiatric studies. 4th edn. Churchill Livingstone, Edinburgh

Fairweather D S 1947 Psychiatric aspects of the post-encephalitic syndrome. Journal of Mental Science 93: 201

Faris P L 1985 Role of cholecystokinin in the control of nociception and food intake. Progress in Clinical and Biological Research 192: 159

Farmer R D T, Pinder R M 1989 Why do fatal overdose rates vary between antidepressants? Acta Psychiatrica Scandinavica 80 (suppl. 354): 25

Fazio R H, Zanna M P, Cooper J 1977 Dissonance and self-perception: an integrative view of each theory's proper domain of application. Journal of Experimental Social Psychology 13: 464

Feighner J F, Boyer W F (editors) 1990 Selective serotonin re-uptake inhibitors: the clinical use of citalopram, fluoxetine, fluvoxamine, paroxetine and sertraline. John Wiley, Chichester

Fenton G 1974 The straightforward EEG in psychiatric practice. Proceedings of the Royal Society of Medicine 67: 911

Fenton G W 1989 The EEG in neuropsychiatry. In: Reynold E H, Trimble M R (eds) The bridge between neurology and psychiatry. Churchill Livingstone, Edinburgh

Ferguson B, Tyrer P 1988 Classifying personality disorder. In: Tyrer P (ed) Personality disorders: diagnosis, management and course. Wright, London

Ferguson B, Tyrer P 1989 Rating instruments in psychiatric research. In: Freeman C, Tyrer P (eds) Research methods in psychiatry: a beginner's guide. Gaskell, London

Ferguson S M, Rayport M, Corrie W S 1986 Brain correlates of aggressive behaviour in temporal lobe epilepsy. In: Doane B K, Livingston K E (eds) The limbic system: functional organization and clinical disorders. Raven Press, New York

Ferraro A 1934 Histopathological findings in two cases clinically diagnosed dementia praecox. American Journal of Psychiatry 90: 883

Ferraro A 1943 Pathological changes in the brain of a case clinically diagnosed dementia praecox. Journal of Neuropathology and Experimental Neurology 2: 84

Ferrier I N, Cotes P M, Crow T J, Johnstone E C 1982 Gonadotropin secretion abnormalities in chronic schizophrenia. Psychological Medicine 12: 263

Ferrier I N, Crow T J, Roberts G W et al, 1984a Alterations in neuropeptides in the limbic lobe in schizophrenia. In: Trimble M R, Zarifian E (eds) Psychopharmacology of the limbic system. Oxford University Press, Oxford

Ferrier I N, Johnstone E C, Crow T J 1984b Hormonal effects of apomorphine in schizophrenia. British Journal of Psychiatry 144 349–357

Festinger L A 1954 A theory of social comparison processes. Human Relations 1: 117

Festinger L 1957 A theory of cognitive dissonance. Row, Peterson, Evanston, Illinois

Festinger L, Carlsmith J M 1959 cognitive consequences of forced compliance. Journal of Abnormal and Social Psychology 58: 203

Fichter M M, Pirke K M, Holsboor E 1986 Weight loss causes neuroendocrine disturbances: experimental study in healthy starving subjects. Psychiatry Research 17: 61

File S E 1985 Tolerance to the behavioral action of benzodiazepines. Neuroscience and Biobehavioral Review 9: 113

Fink M 1979 Convulsive therapy: theory and practice. Raven Press, New York

Fischer M H, Brown R R 1980 Tryptophan and lysine metabolism in alpha-aminoadipic aciduria. American Journal of Medical Genetics 5: 35

Fishman S M, Sheehan D V, Carr D B 1985 Thyroid indices in panic disorder. Journal of Clinical Psychiatry 46: 432

Fitch N, Pinsky L, Lachance R C 1970 A form of bird-headed dwarfism with features of premature senility. American Journal of Diseases of Children 120: 260

Foerster O 1933 The dermatomes in man. Brain 56: 1

Folstein S, Rutter M 1977 Infantile autism: a genetic study of 21 twin pairs. Journal of Child Psychology and Psychiatry 18: 297

Foncin J-F, Salmon D, Bruni A C 1986 Genetics of Alzheimer's Disease: a large kindred with apparent Mendelian transmission; possible implications for a linkage study. In: Briley M, Kato A, Weber M (eds) New concepts in Alzheimer's disease. Macmillan, London

Fontaine R, Chouinard G, Annable L 1984 Rebound anxiety in anxious patients after abrupt withdrawal of benzodiazepine treatment. American Journal of Psychiatry 141: 848

Ford F R 1966 Diseases of the nervous system in infancy, childhood and adolescence 5th edn. Charles C. Thomas, Springfield, Illinois

Fox P, Fox D, Gerrard J W 1980 X-linked mental retardation: Renpenning revisited. American Journal of Medical Genetics 7: 491

Freud S 1986 The essentials of psychoanalysis. Pelican Books, Harmondsworth

Gaffney G R, Berlin F S 1984 Is there hypothalamo–pituitary–gonadal dysfunction in paedophilia? British Journal of Psychiatry 145: 657

Gallhofer B, Trimble M R, Frackowiak R, Gibbs J, Jones T 1985 A study of cerebral blood flow and metabolism in epileptic psychosis using positron emission tomography and oxygen. Journal of Neurology, Neurosurgery and Psychiatry 48: 201

Garattini S, Mennini T, Samanin R 1989 Reduction of food intake by manipulation of central serotonin: current experimental results. British Journal of Psychiatry 155 (suppl. 8): 41

Garbutt J L, Loosen P T, Tipermas A, Prange A J 1983 The TRH test in patients with borderline personality. Psychiatry Research 9: 107

Gardner M J, Altman D G 1986 Confidence intervals rather than P values: estimation rather than hypothesis testing. British Medical Journal 292: 746

Gardner M J, Altman D G 1990 Confidence — and clinical importance — in research findings. British Journal of Psychiatry 156: 472

Geaney D P, Abou-Saleh M T 1990 The use and applications of single-photon emission computerised tomography in dementia. British Journal of Psychiatry 157 (suppl. 9): 66

Gelder M G 1985 Cognitive therapy. In: Granville-Grossman KL (ed) Recent advances in clinical psychiatry, number 5. Churchill Livingstone, Edinburgh

Geocaris K 1957 Psychotic episodes heralding the diagnosis of multiple sclerosis. Bulletin of the Menninger Clinic 21: 107

Gerner R H 1984 Cerebrospinal fluid cholecystokinin and bombesin in psychiatric disorders and normals. In: Post R M, Ballenger J C (ed) Neurobiology of Mood Disorders. Williams and Wilkins, Baltimore

Gerner R H, Gwirtsman H E 1981 Abnormalities of dexamethasone suppression test and urinary M H P G in anorexia nervosa. American Journal of Psychiatry 138: 650

Gerner R H, Yamada T 1982 Altered neuropeptide concentration in cerebrospinal fluid of psychiatric patients. Brain Research 238: 298

Gershon E S, Bunney W E, Leckman J F, van Eerdewegh M, de Bauche B A 1976. The inheritance of affective disorders: a review of data and hypotheses. Behaviour Genetics 6: 227

Gibbons J L, McHugh P R 1962 Plasma cortisol in depressive illness. Journal of Psychiatric Research 1: 162

Gibbs J, Young R C, Smith G P 1973a Cholecystokinin decreases food intake in rats. Journal of Comparative and Physiological Psychology 84: 488

Gibbs J, Young R C, Smith G P 1973b Cholecystokinin elicits satiety in rats with open gastric fistulas. Nature 245: 323

Gil-Ad I, Gurewitz R, Marcovici O, Rosenfeld J, Laron Z 1984 Effects of aging on human plasma growth hormone response to clonidine. Mechanisms of Ageing and Development 27: 97

Gill M, McKeon P, Humphries P 1988 Linkage analysis of manic depression in an Irish family using H-ras 1 and INS DNA markers. Journal of Medical Genetics 25: 634

Gillespie F D 1965 Aniridia, cerebellar ataxia, and oligophrenia in siblings, Archives of Ophthalmology 73: 338

Gilliam T C, Bucan M, MacDonald M E et al 1987 A DNA segment encoding two genes very tightly linked to Huntington's disease. Science 238: 950

Gispen W H 1982 Neuropeptides and behavior: ACTH. Scandinavian Journal of Psychology (suppl. 1): 16

Gitlin M J, Gerner R H 1986 The dexamethasone suppression test and response to somatic treatment: a review. Journal of Clinical Psychiatry 47: 16.

Gloor P 1986 Role of the human limbic system in perception, memory, and affect: lessons from temporal lobe epilepsy. In: Doane B K, Livingston K E (eds) The limbic system: functional organization and clinical disorders. Raven Press, New York

Goa K L, Heel R C 1986 Zopiclone: a review of its pharmacodynamic and pharmacokinetic properties and therapeutic efficacy as a hypnotic. Drugs 32: 48

Goate A M, Haynes A R, Owen M J et al 1989 Predisposing locus for Alzheimer's disease on chromosome 21. Lancet i: 352

Goate A M, Chartier-Harlin M-C, Mullan M et al 1991 A missense mutation in the amyloid precursor protein gene segregates with familial Alzheimer's disease. Nature 349: 704

Goffman E 1961 Asylums: essays on the social situation of mental patients and other inmates. Doubleday, New York

Gold M S, Pearsall H R 1983 Hypothyroidism — or is it depression? Psychosomatics 24: 646

Gold P W, Kaye W, Robertson G L, Ebert M 1983 Abnormalities in plasma and cerebrospinal-fluid arginine vasopressin in patients with anorexia nervosa. New England Journal of Medicine 308: 1117

Gold P W, Chrousos G, Kellner C et al 1984 Psychiatric

implications of basic and clinical studies with corticotropin-releasing factor. American Journal of Psychiatry 141: 619

Goldberg D P 1972 The detection of psychiatric illness by questionnaire. Maudsley monograph 21. Oxford University Press, London

Goldberg E M, Morrison S L 1963 Schizophrenia and social class. British Journal of Psychiatry 109: 785

Goldgaber D, Lerman M I, McBride O W, Sffiotti U, Gajdusek D C 1987 Characterisation and chromosomal localisation of a cDNA encoding brain amyloid of Alzheimer's disease. Science 235: 877

Goltz F 1892 Der Hund ohne Grosshirn; Siebente Abhandlung uber die Verrichtungen des Grosshirns Pflugers. Archiv für die Gesammte Physiologie des Menschen und der Thiere 51: 570

Goodman W K, Price L H, Rasmussen S A, Delgado P L, Heninger G R, Charney D S 1989 Efficacy of fluvoxamine in obsessive–compulsive disorder: a double-blind comparison with placebo. Archives of General Psychiatry 46: 36

Goodwin P K, Bunney W E 1971 Depression following reserpine: a re-evaluation. Seminars in Psychiatry 3: 435

Goodwin D W, Schulsinger F, Hermansen L, Guze S B, Winokur G 1973 Alcohol problems in adoptees raised apart from alcoholic biological parents. Archives of General Psychiatry 28: 238

Gottesman I I, Shields J A 1976 Critical review of recent adoption, twin and family studies. Schizophrenia Bulletin 2: 360

Gottesman I I, Shields J 1982 Schizophrenia, the epigenetic puzzle. Cambridge University Press, Cambridge

Granville-Grossman K L, Turner P 1966 The effect of propranolol on anxiety. Lancet 1: 788

Greden J F, Genero N, Price L 1986 Facial electromyography in depression. Archives of General Psychiatry 43: 269

Green D M, Birdsall T G 1978 Detection and recognition. Psychological Review 85: 192

Green A R, Costain D W 1981 Pharmacology and biochemistry of psychiatric disorders. Wiley, Chichester

Greenbaum J V, Lurie L A 1948 Encephalitis as a causative factor in behavior disorders of children. Journal of the American Medical Association 136: 923

Gregory S, Shawcross C R, Gill D 1985 The Nottingham ECT study. A double-blind comparison of bilateral, unilateral and simulated ECT in depressive illness (Mapperley Hospital Nottingham). British Journal of Psychiatry 146: 520

Greter J, Hagberg B, Steen G, Sodenhjelm U 1978 3-Methylglutaconic aciduria: report on a sibship with infantile progressive encephalopathy. European Journal of Pediatrics 129: 231

Grevert P, Albert L H, Goldstein A 1983 Partial antagonism of placebo analgesia by naloxone. Pain 16: 129

Grossman K, Thank K, Grossman K E 1981 Maternal tactual contact of the newborn after various post-partum conditions of mother–infant contact. Development Psychology 17: 158

Guillemin R 1978 Peptides in the brain: the new endocrinology of the neuron. Science 202: 390

Gusella J F, Wexler N S, Conneally P M et al 1983 A polymorphic DNA marker genetically linked to Huntington's disease. Nature 306: 234

Gwirtsman H E, Roy-Byrne P, Yager J, Gerner R G 1983 Neuroendocrine abnormalities in bulimia. American Journal of Psychiatry 140: 559

Haas R H 1988 The history and challenge of Rett syndrome. Journal of Child Neurology 3: 3

Haber R N 1969 Eidetic images. Scientific American 220: 36

Haggerty J J, Simon J S, Evans D L, Nemeroff C B 1987 Relationship of serum TSH concentration and antithyroid antibodies to diagnosis and DST response in psychiatric inpatients. American Journal of Psychiatry 144: 1491

Hakim S 1964 Algunas Observaciones sobra la Pression del L.C.R. Sindrome Hidrocefalico en al Adulto con 'Pression Normal' del L.C.R. Tesis de Grado. Universidad Javeriana, Bogota, Columbia

Hakim S, Adams R D 1965 The special clinical problem of symptomatic hydrocephalus with normal cerebrospinal fluid pressure: observations on cerebrospinal fluid hydrodynamics. Journal of the Neurological Sciences 2: 307

Halbreich U, Assael M, Ben-David M, Bornstein R 1976 Serum-prolactin in women with premenstrual syndrome. Lancet ii: 654

Hall S B 1929 Mental aspect of epidemic encephalitis. British Medical Journal 1: 444

Hall B D, Riggs F D 1975 A new familial metabolic disorder with progressive osseous changes, microcephaly, coarse facies, flat nasal bridge and severe mental retardation. Birth Defects Original Articles Series XI(5): 79

Hamilton J, Parry B L, Alagna S, Blumenthal S, Herz E 1984 Premenstrual mood changes: a guide to evaluation and treatment. Psychiatric Annals 14: 426

Hanson J W, Myriathopoulos N C, Sedgwick M H A, Smith D W 1976 Risks to the offspring of women treated with hydantoin anticonvulsants, with emphasis on the fetal hydantoin syndrome. Journal of Pediatrics 89: 662

Hardy J A 1990 Genetics of the dementias. Current Opinion in Psychiatry 3: 93

Hardy J, Cowburn R, Barton A 1987 A disorder of cortical GABAergic innervation in Alzheimer's disease. Neuroscience Letters 73: 192

Harlow H F 1932 Social facilitation of feeding in the albino rat. Journal of Genetic Psychology 41: 211

Harlow H F 1959 Love in infant monkeys. Scientific American 200: 68

Harper P S, Youngman S, Anderson M A et al 1985 Genetic linkage between Huntington's disease and the DNA polymorphism G8 in South Wales families. Journal of Medical Genetics 22: 447

Harrisson P, Letchworth A T 1976 Bromocriptine in the treatment of premenstrual tension syndrome. In: Bayliss R I S, Turner P, Maclay W P (eds) Pharmacological and clinical aspects of bromocriptine (Parlodel): proceedings of a symposium held at the Royal College of Physicians, London, 14 May, 1976. Medical Congresses and Symposia Consultants: Tunbridge Wells, Kent

Harrisson P, Strangeway P, McCann J, Catalan J 1989 Paedophilia and hyperprolactinaemia. British Journal of Psychiatry 155: 847

Harvey E N, Loomis A L, Hobart G A 1937 Cerebral states during sleep as studied by human brain potentials. Science Monthly 45: 191

Hatfield E, Traupmann J 1981 Intimate relationships: a perspective from equity theory. In: Duck S, Gilmour R (eds) Personal relationships vol 1. Academic Press, New York

Hauge M, Harvald B, Fischer M et al 1968 The Danish twin register. Acta Geneticae Medicae et Genellologiae 17: 319

Head H 1920 Studies in neurology. Oxford University Press, London

Heider F 1958 The psychology of interpersonal relations. Wiley, New York

Heninger G R, Delgado P, Charney D et al 1989 Behavioral effects of acute tryptophan depletion in depressed patients. Biological Psychiatry 25: 143

Herman M N, Sandok B A 1967 Conversion symptoms in a case of multiple sclerosis. Military Medicine 132: 816

Herrmann W M 1982 Development and critical evaluation of an objective procedure for the electroencephalographic classification of psychotropic drugs. In: Herrmann W M (ed) Electroencephalography in drug research. Gustav Fischer, Stuttgart

Hess R O, Kaveggia E G, Opitz J M 1974 The N syndrome, a 'new' multiple congenital anomaly-mental retardation syndrome. Clinical Genetics 6: 237

Heston L J 1966 Psychiatric disorders in foster home reared children of schizophrenic mothers. British Journal of Psychiatry 112: 819

Heston L J, Mastri A R, Anderson V E, White J 1981 Dementia of the Alzheimer type: clinical genetics, natural history and associated conditions. archives of General Psychiatry 38: 1085

Hindmarch I 1979 A preliminary study of the effects of repeated dosage of clobazam on aspects of the performance, arousal and behaviour in a group of anxiety rated volunteers. European Journal of Clinical Pharmacology 16: 17

Hirsch S R, Gaind R, Rohde P D, Stevens B C, Wing J K 1973 Outpatient maintenance of chronic schizophrenic patients with long-acting fluphenazine: double-blind placebo trial. British Medical Journal i: 633

Hobson J 1974 The cellular basis of sleep cycle control. In: Weitzman E (ed) Advances in sleep research vol 1. Spectrum Publications, New York.

Hodgkinson S, Sherrington R, Gurling H M D et al 1987 Molecular genetic evidence for heterogeneity in manic depression. Nature 325: 805

Hoefnagel D, Pomeroy J, Wurster D, Saxon a 1971 Congenital athetosis, mental deficiency, dwarfism and laxity of skin and ligaments. Helvetica Paediatrica Acta 26: 397

Hogarty G E, goldberg S C, Schooler N R, Ulrich R F 1974 Drugs and sociotherapy in the aftercare of schizophrenic patients. Archives of General Psychiatry 31: 603

Hökfelt T, Skirboll L, Everitt B et al 1985 Distribution of cholecystokinin-like immunoreactivity in the nervous system. co-existence with classical neurotransmitters and other neuropeptides. In: Vanderhaeghen J J, Crawley J (eds) Neuronal cholecystokinin. Annals of the New York Academy of Sciences 448: 255

Holland A J, Hall A, Murray R, Russell G F M, Crisp A H 1984 Anorexia nervosa: a study of 34 twin pairs and a set of triplets. British Journal of Psychiatry 145: 414

Hold E K, Tedeschi C 1943 Cerebral patchy demyelination. Journal of Neuropathology and Experimental Neurology 2:306

Holzman P S, Kringlen E, Levy D L et al 1977 Abnormal pursuit eye movements in schizophrenia: evidence for genetic marker. Archives of General Psychiatry 34: 802

Holzman P S, Kringlen E, Matthysse S et al 1988 A single dominant gene can account for eye tracking dysfunctions and schizophrenia in offspring of discordant twins. Archives of General Psychiatry 45: 641

Homans G C 1961 Social behaviour: its elementary forms. Harcourt Brace, New York

Hudson J I, Pope H G, Jonas J M, Laffer P S, Hudson M S, Melby J C 1983 Hypothalamic–pituitary–adrenal–axis hyperactivity in bulimia. Psychiatry Research 8: 111

Hughes J, Smith T W, Kosterlitz H W, Fothergill L A, Morgan B A, Morris H R 1975 Identification of two related pentapeptides from the brain with potent opiate antagonist activity. Nature 258: 577

Hull C L 1952 A behaviour system. Yale University Press, New Haven, Connecticut

Humphrey M E, Zangwill O L 1951 Cessation of dreaming after brain injury. Journal of neurology, Neurosurgery and Psychiatry 14: 322

Hunter A G W, Jurenka S, Thompson D, Evans J 1982 Absence of the cerebellar granular layer, mental retardation, tapetoretinal degeneration and progressive glomerulopathy: an autosomal recessive oculo-renal-cerebellar syndrome. American Journal of Medical Genetics 11: 383

Huston T L 1973 Ambiguity of acceptance, social desirability and dating choice. Journal of Experimental Social Psychology 9: 32

Inui A, Okita M, Inoue T et al 1988 Mechanism of actions of cholecystokinin octapeptide on food intake and insulin and pancreatic polypeptide release in the dog. Peptides 9: 1093

Itil T M 1974 Quantitative pharmacoelectroencephalography. Use of computerized cerebral biopotentials in psychotropic drug research. In: Itil T M (ed) Modern problems of pharmacopsychiatry, vol 8: psychotropic drugs and the human EEG. Karger, Basel

Itil T M 1977 Qualitative and quantitative EEG findings in schizophrenia. Schizophrenia Bulletin 3: 61

Iversen S D and Iversen L 1981 Substance P - a new CNS transmitter. Hospital Update [May] 497

Jablensky A and Sartorius N 1975 Culture and schizophrenia. Psychological Medicine 5: 113

Jacobs B and Jones B 1978 The role of central monamine and acetylcholin systems in sleep-wakefulness states; mediation or modulation? In: L.L. Butcher (editor) Cholinergic-monaminergic Interactions in the Brain. Academic Press: New York

Jacobs P A, Brunton M, Melville M M, Brittain R P, and McClemont W F 1965 Aggressive behaviour and subnormality. Nature 208: 1351

Jacoby R 1981 Dementia, depression and the CT scan. Psychological Medicine 11: 673

Jacquet Y and Marks N 1976 The c-fragment of β-lipotropin: an endogenous neuroleptic or antipsychotogen. Science 194: 632

James W 1890 The principles of psychology. Holt, Rinehart & Winston, New York

James I M, Pearson R M, Griffith D N W, and Newbury P 1977 The effect of oxprenolol on stage fright in musicians. Lancet ii: 952

Janeway R, Ravens J R, Pearce L A, Odor D L, Suzuki K 1967 Progressive myoclonus epilepsy with Lafora inclusion bodies. I. Clinical, genetic, histopathologic and biochemical aspects. Archives of Neurology 16: 565

Jeffcoate W J 1981 Alcohol and endorphins. Advanced Medicine 17: 189

Jeffcoate W J, Silverstone J T, Edwards C R W, Besser G M 1979 Psychiatric manifestations of Cushing's syndrome: response to lowering of plasma cortisol. Quarterly Journal of Medicine 58: 465

Jefferson J W 1983 Lithium and affective disorders in the elderly. Comprehensive Psychiatry 24: 166

Jefferson J W, Marshall J R 1981 Neuropsychiatric features of medical disorders. Plenum Press, New York

Jenike M A, Shurman O S, Cassem N H et al 1983 Monoamine oxidase inhibitors in obsessive-compulsive disorder. Journal of Clinical Psychiatry 4: 131

Jennett B, Plum F 1972 Persistent vegetative state after brain damage: a syndrome in search of a name. Lancet i: 734

Jensen P K A 1981 Nerve deafness, optic nerve atrophy, and dementia: a new X-linked recessive syndrome? American Journal of Medical Genetics 9: 55

Jilek-Aall L, Jilek W, Miller J R 1979 Clinical and genetic aspects of seizure disorders prevalent in an isolated African population. Epilepsia 20: 613

Johnson C F 1966 Broad thumbs and broad great toes with facial abnormalities and mental retardation. Journal of Pediatrics 68: 942

Johnson J 1969 Organic psychosyndromes due to boxing. British Journal of Psychiatry 115: 45

Johnson D A W 1974 A study of the use of antidepressant medication in general practice. British Journal of Psychiatry 125: 186

Johnson D A W 1975 Observations on the dose regime of fluphenazine decanoate in maintenance therapy of schizophrenia. British Journal of Psychiatry 126: 457

Johnson T 1989 Methodology of clinical trials in psychiatry. In: Freeman C, Tyrer P (eds) Research methods in psychiatry: a beginner's guide. Gaskell, London, pp 12–45

Johnson A M 1990 The comparative pharmacological properties of selective serotonin re-uptake inhibitors in animals. In: Feighner J P, Boyer W F (eds) Selective serotonin re-uptake inhibitors: the clinical use of citalopram, fluoxetine, fluvoxamine, paroxetine and sertraline. John Wiley, Chichester

Johnston A W, McKusick V A 1961 Sex-linked recessive inheritance in spastic paraplegia and Parkinsonism. Proceedings of the Second International Congress of Human Genetics 3: 1652

Johnston A W, McKusick V A 1962 A sex-linked recessive inheritance of spastic paraplegia. American Journal of Medical Genetics 14: 83

Johnstone E C, Crow T J, Frith C D, Carney M W P, Price J S 1978 Mechanism of the antipsychotic effect in the treatment of acute schizophrenia. Lancet i: 848

Johnstone E C, Deakin J F W, Lawler P et al 1980 The Northwick Park ECT trial. Lancet ii: 1317

Jones K L, Smith D W, Harvey M A S, Hall B D, Quan L 1975 Older paternal age and fresh gene mutation: data on additional disorders. Journal of Pediatrics 86: 84

Joseph R, Lefebvre J, Guy E, Job J-C 1958 Dysplasie cranio-diaphysaire progressive. Ses relations avec la dysplasie diaphysaire progressive de Camurati-Engelmann. Annales de Radiologie 1: 477

Jouvet M 1965 Paradoxical sleep — a study of its nature and mechanisms. Progress in Britain Research 18: 20

Jouvet M 1967 Neurophysiology of the states of sleep. In: Quarton G C et al (eds) Neurosciences: a study program. Rockerfeller University Press, New York

Jouvet M 1972 The role of monoamines and acetylcholine-containing neurones in the regulation of the sleep-waking cycle. Ergebnisse der Physiologie, Biologischen Chemie und Experimentellen Pharmakologie 64: 166

Jouvet M 1977 Neuropharmacology of the sleep-waking cycle. In: Iversen L L, Iversen S, Snyder S H (eds) Handbook of psychopharmacology, vol 8: drugs, neurotransmitter and behavior. Plenum Press, New York

Jouvet M, Renault J 1966 Insomnie persistante après lèsions des noyaux du raphé chez le chat. Comptes Rendus des Seances de la Societé de Biologie et de ses Filiales 160: 1461

Juberg R C, Hellman C D 1971 A new familial form of convulsive disorder and mental retardation limited to females. Journal of Pediatrics 79: 726

Judd L L, Risch C, Parker D C, Janowsky D S, Segal D S, Huey L Y 1982 Blunted prolactin response. Archives of General Psychiatry 39: 1413

Kagan J, Kearsley R, Zelazo P 1978 Infancy: its place in human development. Harvard University Press, Cambridge, Massachusetts

Kahana E, Leibowitz U, Alter M 1971 Cerebral multiple sclerosis. Neurology 21: 1179

Kahn R J, McNair D M, Lipman R S et al 1986 Imipramine and chlordiazepoxide in depressive and anxiety disorders: II. Efficacy in anxious out-patients. Archives of General Psychiatry 43: 79

Kaij L 1960 Alcoholism in twins. Almqvist & Wiksell, Stockholm

Kaiya H, Tamura Y, Adachi S et al 1981 Substance P-like immunoreactivity in plasma of psychotic patients and effects of neuroleptics and electroconvulsive therapy. Psychiatry Research 5: 11

Kales A, Kales J D 1974 Sleep disorders: recent findings in the diagnosis and treatment of disturbed sleep. New England Journal of Medicine 290: 487

Kales A, Scharf M B, Kales J D 1978 Rebound insomnia: a new clinical syndrome. Science 201: 1039

Kalivas P W, Miller J S 1984 Substance P modulation of dopamine in the nucleus accumbens. Neuroscience Letters 48: 55

Kamijo K, Saito A, Kato T, Kawasaki K, Yachi A, Wada T 1983 Sex differences in the paradoxical response of serum GH to thyrotropin releasing hormone in cancer patients. Endocrinologica Japonica 30: 777

Kamoun P P, Parvy P, Cathelineau L, Meyer B 1981 Renal histidinuria. Journal of Inherited and Metabolic Disease 4: 217

Kane J M, Rifkin A, Woerner A 1983 Low-dose neuroleptic treatment of chronic schizophrenia. Archives of General Psychiatry 40: 893

Kane J, Honigfeld G, Singer J et al 1988 Clozapine for the treatment of resistant schizophrenics. Archives of General Psychiatry 45: 789

Kane J, Lemaire H-G, Unterbeck A et al 1987 The precursor of Alzheimer's disease amyloid A4 protein resembles a cell surface receptor. Nature 325: 733

Kaplan A S, Garfinkel P E, Brown G M 1989 The D S T and T R H test in bulimia nervosa. British Journal of Psychiatry 154: 86

Kathol R G, Jaeckle R S, Lopez J F, Meller W H 1989 Consistent reduction of ACTH responses to stimulation with CRH, vasopressin and hypoglycaemia in patients with major depression. British Journal of Psychiatry 155: 468

Katschnig H, Merz W A, Alf C 1988 Attack-controlled discontinous administration of Ro 16–6028 as a new strategy in the treatment of panic disorder: results of a placebo-controlled double blind study. Presented at XVIth CINP Congress, Munich 1988

Kaufmann C A, Delisi L E, Lehner T et al 1989 Physical mapping and linkage analysis of a putative schizophrenia locus on chromosome 5q. Schizophrenia Bulletin 15: 441

Keegan J J 1943 Dermatome hypalgesia associated with herniation of intervertebral disc. Archives of Neurology and Psychiatry 50: 67

Keegan J J, Garrett F D 1948 The segmental distribution of the cutaneous nerves in the limbs of man. Anatomical Record 102: 409

Keller M A, Jones K L, Nyhan W L, Francke U, Dixson B 1976 A new syndrome of mental deficiency with craniofacial, limb, and anal abnormalities. Journal of Pediatrics 88: 589

Kelley H H, Woodruff C L 1956 Members' reactions to apparent group approval of a counternorm communication, Journal of Abnormal and Social Psychology 52: 67

Kelly G A 1955 The psychology of personal construction Norton, New York

Kelman H C 1961 Processes of opinion change. Public Opinion Quarterly 25: 57

Kendell R E 1981 The present status of electroconvulsive therapy. British Journal of Psychiatry 139: 265

Kendler K S 1986 Genetics of schizophrenia. In: Frances A J, Hales R E (eds) American Psychiatric Association Annual Review, vol 5. American Psychiatric Press, Washington D.C.

Kendler K S, Davis K L 1981 The genetics and biochemistry of paranoid schizophrenia and other paranoid psychoses. Schizophrenia Bulletin 7: 689

Kennedy J L, Giuffra L A, Moises H W et al 1988 Evidence against linkage of schizophrenia to markets on chromosome 5 in a northern Swedish pedigree. Nature 336: 167

Kety S S, Rosenthal D, Wender P H, Schulsinger F, Jacobsen B 1976 mental illness in the biological and adoptive families of individuals who have become schizophrenic. Behaviour Genetic 6: 219

Kiloh L G, McComas A J, Ossleton J W, Upton A R M 1981 Clinical electroencephalography. 4th edn. Butterworths, London

Kilts C D, Knight D L, Mailman R B, Widerlöv E, Breese G R 1984 Effects of thioridazine and its metabolites on dopaminergic function: drug metabolism as a determinant of the antidopaminergic actions of thioridazine. Journal of Pharmacology and Experimental Therapeutics 231: 334

Kilts C D, Anderson C M, Bissette G, Ely T D, Nemeroff C B 1988 Differential effects of antipsychotic drugs on the neurotensin concentration of discrete rat brain nuclei. Biochemical Pharmacology 37: 1547

King M 1990 Psychological aspects of HIV infection and AIDS. British Journal of Psychiatry 156: 151

Klein D F 1964 Delineation of two drug-responsive anxiety syndromes. Psychopharmacologia 5: 397

Kleinman J E, Iadarola M, Govoni S, Hong J, Gillin J C, Wyatt R J 1983 Postmortem measurements of neuropeptides in human brain. Psychopharmacology Bulletin 19: 375

Kloepfer H W, Platou R V, Hansche W J 1964 Manifestations of a recessive gene for microcephaly in a population isolate. Journal de Genetique Humaine 13: 52

Koch C, Poggio T 1983 A theoretical analysis of electrical properties of spines. Proceedings of the Royal Society of London. Series B: Biological Sciences 218: 455

Koella W P 1984 The limbic system and behaviour. Acts Psychiatrica Scandinavica (suppl. 313) 69: 35

Kolakowska T, Orr M, Gelder M, Heggie M, Wiles D, Franklin M 1979 Clinical significance of plasma drug and prolactin levels during acute chlorpromazine treatment: a replication study. Psychological Medicine 135: 352

Kopelman M D 1985 Multiple memory deficits in alzheimer-type dementia: implications for pharmacotherapy. Psychological Medicine 15: 527

Kousseff B G 1981 Cohen syndrome: further delineatio and inheritance. American Journal of Medical Genetics 9: 25

Kraepelin E 1883 Compendium der Psychiatrie. Abel: Leipzig

Krech D, Crutchfield R S 1948 Theory and problems of social psychology. McGraw-Hill, New York

Kripke D, Lavie P, Parker D, Huey L, Deftos L 1978 Plasma parathyroid hormone and calcium are related to sleep stage cycles. Journal of Clinical Endocrinology and Metabolism 47: 1021

Krishnan K R R 1990 Brain imaging and psychiatric disorders. Current Opinion in Psychiatry 3: 79

Kubler-Rose E 1969 On death and dying. Macmillan, New York

Lader M 1988 The practical use of buspirone. In: Lader M (ed) Buspirone: a new introduction to the treatment of anxiety. Journal of the Royal Society of medicine, International Congress and Symposium Series, Number 133. New York

Lancet 1987 Report with confidence (editorial). Lancet i: 488

Lancet 1988 Klinefelter's syndrome (editorial). Lancet i: 1316

Landis J R, Koch G G 1977 The measurement of observer agreement for categorical data. Biometrics 33: 159

Lange C 1967 The emotions. (Translation of 1885 monograph.) Hafner, New York

Larsson T, Sjögren T, Jacobson G, Sjögren G 1963 Senile dementia: a clinical, socio-medical and genetic study. Acta Psychiatrica Scandinavica Supplement 167

Latcham R W 1985 Familial alcoholism: evidence from 237 alcoholics. British Journal of Psychiatry 147: 54

Lazoff S G, Rybak J J, Parker B R, Luzzatti L 1975 Skeletal dysplasia, occipital horns, intestinal malabsorption, and obstructive uropathy — a new hereditary syndrome. Birth Defects Original Articles Series XI(5): 71

Leake A, Charlton B G, Lowry P J, Jackson S, Fairbairn A, Ferrier In 1990 Plasma N-POMC, ACTH and cortisol concentrations in a psychogeriatric population. British Journal of Psychiatry 156: 676

Lebrecht U, Nowak J Z 1980 Effect of single and repeated electroconvulsive shock on serotonergic system in rat brain. I. Metabolic studies. Neuropharmacology 19: 1049

Leckman J F, Bowers M B, Sturges J S 1981 Relationship between estimated premorbid adjustment and CSF homovanillic acid and 5-hydroxyindoleacetic acid levels. American Journal of Psychiatry 138: 472

Lee T, Seeman P 1980 elevation of brain neuroleptic/dopamine receptors in schizophrenia. American Journal of Psychiatry 137: 191

Lefkowitz R J, Caron M G, Stiles G L 1984 Mechanisms of membrane-receptor regulation. New England Journal of medicine 310: 1570

Leonard B E 1987 A comparison of the pharmacological properties of the novel tricyclic antidepressant drug, lofepramine, with its major metabolite, desipramine: a review. International Clinical Psychopharmacology 2: 281

Leonhard K 1959 Anfteilung der Endogen Psychosen. Akademic Verlag: Berlin

Levental M, Susic V, Rusic M, Rakic L 1975 Rapid-eye-movement (REM) sleep deprivation: effect on acid mucopolysaccharides in rat brain. Archives Internationales de Physiologie et Biochimistrie 83: 221

Levic Z M, Stefanovic B S, Nikolic M Z, and Pisteljic D T 1975 Progressive nuclear ophthalmoplegia associated with mental deficiency, lingua scrotalis, and other neurological and other neurologic and ophthalmologic signs in the family. Neurology 25: 68

Levy R 1990 Controversies in therapeutics: are drugs targeted at Alzheimer's disease useful? 1 Useful for what? British Medical Journal 300: 1131

Levy R, Isaacs A, Behrman J 1971 Neurophysiological correlates of senile dementia. II. The somatosensory evoked response. Psychological Medicine 1: 159

Levy M I, DeNigris Y, Davis K L 1982 Rapid antidepressant activity of melanocyte inhibitory factor: a clinical trial. Biological Psychiatry 17: 259

Lewis S W 1990 Computerised tomography in schizophrenia 15 years on. British Journal of Psychiatry 157 (suppl. 9): 16

Lewy A J 1984 Human melatonin secretion (II): a marker for the circadian system and the effects of light. In: Post R M, Ballenger J C (eds) Neurobiology of mood disorders. Williams and Wilkins, Baltimore

Lewy A J, Kern H E, Rosenthal N E et al 1982 Bright artificial light treatment of a manic-depressive patient with a seasonal mood cycle. American Journal of Psychiatry 139: 1496

Liebowitz M R 1989 Antidepressants in panic disorders. British Journal of Psychiatry 155 (suppl. 6): 46

Liebowitz M R, Fyer A J, Gorman J M et al 1988 Tricyclic therapy of the DSM-III anxiety disorders: a review with implications for further research. Journal of Psychiatric Research 22 (suppl. 1): 7

Lipinski J F, Mallya G, Zimmerman P, Pope H G 1989 Fluoxetine-induced akathisia: clinical and theoretical implications. Journal of Clinical Psychiatry 50: 339

Lishman W A 1968 Brain damage in relation to psychiatric disability after head injury. British Journal of Psychiatry 114: 373

Lishman W A 1987 Organic Psychiatry: The psychological consequences of cerebral disorder. 2nd edn. Blackwell Scientific, Oxford

Livingston K E 1978 The experimental-clinical interface: kindling as a dynamic model of induced limbic system dysfunction. In: Livingston K E, Hornykiewicz O (eds) Limbic mechanisms: the continuing evolution of the limbic system concept. Plenum Press, New York

Ljungberg L 1957 Hysteria: a clinical, prognostic and genetic study. Acta Psychiatrica Scandinavica Supplement 112

Lloyd K G, Thuret F, Pilc A 1985 Upregulation of γ-aminobutyric acid (GABA)B binding sites in rat frontal cortex: a common action of repeated administration of different classes of antidepressants and electroshock. Journal of Pharmacology and Experimental Therapeutics 235: 191

Lock T, Abou-Saleh M T, Edwards R H T 1990 Psychiatry and the new magnetic resonance era. British Journal of Psychiatry 157 (suppl. 9): 38

Login I S, Nagy I, MacLeod R M 1979 Evidence for opiate moderation of hypothalamic dopamine and its influence on prolactin synthesis and release. Physiologist 22(3): 72

Logue V, Durward M, Pratt R T C, Piercy M, Nixon W L B 1968 The quality of survival after rupture of an anterior cerebral aneurysm. British Journal of Psychiatry 114: 137

Loosen P T, Prange A J 1979 TRH in alcoholic men: endocrine responses. Psychosomatic Medicine 41: 584

Loosen P T, Prange A J 1980 T R H: a useful tool for psychoneuroendocrine investigation. Psychoneuroendocrinology 5: 63

Loosen P T, Prange A J 1982 Serum thyrotropin response to thyrotropin-releasing hormone in psychiatric patients: a review. American Journal of Psychiatry 139: 405

Loosen P T, Prange A J 1984 Hormones of the thyroid axis and behavior. In: Nemeroff C B, Dunn A J (eds) Peptides, hormones, and behavior. Spectrum Publications, New York

Lorenz K 1962 King Solomon's Ring. Methuen, London

Lott I T, Williams R S, Schnur J A, Hier D B 1979 Familial amentia, unusual ventricular calcifications, and increased cerebrospinal fluid protein. Neurology 29: 1571

Lowry R B, Miller J R, Fraser F C 1971 A new dominant gene mental retardation syndrome: associated with small stature, tapering fingers, characteristic facies, and possible hydrocephalus. American Journal of Diseases of Children 121: 496

Lowry R B, MacLean R, McLean D M, Tischler B 1978 Cataracts, microcephaly, kyphosis, and limited joint movement in two siblings: a new syndrome. Journal of Pediatrics 79: 282

Maas J W 1975 Biogenic amines and depression. Archives of General Psychiatry 32: 1357

Mackay A V P, Sheppard G P 1979 Pharmacotherapeutic trials in tardive dyskinesia. British Journal of Psychiatry 135: 489

MacLean P D 1962 New findings relevant to the evolution of psychosexual functions of the brain. Journal of Nervous and Mental Disease 135: 289

MacLean P D 1967 The brain in relation to empathy and medical education. Journal of Nervous and Mental Disease 144: 374

MacLean P D 1970 The triune brain, emotion, and scientific bias. In: Schmitt F O (ed) The neurosciences second study program. Rockefeller University Press, New York

MacLean P D 1986 Interview with Paul D. MacLean, M.D. In: Doane B K, Livingston K E (eds) The limbic system: functional organization and clinical disorders. Raven Press, New York

Maj M, Mastronardi P, Cerreta A, 1988 Changes in platelet ^{3}H-imipramine binding in depressed patients receiving electroconvulsive therapy. Biological Psychiatry 24: 469

Malamud N, Cohen P 1958 Unusual form of cerebellar ataxia with sex-linked inheritance. Neurology 8: 261

Manchanda R, Hirsch S R 1981 (Des-tyr)-γ-endorphin in the treatment of schizophrenia. Psychological Medicine 11: 401

Mandel W, Oliver W A 1961 Withdrawal of maintenance antiparkinson drug in the phenothiazine-induced extrapyramidal reaction. American Journal of Psychiatry 118: 350

Mann J J, Aarons S F, Wilner P J et al 1989a A controlled study of the antidepressant efficacy and side effects of (−)–deprenyl. Archives of General Psychiatry 46: 45

Mann J J, Arango V, Marzuk P M, Theccanat S, Reis D J 1989b Evidence for the 5-HT hypothesis of suicide: a review of post-mortem studies. British Journal of Psychiatry 155 (suppl. 8): 7

Mantyh P W, Hunt S P 1984 Evidence for cholecystokinin-like immunoreactive neurons in the rat medulla oblongata which project to the spinal cord. Brain Research 291: 49

Marks I M 1983 Are there anti-compulsive or anti-phobic drugs? Review of the evidence. British Journal of Psychiatry 140: 338

Marks I, O'Sullivan G 1989 Anti-anxiety drug and psychological treatment effects in agoraphobia/panic and obsessive-compulsive disorders. In: Tyrer P (ed)

Psychopharmacology of Anxiety. Oxford University Press, Oxford

Marks I M, Stern R S, Mawson D, Cobb J, McDonald R 1980 Clomipramine and exposure for obsessive–compulsive rituals. British Journal of Psychiatry 136: 1

Marsden C D 1982 Basal ganglia and disease. Lancet ii 1141

Marsden C D, Jenner P 1980 The pathophysiology of extrapyramidal side effects of neuroleptic drugs. Psychological Medicine 10: 55

Marsden C D, Foley T H, Owen D A L, McAllister R G 1967 Peripheral beta-adrenergic receptors concerned with tremor. Clinical Science 33: 53

Martin W R, Eades C G, Thompson J A, Huppler R E, Gilbert P E 1976 The effects of morphine- and nalorphine-like drugs in the nondependent and morphine-dependent chronic spinal dog. Journal of Pharmacology and Experimental Therapeutics 197: 517

Martsolf J T, Hunter A G W, Haworth J C 1978 Severe mental retardation, cataracts, short stature and primary hypogonadism in two brothers. American Journal of Medical Genetics 1: 291

Matthews W B 1979 Multiple sclerosis presenting with acute remitting psychiatric symptoms. Journal of Neurology, Neurosurgery and Psychiatry 42: 859

Matuzas W, Al-Sadir J, Uhlenhuth E H, Glass R M 1987 Mitral valve prolapse and thyroid abnormalities in patients with panic attacks. American Journal of Psychiatry 144: 493

McCarley R, Hobson J 1975 Neuronal excitability modulation over the sleep cycle: a structural and mathematical model. Science 189: 55

McDougall W 1908 Social psychology. G.P. Putnam, New York

McGuffin P, Katz R 1986 Nature, nurture and affective disorder. In: Deakin J F W (ed) The biology of depression. Gaskell, London

McInnes R G 1937 Observations of heredity in neurosis. Proceedings of the Royal Society of Medicine 30: 895

Mechanic D 1978 Medical sociology. 2nd edn. Free Press, Glencoe

Medical Research Council 1965 Report of Clinical Psychiatry Committee. Clinical trial of the treatment of depressive illness. British Medical Journal 1: 881

Meichenbaum D H 1977 Cognitive behavior modification. Plenum Press, New York

Melancon S B, Dallaire L, Lemieux B, Robitaille P, Potier M 1977 Dicarboxylic aminoaciduria: an inborn error of amino acid conservation. Journal of Pediatrics 91: 422

Meldrum B 1985 Excitatory amino acids and anoxic/chaemic brain damage. Trends in Neuroscience 8: 47

Mellman W J, Barness L A, Tedesco T A, Besselman D 1963 Indolylacroyl glycine excretion in a family with mental retardation. Clinica Chimica Acta 8: 843

Meltzer H Y, Lowy M T 1987 The serotonin hypothesis of depression. In: Meltzer H Y (ed) Psychopharmacology, the third generation of progress. Raven Press, New York

Meltzer H Y, Kolakowska T, Fang V S et al 1984 Growth hormone and prolactin response to apomorphine in schizophrenia and the major affective disorder. Archives of General Psychiatry 41: 512

Mendlewicz J, Rainer J D 1977 Adoption study supporting genetic transmission of manic depressive illness. Nature 268: 327

Mendlewicz J, Fleiss J L, Fieve R R 1972 Evidence for X-linkage in the transmission of manic-depressive illness.

Journal of the American Medical Association 222: 1624

Mendlewicz J, Simon P, Sevy S et al 1987 Polymorphic DNA marker on X chromosome and manic depression. Lancet ii: 1230

Menkes J H, Philippart M, Clark D B 1964 Hereditary partial agenesis of corpus callosum. Archives of Neurology 11: 198

Mettler F A 1968 Anatomy of the basal ganglia. In: Vinken P J, Bruyn G W (eds) Handbook of clinical neurology, vol 6. North-Holland, Amsterdam

Mietens C, Weber H 1966 A syndrome characterized by corneal opacity, nystagmus, flexion contracture of the elbows, growth failure, and mental retardation. Journal of Pediatrics 69: 624

Milgram S 1963 Behavioral study of obedience. Journal of Abnormal and Social Psychology 67: 371

Milgram S 1974 Obedience to authority: an experimental view. Harper & Row, New York

Miller G A 1956 The magic number seven, plus or minus two: some limits on our capacity for processing information. Psychological Review 63: 81

Mindham R H S, Howland C, Shepherd M 1973 An evaluation of continuation therapy with tricyclic antidepressants in depressive illness. Psychological Medicine 3: 5

Mirmiran M, Scholtens J, van de Poll N, Uylings H, van der Gugten J, Boer G 1983 Effects of experimental suppression of active (REM) sleep during early development upon adult brain and behavior in the rat. Brain Research 283: 277

Misra P C, Hay G G 1971 Encephalitis presenting as acute schizophrenia. British Medical Journal i: 532

Mitchell J E 1986 Anorexia nervosa: medical and physiological aspects. In: Brownell K D, Foreyt J P (eds) Handbook of eating disorders: physiology, psychology and treatment of obesity, anorexia and bulimia. Basic Books, New York

Mitchell J E, Bantle J P 1983 Metabolic and endocrine investigations in women of normal weight with the bulimia syndrome. Biological Psychiatry 18: 355

Mohrenweiser H W 1981 Frequency of enzyme deficiency variants in erythrocytes of newborn infants. Proceedings of the National Academy of Sciences 78: 5046

Mollica F, Pavone L, Antener I 1972 Short stature, mental retardation and ocular alterations in three siblings. Helvetica Paediatrica Acta 27: 463

Montgomery S A 1980 Clomipramine in obsessional neurosis: a placebo controlled trial. Pharmaceutical Medicine 1: 189

Montgomery S A 1988a The benefits and risks of 5-HT uptake inhibitors in depression. British Journal of Psychiatry 153 (suppl. 3): 7

Montgomery S A 1988b The risk of suicide with antidepressants. International Clinical Psychopharmacology 3: 15

Montgomery S A, Green M D 1988 The use of cholecystokinin in schizophrenia: a review. Psychological Medicine 18: 593

Montgomery S A, Dufour H, Brion S et al 1988 The prophylactic efficacy of fluoxetine in unipolar depression. British Journal of Psychiatry 153 (suppl. 3): 69

Moore R H 1979 The effects on blood pressure, heart rate and behavior of chronically-administered dopamine into the limbic system. Federation Proceedings 38: 858

Moore R Y 1982 The suprachiasmatic nucleus and the organization of a circadian system. Trends in Neuroscience 5: 404

Morgan K, Oswald I 1982 Anxiety caused by a short-life hypnotic. British Medical Journal 284: 942

Morihisa J M 1987 Functional brain imaging techniques. In: Hales R, Frances A (eds) American Psychiatric Association annual review number 6. American Psychiatric Press, Washington D.C.

Morihisa J M, McAnulty G B 1985 Structure and function: brain electrical activity mapping (BEAM) and computed tomography (CT scan) in schizophrenia. Biological Psychiatry 20: 3

Morley S, Snaith P 1989 Principles of psychological assessment. In: Freeman C, Tyrer P (eds) Research methods in psychiatry: a beginner's guide. Gaskell, London

Morris J B, Beck A T 1974 The efficacy of antidepressant drugs. A review of research (1958 to 1972). Archives of General Psychiatry 30: 667

Moroji T, Watanabe N, Aoki N, Itoh S 1982 Antipsychotic effects of ceruletide (caerulein) on chronic schizophrenia. Archives of General Psychiatry 39: 485

Morton N E 1955 Sequential tests for the detection of linkage. American Journal of Human Genetics 7: 277

Moynahan E J 1962 Familial congenital alopecia, epilepsy, mental retardation with unusual electroencephalograms. Proceedings of the Royal Society of Medicine 55: 411

Mullen P, James V, Lightman S, Linsell C, Peart W 1980 A relationship between plasma renin activity and the rapid eye movement phase of sleep in man. Journal of Clinical Endocrinology and Metabolism 50: 466

Murdock B B 1961 The retention of individual items. Journal of Experimental Psychology 62: 618

Murdock B B 1962 The serial position effect of free recall. Journal of Experimental Psychology 64: 482

Murphy S, Tyrer P 1989 Rating scales for special purposes. I: Psychotherapy. In: Freeman C, Tyrer P (eds) Research methods in psychiatry: a beginner's guide. Gaskell, London

Murphy M R, MacLean P D, Hamilton S C 1981 Species-typical behavior of hamsters deprived from birth of the neocortex. Science 213: 459

Murray H A 1938 Explorations in personality. Oxford University Press, London

Murray R M, McGuffin P 1988 Genetic aspects of mental disorders. In: Kendell R E, Zealley A K (eds) Companion to psychiatric studies. 4th edn. Churchill Livingstone, Edinburgh

Murray R M, Oon M C H, Rodnight R, Birley J L T, Smith A 1979 Increased excretion of dimethyltryptamine and certain features of psychosis. Archives of General Psychiatry 36: 644

Murray R M, Clifford C A, Gurling H M D 1983 Twin and alcoholism studies. In: Gallanter M (ed) Recent developments in alcoholism vol I. Gardner Press, New York

Mutchinick O 1972 A syndrome of mental and physical retardation, speech disorders, and peculiar facies in two sisters. Journal of Medical Genetics 9: 60

Naiman J L, Fraser F C 1955 Agenesis of the corpus callosum. A report of two cases in siblings. Archives of Neurology and Psychiatry 74: 182–185

Nebes R D, Sperry R W 1971 Cerebral dominance in perception. Neuropsychologia 9: 247

Nee L E, Polinsky R J, Eldridge R, Weingartner H, Smallberg S, Ebert M 1983 A family with histologically confirmed Alzheimer's disease. Archives of Neurology 40: 203

Nemeroff C B 1990 Neuropeptide involvement in affective disorders. Current Opinion in Psychiatry 3: 108

Nemeroff C B, Youngblood W W, Manberg P J, Prange A J, Kizer J S 1983 Regional brain concentrations of neuropeptides in Huntington's chorea and schizophrenia. Science 221: 972

Neuhauser G, Kaveggia E G, France T D, Opitz J M 1975 Syndrome of mental retardation, seizures, hypotonic cerebral palsy and megalocorneae, recessively inherited. Zeitschrift für Kinderheilkunde 120: 1

Neuhauser G, Kaveggia E G, Opitz J M 1976 A craniosynostosis–craniofacial dysostosis syndrome with mental retardation and other malformations: 'craniofacial dyssynostosis'. European Journal of Pediatrics 123:15

Newcomb T 1956 The prediction of interpersonal attraction. American Psychologist 11: 575

Newcombe F 1983 The psychological consequences of closed head injury: assessment and rehabilitation. Injury 14: 111

Niikawa N, Matsuda I, Ohsawa T, Kajii T 1978 Familial occurrence of a syndrome with mental retardation, nasal hypoplasia, peripheral dysostosis, and blue eyes in Japanese siblings. Human Genetics 42: 227

Nishino N, Koizumi K, Brooks C Mc 1976 The role of the suprachiasmatic nucleus of the hypothalamus in the production of circadian rhythm. Brain Research 112: 45

Norman R M, Urich H 1958 Cerebellar hypoplasia associated with systemic degeneration in early life. Journal of Neurology, Neurosurgery and Psychiatry 21: 159

Noyes R, Clancy J, Crowe R, Hoenk P R, Slymen D J 1978 The familial prevalence of anxiety neurosis. Archives of General Psychiatry 35: 1057

O'Brien P M S, Symonds E M 1982 Prolactin levels in the premenstrual syndrome. British Journal of Obstetrics and Gynaecology 89: 306

Ockerman P A 1967 A generalized storage disorder resembling Hurler's syndrome. Lancet ii: 239

Oleson D R, Johnson D R 1988 Regulation of human natural cytotoxicity by enkephalins and selective opiate agonists. Brain, Behavior and Immunity 2: 171

Oliver C, Holland A J 1986 Down's syndrome and Alzheimer's disease: a review. Psychological Medicine 16: 307

Omura K, Yamanaka N, Higami S et al 1976 Lysine malabsorption syndrome: a new type of transport defect. Pediatrics 57: 102

Opitz J M, Kaveggia E G, Durkin-Stamm M V, Pendelton E 1978 Diagnostic-genetic studies in severe mental retardation. Birth Defects Original Article Series XIV(6B): 1

Oram J J, Edwardson J, Millard P H 1981 Investigation of cerebrospinal fluid neuropeptides in idiopathic senile dementia. Gerontology 27: 216

Ota Y 1969 Psychiatric studies on civilian head injuries. In: Walker A E, Caveness W F, Critchley M (eds) The late effects of head injury. Thomas, Springfield, Illinois

Overstreet D H 1986 Selective breeding for increased cholinergic function: development of a new animal model of depression. Biological Psychiatry 21: 49

Owen F, Cross A J, Crow T J, Longden A, Poulter M, Riley G J 1978 Increased dopamine-receptor sensitivity in schizophrenia. Lancet ii: 223

Paddison R M, Moossy J, Derbes V J, Kloepfer H W 1963 Cockayne's syndrome. A report of five new cases with biochemical, chromosomal, dermatologic, genetic and neuropathologic observations. Dermatologica Tropica 2: 195

Paine R S 1960 Evaluation of familial biochemically determined mental retardation in children, with special reference to aminoaciduria. New England Journal of Medicine 262: 658

Pallister P D, Herrmann J, Spranger J W, Gorlin R J, Langer L O, Opitz J M 1974 The W syndrome. Birth Defects Original Article Series X(7): 51

Pandey G N, Garver D L, Tamminga C, Eriksen S, Ali S I, Davis J M 1977 Postsynaptic supersensitivity in schizophrenia. American Journal of Psychiatry 134: 518

Pinder R M 1988 The benefits and risks of antidepressant drugs. Human Psychopharmacology 3: 73

Paprocki J, Barcala Peixoto M P, Mendes Andrade N 1975 A controlled double-blind comparison between loxapine and haloperidol in acute newly hospitalized schizophrenic patients. A Folha Médica 70: 533

Pare C M B, Kline N, Hallstrom C, Cooper T B 1982 Will amitriptyline prevent the 'cheese' reaction of monoamine oxidase inhibitors? Lancet ii: 183

Parker N 1957 Manic-depressive psychosis following head injury. Medical Journal of Australia 2: 20

Parker G, Roberts C J C 1983 Zopiclone in chronic liver disease. British Journal of Clinical Pharmacology 16: 259

Parrish J M, Wilroy R S 1980 The Dubowitz syndrome: the psychological status of ten cases at follow-up. American Journal of Medical Genetics 6: 3

Parsons T 1937 The structure of social action. McGraw-Hill, New York

Parsons T 1951 The social system. Free Press, Glencoe

Partanen J K, Bruun K, Markkanen T 1966 Inheritance of drinking behavior: a study on intelligence, personality and use of alcohol in adult twins. In: Finnish Foundation for Alcohol Studies publication number 14. Finnish Foundation for Alcohol Studies: Helsinki

Passwell J H, Goodman R M, Ziprkowski M, Cohen B E.1975 Congenital ichthyosis, mental retardation, dwarfism and renal impairment: a new syndrome. Clinical Genetics 8: 59

Patel V, Watanabe I, Zeman W 1972 Deficiency of alpha-L-fucosidase. Science 176: 426

Pavlov I P 1927 Conditioned reflexes. Oxford University Press, London

Paykel E S 1979 Predictors of treatment response. In: Paykel E S, Coppen A (eds) Psychopharmacology of affective disorders. Oxford University Press, Oxford

Paykel E S 1989 Treatment of depression. The relevance of research for clinical practice. British Journal of Psychiatry 155: 754

Pearson C M, Kalyanaraman K 1972 The periodic paralyses. In: Stanbury J B, Wyngaarden J B, Fredrickson D S (eds) The metabolic basis of inherited disease. 3rd edn. McGraw-Hill, New York

Peck D F 1989 Research with single (or few) patients. In: Freeman C, Tyrer P (editors) Research methods in psychiatry: A beginner's guide. Gaskell, London

Peck D F, Shapiro C M (eds) 1990 Measuring human problems: a practical guide. John Wiley, Chichester

Peet M, Harvey N S 1991 Lithium maintenance: 1. A standard education programme for patients. British Journal of Psychiatry 158: 197

Penfield W 1938 The cerebral cortex in man. I. The cerebral cortex and consciousness. Archives of Neurology and Psychiatry 40: 417

Penfield W 1951 Memory mechanisms. Archives of Neurology and Psychiatry 67: 178

Penfield W 1955 The permanent record of the stream of consciousness. Proceedings of the 14th International Congress of Psychology. Montréal 1954. Acta Psychologica 11: 47

Pericak-Vance M A, Yamaoka L H, Haynes C S et al 1988 Genetic linkage studies in Alzheimer's disease families. Experimental Neurology 102: 271

Perry E K, Perry R H 1982 Neurotransmitter and neuropeptide systems in Alzheimer-type dementia. In: Hoyer S (ed) The ageing brain. Springer-Verlag, Berlin

Perry T L, Hansen S, Klaster M 1973 Huntington's chorea, deficiency of gamma-aminobutyric acid in the brain. New England Journal of Medicine 288: 337

Perry E, Gibson P, Blessed G, Perry R 1977 Neurotransmitter enzyme abnormalities in senile dementia. Science 34: 247

Perry E, Tomlinson B, Blessed G, Bergmann K, Gibson P, Perry R 1978 Correlation of cholinergic abnormalities with senile plaques and mental test scores in senile dementia. British Medical Journal ii: 1457

Perry E K, Curtis M, Dick D J et al 1985 Cholinergic correlates of cognitive impairment in Parkinson's disease: comparisons with Alzheimer's disease. Journal of Neurology, Neurosurgery and Psychiatry 48: 412

Petty R E, Cacioppo J T 1981 Attitudes and persuasion: classic and contemporary approaches. William C Brown, Dubuque, Iowa

Phelan M C, Pellock J M, Nance W E 1981 Discordant expression of fetal hydantoin syndrome in a pair of dizygotic twins with different fathers. (Abstract) American Journal of Human Genetics 33: 67A

Piaget J 1963 The origins of intelligence in children. Norton, New York

Piaget J 1971 The construction of reality in the child. Ballantine, New York

Pickar D, Davis G C, Schulz C et al 1981 Behavioral and biological effects of acute β-endorphin injection in schizophrenic and depressed patients. American Journal of Psychiatry 138: 160

Pickar D, Vartanian F, Bunney W E et al 1982 Short-term naloxone administration in schizophrenic and manic patients. Archives of General Psychiatry 39: 313

Pickar D, Breier A, Kelso J 1988 Plasma homovanillic acid as an index of central dopaminergic activity: studies in schizophrenic patients. Annals of the New York Academy of Sciences 537: 339

Plutchik R 1980 Emotion: A psychoevolutionary synthesis. Harper & Row, New York

Pocock S J 1977 Group sequential methods in the design and analysis of clinical trials. Biometrika 64: 191

Pocock S J 1983 Clinical trials: A practical approach. John Wiley, Chichester

Pollin W, Cardon P V, Kety S S 1961 Effects of aminoacid feedings in schizophrenic patients treated with iproniazid. Science 133: 104

Pomeranz B, Chiu D 1976 Naloxone blockade of acupuncture analgesia: endorphin implicated. Life Sciences 19: 1757

Popper K 1963 The Logic of Scientific Discovery. Hutchinson, London

Post F 1971 Schizo-affective symptomatology in late life. British Journal of Psychiatry 118: 437

Post R M, Jimerson D C, Ballenger J C, Lake C R, Uhde T W, Goodwin F K 1984 CSF norepinephrine and its metabolites in manic depressive illness. In: Post R M, Ballenger J G (eds) Neurobiology of mood disorders. Williams & Wilkins, Baltimore

Post R M, Uhde T W, Roy-Byrne P P, Joffe R T 1987

Correlates of antimanic response to carbamazepine. Psychiatry Research 21: 71

Postmes J, Coengracht J 1983 Journal of Drug Research 8: 2051

Potamianos G, Kellett J M 1982 Anticholinergic drugs and memory: the effects of benzhexol on memory in a group of geriatric patients. British Journal of Psychiatry 140: 470

Power K G, Jerrom D W A, Simpson R J, Mitchell M 1985 Controlled study of withdrawal symtoms and rebound anxiety after six week course of diazepam for generalised anxiety. British Medical Journal 290: 1246

Prange A J, Wilson I C 1972 Thyrotropin releasing hormone (TRH) for the immediate relief of depression: a preliminary report. Psychopharmacologia 26: 82

Prange A J, Wilson I C, Lara P P, Alltop L B, Breese G R 1972 Effects of thyrotropin releasing hormone in depression. Lancet ii: 999

Prange A J, Loosen P T, Wilson I C, Meltzer H Y, Fang V S 1979 Behavioural and endocrine responses of schizophrenic patients to TRH. Archives of General Psychiatry 36: 1086

Prange A J, Loosen P T, Wilson I C, Lipton M A 1984 The therapeutic use of hormones of the thyroid axis in depression. In: Post R M, Ballenger J C (eds) Neurobiology of mood disorders. Williams & Wilkins, Baltimore, Maryland

Priest R G, Montgomery S A 1988 Benzodiazepines and dependence: a College statement. Psychiatric Bulletin of the Royal College of Psychiatrists 12: 107

Proops R, Taylor A M R, Insley J 1981 A clinical study of a family with Cockayne's syndrome. Journal of Medical Genetics 18: 288

Quabbe H 1977 Chronobiology of growth hormone secretion. Chronobiologia 4: 217

Quality Assurance Project 1985 Treatment outlines for the management of anxiety states. Australian and New Zealand Journal of Psychiatry 19: 138

Quetsch R M, Achor R W P, Litin E M, Fauchett R L 1959 Depressive reactions in hypotensive patients: a comparison of those treated with Rauwolfia and those receiving no specific antihypotensive treatment. Circulation 19: 366

Quinsey V L, Chaplin T C, Upfold D 1984 Sexual arousal to nonsexual violence and sadomasochistic themes among rapists and non-sex-offenders. Journal of Consulting and Clinical Psychology 52: 651

Rahmani Z, Blouin J L, Creau-Goldberg N et al 1989 Critical role of the D21S55 region on chromosome 21 in the pathogenesis of Down syndrome. Proceedings of the National Academy of Sciences 86: 5958

Raitta C, Santavuori P, Lamminen M, Leisti J 1978 Ophthalmological findings in a new syndrome with muscle, eye and brain involvement. Acta Ophthalmologica 56: 465–472

Ramani S V 1981 Psychosis associated with frontal lobe lesions in Schilder's cerebral sclerosis. Journal of Clinical Psychiatry 42: 250

Ramani V, Gumnit R J 1981 Intensive monitoring of epileptic patients with a history of episodic aggression. Archives of Neurology 38: 570

Rasmussen K, Morilak D A, Jacobs B L 1986 Single unit activity of locus coeruleus neurons in the freely moving cat. I. During naturalistic behaviors and in response to simple and complex stimuli. Brain Research 371: 324

Razani J, White K, White J et al 1983 The safety and efficacy of combined amitriptyline and tranylcypromine antidepressant treatment. Archives of General Psychiatry 40: 657

Reilly E L, Halmik A, Noyes R 1973 Electro-encephalographic responses to lithium. International Pharmacopsychiatry 8: 208

Reimherr F W, Wood D R, Byerley B, Brainard J, Grosser B I 1984 Characteristics of responders to fluoxetine. Psychopharmacology Bulletin 20: 70

Rémillard G, Andermann F, Testa G F et al 1983 Sexual ictal manifestations predominate in women with temporal lobe epilepsy: a finding suggesting sexual dimorphism in the human brain. Neurology 33: 323

Riccardi V M, Hassler E, Lubinsky M S 1977 The FG syndrome: further characterization, report of a third family, and of a sporadic case. American Journal of Medical Genetics 1: 47

Rimoin D L, McAlister W H 1971 Metaphyseal dysostosis, conductive hearing loss, and mental retardation: a recessively inherited syndrome. Birth Defects Original Article Series VII(4): 116

Robinson D S, Cooper T B, Jindal S P, Corcella J, Lutz T 1985 Metabolism and pharmacokinetics of phenelzine: lack of evidence for acetylation pathway in humans. Journal of Clinical Psychopharmacology 5: 333

Roe A 1944 The adult adjustment of children of alcoholic parents raised in foster homes. Journal of Studies on Alcohol 5: 378

Rogers R L, Meyer J S, Mortel K F, Mahurin R K, Judd B W 1986 Decreased cerebral blood flow precedes multi-infarct dementia, but follows senile dementia of Alzheimer type. Neurology 36: 1

Ron M 1983 The alcoholic brain: CT scan and psychological findings. Psychological Medicine Monograph Supplement 3: 1

Rosenthal D, Wender P H, Kety S S, Schulsinger F, Welner J, Reider R 1975 Parent-child relationships and psychopathological disorder in the child. Archives of General Psychiatry 32: 466

Rosenthal N E, Sack D A, Gillin J C et al 1984 Seasonal affective disorder. A description of the syndrome and preliminary findings with light therapy. Archives of General Psychiatry 41: 72

Rossor M N, Garrett N J, Johnson A J, Mountjoy C Q, Roth M, Iversen L L 1982. A post-mortem study of the cholinergic and GABA systems in senile dementia. Brain 105: 313

Roth M 1951 Changes in the EEG under barbiturate anaesthesia produced by ECT and their significance for the theory of ECT action. Electroencephalography and Clinical Neurophysiology 3: 261

Rothblat L A, Schwarz M L 1979 The effect of monocular deprivation on dendritic spines in visual cortex of young and adult albino rats: evidence for a sensitive period. Brain Research 161: 156

Rothman K 1978 A show of confidence. New England Journal of Medicine 299: 1362

Rotrosen J, Angrist B M, Gershon S, Sachar E J, Halpern F 1976 Dopamine receptor alteration in schizophrenia: neuroendocrine evidence. Psychopharmacology 51: 1

Routtenberg A, Santos-Anderson R 1977 The central role of prefrontal cortex in intracranial self-stimulation: a case history of anatomical localization of motivational substrates. In Iversen L L, Iversen S D, Snyder S H (eds)

Handbook of psychopharmacology vol 8: drugs, neurotransmitters and behavior. Plenum Press, New York

Roy A, DeJong J, Gold P et al 1990 Cerebrospinal fluid levels of somatostatin, corticotropin-releasing hormone and corticotropin in alcoholism. Acta Psychiatrica Scandinavica 82: 44

Roy-Byrne P P, Uhde T W, Rubinow D R, Post R M 1986 Reduced TSH and prolactin responses to TRH in patients with panic disorder. American Journal of Psychiatry 143: 503

Ruberg M, Ploska A, Javoy-Agid Y 1982 Muscarinic binding and choline acetyltransferase activity in Parkinsonian subjects with reference to dementia. Brain Research 232: 129

Rubin P C 1990 Editorial comment. British Medical Journal 300: 1131

Rubinow D R 1986 Cerebrospinal fluid somatostatin and psychiatric illness. Biological Psychiatry 21: 341

Rubinstein J H, Taybi H 1963 Broad thumbs and toes and facial abnormalities. American Journal of Diseases of Children 105: 588

Rufer V, Bauer J, Soukup F 1970 On the heredity of eye colour. Acta Universitatis Carolinae Medica 16: 429

Rundus D 1971 Analysis of rehearsal processes in free recall. Journal of Experimental Psychology 89: 63

Russell W R, Smith A 1961 Post-traumatic amnesia in closed head injury. Archives of Neurology 5: 4

Ruther E, Jungkunz G, Nedopil N 1980 Clinical effects of the synthetic analogue of methionine enkephalin FK 33–824. Progress in Neuropsychopharmacology 4: 304

Rutter M L 1972 Maternal deprivation reassessed. Penguin, Harmondsworth

Sachar E J, Hellman L, Fukushima D K, Gallagher T F 1972 Cortisol production in mania. Archives of General Psychiatry 26: 137

Sachar E J, Hellman L, Roffwarg H P et al 1973 Disrupted 24 hour patterns of cortisol secretion in psychotic depression. Archives of General Psychiatry 28: 19

Sachar E J, Asnis G, Nathan R S et al 1980 Dextroamphetamine and cortisol in depression. Archives of General Psychiatry 37: 755

Sachs J D S 1967 Recognition memory for syntactic and semantic aspects of connected discourse. Perception and Psychophysics 2: 437

St Clair D, Blackwood D, Muir W et al 1989 No linkage of chromosome 5q11-q13 markers to schizophrenia in Scottish families. Nature 339: 305

St George-Hyslop P H, Tanzi R E, Polinsky R J et al 1987 The genetic defect causing familial Alzheimer's disease maps on chromosome 21. Science 235: 885

St Hilaire J M, Gilbert M, Bouvier G, Barbeau A 1980 Epilepsy and aggression: two cases with depth electrode studies. In: Robb P (ed) Epilepsy updated: causes and treatment. Yearbook Medical Publishers, Chicago

Sandman C A, George J, Nolan J, van Riezen H, Kastin A 1975 Enhancement of attention in man with $ACTH/MSH_{4-10}$. Physiology and Behavior 15: 427

Sassin J, Frantz A, Weitzman E, Kapen S 1972 Human prolactin: 24-hour patterns with increased release during sleep. Science 17: 1205

Sastre J, Sakai K, Jouvet M 1979 Persistence of paradoxical sleep after destruction of the pontine gigantocellular tegmental field with kainic acid in the cat. Comptes Rendus des Seances de l'Academie des Sciences 289: 959

Satterfield J H 1972 Auditory evoked cortical response studies in depressed patients and normal control subjects. In: Williams T A, Katz M M, Sheild J A (eds) Recent advances in the psychobiology of the depressive illnesses. U.S. Government Printing Office, Washington D.C.

Sauk J J, Litt R, Espiritu C E, Delaney J R 1973 Familial bird-headed dwarfism (Seckel's syndrome). Journal of medical Genetics 10: 196

Schachat A P and Maumenee I H 1982 The Bardet–Biedl syndrome and related disorders. Archives of Ophthalmology 100: 285

Schachter S 1971 Emotion, obesity and crime. Academic Press, New York

Schachter S, Singer J E 1962 Cognitive, social and physiological determinants of emotional state. Psychological Review 69: 379

Schally A V 1978 Aspects of hypothalamic regulation of the pituitary gland. Science 202: 18

Schellenberg G D, Bird T D, Wisjman E M et al 1988 Absence of linkage of chromosome 2lq21 markers to familial Alzheimer's disease. Science 241: 1507

Schenk V W D 1959 Re-examination of a family with Pick's disease. Annals of Human Genetics 23: 325

Schiavi R C, Owen D, Fogel M et al 1978 Pituitary gonadal function in XXY and XYY men identified in a population survey. Clinical Endocrinology 9: 223

Schildkraut J J 1965 The catecholamine hypothesis of affective disorders: a review of supporting evidence. American Journal of Psychiatry 122: 509

Schinzel A, Schmid W 1980 Hallux duplication, postaxial polydactyly, absence of the corpus callosum, severe mental retardation and additional anomalies in two unrelated patients: a new syndrome. American Journal of Medical Genetics 6: 241

Schmidtke A, Fleckenstein P, Beckman N H 1989 The dexamethasone suppression test and suicide attempts. Acta Psychiatrica Scandinavica 79: 276

Schuckit M, Robins E, Feighner J 1971 Tricyclic antidepressants and monoamine oxidase inhibitors. Archives of General Psychiatry 24: 509

Schurig V, van Orman A, Bowen P 1981 Nonprogressive cerebellar disorder with mental retardation and autosomal recessive inheritance in Hutterites. American Journal of Medical Genetics 9: 43

Schwartz M F, Bauman J E, Masters W H 1982 Hyperprolactinaemia and sexual disorders in men. Biological Psychiatry 17: 861

Scott A I F, Whalley L J, Bennie J, Bowler G 1986 Oestrogen-stimulated neurophysin and outcome after electroconvulsive therapy. Lancet i: 1411

Scott-Emuakpor A B, Heffelfinger J, Higgins J V 1977 A syndrome of microcephaly and cataracts in four siblings: a new genetic syndrome? American Journal of Diseases of Children 131: 167

Sedvall G C, Wode-Helgodt B 1980 Aberrant monoamine metabolite in CSF and family history of schizophrenia. Archives of General Psychiatry 37: 1113

Seemanova E, Lesny I, Hyanek J, Brachfeld K, Rossler M, Proskova M 1973 X-chromosomal recessive microcephaly with epilepsy, spastic tetraplegia and absent abdominal reflex. New variety of 'Paine syndrome'? Humangenetik 20: 113

Seligman M E P 1975 Helplessness: on depression, development and death. Freeman, San Francisco

Serra G, Gessa G L 1984 Role of brain monoamines and peptides in the regulation of male sexual behavior. In: Shah N S, Donald A G (eds) Psychoneuroendocrine dysfunction. Plenum Press, New York

Seutin V, Mossotte L, Dresse A 1989 Electrophysiological effects of neurotensin on dopaminergic neurones of the ventral tegmental area of the rat in vitro. Neuropharmacology 28: 949

Shagass C 1983 Evoked potentials in adult psychiatry. In: Hughes J R, Wilson W P (eds) EEG and evoked potentials in psychiatry and behavioral neurology. Butterworths, Boston

Shagass C, Roemer R A, Straumanis J J, Amadeo M 1978 Evoked potential correlates of psychosis. Biological Psychiatry 13: 163

Shawcross C R, Tyrer P 1987 The place of monoamine-oxidase inhibitors in the treatment of resistant depression. In: Zohar J, Belmaker R H (eds) Treating resistant depression. PMA Publishing Corporation, New York

Sheard M H, Marini J L, Bridges C I, Wagner E 1976 The effect of lithium on impulsive aggressive behavior in man. American Journal of Psychiatry 133: 1409

Sheehan D V, Ballenger J, Jacobsen G 1980 Treatment of endogenous anxiety with phobic, hysterical, and hypochondriacal symptoms. Archives of General Psychiatry 37: 51

Sheehan D V, Zak J P, Miller J A, Junior J R, Fanous BS 1988 Panic disorder: the potential role of serotonin reuptake inhibitors. Journal of clinical Psychiatry 49 (suppl. 1): 30

Sherif M 1936 The psychology of social norms. Harper & Row, New York

Sherrington R, Brynjolfsson J, Petersson H et al 1988 Localization of a susceptibility locus for schizophrenia on chromosome 5. Nature 336: 164

Shih V E, Efron M L, Moser H W 1969 Hyperornithinemia, hyperammonemia, hyperammonemia, and homocitrullinemia: a new disorder of amino acid metabolism associated with myoclonic seizures and mental retardation. American Journal of Diseases of Children 117: 83

Shokeir M H K 1970 X-linked cerebella ataxia. Clinical Genetics 1: 225

Shokeir M H K 1977 Universal permanent alopecia, psychomotor epilepsy, pyorrhea and mental subnormality. Clinical Genetics 11: 13

Shopsin B, Friedman E, Gershon S 1976 Parachlorphenylalanine reversal of tranylcypromine effects in depressed patients. Archives of General Psychiatry 33: 811

Silberman E K, Vivaldi E, Garfield J, McCarley R W, Hobson J A 1980 Carbachol triggering of desynchronized sleep phenomena: enhancement via small volume infusions. Brain Research 191: 215

Silverstein M N, Ellefson R D 1972 The syndrome of sea-blue histiocyte. Seminars in Hematology 9: 299

Silverstone T 1984 Response to bromocriptine distinguishes bipolar from unipolar depression. Lancet i: 903

Simpson G M, Frederickson E, Palmer R, Pi E, Sloane R B, White K 1985 Platelet monoamine oxidase inhibition by deprenyl and tranylcypromine: implications for clinical use. Biological Psychiatry 20: 684

Singer S J, Nicolson G L 1972 The fluid mosaic model of the structure of cell membranes. Science 175: 720

Sitaram N, Weingartner H, Gillin M C 1979 Choline chloride and arecoline: effects on memory and sleep in man. In: Barbeau A, Growdon J W, Wurtman R W (eds) Nutrition and the brain vol 5: choline and lecithin in brain disorders. Raven Press, New York

Sjaastad O, Berstad J, Gjesdahl P, Gjessing L R 1976 Homocarnosinosis. 2. A familial metabolic disorder associated with spastic paraplegia, progressive mental deficiency, and retinal pigmentation. Acta Neurologica Scandinavica 53: 275

Sjogren T 1950 Hereditary congenital spinocerebellar ataxia accompanied by congenital cataract and oligophrenia. A genetic and clinical investigation. Confinia Neurologica 10: 293

Sjoland B, Terenius L, Erikisson M 1977 Increased cerebrospinal fluid levels of endorphin after electro-acupuncture. Acta Physiologica Scandinavica 100: 382

Skinner BF 1938 The behavior of organisms. Appleton–Century–Crofts, New York

Skre H, Loken A C 1970 Myoclonus epilepsy and subacute presenile dementia in heredo-ataxia. A clinical, electroencephalographic, and pathological study with a discussion of classification and etiology. Acta Neurologica Scandinavica 46: 18

Slade P D 1982 Towards a functional analysis of anorexia nervosa and bulimia nervosa. British Journal of Clinical Psychology 21: 167

Slater E 1961 Hysteria 311. Journal of Mental Science 107: 359

Small R G 1968 Coats' disease and muscular dystrophy. Transactions of the American Academy of Ophthalmology and Otolaryngology 72: 225

Small J G, Milstein V, Perez H C, Small I F, Moore D F 1972 EEG and neurophysiological studies of lithium in normal volunteers. Biological Psychiatry 5: 65

Smith A 1961 duration of impaired consciousness as an index of severity in closed head injuries: a review. Diseases of the Nervous System 22: 69

Smith A J, Strang L B 1958 An inborn error of metabolism with the urinary excretion of alpha-hydroxy-butyric acid and phenylpyruvic acid. Archives of Disease in Childhood 33: 109

Smith R C, Baumgartner R, Ravichandran G K et al 1984 Plasma and red cell levels of thioridazine and clinical response in schizophrenia. Psychiatry Research 12: 287

Snell R S 1987 Clinical neuroanatomy for medical students. 2nd edn. Little Brown, Boston

Snyder S H 1986 Neuronal receptors. Annual Review of Physiology 48: 461

Snyder F, Hobson J A, Morrison D F, Goldfrank F 1964 Changes in respiration, heart rate, and systolic blood pressure in human sleep. Journal of Applied Physiology 19: 417

Snyder S H, Banerjee S P, Yamamura H I, Greenberg D 1974 Drugs, neurotransmitters and schizophrenia. Science 184: 1243

Soloff P H, George A, Nathan R S, Shulz P M, Ulrich R F, Perel J M 1986 Progress in pharmacotherapy of borderline disorders: a double blind study of amitriptyline, haloperidol and placebo. Archives of General Psychiatry 43: 691

Solomon C M, Holzman P S, Levin S et al 1987 The association between eye tracking dysfunctions and thought disorder in psychosis. Archives of General Psychiatry 44: 31

Sommer A, Rathbun M A, Battles M L 1974 A syndrome of

partial aniridia, unilateral renal agenesis, and mild psychomotor retardation in siblings. Journal of Pediatrics 85: 870

Somogyi P, Hodgson A J, Smith A D, Nunzi M G, Gorio A, Wu J-Y 1984 Different populations of GABAergic neurons in the visual cortex and hippocampus of cat contain somatostatin- or cholecystokinin-immunoreactive material. Journal of Neuroscience 4: 2590

Sørensen K V, Christensen S E, Dupont E, Hansen A P, Pedersen E, Orskov H 1980 Low somatostatin content in cerebrospinal fluid in multiple sclerosis. An indicator of disease activity? Acta Neurologica Scandinavica 61: 186

Sørensen P S, Bolwig T G, Lauritsen B, Bengtson O 1981 Electroconvulsive therapy: a comparison of seizure duration as monitored with electroencephalograph and electromyograph. Acta Psychiatrica Scandinavica 64: 1193

Sperling G 1960 The information available in brief visual presentations. Psychological Monographs 74: 1

Spokes E G S 1980 Neurochemical alterations in Huntington's chorea: a study of post-mortem brain tissue. Brain 103: 179

Standish-Barry H M A S, Bouras N, Bridges P K, Bartlett J R 1982 Pneumoencephalographic and CAT scan changes in affective disorder. British Journal of Psychiatry 141: 614

Stanley M, Brown G M 1988 Melatonin levels are reduced in the pineal glands of suicide victims. Psychopharmacology Bulletin 24: 484

Steadman J H, Graham J G 1970 Head injuries: an analysis and follow-up study. Proceedings of the Royal Society of Medicine 63: 23

Stevens J H, Mathews M (eds) 1978 Mother/child, father/child relationships. National Association for the Education of Young Children, Washington D.C.

Stimmler L, Jensen N, Toseland P 1970 Alaninuria, associated with microephaly, dwarfism, enamel hypoplasia, and diabetes mellitus in two sisters. Archives of Disease in Childhood 45: 682

Stokes P E, Stoll P M, Koslow S H et al 1984 Pretreatment DST and hypothalamic-pituitary-adrenocortical function in depressed patients and comparison groups. Archives of General Psychiatry 41: 257

Storey P B 1967 Psychiatric sequelae of subarachnoid haemorrhage. Journal of Psychosomatic Research 13: 175

Storey P B 1970 Brain damage and personality change after subrachoid haemorrhage. British Journal of Psychiatry 117: 129

Stransky E, Bayani-Sioson P S, Lee W 1962 A peculiar type of familial mental deficiency, probably due to metabolic disturbance. A preliminary report. Journal of the Philippine Medical Association 38: 903

Straumanis J J, Shagass C, Schwartz M 1965 Visually evoked cerebral response changes associated with chronic brain syndrome and aging. Journal of Gerontology 20: 498

Stromgren L S, Juul-Jensen P 1975 EEG in unilateral and bilateral electroconvulsive therapy. Acta Psychiatrica Scandinavica 51: 340

Strong W B 1968 Familial syndrome of right-sided aortic arch, mental deficiency, and facial dysmorphism. Journal of Pediatrics 73: 882

Suda M, Hayaishi O, Nakagawa H (editors) 1979 Biological rhythms and their central mechanisms. North-Holland Biomedical Press, Amsterdam

Sulser F 1984 Regulation and function of noradrenaline receptor systems in brain. Neuropharmacology 23: 255

Sulser F, Watts J, Brodie B B 1962 On the mechanism of action of antidepressant action of imipramine-like drugs. Annals of the New York Academy of Science 96: 270

Sulser F, Vetulani J, Mobley P L 1978 Mode of action of antidepressant drugs. Biochemical Pharmacology 27: 257

Surridge D 1969 An investigation into some psychiatric aspects of multiple sclerosis. British Journal of Psychiatry 115: 749

Sutherland G R 1979 Hereditable fragile sites on human chromosomes. II. Distribution, phenotypic effects and cytogenetics. Human Genetics 53: 136

Swann A C, Secunda S, Davis J M et al 1983 CSF monoamine metabolites in mania. American Journal of Psychiatry 140: 396

Swigar M E, Lankenhal H, Bhimani S, Conway B, Benes F 1983 CT scan abnormalities in eating disorders. American Psychiatric Association Annual Meeting. New Research Abstracts 62. New York

Symonds C P 1937 mental disorder following head injury. Proceedings of the Royal Society of Medicine 30: 1081

Takahashi Y, Kipnis D, Daughaday W 1968 Growth hormone secretion during sleep. Journal of Clinical Investigation 47: 2079

Takeuchi K, Uematsu M, Ofuji M, Morikiyo M, Kaiya H 1988 Substance P involved in mental disorders. Progress in Neuropsychopharmacology and Biological Psychiatry 12 (suppl.): S157

Tamminga C A, Littman R L, Alphs L D, Chase T N, Thaker G K, Wagman A M 1986 Neuronal cholecystokinin and schizophrenia: pathogenic and therapeutic studies. Psychopharmacology 88: 387

Tancer M E, Stein M B, Gelernter C S, Uhde T W 1990 The hypothalamic–pituitary–thyroid axis in social phobia. American Journal of Psychiatry 147: 929

Tandon R, Greden J F 1989 Cholinergic hyperactivity and negative schizophrenic symptoms. A model of cholinergic/dopaminergic interactions in schizophrenia. Archives of General Psychiatry 46: 745

Tanzi R E, Gusella J F, Watkins P C et al 1987a Amyloid B protein gene: cDNA, mRNA distribution, and genetic linkage near the Alzheimer locus. Science 235: 880

Tanzi R E, St George-Hyslop P H, Haines J L et al 1987b The genetic defect in Alzheimer's disease is not tightly linked to the amyloid beta-protein gene. Nature 329: 156

Tarachow S 1939 The Korsakoff psychosis in spontaneous subarachnoid haemorrhage. American Journal of Psychiatry 95: 887

Targum S D, Rosen L N, Delisi L E, Weinberger D R, Citrin C M 1983 Cerebral ventricular size in major depressive disorder: association with delusional symptoms. Biological Psychiatry 18: 329

Tay C H 1970 Ichthyosiform erythroderma, hair shaft abnormalities, and mental and growth retardation: a new recessive disorder. Archives of Dermatology 104: 4

Thaker G K, Nguyen J A, Tamminga C A 1989 Increased saccadic distractibility in tardive dyskinesia: functional evidence for subcortical GABA dysfunction. Biological Psychiatry 25: 49

Theander S 1970 Anorexia nervosa: a psychiatric investigation of female patients. Acta Psychiatrica Scandinavica Supplement 214

Theander S, Granholm L 1967 Sequelae after spontaneous subarachnoid haemorrhage, with special reference to hydrocephalus and Korsakoff's syndrome. Acta Neurologica Scandinavica 43: 479

Thompson C (ed) 1989 The instruments of psychiatric research. John Wiley, Chichester

Thompson C, Franey C, Arendt J, Checkley S A 1988 A comparison of melatonin secretion in depressed patients and normal subjects. British Journal of Psychiatry 152: 260

Thomsen K, Schou M 1968 American Journal of Physiology 215: 823

Thorén P, Asberg M, Bertilsson L et al 1980 Clomipramine treatment of obsessive-compulsive disorder. II. Biochemical aspects. Archives of General Psychiatry 37: 1289

Thorndike E L 1911 animal intelligence. Macmillan, New York

Tiffany S T, Martin E M, Baker T B 1986 Treatments for cigarette smoking: an evaluation of the contributions of aversion and counselling procedures. Behaviour Research and Therapy 24: 437

Tomlinson B E, Blessed G, Roth M 1970 Observations on the brains of demented old people. Journal of the Neurological Sciences 11: 205

Tricklebank M D 1987 Subtypes of 5-HT receptors. Journal of Psychopharmacology 1: 222

Trimble M R 1978 Serum prolactin in epilepsy and hysteria. British Medical Journal 11: 1682

Trimble M R 1981 Neuropsychiatry. Wiley, Chichester

Trimble M R 1985 New brain imaging techniques and psychiatry. In: Granville-Grossman K (ed) Recent advances in clinical psychiatry, number 5. Churchill Livingstone, Edinburgh

Tuomaala P, Haapanen E 1968 Three siblings with similar anomalies in the eyes, bones and skin. Acta Ophthalmologica 46: 365

Turcot J, Despres J P, St Pierre F 1959 Malignant tumors of the central nervous system associated with familial polyposis of the colon: report of two cases. Diseases of the Colon and Rectum 2: 465

Turkington R W 1972 Phenothiazine stimulation test for prolactin reserve: the syndrome of isolated prolactin deficiency. Journal of Clinical Endocrinology 34: 247

Tyler H R 1958 Pelizaeus-Merzbacher disease: a clinical study. Archives of Neurology and Psychiatry 80: 162

Tyrer P 1976 The role of bodily feelings in anxiety. Maudsley monograph number 23. Oxford University Press, London

Tyrer P 1979 Clinical use of monoamine oxidase inhibitors. In: Paykel E S, Coppen A (eds) Psychopharmacology of affective disorders. Oxford University Press, Oxford

Tyrer P 1982a Antianxiety drugs and monoamine oxidase inhibitors. In: Tyrer PJ (ed) Drugs in psychiatric practice. Butterworths, London

Tyrer P 1982b The concept of somatic anxiety. British Journal of Psychiatry 140: 325

Tyrer P 1984a Clinical effects of abrupt withdrawal from tricyclic antidepressants and monoamine-oxidase inhibitors after long-term treatment. Journal of Affective Disorders 6: 1

Tyrer P J 1984b Benzodiazepines on trial. British Medical Journal 288: 1101

Tyrer S P 1985 Lithium in the treatment of mania. Journal of Affective Disorders 8: 251

Tyrer P 1988 Dependence as a limiting factor in the clinical use of minor tranquillizers. Pharmacology and Therapeutics 36: 173

Tyrer P 1989a Classification of neurosis. John Wiley, Chichester

Tyrer P 1989b Choice of treatment in anxiety. In: Tyrer P (ed) Psychopharmacology of anxiety. Oxford University Press, Oxford

Tyrer S P 1989 Assessment and physical treatment of affective disorder. British Journal of Hospital Medicine 42: 184

Tyrer P, 1991 The benzodiazepine post-withdrawal syndrome. Stress Medicine 1: 1

Tyrer P, Harrison-Read P 1990 New perspectives in treatment with monoamine oxidase inhibitors. International Review of Psychiatry 2: 331

Tyrer P J, Lader M H 1974 Physiological and psychological effects of propranolol, (+)– propranolol and diazepam in induced anxiety. British Journal of Clinical Pharmacology 1: 379

Tyrer P, Marsden C 1985 New antidepressant drugs: is there anything new they tell us about depression? Trends in Neurosciences 8: 427

Tyrer P, Murphy S 1987 The place of benzodiazepines in psychiatric practice. British Journal of Psychiatry 151: 719

Tyrer P, Murphy S 1990 Efficacy of combined antidepressant therapy in resistant neurotic disorder. British Journal of Psychiatry 156: 115

Tyrer P, Shawcross C 1988 Monoamine-oxidase inhibitors in anxiety disorders. Journal of Psychiatric Research 22 (suppl. 1): 87

Tyrer P, Seivewright N, Murphy S et al 1988 The Nottingham study of neurotic disorder: comparison of drug and psychological treatments. Lancet ii: 235

Ugedo L, Grenhoff J, Svensson T H 1989 Ritanserin, a 5-HT2 receptor antagonist, activates midbrain dopamine neurons by blocking serotonergic inhibition. Psychopharmacology 98: 45

Uhde T W, Vittone B J, Siever L J, Kaye W H, Post R M 1986 Blunted growth hormone responses to clonidine in panic disorder patients. Biological Psychiatry 21: 1081

Valmier J, Touchon J, Daures P, Zanca M, Baldy-Moulinier M 1987 Correlations between cerebral blood flow variations and clinical parameters in temporal lobe epilepsy: an interictal study. Journal of Neurology, Neurosurgery and Psychiatry 50: 1306

van Bogaert L, Moreau M 1939–1941 Combination de l'amyotrophie de Charcot-Marie-Tooth et de la maladie de Friedreich chez plusieurs membres d'une famille. Encephale 34: 312

van Broeckhoven C L, Genthe A M, Vanderberghe A et al 1987 Failure of familial Alzheimer's disease to segregate with the A4 amyloid gene in several European families. Nature 329: 153

van Broeckhoven C, Backhovens H, van Hul W et al 1989 Linkage analysis of two extended Alzheimer families with chromosome 21 DNA markers. HGMX A2257

van Cauter E, Desir D, Refetoff S et al 1982 The relationship between episodic variations of plasma prolactin and REM-non-REM cyclicity is an artifact. Journal of Clinical Endocrinology and Metabolism 54: 70

van den Bosch J 1959 A new syndrome in three generations of a Dutch family. Ophthalmologica 137: 422

van der Welde C D 1983 Rapid clinical effectiveness of MIF-1 in the treatment of major depressive illness. Peptides 4: 297

van Hoof F, Hers H G 1968 Mucopolysaccharidosis by absence of alpha-fucosidase. Lancet i: 1198

van Praag H M, Korf J 1971 Retarded depression and dopamine metabolism. Psychopharmacologia 19: 199

van Praag H M, Kahn R, Asnis G M, Lemus C Z, Brown S L 1987 Therapeutic indications for serotonin-potentiating compounds: a hypothesis. Biological Psychiatry 22: 205

Vanderhaeghen J J, Crawley J N (eds) 1985 Neuronal cholecystokinin. Annals of the New York Academy of Science, New York

Vauhkonen K 1959 Suicide among the male disabled with war injuries to the brain. Acta Psychiatrica et Neurologica Scandinavica (suppl. 137): 90

Vekemans M, Delvoye P, L'Hermite M, Robyn C 1977 Serum prolactin levels during the menstrual cycle. Journal of Clinical Endocrinology and Metabolism 44: 989

Vining E, Kosten T R, Kleber H D 1988 Clinical utility of rapid clonidine-naltrexone detoxification for opioid abusers. British Journal of Addiction 83: 567

Vitek M P, Rasool C, de Sauvage F et al 1988 absence of mutation in the β amyloid cDNAs cloned from the brains of three patients with sporadic Alzheimer's disease. Molecular Brain Research 4: 121

von Knorring L, Espvall M, Perris C 1974 Averaged evoked responses, pain measures, and personality variables in patients with depressive disorders. Acta Psychiatrica Scandinavica (suppl. 255): 99

von Knorring A, Cloninger C R, Bohman M, Sigvardsson S 1983 An adoption study of depressive disorders and substance abuse. Archives of General Psychiatry 40: 943

Waggoner R W, Lowenberg-Scharenberg K, Schilling M E 1942 Agenesis of white matter with idiocy. American Journal of Mental Deficiency 47: 20

Walster E, Aronson E, Abrahams D 1966 On increasing the persuasiveness of a low prestige communicator. Journal of Experimental Social Psychology 2: 325

Walton J N 1953 The Korsakoff syndrome in spontaneous subarachnoid haemorrhage. Journal of Mental Science 99: 521

Warkany J, Monroe B B, Sutherland B S 1961 Intrauterine growth retardation. American Journal of Diseases of Children 102: 249

Wasmuth J J, Hewitt J, Smith B et al 1988 A highly polymorphic locus very tightly linked to the Huntington's disease gene. Nature 332: 734

Watson J D, Crick F H C 1953 Molecular structure of nucleic acids: a structure for deoxyribose nucleic acid. Nature 171: 737

Watson J B, Rayner R 1920 Conditioned emotional reactions. Journal of Experimental Psychology 3: 1

Webb W B 1981 Some theories about sleep and their clinical implications. Psychiatric Annals 11: 415

Weeke A, Weeke J 1978 Disturbed circadian variation of serum thyrotropin in patients with endogenous depression. Acta Psychiatrica Scandinavica 57: 281

Wehr T A, Jacobsen F M, Sack D A et al 1986 Timing of phototherapy and its effects on melatonin secretion do not appear to be critical for its antidepressant effect in seasonal affective disorders. Archives of General Psychiatry 43: 870

Weiner R D 1980 Persistence of ECT-induced EEG changes. Journal of Nervous and Mental Disease 168: 224

Weissman M M, Merikangas K R, John K, Wickramaratne P, Prusoff B A, Kidd K K 1986 Family-genetic studies of psychiatric disorders. Archives of General Psychiatry 43: 1104

Weizman R, Weizman A, Gil-Ad I, Tyano S, Laron Z 1982 Abnormal growth hormone response to TRH in chronic adolescent schizophrenic patients. British Journal of Psychiatry 141: 582

Wender P H, Rosenthal D, Kety S S, Schulsinger F, Welner J 1973 Social class and psychopathology in adoptees: a natural experimental method for separating the roles of genetic and experiential factors. Archives of General Psychiatry 28: 318

Westlake R J, Rastegar A 1973 Hyperpyrexia from drug combinations. Journal of the American Medical Association 225: 1250

Wetterberg L, Beck-Friis J, Aperia B, Petterson U 1979 Melatonin/cortisol ratio in depression. Lancet ii: 1361

Whaley W L, Michiels F, MacDonald M E et al 1988 Mapping of D4S98/S114/S113 confines the Huntington's defect to a reduced physical region at the telomere of chromosome 4. Nucleic Acids Research 16: 11769

Whalley L J 1987 Causal models of Alzheimer's disease. In: Malkin J C, Rossor M N (eds) Alzheimer's disease — current approaches. Duphar Medical Relations, Southampton

Whalley L J, Rosie R, Dick H et al 1982 Immediate increased in plasma prolactin and neurophysin but not other hormones after electroconvulsive therapy. Lancet ii: 1064

Whalley L J, Christie J E, Brown S, Arbuthnott G W 1984 Schneider's first rank symptoms of schizophrenia: an association with increased growth hormone response to apomorphine. Archives of General Psychiatry 41: 1040

Will R G, Masters W B 1982 Evidence for case-to-case transmission of Creutzfeldt–Jakob disease. Journal of Neurology, Neurosurgery and Psychiatry 45: 235

Williams C A, Frias J L 1982 The Angelman ('happy puppet') syndrome. American Journal of Medical Genetics 11: 453

Williams J J, Sandlin C S, Dumars K W 1978 New syndrome: microcephaly associated with achalasia. (Abstract) American Journal of Human Genetics 30: 106A

Williams P L, Warwick R, Dyson M, Bannister L H 1989 Gray's anatomy. 37th edn. Churchill Livingstone, Edinburgh

Wilson I C, Prange A J, Lara P P, Alltop L B, Stikeleather R A, Lipton M A 1973 THR (lopremone): psychobiological responses to normal women. I. Subjective experiences. Archives of General Psychiatry 29: 15

Wing J K 1967 Social treatment, rehabilitation and management. In Coppen A, Walker A (eds) Recent developments in schizophrenia. British Journal of Psychiatry Special Publication no. 1

Wing J K 1978 Schizophrenia: Towards a new synthesis. Academic Press, London

Wing J K, Sturt E 1978 The PSE-ID-CATEGO system: a supplementary manual. Institute of Psychiatry, London

Wing J K, Cooper J E, Sartorius N 1974 The measurement and classification of psychiatric symptoms. Cambridge University Press, London

Wing J K, Mann S A, Leff J P, Nixon J M 1978 The concept of a 'case' in psychiatric population surveys. Psychological Medicine 8: 203

Winokur G, Clayton P, Reich T 1969 Manic depressive illness. C.V. Mosby, St Louis

Wise C D, Berger B D, Stein L 1972 Benzodiazepines: anxiety-reducing activity by reduction of serotonin turnover in the brain. Science 17: 180

Witkin H A, Mednick S A, Schulsinger F 1976 Criminality and XYY and XXY man. Science 193: 547

Wolpe J 1950 The genesis of neurosis. South African Medical Journal 24: 613

Wolpe J 1958 Psychotherapy by reciprocal inhibition.

Stanford University Press, Stanford

World Health Organization 1990 ICD-10 chapter V: mental and behavioural disorders (including disorders of psychological development). Diagnostic criteria for research (February 1990 draft for field trials). World Health Organization, Geneva

Wyatt R J, Termini B A, Davis J 1971 Biochemical and sleep studies of schizophrenia: a review of the literature 1960–1970. Schizophrenia Bulletin 4: 10

Yassa R 1985 The Pisa syndrome: a report of two cases. British Journal of Psychiatry 146: 93

Yehuda R, Southwick S M, Nussbaum G, Wahby V, Giller E L, Mason J W 1990 Low urinary cortisol excretion in patients with posttraumatic stress disorder. Journal of Nervous and Mental Disease 178: 366

Yoshida T, Tada K, Honda Y, Arakawa T 1971 Urochanic aciduria: a defect in the urocanase activity in the liver of a mentally retarded. Tohoku Journal of Experimental Medicine 104: 305

Young J P R, Hughes W C, Lader M H 1976 A controlled comparison of flupenthixol and amitriptyline in depressed outpatients. British Medical Journal i: 1116

Young J P R, Lader M H, Hughes W C 1979 Controlled trial of trimipramine, monoamine oxidase inhibitors, and combined treatment in depressed outpatients. British Medical Journal ii: 1315

Young S N, Pihl R O, Ervin F R 1986 The effect of altered tryptophan levels on mood and behavior in normal human males. Clinical Neuropharmacology 9 (suppl. 4): 516

Index